READING THE STORY IN DNA

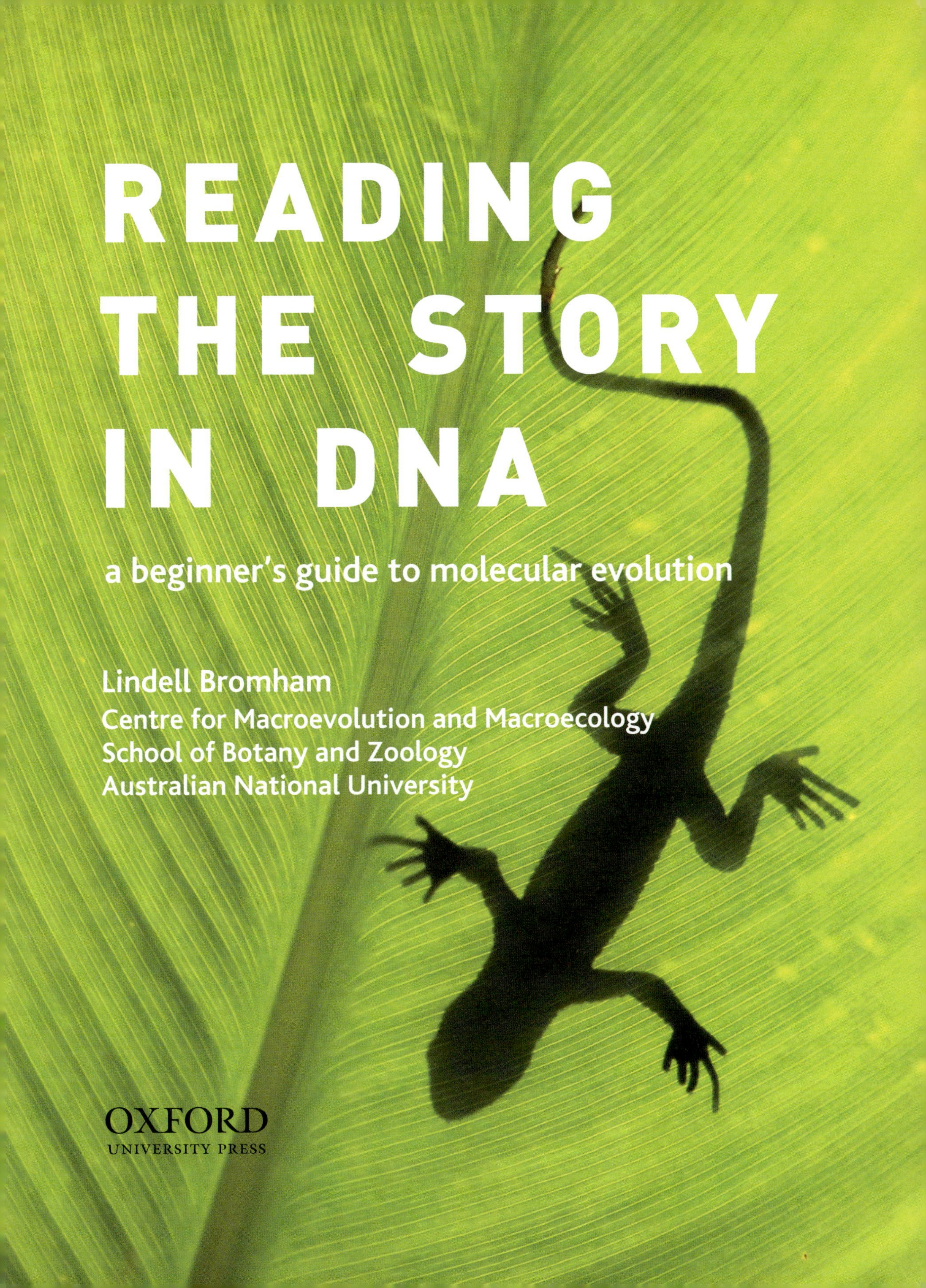

READING THE STORY IN DNA

a beginner's guide to molecular evolution

Lindell Bromham
Centre for Macroevolution and Macroecology
School of Botany and Zoology
Australian National University

OXFORD
UNIVERSITY PRESS

OXFORD
UNIVERSITY PRESS

Great Clarendon Street, Oxford OX2 6DP

Oxford University Press is a department of the University of Oxford.
It furthers the University's objective of excellence in research, scholarship,
and education by publishing worldwide in

Oxford New York

Auckland Cape Town Dar es Salaam Hong Kong Karachi
Kuala Lumpur Madrid Melbourne Mexico City Nairobi
New Delhi Shanghai Taipei Toronto

With offices in

Argentina Austria Brazil Chile Czech Republic France Greece
Guatemala Hungary Italy Japan Poland Portugal Singapore
South Korea Switzerland Thailand Turkey Ukraine Vietnam

Oxford is a registered trade mark of Oxford University Press
in the UK and in certain other countries

Published in the United States
by Oxford University Press Inc., New York

First published 2008

British Library Cataloguing in Publication Data
Data available

Library of Congress Cataloging in Publication Data
Bromham, Lindell.
Reading the story in DNA : a beginner's guide to molecular evolution / Lindell Bromham.
p. ; cm.
Includes bibliographical references and index.
ISBN 978-0-19-929091-8
1. DNA. 2. Molecular evolution. 3. Evolutionary genetics. I. Title.
[DNLM: 1. DNA. 2. Evolution, Molecular. QU 58.5 B868r 2008]
QP624.B74 2008
572.8′6—dc22
2008017073

Typeset by Graphicraft Limited, Hong Kong
Printed in Italy
by L.E.G.O. S.p.A.

ISBN 978–0–19–929091–8

1 3 5 7 9 10 8 6 4 2

For Marcel,

without whom nothing would be possible.

Preface

The aims of this book, and the features it contains, are covered more fully in Chapter 1. Here is a brief summary with references to pages where each point is discussed in more detail.

Who is this book for?

This book is aimed at biology students who need to learn how to apply molecular data to problems in evolution and ecology. The book assumes no more than a basic grounding in biology and genetics, so should be suitable for entry-level undergraduate students (see Chapter 1, page 24). However, this book may also be useful to bioinformaticians whose background is in maths, statistics, or computation, who need to learn evolutionary and genetic principles in order to interpret patterns in biological data.

What is in this book?

This book is designed to be read in a variety of ways, depending on your interests or background knowledge (see page 27). To facilitate getting what you need from this book, it is divided into several intercalated sections.

The **main text** provides the background to key issues in molecular evolution, and should be accessible to entry-level students with no previous knowledge of the subject.

TechBoxes go into more detail on specific methods or ideas. The language is more technical and they may be challenging to entry-level students, but more informative for students with some previous knowledge of molecular genetics.

Case Studies illustrate recent scientific research related to the principles covered in the chapter, cross-referenced to TechBoxes that explain some of the techniques used.

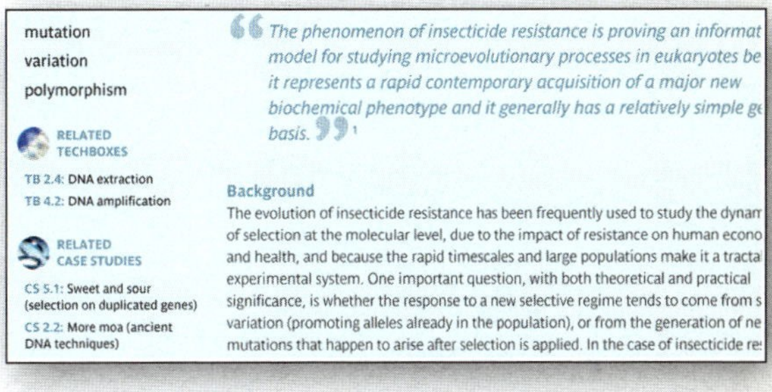
mutation
variation
polymorphism
RELATED TECHBOXES
TB 2.4: DNA extraction
TB 4.2: DNA amplification
RELATED CASE STUDIES
CS 5.1: Sweet and sour (selection on duplicated genes)
CS 2.2: More moa (ancient DNA techniques)

"The phenomenon of insecticide resistance is proving an informat model for studying microevolutionary processes in eukaryotes be it represents a rapid contemporary acquisition of a major new biochemical phenotype and it generally has a relatively simple ge basis." 1

Background

The evolution of insecticide resistance has been frequently used to study the dynam of selection at the molecular level, due to the impact of resistance on human econo and health, and because the rapid timescales and large populations make it a tracta experimental system. One important question, with both theoretical and practical significance, is whether the response to a new selective regime tends to come from s variation (promoting alleles already in the population), or from the generation of ne mutations that happen to arise after selection is applied. In the case of insecticide re

Heroes of the Genetic Revolution illustrate scientists who have worked in the area (see page 25).

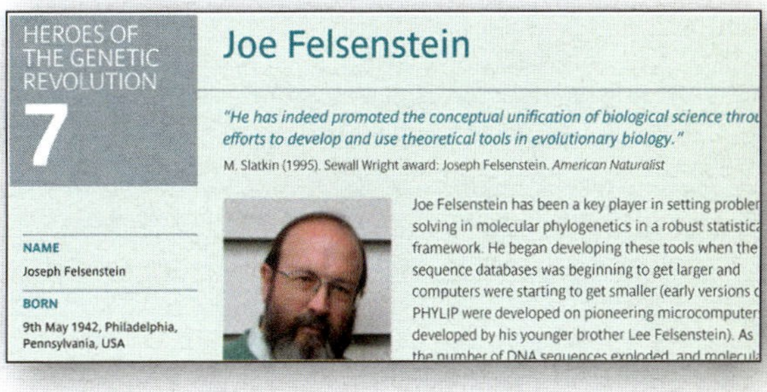
HEROES OF THE GENETIC REVOLUTION
7
Joe Felsenstein

"He has indeed promoted the conceptual unification of biological science thro efforts to develop and use theoretical tools in evolutionary biology."

M. Slatkin (1995). Sewall Wright award: Joseph Felsenstein. *American Naturalist*

NAME
Joseph Felsenstein
BORN
9th May 1942, Philadelphia, Pennsylvania, USA

Joe Felsenstein has been a key player in setting proble solving in molecular phylogenetics in a robust statistic framework. He began developing these tools when the sequence databases was beginning to get larger and computers were starting to get smaller (early versions o PHYLIP were developed on pioneering microcompute developed by his younger brother Lee Felsenstein). As the number of DNA sequences exploded and molecul

Other features of the text include:

Cross-references to information in other chapters, to help you navigate through the book and build up a knowledge base.

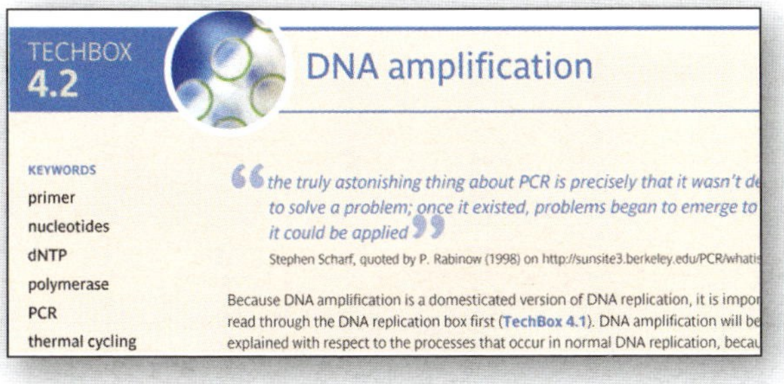
TECHBOX 4.2
DNA amplification

KEYWORDS
primer
nucleotides
dNTP
polymerase
PCR
thermal cycling

"the truly astonishing thing about PCR is precisely that it wasn't de to solve a problem; once it existed, problems began to emerge to it could be applied"

Stephen Scharf, quoted by P. Rabinow (1998) on http://sunsite3.berkeley.edu/PCR/whatis

Because DNA amplification is a domesticated version of DNA replication, it is impor read through the DNA replication box first (TechBox 4.1). DNA amplification will be explained with respect to the processes that occur in normal DNA replication, becau

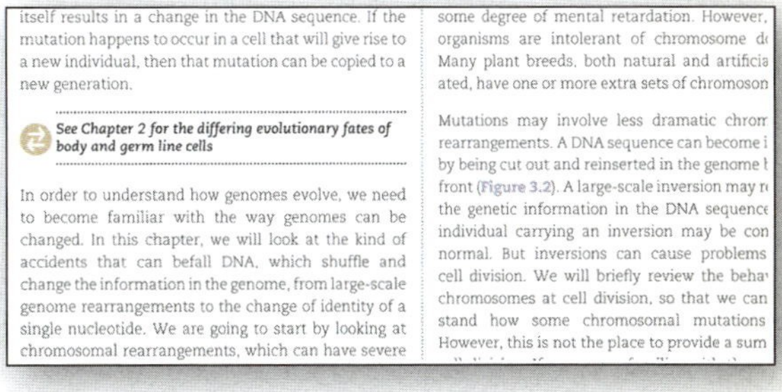
itself results in a change in the DNA sequence. If the mutation happens to occur in a cell that will give rise to a new individual, then that mutation can be copied to a new generation.

See Chapter 2 for the differing evolutionary fates of body and germ line cells

In order to understand how genomes evolve, we need to become familiar with the way genomes can be changed. In this chapter, we will look at the kind of accidents that can befall DNA, which shuffle and change the information in the genome, from large-scale genome rearrangements to the change of identity of a single nucleotide. We are going to start by looking at chromosomal rearrangements, which can have severe

some degree of mental retardation. However, organisms are intolerant of chromosome d Many plant breeds, both natural and artificia ated, have one or more extra sets of chromoson

Mutations may involve less dramatic chrom rearrangements. A DNA sequence can become i by being cut out and reinserted in the genome t front (Figure 3.2). A large-scale inversion may r the genetic information in the DNA sequence individual carrying an inversion may be con normal. But inversions can cause problems cell division. We will briefly review the beha chromosomes at cell division, so that we can stand how some chromosomal mutations However, this is not the place to provide a sum

The **Glossary** contains some basic biological terms and explains some tricky concepts in evolution and genetics.

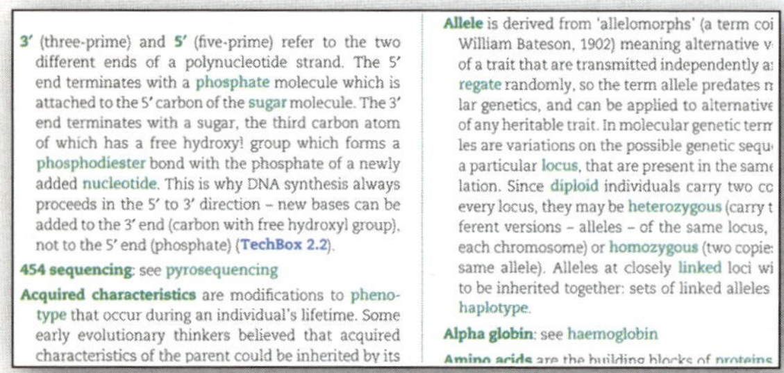
3′ (three-prime) and **5′** (five-prime) refer to the two different ends of a polynucleotide strand. The 5′ end terminates with a phosphate molecule which is attached to the 5′ carbon of the sugar molecule. The 3′ end terminates with a sugar, the third carbon atom of which has a free hydroxyl group which forms a phosphodiester bond with the phosphate of a newly added nucleotide. This is why DNA synthesis always proceeds in the 5′ to 3′ direction – new bases can be added to the 3′ end (carbon with free hydroxyl group), not to the 5′ end (phosphate) (TechBox 2.2).

454 sequencing: see pyrosequencing

Acquired characteristics are modifications to phenotype that occur during an individual's lifetime. Some early evolutionary thinkers believed that acquired characteristics of the parent could be inherited by its

Allele is derived from 'allelomorphs' (a term coi William Bateson, 1902) meaning alternative v of a trait that are transmitted independently a regate randomly, so the term allele predates n lar genetics, and can be applied to alternative of any heritable trait. In molecular genetic term les are variations on the possible genetic sequ a particular locus, that are present in the same lation. Since diploid individuals carry two cc every locus, they may be heterozygous (carry t ferent versions – alleles – of the same locus, each chromosome) or homozygous (two copie same allele). Alleles at closely linked loci wi to be inherited together: sets of linked alleles haplotype.

Alpha globin: see haemoglobin

Amino acids are the building blocks of proteins

The **figure legends** sometimes have interesting facts not directly related to the subject of the chapter, because that's what makes biology fun.

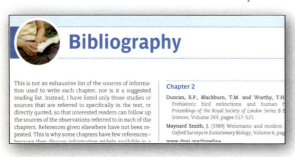

Figure 5.6 Males of the side-blotched lizard *Uta stansburiana* come in three kinds, each characterized by a mating strategy and a throat colour. The selective advantage of each morph depends on the relative frequency of the others: in a population with lots of orange-throats it's better to be a yellow than a blue, but when there are a lot of blues around, it is better to be an orange than a yellow. Frequencies of the three morphs fluctuate on a 4 to 5-year cycle. When Barry Sinervo and Curtis Lively discovered this, they 'looked at each other and . . . said "Dude! It's the rock–scissors–paper game!"' (see

The **Bibliography** contains the publication details of some studies referred to in the text (usually referred to by first author and date, e.g. 'Bromham 2008'), plus the sources of some of the studies cited in each chapter.

Bibliography

This is not an exhaustive list of the sources of information used to write each chapter, nor is it a suggested reading list. Instead, I have listed only those studies or sources that are referred to specifically in the text, or directly quoted, so that interested readers can follow up the sources of the observations referred to in each of the chapters. References given elsewhere have not been repeated. This is why some chapters have few references – because they discuss information widely available in a

Chapter 2

Duncan, R.P., Blackburn, T.M. and Worthy, T.H. Prehistoric bird extinctions and human Proceedings of the Royal Society of London Series B Sciences, Volume 269, pages 517–521.

Maynard Smith, J. (1989) Weismann and modern Oxford Surveys in Evolutionary Biology, Volume 6, pag

www.dnai.org/timeline

What is not in this book?

Instructions for computer programs: The aim of this book is to give you an understanding of the evolutionary principles that are the basis of molecular genetic analysis, rather than a training in the use of particular programs. The reason that this book sticks to general principles is that new methods are constantly being developed. Any programs or methods I describe today are likely to be out of date within a few years. But by giving you the general principles underlying those methods, I hope to equip you with the intellectual tools you need to learn how to use particular methods, and to adapt your analysis when new methods become available.

Equations: Unlike most molecular evolution textbooks, there is no maths in this book. This book is equation-free in an attempt to make the field of molecular evolution accessible to all biologists, including those who prefer explanations based on words or images rather than algebra (see page 28). You might choose to read this book alongside a traditional molecular evolution text which will provide all the population genetic equations for the principles discussed here.

What is in the Online Resource Centre?

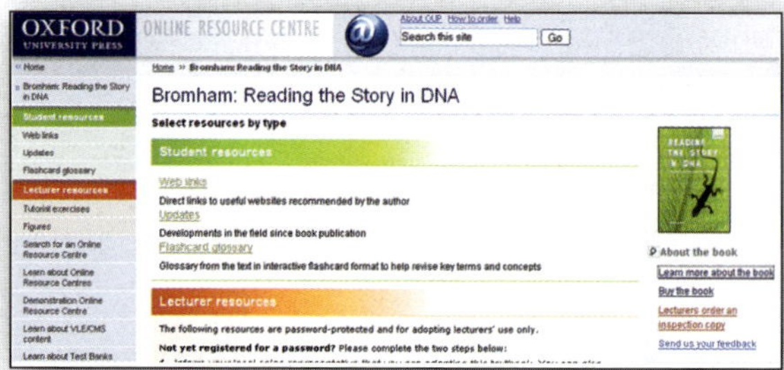

For lecturers: downloadable figures, tutorial exercises, practical projects.

For students: flashcard glossary, links to relevant resources, scientific updates.

For everyone: the chance to give feedback on the book.

Go to **www.oxfordtextbooks.co.uk/orc/bromham**

Acknowledgements

Thanks to Jonathan Crowe for initiating this project and for his cheerful optimism throughout. I owe a great debt to Meg Woolfit, whose tenacity and tireless attention to detail have greatly improved this book. I am grateful to the many people who have provided helpful comments on draft chapters, including Rob Lanfear, John Welch, Eric Fontanillas, Jessica Thomas, Marcel Cardillo, Matt Phillips, Kim Sterelny, Brett Calcott, Patrick Forber, David Bryant, Andrew Rambaut, Toni Craig, Katherine Bache, Ted Phelps, Heather Barnett, and Yubelsy Bello. Thanks to Tempo and Mode people for encouragement and collaborative glossary writing (especially Brett, Kim, John, Jochen, Marcel, and Holly). The following people generously allowed their images to be used in this book: Scott Baker, Marcel Cardillo, Raj and Ruby Akinsanya, Dan Edwards, Theo Evans, Ben Kerr, Asha Cardillo, Tom and Ben Dickinson, Carol Hartley, Ingrid and Anita Cardillo, Tom Cardillo, Rod Peakall, Conrad Hoskin, Meg Woolfit, Ted Phelps, Andrew and Hamish Rambaut, and Jo Kelly. Thanks to Beverley and Barry Bromham, Tom and Annemarie Cardillo, and David and Lysbeth Gibson for their support and encouragement, and to Marcel and Asha Cardillo for their heroic support and stoic tolerance.

Contents

Descent with modification

Or: how do I detect natural selection?

page 137

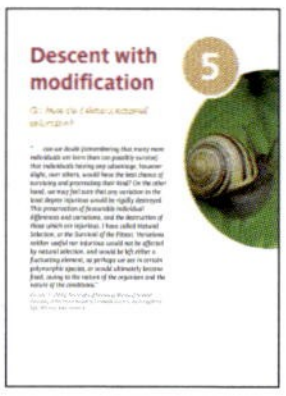

Origin of species

Or: how do I align DNA sequences?

page 185

Tree of life

Or: how do I construct a phylogeny?

page 224

Tempo and mode

Or: how do I estimate molecular dates?

page 279

You are a scientist

Or: what do I do now?

List of TechBoxes

List of Case Studies

The story in DNA

Or: what kind of information can I get from DNA?

"In nature's infinite book of secrecy a little I can read."

William Shakespeare (1623) *Anthony and Cleopatra*

What this chapter is about

This chapter gives an overview of the kind of information that can be gained from analysing DNA sequences, including identifying individuals, unravelling social interactions, understanding the evolution of major adaptations, tracing the evolutionary origins of lineages, and investigating the tempo and mode of evolution over all time scales. By taking a broad look at the use of DNA sequences in evolution and ecology, we will set the scene for later chapters; topics only briefly mentioned here will be covered in more detail later in the book.

Key concepts

Evolutionary biology: evolution connects individual lives to population-level processes to large-scale evolutionary change.

Molecular evolution: information on all of these levels is available in DNA.

Techniques: DNA sequencing

The mystery of the Chilean blob

In July 2003, a 13-tonne blob washed up on a Chilean beach (**Figure 1.1**). With no bones, no skin, not even any cells, there was nothing obvious to identify the origin of the mystery blob. Theories abounded: it was a giant squid, a new species of octopus, an unknown monster from the deep, an alien. This was not the first appearance of a giant 'globster'. In 1896, an 18-metre blob washed up on St Augustine beach in Florida, USA. The identification of the blob varied from giant squid to whale blubber. However, when it was formally described in the scientific literature (albeit sight unseen), it was given the scientific name *Octopus giganteus*. There have been at least half a dozen other reported globsters, including the 1960 Tasmanian West Coast Monster, 9 m long and 2.5 m tall, which, although badly decomposed, was described as being hairy.

The mystery of the Chilean blob was solved in 2004 when researchers sequenced DNA samples from the globster, and compared them to DNA sequences held in the gigantic public database GenBank (see **TechBox 1.1**). The sequences matched that of a sperm whale (scientific name *Physeter catadon*: **Figure 1.2**, p. 7), The globster was nothing but blubber (see **Case Study 1.1**, p. 5). The team also retrospectively solved previous sea monster mysteries. For example, DNA extracted from samples of the 'Nantucket Blob' of 1996 showed that it had been the remains of a fin whale (*Baleonoptera physalus*). As reported on Unexplained-Mysteries.com: 'One of the myths of the sea has been skewered by gene researchers'.

Figure 1.1 The blob that washed up on a Chilean beach had people guessing: squid, octopus, kraken? Similar blobs have been found on beaches all over the world.

Image courtesy of Elsa Cabrera/ CCC.

The DNA sample taken from the Chilean blob was enough to unambiguously identify it as the remains of a sperm whale. But that DNA sample could do far more. From that one tiny sample, we could identify not only the species it came from, but also, given enough data, which individual whale. We could use that DNA sample to predict where that individual whale was born, and to understand the relationships it had during its lifetime, both with its immediate family and members of its social group. That sample of DNA could allow us to track whale movements across the globe, both in space and in time. We could use it to trace this whale's family history back through the ages, exploring how the whales responded to a changing world as ice ages came and went. The DNA sample could help us reconstruct the evolution of important whale adaptations such as echolocation, and identify the nearest mammalian relatives of the whales, a question that has perplexed biologists for centuries. By comparing this whale DNA to the DNA of other mammals, we could ask whether the rise of modern mammals was contingent on the extinction of the dinosaurs. Deeper still, this DNA gives the chance to look beyond the fossil record, potentially shedding light on one of the greatest biological mysteries, the explosive beginnings of the animals, and back further still to the origin of the kingdoms.

The reason the DNA sequences from the genomes of different species can provide information at all of these different levels of evolutionary history – from individuals to populations to species, families, phyla, and kingdoms of life – is that different parts of the genome evolve at

TECHBOX
1.1

GenBank

KEYWORDS

Entrez

database

NCBI

accession

annotation

FURTHER INFORMATION

The first chapter of the NCBI handbook (freely available online) explains their databases, including GenBank (*www.ncbi.nlm.nih.gov/books/bv.fcgi?rid=handbook*).

RELATED TECHBOXES

TB 6.1: Taxonomy

TB 3.4: BLAST

RELATED CASE STUDIES

CS 1.1: Chilean blob (identifying species)

CS 6.1: Barcoding nematodes (DNA taxonomy)

GenBank is a golden example of the international science community sharing data freely. It is based at the National Center for Biotechnology Information (NCBI) in the US, but is synchronized with European (EMBL) and Japanese (DDBJ) molecular databases so that they all share the same data. GenBank contains most of the DNA sequences that have ever been produced. Whenever a scientist sequences a section of DNA, they should submit the sequence to GenBank so that anyone else can access the sequence and use it in their own research. Submission to GenBank is usually a requirement of publication in the scientific press, in line with the ethos of repeatability of scientific experiments (any scientist should have access to the data and materials needed to check published results). In addition, there are many sequences on GenBank that have never been formally published, but are available for anyone to use. At the time of writing, GenBank contains over 77 billion nucleotides of DNA sequences, from over 200,000 species.

When a researcher submits a DNA sequence, they provide information about the organism it was sampled from, what kind of sequence it is, and other features of the sequence (this information is broadly known as sequence annotation). For example, if the sequence contains part of a protein-coding gene, the information given might include the location of the beginning and end of the coding sequence, the amino acid sequence of the protein product generated from it and the likely role of the protein product. Genome sequencing projects often rely on automated annotation to identify and label features such as coding regions or gene regulation elements.

Each GenBank submission is assigned a unique accession (identification) number that can be used to retrieve that sequence from the database. Sequences can also be accessed by searching for an organism, gene, author name, or key word. For example, if you type 'Chilean blob' into the GenBank search engine, Entrez, you retrieve three sequences from the study described in **Case Study 1.1** with the accession numbers AY582746, AY582747, and AY582748. If you type these accession numbers into query box on the Entrez search engine, it will locate the records of these sequences in the database so that you can access them. **Figure TB1.1** shows the GenBank entry you will retrieve if you enter the accession number AY582746.

Accepting submissions from any individual or laboratory is one of the strengths of GenBank, allowing it to rapidly expand to cover more species and more genes. But it is also a weakness, as it is difficult to guarantee submission quality. It is inevitable that some sequences contain mistakes made in the sequencing process. Worse, some sequences may represent contaminants, rather than the target sequence reported. Sometimes there are errors in annotation (sequences may be listed as the wrong gene or from an incorrect species). So while sequences from GenBank are a boon to biological research, they should not be used uncritically: check your own analyses for aberrant results that could be caused by a mistake in GenBank, and look for alternative sequences if you suspect any problems.

GenBank can be accessed using any internet browser. The easiest approach is to go to the Entrez search engine (www.ncbi.nlm.nih.gov/gquery/) which allows you to search a large collection of databases at once. In addition to the DNA sequence database GenBank, the search engine Entrez allows you to search a range of other databases, including whole genome sequences, single nucleotide polymorphisms (SNPs: **TechBox 3.3**), population-based datasets, functional and structural information on gene products, and taxonomic information.

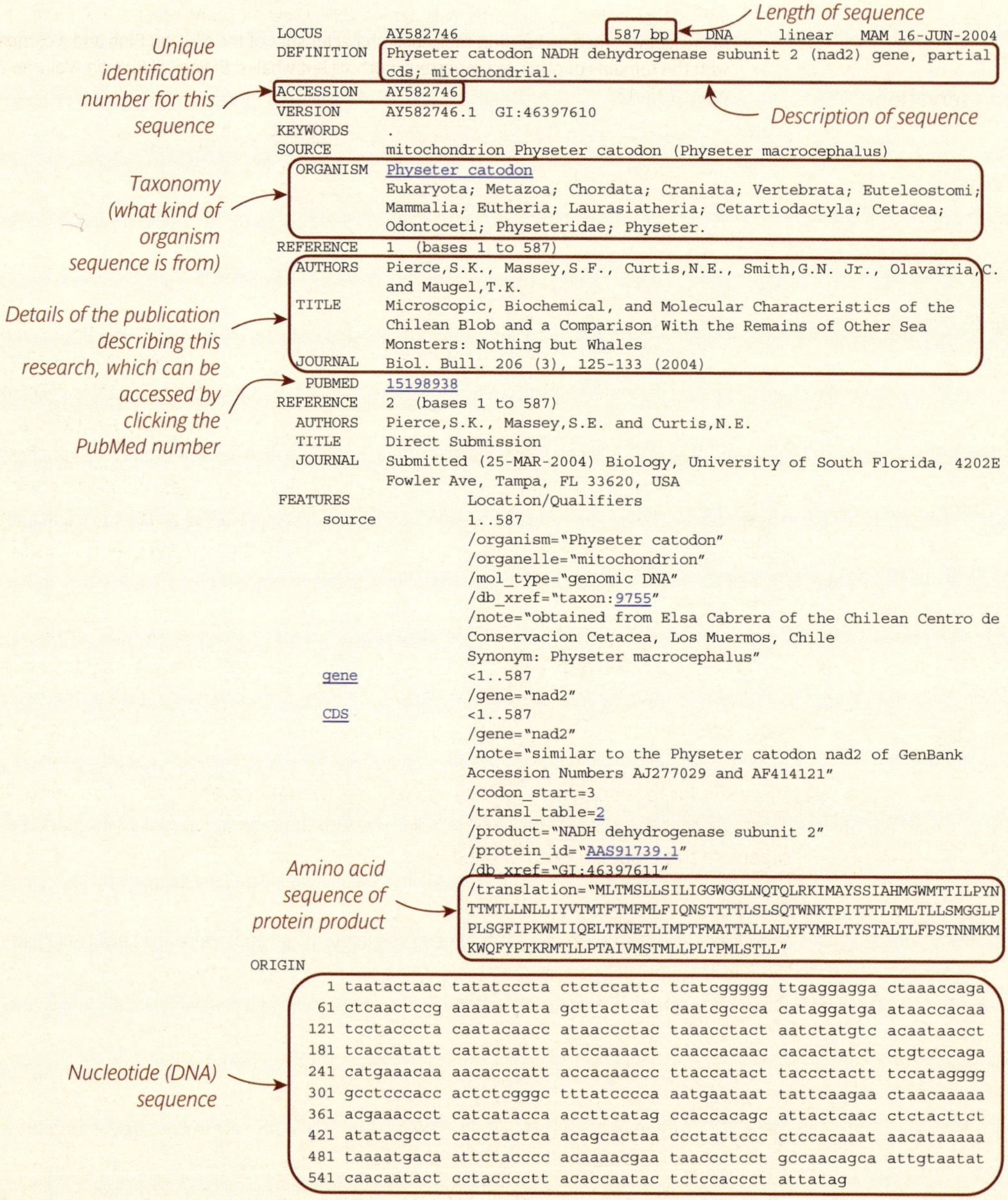

```
LOCUS       AY582746                 587 bp    DNA     linear   MAM 16-JUN-2004
DEFINITION  Physeter catodon NADH dehydrogenase subunit 2 (nad2) gene, partial
            cds; mitochondrial.
ACCESSION   AY582746
VERSION     AY582746.1  GI:46397610
KEYWORDS    .
SOURCE      mitochondrion Physeter catodon (Physeter macrocephalus)
  ORGANISM  Physeter catodon
            Eukaryota; Metazoa; Chordata; Craniata; Vertebrata; Euteleostomi;
            Mammalia; Eutheria; Laurasiatheria; Cetartiodactyla; Cetacea;
            Odontoceti; Physeteridae; Physeter.
REFERENCE   1  (bases 1 to 587)
  AUTHORS   Pierce,S.K., Massey,S.F., Curtis,N.E., Smith,G.N. Jr., Olavarria,C.
            and Maugel,T.K.
  TITLE     Microscopic, Biochemical, and Molecular Characteristics of the
            Chilean Blob and a Comparison With the Remains of Other Sea
            Monsters: Nothing but Whales
  JOURNAL   Biol. Bull. 206 (3), 125-133 (2004)
   PUBMED   15198938
REFERENCE   2  (bases 1 to 587)
  AUTHORS   Pierce,S.K., Massey,S.E. and Curtis,N.E.
  TITLE     Direct Submission
  JOURNAL   Submitted (25-MAR-2004) Biology, University of South Florida, 4202E
            Fowler Ave, Tampa, FL 33620, USA
FEATURES             Location/Qualifiers
     source          1..587
                     /organism="Physeter catodon"
                     /organelle="mitochondrion"
                     /mol_type="genomic DNA"
                     /db_xref="taxon:9755"
                     /note="obtained from Elsa Cabrera of the Chilean Centro de
                     Conservacion Cetacea, Los Muermos, Chile
                     Synonym: Physeter macrocephalus"
     gene            <1..587
                     /gene="nad2"
     CDS             <1..587
                     /gene="nad2"
                     /note="similar to the Physeter catodon nad2 of GenBank
                     Accession Numbers AJ277029 and AF414121"
                     /codon_start=3
                     /transl_table=2
                     /product="NADH dehydrogenase subunit 2"
                     /protein_id="AAS91739.1"
                     /db_xref="GI:46397611"
                     /translation="MLTMSLLSILIGGWGGLNQTQLRKIMAYSSIAHMGWMTTILPYN
                     TTMTLLNLLIYVTMTFTMFMLFIQNSTTTTLSLSQTWNKTPITTTLTMLTLLSMGGLP
                     PLSGFIPKWMIIQELTKNETLIMPTFMATTALLNLYFYMRLTYSTALTLFPSTNNMKM
                     KWQFYPTKRMTLLPTAIVMSTMLLPLTPMLSTLL"
ORIGIN
        1 taatactaac tatatcccta ctctccattc tcatcggggg ttgaggagga ctaaaccaga
       61 ctcaactccg aaaaattata gcttactcat caatcgccca cataggatga ataaccacaa
      121 tcctacccta caatacaacc ataaccctac taaacctact aatctatgtc acaataacct
      181 tcaccatatt catactattt atccaaaact caaccacaac cacactatct ctgtcccaga
      241 catgaaacaa aacacccatt accacaaccc ttaccatact taccctactt tccatagggg
      301 gcctcccacc actctcgggc tttatcccca aatgaataat tattcaagaa ctaacaaaaa
      361 acgaaaccct catcatacca accttcatag ccaccacagc attactcaac ctctacttct
      421 atatacgcct cacctactca acagcactaa ccctattccc ctccacaaat aacataaaaa
      481 taaaatgaca attctacccc acaaaacgaa taaccctcct gccaacagca attgtaatat
      541 caacaatact cctacccctt acaccaatac tctccaccct attatag
```


Figure TB1.1 The GenBank entry for a sequence from the Chilean Blob described in **Case Study 1.1**. Important features of the GenBank entry have been labelled in red: all GenBank entries should have the same basic layout.

CASE STUDY **1.1**

Solving the mystery of the Chilean blob: identifying species using DNA sequences

KEYWORDS

GenBank

sample preservation

ethanol

formaldehyde

contamination

mitochondrial

DNA barcoding

cryptozoology

RELATED TECHBOXES

TB 1.1: GenBank

TB 6.3: Multiple alignment

RELATED CASE STUDIES

CS 2.1: Origin of faeces (identifying species from remote samples)

CS 2.2: More moa (identifying species from ancient DNA)

Pierce, S.K., Massey, S.E., Curtis, N.E., Smith, G.N., Olavarri, C. and Maugel, T.K. (2004) Microscopic, biochemical, and molecular characteristics of the Chilean Blob and a comparison with the remains of other sea monsters: nothing but whales. *Biological Bulletin*, Volume 206, pages 125–133

> “*. . . to our disappointment, we have not found any evidence that any of the blobs are the remains of gigantic octopods, or sea monsters of unknown species.*” [1]

Background

Strange gelatinous material occasionally washes up on beaches around the world. The large mass that washed up on a beach in Florida in 1896 was too heavy to be removed using a team of horses and so rubbery that axes bounced off it. The material generally lacks cells that would aid its identification, and so these ‘globsters’ have variously been described as the remains of giant squid, unknown species of octopus, or whale blubber. The lack of a clear species identification has resulted in globsters being listed on websites that discuss unexplained phenomena: the blob that washed up on a beach in Chile in 2003 made international headlines and was discussed by biologists and conspiracy theorists all over the world.

Aim

DNA sequences provide the means to unambiguously identify a biological sample, through comparison of the sequence to that from known species. Even if the sample is from a previously undescribed species, then comparing the sequence to a database of known sequences can show which species is the closest relative (for example, if a globster is a previously unknown species of giant octopus, then DNA sequences from the sample will be more similar to sequences from other octopuses than they are to those of squid or whales). These researchers aimed to use DNA from the Chilean blob to work out what kind of organism the blob had been derived from.

Methods

Small pieces of the blob were frozen, or preserved in ethanol: both procedures protect the DNA in the sample from decaying. DNA is everywhere, so contamination in sequencing labs can be a problem (Chapter 4). These researchers reduced the chance of their sequences being contaminants by analysing two different samples in independent laboratories, to check that they get the same result (this can rule out incidental contamination from the lab, but not contamination of the source of the samples). One lab in the US analysed a frozen sample of the blob, and a lab in New Zealand analysed an ethanol-preserved sample.
DNA was extracted from the samples and sequenced. The US team sequenced part of mitochondrial gene called NADH2, which codes for the subunit of a fundamental metabolic enzyme. The NZ team sequenced the mitochondrial control region. These sequences were chosen because they tend to vary between species, and because they are relatively easy to

sequence, so there are a lot of comparable sequences in the international sequence database, GenBank (**TechBox 1.1**).

Results

To identify the origin of the blob, the DNA sequences were compared to existing sequences in GenBank. The sequences of NADH2 from the blob were identical to two NADH2 sequences in the database, which had both been taken from sperm whales (*Physeter catadon*, also known by the species name *Physeter macrocephalus*: **Figure 1.2**). Because this sequence normally varies between species, this is convincing evidence that the blob was from a sperm whale. The mitochondrial control region sequence from the Chilean blob was 99% identical to *P. catadon* sequences in GenBank. The control region tends to evolve faster than NADH2, and may vary between individuals within a species, so it is often used for population-level studies. Although the control region sequence differed from the sperm whale sequence on GenBank, it was only as different as you would expect from two different individuals in the same species.

```
                   1                                                          60
Physeter catadon   TAATACTAACTATATCCCTACTCTCCATTCTCATCGGGGGTTGAGGAGGACTAAACCAGA
Chilean blob       TAATACTAACTATATCCCTACTGTCCATTCTCATCGGGGGTTGAGGAGGACTAAACCAGA

                   61                                                        120
Physeter catadon   CTCAACTCCGAAAAATTATAGCTTACTCATCAATCGCCCACATAGGATGAATAACCACAA
Chilean blob       CTCAACTCCGAAAAATTATAGCTTACTCATCAATCGCCCACATAGGATGAATAACCACAA

                   121                                                       180
Physeter catadon   TCCTACCCTACAATACAACCATAACCCTACTAAACCTACTAATCTATGTCACAATAACCT
Chilean blob       TCCTACCCTACAATACAACCATAACCCTACTAAACCTACTAATCTATGTCACAATAACCT

                   181                                                       240
Physeter catadon   TCACCATATTCATACTATTTATCCAAAACTCAACCACAACCACACTATCTCTGTCCCAGA
Chilean blob       TCACCATATTCATACTATTTATCCAAAACTCAACCACAACCACACTATCTCTGTCCCAGA

                   241                                                       300
Physeter catadon   CATGAAACAAAACACCCATTACCACAACCCTTACCATACTTACCCTACTTTCCATAGGGG
Chilean blob       CATGAAACAAAACACCCATTACCACAACCCTTACCATACTTACCCTACTTTCCATAGGGG

                   301                                                       360
Physeter catadon   GCCTCCCACCACTCTCGGGCTTTATCCCCAAATGAATAATTATTCAAGAACTAACAAAAA
Chilean blob       GCCTCCCACCACTCTCGGGCTTTATCCCCAAATGAATAATTATTCAAGAACTAACAAAAA

                   361                                                       420
Physeter catadon   ACGAAACCCTCATCATACCAACCTTCATAGCCACCACAGCATTACTCAACCTCTACTTCT
Chilean blob       ACGAAACCCTCATCATACCAACCTTCATAGCCACCACAGCATTACTCAACCTCTACTTCT

                   421                                                       480
Physeter catadon   ATATACGCCTCACCTACTCAACAGCACTAACCCTATTCCCCTCCACAAATAACATAAAAA
Chilean blob       ATATACGCCTCACCTACTCAACAGCACTAACCCTATTCCCCTCCACAAATAACATAAAAA

                   481                                                       540
Physeter catadon   TAAAATGACAATTCTACCCCACAAAACGAATAACCCTCCTGCCAACAGCAATTGTAATAT
Chilean blob       TAAAATGACAATTCTACCCCACAAAACGAATAACCCTCCTGCCAACAGCAATTGTAATAT

                   541                                          587
Physeter catadon   CAACAATACTCCTACCCCTTACACCAATACTCTCCACCCTATTATAG
Chilean blob       CAACAATACTCCTACCCCTTACACCAATACTCTCCACCCTATTATAG
```

Figure CS1.1 Part of an alignment of a sequence from the mysterious Chilean blob with that of a sperm whale (*Physeter catadon*), showing that the sequences are identical for this part of their mitochondrial genome.

Conclusions

The identical (or near-identical) match between the sequences from the blob and whale sequences in GenBank prove beyond doubt that the blob is the remains of a sperm whale.

Limitations

Selection of an appropriate gene to sequence is critical – there has to be enough difference between species to allow discrimination, yet not so much difference that the relationship to other species is unclear (see Chapter 6). Sample quality and nature of preservation are important for DNA extraction. They could not extract DNA from earlier 'blob' samples that had been preserved in formaldehyde (including samples of the relatively recent Tasmanian West Coast Monster), because formaldehyde destroys DNA. This is a shame, because many museum specimens are pickled in formaldehyde.

Future work

Because DNA can be extracted from samples such as hair, faeces, or saliva, molecular analysis has the potential to solve many mysteries of cryptozoology (the field that seeks to establish whether apparently mythical creatures are in fact real organisms). For example, DNA from purported Yeti hair samples was surprisingly similar to DNA sequences from horses[2]. The DNA barcoding movement seeks to catalogue DNA sequences of all species so that any unknown sample can be identified.

References

1. Pierce, S.K., Massey, S.E., Curtis, N.E., Smith, G.N., Olavarri, C. and Maugel, T.K. (2004) Microscopic, biochemical, and molecular characteristics of the Chilean Blob and a comparison with the remains of other sea monsters: nothing but whales. *Biological Bulletin*, Volume 206, pages 125–133.
2. Milinkovitch, M.C., Caccone, A. and Amato, G. (2004) Molecular phylogenetic analyses indicate extensive morphological convergence between the 'yeti' and primates. *Molecular Phylogenetics and Evolution*, Volume 31, pages 1–3.

Figure 1.2 Sperm whales (*Physeter catadon*, also known as *Physeter macrocephalus*) derive their common name from spermaceti, a waxy, milky white substance that is found in abundance in sperm whales' heads. The exact function of spermaceti is not known: it might act as a sounding medium for echolocation, or to aid buoyancy, or possibly to give the head extra heft for head-buts in male-to-male combat. When whaling was a global industry, whale oil, derived from spermaceti, had variety of industrial uses, including the production of candles.

Reproduced by permission of Christian Darkin/ Science Photo Library.

different rates. This means that different parts of the genome can be selected to tell different stories. This chapter will use examples from the scientific literature to briefly illustrate what kind of information you can get from a tiny DNA sample. The rest of the book will show you how to do it.

Individuals, families, and populations

The genome of a sperm whale contains over 3,000,000,000 nucleotides of DNA. The nucleotides come in four types, which are given the single-letter codes A, C, G, and T. These four letters make up the DNA alphabet. All of the information needed to make the essential parts of a whale, such as the skin, the eyes, the blubber, and the blood, is coded in these four letters.

The genetic code is covered in Chapter 2

Most of the genome is exactly the same for all sperm whales, because all whales must be able to make functional skin, eyes, blubber, and blood in order to survive. But some of the DNA letters can change without destroying the information needed to make a working whale. Because of this, some DNA sequences vary slightly between individual sperm whales. Most of these differences between genomes arise when DNA is copied from the parent's genome to make the eggs or sperm that will go on to form a new individual. DNA copying is astoundingly accurate, but it is not perfect. In fact, you could probably expect around 100 differences in the nucleotide sequence of the sperm whale genome between a parent whale and its calf, due to mutation. The upshot of this is that, although most DNA sequences are exactly the same in all sperm whales, every individual whale has some changes to the genome that make it unique. So given enough DNA sequence data it would be possible to tell not only which species the Chilean blob had come from, but also which individual whale. The same rationale applies to forensic DNA analysis to identify biological samples left at crime scenes: because each individual has a unique genome, if the DNA from the victim matches the blood on the accused's clothes, then the two must be linked.

Chapter 3 explains how mutation makes individual genomes unique

The sperm whale's genome is copied when it reproduces, and any changes to its genome will also be inherited by the whale's offspring. Therefore, by asking which individuals share particular sequence changes, biologists can use DNA sequences to reveal the relationships between individual whales. Since DNA is inherited from both father and mother with relatively few changes, it is possible to use DNA sequences to conclusively identify an individual's parents (hence the growing number of companies offering 'paternity tests'). The inheritance of specific DNA differences can also be traced back through a whale's family tree, to its parents, grandparents, and great grandparents, and so on. So, in addition to identifying specific individuals, if you take DNA sequences from a whole group of whales you can tell who is whose mother or sister or cousin. More generally, you can start to understand how populations of sperm whales interact and interbreed.

Chapter 4 explains how DNA replication results in related individuals being more genetically similar

Sperm whale social groups

Sperm whales travel in social groups that co-operate to defend and protect each other, and may even share suckling of calves. It is difficult to determine the membership of these groups from sightings alone, because of the practical difficulties of observing whale behaviour, most of which happens underwater. To make things even more difficult, sperm whales can travel across entire oceans and can dive to a depth of a kilometre. Biologists who study whale behaviour generally have to be content with hanging around in boats, waiting for their subjects to surface. But when they do surface, in addition to taking photos which allow individual whales to be identified, biologists can zip over in

Figure 1.3 Research groups that use molecular data to study whales typically wait for whales to surface, then either scoop sloughed-off skin cells from the surface, or shoot biopsy darts that take small tissue samples. This photograph shows Scott Baker and his research student Carlos Olavarria in Tonga approaching a humpback whale in order to take a biopsy.

Reproduced by permission of C. Scott Baker, Marine Mammal Institute, Oregon State University and School of Biological Sciences, University of Auckland.

worryingly small boats and pick up the bits of skin that the whales leave behind on the surface when they resubmerge (**Figure 1.3**). The DNA extracted from these bits of whale skin not only identifies the individuals in the group, but also reveals their relationships to each other. This has allowed researchers to described sperm whale social groups in detail.

DNA analysis from these skin samples shows that sperm whale social groups are made up of 'matrilines', or female family groups (mothers, daughters, sisters, and so on). Males leave the group before they mature. But not all the individuals in the group are related to each other – there are members of several different matrilines in each group. This suggests that, while adult males come and go and rarely stay with the group longer than it takes to father more offspring, female sperm whales form long-term, and possibly life-long, relationships. By sequencing sperm whale skin samples (**TechBox 1.2**), it is possible to get an insight into the private lives of animals that were previously hard to observe, yet without disrupting their natural behaviour.

Whale populations in space and time

Sperm whales are a global species, found in every ocean. Female social groups inhabit relatively warm temperate and tropical waters and do not tend to move between oceans. But males, who leave the matrilineal social groups at a young age to join roaming 'bachelor schools', travel more widely, going to cooler polar waters to feed, and returning to lower latitudes to mate. Because males move away from their natal group and can travel between oceans, a calf's father could have been born in any ocean of the world. But if a sperm whale calf is born in the Pacific Ocean, then we can be almost certain that its mother was also born in the Pacific and so was its grandmother.

This behavioural difference between males and females is reflected in sperm whale DNA. A sperm whale's nuclear DNA is carried on 44 chromosomes, located in the nucleus of each cell. Each individual whale inherits half of its chromosomes from its mother and half from its father. As males move around the world and mate, they spread their nuclear DNA around.

TECHBOX 1.2 DNA sequencing

KEYWORDS

chain termination

Sanger sequencing

dideoxy

dNTP

primer

gel

pyrosequencing

FURTHER INFORMATION

An explanation of sequencing using an automated sequencing machine, with helpful diagrams, can be found at *http://allserv.rug.ac.be/~avierstr/principles/seq.html*

RELATED TECHBOXES

TB 2.4: DNA extraction

TB 4.2: DNA amplification

RELATED CASE STUDIES

CS 1.1: Chilean blob (identifying species)

CS 1.2: Whale meat (DNA surveillance)

To understand DNA sequencing, it is helpful to have a grasp of DNA structure (**TechBox 2.2**), DNA replication (**TechBox 4.1**), and DNA amplification (**TechBox 4.2**). This means that you may find this box rather challenging if you do not already have a basic understanding of molecular genetics. So why is DNA sequencing introduced in the first chapter? Because it is fundamental to all other topics covered in this book. We have to start somewhere, so let's start with sequencing. Just as this introductory chapter serves as a road map to the topics we will cover in later chapters, this TechBox will briefly touch upon many topics that we will cover in later TechBoxes. My advice is to read through it now without worrying too much about the details, then come back to it later when you have built up a basic understanding of the concepts involved.

1 DNA extraction

Most biological samples, such as a drop of dried blood, a fresh leaf, or the marrow from an old bone, contain DNA. Some tissues are easier to extract DNA from than others, and fresh material is usually easier to work with than old specimens (and some preservation techniques destroy DNA: **Case Study 1.1**). Chemicals, such as phenol and chloroform, are added to the sample to break down the cell membranes and precipitate the DNA so that it is floating in an identifiable layer of liquid that can be pipetted out.

DNA extraction is explained in **TechBox 2.4**

2 Amplification

You need to have a vast number of copies of the DNA you wish to sequence. So you need to take the DNA in your sample and amplify (make many copies of) the sequence that you are interested in. The most common way of doing this is following a series of reactions called the polymerase chain reaction (PCR). The DNA sample is heated (to separate the double strands), mixed with a primer (a short string of RNA that matches the start of the sequence: **TechBox 4.3**), cooled (so that the primer bonds with the DNA from the sample), then warmed again with polymerase (DNA-copying enzyme) which attaches to the primers so that only the target sequence is amplified.

DNA amplification is explained in **TechBox 4.2**

3 Sequencing

Here is an overview of the most common kind of sequencing reaction, which will be described in more detail below. The amplified DNA is heated (to separate the strands) and mixed with polymerase (DNA-copying enzyme), nucleotides (DNA 'letters'), and primers (short DNA sequences that serve as starting blocks for DNA synthesis). Some of the nucleotides are labelled with something that makes them detectable, such as fluorescent dyes or radioactive labels. And some of the nucleotides are modified so that they stop the synthesis of DNA whenever they are added to the new strand. The polymerase makes copies of the DNA in the sample, by constructing chains of nucleotides that are complementary to the template DNA (**TechBoxes 2.2** and **4.1**). While making these nucleotide chains, the reaction occasionally incorporates a labelled nucleotide. And sometimes incorporates a reaction-stopping nucleotide, at which point the growth of that DNA chain stops. So throughout the reaction mixture, copies of the template DNA are being made by polymerase, and those copies are stopping at different points, whenever the polymerase happens to incorporate a reaction-stopping nucleotide. The result is a solution of DNA sequences of different lengths, each ending at a different 'letter' in the DNA sequence. If these sequences are 'read' in order of length, then the sequence of bases at the ends of the fragments provides the sequence for the template they were all copied from.

Chain-termination sequencing in more detail: The Sanger sequencing method (see **Heroes 1**) is known as the chain-termination method because it uses modified nucleotides to halt DNA synthesis at different points to produce an array of DNA fragments, each of which stops at a different nucleotide in the sequence. This method is also known as dideoxy sequencing because the modified nucleotides are missing their OH group. Since nucleotides are added to the OH group of the last nucleotide in the chain (see **TechBox 4.1**), once a dideoxynucleotide is added, no more nucleotides can be added to that chain, so synthesis stops.

The growth of a DNA chain is described in TechBoxes 4.1 and 4.2

The specific details differ between methods, but the basic approach of Sanger sequencing is:

(i) Denature: The amplified DNA is heated, to separate the double-stranded DNA into single strands.

The chemical bonds that hold the DNA helix together are described in TechBox 2.2

Case Study 4.1 describes an application of DNA hybridization (separating and rejoining DNA helices)

(ii) Prime: short sequences that have a complementary sequence to the target sequence are added to the single stranded DNA, which is cooled so that the primers can bind to the template (**TechBox 4.2**). The primer provides a starting block for synthesizing a new DNA strand, because it provides a free OH group for newly added bases to bind to (**TechBox 4.3**).

Primers are explained in TechBox 4.3

DNA synthesis in natural systems is covered in TechBox 4.1 and the laboratory methods for DNA amplification are covered in TechBox 4.2

(iii) Nucleotides: the amplified DNA with attached primers is split into four samples. To each of these samples is added the four DNA bases (in the form of deoxynucleotides: dATP + dTTP + dCTP + dGTP) and DNA polymerase (the enzyme that makes a new DNA strand to match the template: see **TechBox 4.1**). In addition, each sample has a different chain-terminating dideoxynucleotide (ddATP or ddTTP or ddCTP or ddGTP).

Nucleotides are covered in TechBox 2.2 on DNA structure

(iv) Grow: As the polymerase moves along the template, it picks up and adds nucleotides to make a matching DNA strand. If it picks up a deoxynucleotide (dNTP), it continues to add nucleotides to the growing strand, but if it incorporates a dideoxynucleotide (ddNTP), chain elongation stops at that point. Using the example sequence given in the diagram, in the ddCTP sample, the polymerase will add nucleotides to the growing strand.

When it gets to a C it might incorporate a normal dCTP and keep going, or it might incorporate a chain terminating ddCTP and stop.

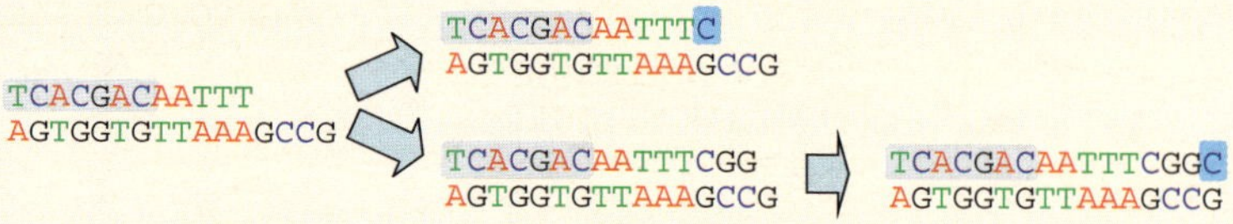

The upshot of this is that, in the ddCTP sample, there will be fragments of many different lengths, but they will all stop at a C. Similarly, the ddTTP sample will have fragments of different lengths that all end at a place in a sequence where there is a T.

DNA synthesis is covered in Chapter 4

(v) Run: To read the sequence, it is necessary to order the fragments with respect to size and report which nucleotide is at the end of each fragment. In traditional Sanger sequencing, the fragments are labelled radioactively (usually by labelling either the dNTPs or ddNTPs). The fragments are run through a gel, with each of the four samples run in a different lane. Since longer sequences are larger and travel more slowly through the gel, the bands will appear on the gel in order of length, so the nucleotides at the end of each fragment can be read in the order of the bands along the gel. In some automated sequencing reactions fluorescent dyes are added to the primer sequences, so that the fragments can be read by being passed through an optical reader. Alternatively, the ddNTPs can be labelled with different coloured fluorescent dyes: one advantage of this approach is that all labelled dNTPS can be added together, rather than in four separate reactions.

The movement of DNA fragments through a gel is described in Chapter 5

Pyrosequencing: this alternative approach to Sanger sequencing is gaining popularity for large-scale sequencing projects. In standard sequencing reactions, the reporter is some kind of label attached to the products of DNA synthesis. In pyrosequencing, the chain-elongation reaction itself is the reporter. Luciferase (a type of enzyme that produces bioluminescence, such as the enzyme that makes fireflies light up) is included in the reaction mixture, along with DNA, primers, and polymerase. Each dNTP is washed over the reaction mixture in turn. The incorporation of each new nucleotide releases a flash of light, which identifies the position of that particular base in the DNA sequence. The sequence can be read off by monitoring whether light is released as each type of nucleotide is added. 454 sequencing technology, which enables rapid, large-scale sequencing, is based on pyrosequencing.

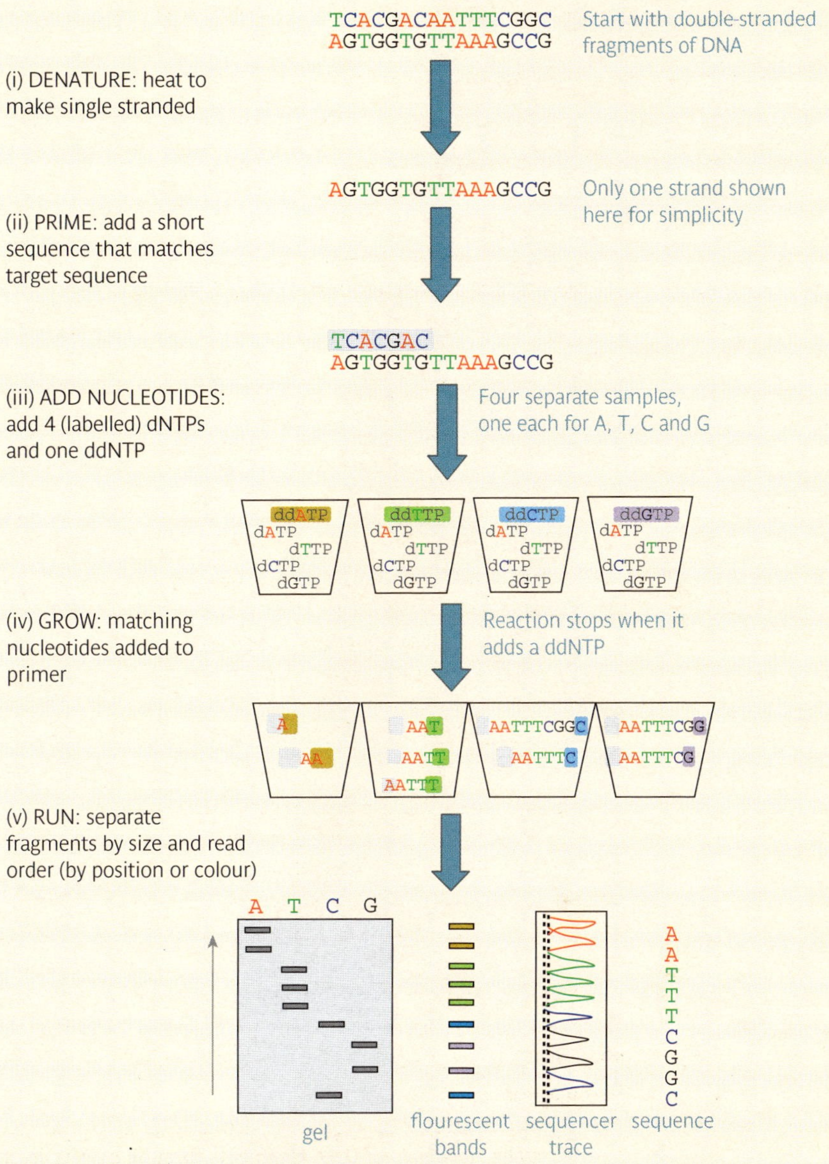

Figure TB1.2 A rough guide to DNA sequencing using the chain-termination (Sanger) method, including several different ways of reading the order of the nucleotides.

But most cells contain another source of DNA, in addition to the chromosomes in the nucleus. Mitochondria – energy-producing organelles found in the cellular cytoplasm – contain their own tiny genomes, less than 0.01% the size of the nuclear genome. In sperm whales, as in most vertebrate species, mitochondria are passed from generation to generation in the egg cells supplied by the mother. Males do not pass their mitochondrial DNA to their offspring, because the mitochondria carried in sperm cells are jettisoned from the fertilized embryo. Therefore, mitochondrial DNA is inherited through the female line. Since females do not roam as widely as males, each whale will tend to have the mitochondrial sequence typical of the ocean in which it was born. This means that if you were to give a sample of sperm whale skin (or blood, blubber, or tooth) to a biologist, by sequencing the mitochondrial DNA, they could probably tell you in which region of the world that whale was born (see **Case Study 1.2**).

DNA sequences can reveal not only the current global distribution of sperm whales, but also how the population has changed over time. Changes accumulate in the mitochondrial genome as it is copied and passed from mother to daughter. With every generation, more DNA differences accumulate, so the number of differences between individual genomes increases. When measures of genetic similarity are taken for a whole population, it gives an indication of population size and mating structure. In a small population, there is an increased chance of mating with a relative. This means that in a small, inbred population, you would generally not have to trace two individuals' family trees back many generations before you found a shared ancestor. Since these two related individuals will have both inherited some of the same genetic variants from their common ancestor, you would expect their genomes to be more similar than when you compared two unrelated individuals. In a large population where mating between unrelated individuals is the rule, you would have to go much further back to find an ancestor shared by any two individuals. So two individuals in a large population probably have less of their genome in common than two individuals in a small population. Therefore, the average number of genetic differences between individuals gives an indication of population size. This can be very useful for estimating the numbers of individuals in species that are difficult to survey directly, such as sperm whales.

The effect of population size on DNA evolution is covered in Chapter 5

Current estimates suggest that the size of the global sperm whale population is probably around 360,000 individuals. But the genetic diversity of this population is surprisingly low. The low number of differences between the mitochondrial DNA sequences of sperm whales from around the globe suggests that they are all descended from a small number of founding mothers who survived the last ice age. Since sperm whale females prefer warmer waters, their distribution may have shrunk towards the equator as the world's oceans cooled, reducing their population size. As the ice age ended and the climate warmed, the sperm whale population might have expanded and spread out around the globe once more.

Uncovering the evolutionary past

Although the population may have reduced and expanded with the changing climate, sperm whales have swum in the oceans for millions of years. Sperm whales hunt using echolocation: they emit bursts of ultrasonic sound and use the sound reflected back from their surroundings to locate prey (which, in the case of sperm whales, includes giant squid). Echolocation is a characteristic that sperm whales share with other members of the Odontoceti, the group of predatory toothed whales that includes the dolphins and orcas. The other main group of cetaceans, the Mysticeti, have no need for echolocation. They are the baleen whales that use huge filter plates in their mouths (the baleen) to sieve plankton out of the water as they swim. The Mysticeti includes the gigantic blue whale, which, weighing up to 150 tonnes, is the largest animal that has ever lived. It was previously assumed that these two very different types of cetacean, odontocetes and mysticetes, evolved from a primitive whale species over 30 million years ago, and then each lineage developed special adaptations to their different lifestyles: the odontocetes evolved echolocation for hunting prey, and the mysticetes evolved baleens for sieving plankton.

The genome of the sperm whale tells a different story. Sperm whales are currently classified as odontocetes because they have classic toothed whale characteristics, such as teeth and echolocation (**Figure 1.4**, p. 17). But DNA sequences from sperm whales are more similar to sequences from baleen whales, not toothed whales.

CASE STUDY 1.2

DNA surveillance: using DNA species identification to trace illegal trade in whale meat

KEYWORDS

PCR

conservation

CITES

database

geographic origin

identifying individuals

DNA barcoding

FURTHER INFORMATION

www.dna-surveillance.auckland.ac.nz is the web-based species-identification site that takes a DNA sequence and aligns it against a curated database of reference sequences, and calculates genetic distance to known species.

RELATED TECHBOXES

TB 1.1: GenBank

TB 4.2: DNA amplification

RELATED CASE STUDIES

CS 2.1: Origin of faeces (identifying species from remote samples)

CS 6.2: Keeping the pieces (DNA and conservation)

Baker, C.S. and Palumbi, S.R. (1994) Which whales are hunted? A molecular-genetic approach to monitoring whaling. *Science*, Volume 265, pages 1538–1539

> “*These results confirmed the power of molecular methods in monitoring retail markets and pointed to the inadequacy of the current moratorium for ensuring the recovery of protected species.*”[1]

Background

Following dramatic falls in global whale populations, commercial hunting of whales was banned by international treaty in 1986. Hunting of whales is now only conducted by a small number of countries. While some whale populations are increasing, there is a great deal of concern that many species are threatened with extinction. Protection for vulnerable whale species is therefore a conservation priority.

Aim

Japan continues to hunt minke whales (*Balaeonoptera acutorostrata* and *Balaeonoptera bonaerensis*) for scientific research, and, while import of whale meat is prohibited, there is no law against the sale of whale meat on the domestic market. So whale meat that is a by-product of the scientific catch can be legally sold in Japan. These scientists set out to test if all of the whale meat available in Japanese markets was sourced from the reported scientific catch of minke whales[2].

Methods

The UN Convention on Trade in Endangered Species (CITES), which prohibits international trade in rare animal products, does not allow whale meat to be taken across national boundaries without a permit. But amplified DNA (PCR product: **TechBox 4.2**) does not come under CITES legislation, because it is a synthetic copy of the DNA from the original sample. So the researchers developed a mobile PCR kit that they took to Japan. They would buy whale meat products from the market, then surreptitiously extract and PCR the DNA in their hotel room (**Figure CS1.2**). They took the PCR products back to universities in the US and New Zealand and sequenced them, and compared the whale meat sequences to sequences from known cetacean species.

Results

In the past 8 years, the team have identified the species and in some cases the geographic origin of more than 1100 whale meat products from at least 28 different species of whales and dolphins, including protected humpback, western gray, fin, sei, and Bryde’s whales. Some whale meat was from species that could not have been caught in Japanese waters, such as a type of humpback whale found only in Mexican coastal waters, implying that whale meat was being moved between countries. They have even identified a particular individual, a rare fin whale/blue whale hybrid, killed off Iceland in 1989 and purchased in an Osaka market in 1993[3].

Figure CS1.2 Scott Baker using a 'portable laboratory', consisting of a PCR machine plus chemical reagents which were capable of being carried in a suitcase, to extract and amplify DNA in a hotel room. New DNA-amplification methods that do not rely on cycles of heating and cooling may make mobile DNA testing much easier (see **TechBox 4.2**).

Reproduced by permission of C. Scott Baker, Marine Mammal Institute, Oregon State University and School of Biological Sciences, University of Auckland.

Conclusions

Whale meat purchased in Japan came from a variety of species, including species that have been banned from hunting for over three decades, and species that cannot have been caught in Japanese waters. It was therefore not entirely by-products of the scientific catch.

Limitations

The DNA surveillance database is currently limited to two mitochondrial sequences for cetacean species, but could be expanded to other genes and other taxa (for example, it has recently been extended to include 'What Rat is That?', to identify rat species which can be hard to identify on morphological grounds alone). DNA surveillance reports the closest match between the database and the sample DNA: in the case of a poor match to the database, the sample should be checked against GenBank (**TechBox 1.1**, p. 3).

Future work

The team aims to develop 'Same-Day Whale Identification by DNA': using species-specific primers to test whether a sample is from the target species (**TechBox 4.3**; **Case Study 2.1**). This approach is being extended from identification of protected species to monitoring the demographic impact of hunting on species such as the minke whale[1]. The researchers have also used microsatellites to track multiple products from the same whale, in order to estimate the number of individual whales ending up in the market[4].

References

1. Baker, C.S., Lento, G.M., Cipriano, F. and Palumbi, S.R. (2000) Predicted decline of protected whales based on molecular genetic monitoring of Japanese and Korean markets. *Proceedings of the Royal Society London B*, Volume 267, pages 1191–1199.
2. Baker, C.S., Cipriano, F. and Palumbi, S.R. (1996) Molecular genetic identification of whale and dolphin products from commercial markets in Korea and Japan. *Molecular Ecology*, Volume 5, pages 671–685.
3. Cipriano, F. and Palumbi, S.R. (1999) Genetic tracking of a protected whale. *Nature*, Volume 397, pages 307–308.
4. Baker, C.S., Cooke, J.G., Lavery, S., Dalebout, M.L., Ma, Y.-U., Funahashi, N., Carraher, C. and Brownell, R.L. Jnr (2007) Estimating the number of whales entering trade using DNA profiling and capture-recapture analysis of market products. *Molecular Ecology*, Volume 16, pages 2617–2626.

Figure 1.4

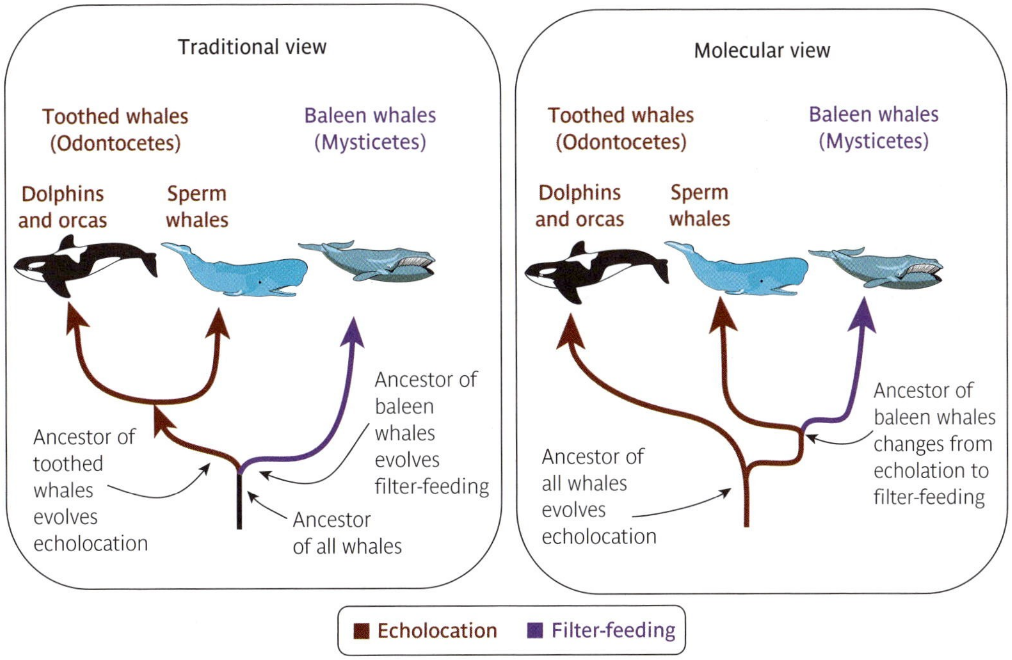

This suggests that sperm whales are more closely related to the filter-feeding odontocetes than they are to the other echolocating mysticetes. This may seem like a fairly unimportant distinction, but it has important implications for the interpretation of whale evolution. If sperm whales belong with the mysticetes, then both major branches of the whale tree contain classic odontocete characters such as echolocation (**Figure 1.4**).

If echolocation is present in both major branches of the whale lineage, then the most likely explanation is that both lineages inherited echolocation from their common ancestor. This means that the ancestral whale must have had echolocation, but it was lost in the baleen whales as they adapted to a new way of life. This hypothesis, built upon DNA evidence, has gained some support from morphological studies, such as the evidence of vestigial 'melons' (echolocation sounding chambers) in baleen whales, remnants left from their predatory ancestor.

DNA sequences are valuable sources of information for ascertaining the relationships between living species. And these relationships can tell us a lot about the evolutionary past. The relationships between whale lineages, revealed by analysis of sperm whale DNA, shows that evolution both adds and takes away complex characteristics. In this way, evolutionary changes can sometimes obscure the history of species. But while a species' appearance changes, the genome continues to record evolutionary history. Because of this, molecular data can sometimes give a clearer view of a species' evolutionary past than can its highly modified morphology.

The importance of inherited similarities (homologies) for uncovering evolutionary relationships is covered in Chapter 6

Four-legged ancestors

Whales may look like giant fish, but inside they are typical mammals, with mammalian blood, bones, and organs. So although adaptations to life at sea have, in many ways, erased the signal of the whale's past, traces of the whales' ancestry can be seen in the way that the mammalian finger bones have been modified into flippers. Some whale species even have the remains of a pelvis left over from an ancestor that walked on four legs. But what exactly was this four-legged ancestor like? While fossil data are the ultimate source of information on the morphology of ancestral species, molecular data can provide important clues by revealing which living mammals are the whales' closest relatives.

The whale skeleton has been so highly modified by evolution that biologists have argued over whether the whale is most closely related to artiodactyls (such as cows, camels, pigs), perissodactyls (horses, rhinos, tapirs), or carnivorans (cats, dogs, bears, etc.). But even when morphology changes dramatically, the genome continues, by and large, to steadily accumulate changes. Just as changes to the genome every generation allow the relationships between individual whales to be traced, so the sum of these changes over longer time periods allows the evolutionary relationships between species to be uncovered. By comparing the similarities and differences in the DNA sequences of different species it is possible to reconstruct their history as an evolutionary tree.

Chapter 7 explains how to estimate phylogenies (evolutionary trees) from DNA sequence data

When DNA sequences from whales are compared with those from other mammals, it is clear that they are most similar to artiodactyls, something that had been suspected for some time. But, more surprisingly, the DNA suggested that the whales' closest living relative is the hippo (**Figure 1.5**), and that whales and hippos share a more recent common ancestor that either do with the rest of the artiodactyls, such as pigs, goats, and camels. This initially startling idea came to be known as the Whippo Hypothesis. Although whales and hippos share some unusual characteristics, such as thick hairless skin insulated with layers of fat, these were previously considered to be convergent adaptations. That is, it was assumed that both whales and hippos independently evolved the same solutions to the shared problems of being warm-blooded mammals living in cold water. Molecular phylogenies suggest that these shared traits are not coincidental, but may have been inherited by both whales and hippos from a shared, semiaquatic ancestor. The DNA evidence suggests that hippos might provide clues to the evolution of fish-like whales from their hairy, four-legged, land-dwelling relatives.

The first whales

The fossil record of whales has improved dramatically in the last decade, with new finds providing more information on stages in the evolutionary series. The oldest whale fossils are 'walking whales' from the beginning of the Cenozoic period (which runs from 65 million years (Myr) ago to the present, and is sometimes given the romantic name 'Age of Mammals'). These four-legged mammals took to the water not long after the oceans had been vacated by the great marine reptiles, the fish-like ichthyosaurs, the serpent-like mososaurs, and the Loch Ness monster-like plesiosaurs. Fossils of these great aquatic reptiles are known from throughout the Mesozoic (from 250 to 65 Myr ago, known as the 'Age of Reptiles'). But, along with the dinosaurs, the icthyosaurs, mososaurs, and pleiosaurs all disappear from the fossil record by the beginning of Cenozoic era.

You can find the names of evolutionary eras on the geological timescale in Appendix III

The mammals that took over from the great sea-dwelling reptiles evolved similar adaptations to aquatic life, such as streamlined bodies and flippers rather than legs. Because of this, many odontocete whales (such as dolphins) look remarkably similar to the fish-like reptilian ichthyosaurs. The independent acquisition of similar traits is called evolutionary convergence (Chapter 6). The fossil record tells a similar story of convergence for other major groups of mammals which appear after the dinosaurs disappeared: hoofed mammals replaced browsing sauropods, carnivorans replaced predatory dinosaurs.

This picture of an evolutionary scramble to fill a world vacated by the dinosaurs has strongly influenced biologists' views of the relationships between the major groups of mammals, such as primates, artiodactyls, and bats. If, on being released from the tyranny of the giant reptiles, mammal groups all evolved simultaneously from a common ancestral stock, then rather than a serially branching evolutionary tree, mammalian relationships may resemble a 'bush', with all branches arising at once from a common root. This conclusion was supported by morphological studies that often failed to resolve any clear relationships between the mammalian orders.

But as the amount of DNA sequence data increases, the phylogeny of mammals is being resolved, and surprising new relationships have been suggested. For example, the base of the new molecular tree for mammals seems to be firmly rooted in Africa. DNA sequences

Figure 1.5 Hippos *(Hippopotamus amphibious)*, equally at home in the water and on the land, may provide clues to the origins of the fully aquatic Cetacea (whales and dolphins). Although hippos spend most of their time in shallow water, and can dive for 5 minutes or more, on land they can run faster than humans.

Reproduced by permission of Paul Martitz.

analysis has united an unlikely group of mammals, including aardvarks, elephants, and tenrecs, into a group now known as the Afrotheria (**Figure 1.6**, p. 20). When the evolutionary tree of mammals is constructed from molecular data, it is these Afrotherian lineages that are amongst the earliest lineages to branch off from the other mammals. This has been used to suggest that the early diversification of placental mammals occurred in ancient Africa, when the continent was isolated from the rest of the world, then spread out from there in the Cenozoic. There is currently little available fossil evidence that could shed light on land vertebrates in Africa during the Mesozoic, so an African genesis of mammals would essentially hide their early evolutionary history from the known fossil record.

Furthermore, the DNA evidence suggests that many of the major branches of the mammalian evolutionary tree stretch back into the time of the dinosaurs. Because changes to the genome accumulate continuously, the longer two lineages have been evolving separately the more differences you expect to see between their genomes. For example, the genomes of two species of baleen whale are more similar than either is to a sperm whale's genome, because the baleen whales' genomes were more recently copied from the same shared ancestor. If changes to DNA accumulate at a predictable rate in all species, then we can use a measure of genetic difference to estimate when two species last shared a common ancestor. When DNA from mammals such as whales, cats, monkeys, and rats are compared, the results are surprising: there are far more DNA differences between the major mammal groups than you would expect if their common ancestor had lived less than 65 Myr ago. Instead, the molecular data suggest that major branches of mammalian

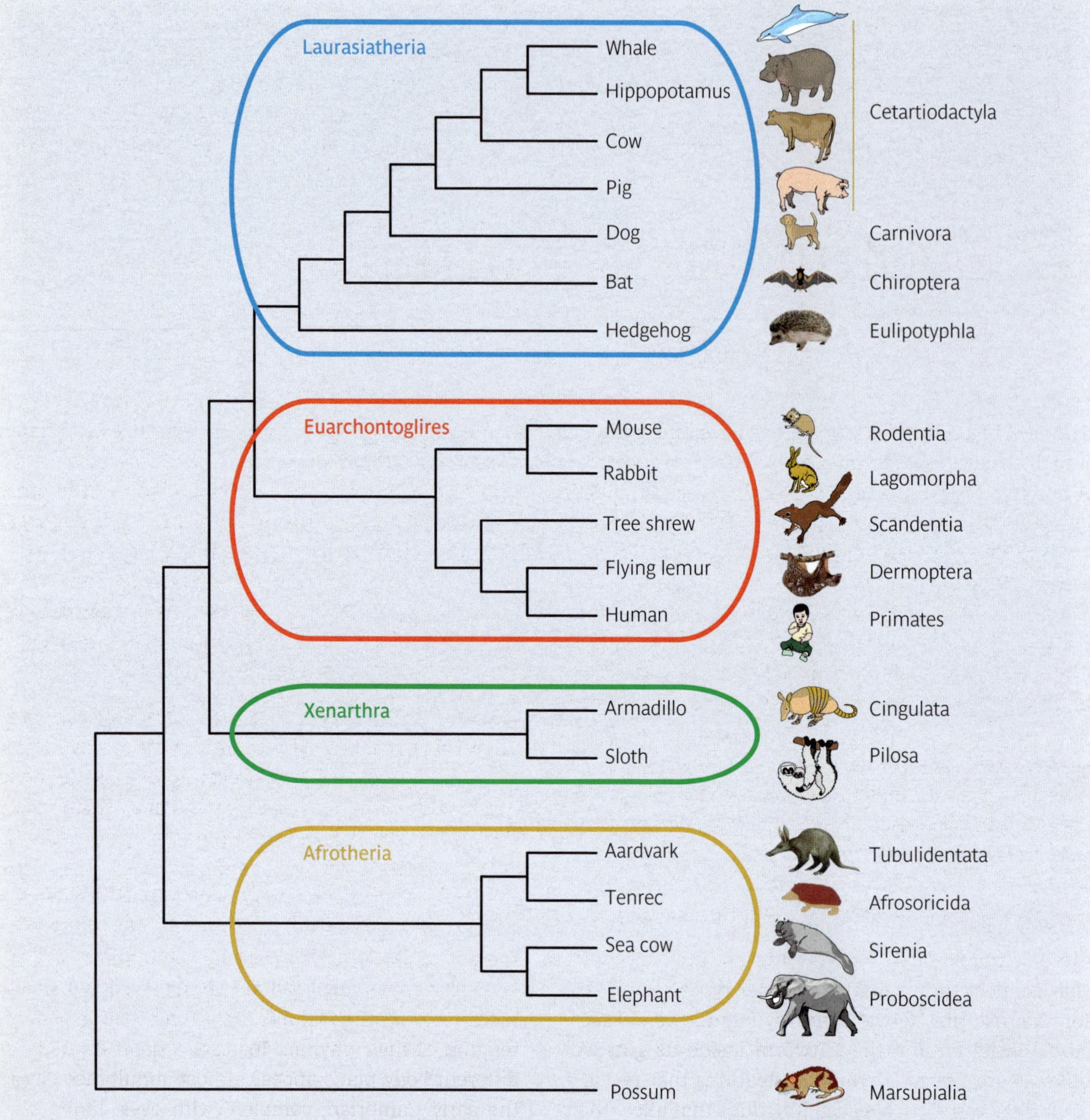

Figure 1.6 The phylogeny of placental mammals has been revised in the last decade, reshaping ideas about mammalian evolution. In particular, molecular data suggest a new grouping of 'Afrotherians' as one of the earliest lineages to emerge from the mammalian radiation.

evolutionary tree arose deep in the Mesozoic, long before the final extinction of the dinosaurs.

Estimating evolutionary time from DNA sequences is covered in Chapter 8

So analyses of DNA sequences paint a very different picture of mammal evolution, not an explosive post-dinosaur radiation, but a gradual Mesozoic diversification. But these molecular date estimates are controversial. If changes to DNA happen when it is copied each generation, then the gigantic sperm whales that take a decade to mature might accumulate DNA differences at a slower rate than their diminutive artiodactyl ancestors that might have bred every year. If that was the case, then assuming that DNA differences accumulated at the same rate in the ancient walking whales as they do in their gigantic sperm whale descendants could lead to incorrect estimates of the time of origin of the whale lineage. As our understanding of molecular evolution grows, as statistical methods become more sophisticated, and as computers get more powerful, we will become better at deciphering the story in the DNA.

Evolution of animal body plans

The DNA of a sperm whale can reach even further back into the whales' history, back to its vertebrate ancestors. A sperm whale's flipper looks nothing like a sheep's hoof, a monkey's hand, or a bat's wing, and yet we can recognize the same underlying structure, not only in all mammals but also in a gecko's toes and a parrot's wing. These disparate animals also share the 'head and tail' body plan that unites the members of the phylum Chordata, which includes the mammals, birds, reptiles, and fish. The group gets its name from the central nervous cord which runs from the head (which holds the sense organs, mouth, and respiratory equipment) to the muscular tail. It is possible to piece together a more-or-less continuous series of forms that illustrate the evolution of the chordate body plan, linking gradual transitions along the evolutionary paths that connect the sperm whale to the monkey, the parrot, and the gecko.

But it is not so easy to follow the evolutionary paths that connect the chordates to other types of animals, for example linking the sperm whale to the barnacle stuck on its fin and the flatworm inside its guts. All animals are united by common features that reveal a shared ancestry, such as cell junctions that allow their multicellular bodies to be co-ordinated, absence of cell walls which permits movement and flexibility, and heterotrophy (consuming biological material as food, rather than producing their own energy from light or chemicals). But beyond the basic shared features of multicellularity, locomotion, and heterotrophy, the main groups of animals are strikingly different. The major divisions of the animal kingdom – the phyla – are often considered to represent different body plans, or basic ways of constructing animals. For example, the arthropod body plan consists of a segmented body with jointed appendages, all clothed in a jointed exoskeleton of protein and chitin (which is what makes bugs crunch underfoot). The echinoderm body plan, on the other hand, has pentameral symmetry (like a five-pointed star), an exoskeleton made of hard calcite, and a water vascular system that, in addition to transporting nutrients around the body, can power locomotion through hundreds of soft, hydraulically operated feet (turn a starfish over to see these tube feet in action).

These animal body plans are so different from each other that there has been an intense (and often rather lively) debate about how they evolved. We can trace the whale and barnacle lineages back half a billion years, to Cambrian-age rocks that contain fossilized animals with recognizable chordate and arthropod body plans (**Figure 1.7**: alas, flatworms do not tend to leave fossils). But there the trail ends rather abruptly. There is no continuous series of fossils showing the different animal body plans diverging gradually from each other, slowly modifying existing features to form the special characteristics of their phylum. Instead, a great diversity of different body plans appear almost simultaneously in the early Cambrian, complete with eyes, limbs, segments, or armour, in an explosion of animal forms. Some interpret this Cambrian explosion as the result of an imperfect fossil record, that failed to preserve the earlier animal ancestors, perhaps because they were small and squishy and lacked the skeletons, shells, and spikes that would have granted them geological immortality. DNA evidence has been used to support this interpretation. When genes shared by whales, barnacles, and flatworms are compared, there are more differences in the DNA sequences than expected from half a billion years of evolution, suggesting that these major

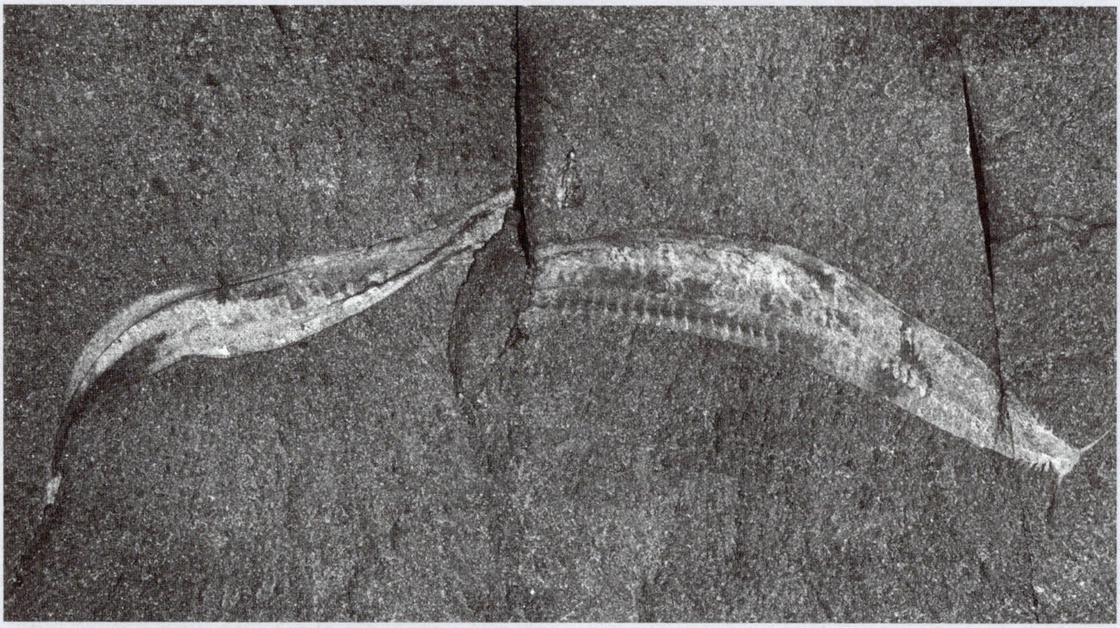

Figure 1.7 The arthropod and chordate 'body plans' have a long history and can be found in Cambrian age rocks, over half a billion years old. Fossil beds such as the Burgess Shale in Canada and Chenjiang in China contain exquisitely preserved examples of these aquatic animals, such as *Marrella* (tiny arthropod 'lace crabs') and *Pikaia* (an early chordate, something like the modern amphioxus).
© 2007 Smithsonian Institution

branches of the evolutionary tree of animals stretch back beyond the first fossils, deep into the Precambrian era. If the beginnings of the animal kingdom are before the fossil record starts, then DNA evidence will be essential in revealing its ancient history.

The role of DNA sequences in understanding the diversification of animals is covered in Chapter 8

But there is an alternative interpretation that is gaining ground, fuelled by the growth in developmental biology.

Could it be possible that body plans as different as chordates, arthopods, and flatworms evolved so rapidly that it requires a radically different evolutionary explanation than the gradual Darwinian plodding that transformed the ancestral chordate into lungfish, parrots, geckos, and whales? There has been a great deal of excitement surrounding the discovery of potential 'body-plan genes', which, with a single mutation, can have dramatic effects such as causing a second pair of wings to grow, or switching on the formation of extra eyes. Could the Cambrian explosion represent a remarkable period when changes to body-plan genes created animal diversity in evolutionary leaps, rather than taking a long series of tiny evolutionary steps? We cannot directly study the genes that were present in long-extinct animals, but we can use DNA sequence analysis to reconstruct the likely state of ancient genomes. The DNA sequences of body-plan genes from animals alive today can be used to trace the history of the genes themselves, as they were copied and changed and shaped to take on new functions.

Back to the beginning

Some sequences in the sperm whale genome code for such fundamental properties of life that they are shared not only with other animals, but with all living things, including oak trees, mushrooms, seaweed, and bacteria. The DNA sequences of these basic genes reveal the deep history of all life. These shared parts of our genome demonstrate that all life on Earth has a single common origin. The original genome, present in the last common ancestor of all plants, animals, fungi, algae, and bacteria, has been modified and expanded, but there is a part of that ancestral genome in all of us. The DNA from the globster was enough to prove that it was the remains of a sperm whale, and not (as some hoped) an extraterrestrial visitor. But if biologists ever do get their hands on a sample of alien life, the first thing they will do is check to see if it has DNA. If it does, it is almost certainly our distant relative. And using the techniques described in this book, we would be able to use the alien DNA sequence to investigate the history and biology of our cousins from space.

A small piece of the mystery blob that washed up on a beach in Chile contains enough DNA to tell a long and wonderful story. The story begins with a sperm whale, born in the same ocean that its mother had been, and its grandmother, and great grandmother, and so on back into history. Our sperm whale's ancestor was one of the founding mothers who survived the last ice age, although she may have had to retreat to tropical waters to do so. The story reaches back in time to tell of the origins of the sperm whales, whose predatory ancestors used echolocation to swim after their prey, leaving their four-legged relatives behind, wallowing in the mud. These ancestors were part of a radiation that exploded onto the world as the dinosaurs left the stage, and yet the roots of this radiation were planted firmly in the time of the reptiles. Both the reptiles and the mammals were themselves products of the diversification of the successful chordate body plan, which appeared in the fossil record half a billion years ago, but may have more ancient beginnings. And using the DNA from the blob, we can follow the story right back, beyond the limits of the fossil record, to the origins of the animal kingdom and, ultimately, back to the last living ancestor of all life. Not bad for a piece of blubber.

About this book

What is this book for?

The aim of this book is to provide a from-the-ground-up introduction to the use of DNA sequence data in evolutionary biology. Obviously, it is not exhaustive: there are many fascinating and useful ideas and techniques that are not covered here. And, in such a fast moving field, it is inevitable that this book will be out-of-date almost as soon as it is printed. So rather than trying to give an exhaustive introduction to the field, the aim is to give you the kind of fundamental information and intellectual tools you need to understand the way DNA sequences are used in biology.

That is why you won't find instructions for specific programs or particular laboratory protocols in this book. A program that is all the rage today is likely to be superseded by improved methods next year. Lab protocols are replaced on a regular basis by new techniques. The aim of this book is to give you the background knowledge you need to understand not only the techniques available today, but hopefully, to lay the groundwork for the new methods of the future. If you are taught to follow instructions to use a particular phylogenetic program to produce a particular result, you may not have the background knowledge you need to modify your approach if you get strange results, or to reshape your analysis plan in response to new ideas, or to expand what you know to undertake different analyses. But, if you have learned the general principles on which the programs are based, you will be better able to pick up the program's instruction book and make decisions for yourself about your own analysis. Therefore, the aim of this book is to introduce the basic principles of molecular evolution, in order to help you understand why there is information about evolution in DNA, how that information can be 'decoded', and what kinds of questions we can use that information to answer.

Who is this book for?

This book should be suitable for university or college students who wish to gain a basic grounding in the application of DNA sequence data to answering questions in evolutionary biology. The reader I have in mind has a basic background in biology (say, to high school level), but no specialist knowledge of genetics, evolution, or ecology. Importantly, the focus of this book is on the 'whole-organism' biologist: someone who is primarily interested in how species persist and coexist and evolve, whether they are interested in a particular group (say, chameleons) or a particular topic (such as sexual selection). Whole-organism biologists are increasingly using molecular techniques in their research: for example, using microarrays to uncover differences in gene expression between individuals in a social hierarchy, phylogenetics to judge the importance of biogeographic patterns of species richness, or molecular dating to understand the evolution of animal body plans. However, their interest remains, by and large, at the level of the organism (or species or higher group). They use molecular techniques for the information they can gain about their organisms, not to illuminate the genetics or biochemistry of their subjects.

While there are a great number of bioinformatics textbooks available, most are concerned primarily with the informatics, and have rather less focus on the bio. Many biologists do not find such an approach very inviting or comfortable. Here, I have attempted to take a complementary approach: starting with evolutionary principles, and illustrated throughout with biological examples, I ask how we can make use of the information coded in DNA to answer the kinds of questions that a whole-organism biologist is interested in.

But I also hope that students coming from other backgrounds might benefit from this book. Many of the brightest minds in computational and statistical fields are attracted to the challenges posed by the analysis of biological data. But some bioinformatics books simplify the biology in order to emphasize the techniques. While it may be convenient to treat DNA simply as a string of letters, the genome is actually a complex, fascinating, and intricate adaptation. DNA and protein sequences are the products of evolution. Therefore, a grounding in the principles of evolution is needed to underpin advances in bioformatics.

How the book is structured

This book consists of four elements: the main text and three types of box. Any of these elements can be read without the others, though they will make most sense when read together. I have chosen this structure to make it easier for you to get what you want from this book. Below, I will explain how the way you use the book might differ depending on what you want to find out. First, though, I will briefly outline each of the elements of the book:

- **Main text** aims to introduce the important concepts needed to build up a foundation for the application of molecular data to evolutionary biology. I have attempted to make the main text as untechnical as possible, and to illustrate key concepts with examples from whole-organism biology. Wherever possible I have moved details of methods, procedures, or biochemical information to the TechBoxes. The main text can stand alone without the details given in any of the boxes, but the boxes will give a fuller appreciation of the techniques and applications of molecular data.

TechBoxes give more detailed information about methods, or take a more in-depth look at particular topics. They can be seen as 'optional extras', but each one contributes to an understanding of the topic as a whole. The main text and TechBoxes are complementary and can be read together to give a fuller account of the topic than the main text alone. However, readers may choose to ignore TechBoxes, or only read specific boxes on particular topics. TechBoxes are cross-referenced, and linked to Case Studies where possible. Each TechBox gives references to further reading and, where possible, freely available resources such as software or online databases.

Case Studies provide a summary of a scientific study that has relevance to the topics covered in that chapter. I have mainly used studies published in the last 5 years, rather than reporting 'classic' papers. The Case Studies are chosen for the diversity of topics and techniques used, and do not represent the best possible studies in this field. Due to space limitations, Case Studies may contain terms or techniques that are not fully explained, so, if you are interested, you should follow up the references given in the box. Where possible, abstracts or articles will be provided on the companion website.

Heroes of the Genetic Revolution: The title is tongue-in-cheek, but the purpose is a noble one: to put some faces to the science (**Heroes 1**). For each chapter, I have highlighted the work of one scientist who contributed to the area covered in that chapter (ideally, someone who has made a broad contribution to the field, rather than a single discovery). It hardly need be said that this is not an exhaustive list, nor should it be read as a 'best of' list, for there are a great many other scientists whose contribution has been as great or greater than those featured here. So please do not feel insulted if your favourite scientist (or yourself) has been left out. It is also an evidently biased list, including some of my friends and colleagues, but anybody's list of heroes would be similarly biased, so I make no apology. These boxes are intended merely as illustrations of the kind of work that has been done, and the type of people who have done it. The Heroes listed in this book range from illustrious scientists of the past, replete with fulsome beards, to younger scientists becoming established in the field today. I would be happy to receive nominations for Heroes to appear in subsequent editions.

Glossary: I am expecting that most people reading this book have a basic knowledge of biology. Because of this, many basic concepts are assumed, such as a passing familiarity with cell division and Mendelian inheritance. However, I hope that the book will not be impossible to read even if you are unfamiliar with these basic concepts. To help with this, I have tried to include all biological terms in the Glossary. If you can't remember what mitosis is, don't know what I mean when I say 'genotype', or have no idea what a moa is, then you can turn to the glossary for some pointers. The Glossary also reiterates terms introduced in the text, so it should also help if you are skipping through the book to relevant passages, rather than reading the book in order.

Further information contains some suggested reading for those interested in exploring the topic further. I have tried to focus on books, rather than journal articles, and attempted to list fairly accessible works, though you may find that some of them are more technical than this book.

Bibliography contains some of the studies referred to in the text, though it is not an exhaustive list of all the sources used in researching each chapter. You may find references to these publications in the figure legends, given in the format Author-Year (e.g. Bromham 2008), in which case you can find the full publication details in the bibliography.

How to use this book

The ideal way to read this book is, not surprisingly, taking each chapter in turn, reading all of the main text and the boxes, as each chapter is designed to build on the concepts introduced in the preceding chapters. Each chapter is structured around a theoretical aspect of molecular evolution, a key concept in evolutionary biology, and a practical application of these principles. For example, Chapter 4 describes DNA amplification techniques (practice), but in order to understand these you need to grasp the basics of DNA replication (molecular evolutionary theory), which also illuminates the hierarchical nature of biological organization (evolutionary principles). Similarly, in Chapter 7, in order to be able to estimate phylogenies (practice), you need to understand the process of genetic divergence (molecular evolutionary theory), which sheds light on diversification (evolutionary principles). Because each chapter has many roles, there are several different ways I could have ordered this book. I had originally organized it along practical lines: how to extract DNA (Chapter 2), then amplify DNA (Chapter 4), then align DNA sequences (Chapter 6), then use the alignment to measure individuals differences (Chapter 3), to detect population processes (Chapter 5), reconstruct phylogenies (Chapter 7), and measure evolutionary rates (Chapter 8). But then I

HEROES OF THE GENETIC REVOLUTION 1

Fred Sanger

"It is like a voyage of discovery into unknown lands, seeking not for new territory but for new knowledge. It should appeal to those with a good sense of adventure."

Fred Sanger (quoted on www.brainyquote.com)

NAME

Frederick Sanger

BORN

13 August 1918, Rendcombe, Gloucestershire, UK

KEY PUBLICATIONS

Sanger, F., Nicklen, S. and Coulson, A.R. (1977) DNA sequencing with chain-terminating inhibitors. *Proceedings of the National Academy of Sciences USA*, Volume 74, pages 5463–5467.

Sanger, F., Coulson, A.R., Friedmann, T., Air, G.M., Barrell, B.G., Brown, N.L., Fiddes, J.C., Hutchison III, C.A., Slocombe, P.M. and Smith, M. (1978) The nucleotide sequence of bacteriophage X174. *Journal of Molecular Biology*, Volume 125, pages 225–246.

FURTHER INFORMATION

You can watch a video of an interview with Frederick Sanger at *www.vega.org.uk/video/programme/18*, or read both of his Nobel lectures at *http://nobelprize.org/chemistry/laureates/*

Figure Hero1a Fred Sanger in front of the research institute that now bears his name, the Sanger Institute near Cambridge, UK. From *www.sanger.ac.uk/Info/Intro/sanger.shtml* with permission of the Sanger Institute.

Courtesy of Wellcome Library, London

Fred Sanger's interest in biology was started at an early age, though he was not a high-achieving student. He chose science when he went to university, rather than his father's profession of medicine, because the focus of the scientific method appealed to him. He loved biochemistry, and it was in this field that he first began to distinguish himself as a student. A conscientious objector during the second world war, he studied for his PhD at Cambridge, on 'lysine metabolism and a more practical problem concerning the nitrogen of potatoes'. During his career he made almost countless advances in the techniques of protein sequencing and DNA sequencing. But perhaps more profoundly he put the idea of the sequence at the heart of biology. By showing that proteins consisted of a particular sequence of amino acids, he showed that the gene, the unit of hereditary information, must have a similar sequential arrangement of units, then provided a means for reading that sequence from the genes themselves.

In the early 1940s, there was a great deal of interest in protein structure. It was known that proteins contained characteristic proportions of the 20 amino acids, and it was assumed that this somehow gave proteins their specific functions and structures. But there was a debate between those who thought that the amino acids formed a particular sequence that gave the protein its function, and those who thought that different amino acids occurred at regular intervals throughout the protein, giving proteins a structural periodicity. Sanger and colleagues first determined that insulin was made up of four polypeptide chains, two each of two types. They then used a series of biochemical techniques, such as shearing particular chemical bonds then fractionating the portions of the chains using chromatography, to gradually deduce the amino acid sequence of each of the chains. The complete sequence demonstrated the 'classical peptide hypothesis' that the sequence of a protein was unique and consistent. Sanger also laid the groundwork for sequence comparisons, showing that most of the insulin sequence was identical between several mammal species, except for three residues in one part of the A chain. He pre-empted the neutral theory in suggesting that the exact residue in these variable sites was relatively unimportant to the functioning of the protein (see **TechBox 5.2**).

Sanger then moved to the Laboratory of Molecular Biology in Cambridge, where he became interested in gene sequencing. Although it was widely believed that DNA carried a linear code, corresponding to the amino acid sequence in proteins, there was no method for reading the sequence of bases in a DNA molecule. Sanger developed methods for sequencing both RNA and DNA, and was the first to produce a whole-genome sequence, that of a bacteriophage (small viruses that parasitize bacteria). The DNA sequencing method developed by Sanger is known formally as the dideoxy method, or the chain-termination method, or simply as 'the Sanger method' (however, since all DNA sequencing is now done in this way, the term 'Sanger method' is rapidly becoming defunct). Sanger and his colleagues went on to invent 'shotgun sequencing', where the genome is chopped up into fragments and the whole genome sequence inferred by looking for regions of overlap between the fragments. In this way, they produced the first bacterial genome sequence, the first human mitochondrial genome sequence, and paved the way for the human genome sequencing project (which is in part led by the research institute that bears Sanger's name).

Fred Sanger's work has underpinned the whole field of molecular evolution. He is currently the only person to have gained two Nobel prizes in Chemistry (there is no Nobel Prize for biology, so the study of biomolecules is considered chemistry, and any other biological advance has to fit under the 'physiology or medicine' prize). When he collected the second Nobel medal, he was quoted as saying he couldn't rule out getting a third at some point. I don't think anyone would be surprised if he did.

> 'Often if one takes stock at the end of a day or a week or a month and asks oneself what have I actually accomplished during this period, the answer is often 'nothing' or very little and one is apt to be discouraged and wonder if it is really worth all the effort that one devotes to some small detail of science that may in fact never materialize. It is at times like the present that one knows that it is always worth-while...'
>
> Frederick Sanger, Nobel prize acceptance speech, 1958

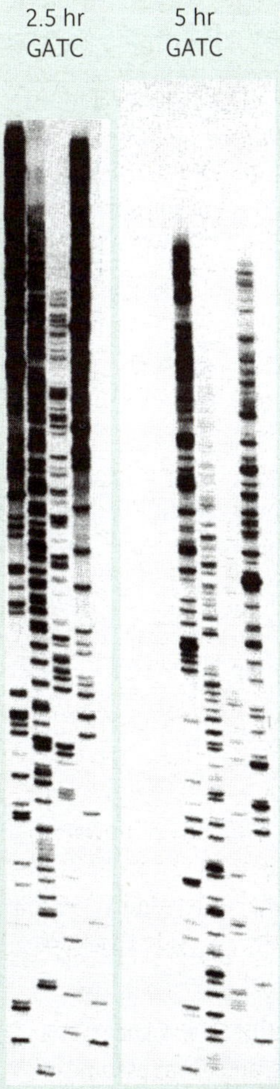

Figure H1b An autoradiograph produced by the dideoxy method of sequencing, taken from Fred Sanger's 1980 Nobel Prize lecture (*http://nobelprize.org/chemistry/laureates/1980/sanger-lecture.html*). See **TechBox 1.2** for a explanation of how this autoradiograph was produced. Most sequencing today is done by automatic sequencers using fluorescent dyes, making autoradiographs like this one virtually obsolete (but still of nostalgic value to biologists over a certain age).

decided that the point of this book is that to apply these techniques intelligently, you need a basic grounding in the principles of molecular evolution and evolutionary biology. So then I restructured the material in the order of key concepts: DNA structure (Chapter 2), mutation (Chapter 3), DNA replication (Chapter 4), substitution (Chapter 5), homology (Chapter 6), and divergence (Chapter 7 and 8). I hope this works.

In any case, the structure of main text plus boxes is designed to allow you to approach the book in several different ways, depending on what you need to know:

If you want a general background in evolutionary biology, then you can read through the main part of the text, without bothering with the case studies and the technical details. The main text is designed to stand alone from the boxes, so you need only dip into the TechBoxes, Case Studies, or Heroes that interest you.

If you want a basic introduction to molecular evolution, you may find much of the main text useful, but there will be many digressions into whole-organism biology that you may find less interesting. You will find many of the TechBoxes useful, as they focus on specific aspects of theory (such as neutral theory: **TechBox 5.2**) and techniques (such as DNA amplification: **TechBox 4.2**). As you read the text, you can decide which ideas you would like to know more about, and following the cross-references to the relevant boxes.

If you are after information on specific topics, you might use the index to track down relevant TechBoxes. So, for example, if you are interested in designing a study to track cross-species transfers in a family of viruses, you might start by looking at how to use BLAST (**TechBox 3.4**) to get sequence data from GenBank (**TechBox 1.1**), then how to align those sequences (**TechBox 6.3**), how to estimate a phylogeny (**TechBoxes 7.1**, **7.2**, **7.3**) and molecular dates (**TechBox 8.4**), and test how well your data support a particular hypothesis (**TechBox 7.4**). Cross-referencing between these boxes should make it easier for you to hop between relevant topics.

If you are after guidance for a particular research project: If you know what kind of techniques you wish to use, then the TechBoxes may help you work out what you need to know to apply these tests yourself. If you are not sure what you need to do, the Case Studies may provide hints. Each Case Study has keywords that draw attention to the subjects, questions, and techniques used, so you can use these to find a Case Study relevant to your research. When you locate a TechBox or Case Study that's relevant to your work, you will find them in the chapters that give the background to those topics. For example, if you are interested in using single nucleotide polymorphisms (SNPs) to find the genetic basis of disease, you might go to **Case Study 3.2** (biobanking). You will find it in a chapter on mutation, which is a key concept in studying SNPs.

Where to go if you want more information

This book is a very brief introduction to molecular evolution. You will find a few reasonably accessible, but more detailed, works listed in the Further information section at the end of each chapter. The companion website for this book contains some links that will take you to further reading for the topics covered in each chapter, and may also point to new studies published after this book went to press. If you are interested in the examples mentioned in the chapter, a few key references are given at the end of the book in the Bibliography. I hope that the TechBoxes, Case Studies, and examples given in the book will act as a springboard for interested readers to begin wider reading on the topic. Because the field changes rapidly, the best way to do this is to use the internet. A quick search will reveal a great many freely available laboratory and teaching resources where the latest techniques are explained. There are also several freely accessible databases of scientific papers – such as Google Scholar or PubMed central – where you can access original research. If you are a member of an institution that has paper or electronic subscriptions to scientific journals, then you should not find it too difficult to get access to primary research papers. However, if you are unable to obtain a particular scientific paper because you do not have access to the journal, try visiting the homepage of the author of the paper, where there may be downloadable versions of their publications. If there aren't, consider writing to the author: most scientists, particularly university academics, take their obligation to make their work accessible seriously, and will be happy to send a copy of their papers to you.

Where are the equations?

Many scientists in the field of molecular evolution will be shocked, and quite probably appalled, that there are no equations in this book. This is not because equations are bad. On the contrary, they are the most succinct and useful way of encapsulating many statements about molecular evolution. But equations are not everybody's cup of tea. If you are one of those whose eyes go blurry as they pass over equations, you may be disheartened by the statement by the great evolutionary biologist John Maynard Smith who said: 'if you can't stand algebra, stay away from evolutionary biology'. But consider for a moment how fortunate we are that Charles Darwin could not retrospectively heed this advice. Darwin never did get the hang of mathematics, and produced his vast catalogue of work without recourse to any equations. Being handy with equations is a skill worth developing if you wish to understand molecular evolution. But it is, in my opinion, by no means an essential prerequisite. Far more important is the ability to think in terms of the evolutionary principles underlying molecular data, whether you phrase those principles in words or algebra.

There are plenty of texts on population genetics and molecular evolution containing equations but none, that I know of, with none. My aim in explaining molecular evolution without recourse to algebra is to ease the 'maths panic' felt by many biologists when they are learning to analyse molecular data. By describing the basic principles of molecular evolution in words rather than equations, I hope to give the mathematically nervous sufficient background, and courage, to read and understand the classic texts. If you do feel comfortable with equations, you may wish to read a traditional molecular evolution text alongside this one, so you can see both the verbal and algebraic statements of the basic principles.

Conclusions

DNA can be extracted and sequenced from a wide range of biological samples, providing a wealth of information about evolution and ecology. The analysis of DNA sequences contributes to evolutionary biology at all levels, from dating the origin of the biological kingdoms to untangling family relationships. In this chapter, the information that can be gained from the analysis of DNA sequences has been illustrated by considering a single DNA sample, and how it can shed light on the evolutionary history of individuals, families, social groups, populations, species, lineages, and kingdoms.

The aim of this book is not to provide you with protocols for lab procedures or instructions for software packages used in the production and analysis of DNA sequences. Instead, this book should provide you with the background knowledge you need to understand these techniques. The most important place to start is an understanding of the material basis of heredity. In Chapter 2, we will take a brief look at the history of the discovery that DNA carries genetic information. Outlining some of the important steps in the development of the field of genetics will serve to illustrate some of the key features that make DNA sequences so useful to evolutionary biologists.

Further information

The Ancestor's Tale takes a similar journey to the one in this chapter, following a lineage back through time, and using the journey to discuss evolutionary principles and research on the way.

Dawkins, R. (2004) *The Ancestor's Tale: a pilgrimage to the dawn of life*. Weidenfeld & Nicholson.

This popular-science book takes the opposite path – from the early history of life towards the future – with tales of biologists and research along the way.

Fortey, R (1998) *Life: an Unauthorised Biography. A natural history of the first 4,000,000,000 years of life on earth*. HarperCollins.

2 The immortal germline

Or: how do I get DNA samples?

"If you could use a big enough magnifying glass you would find that there is really only one kind of life on the Earth: the most central machinery in all organisms is built out of the same set of micro-components, the same set of small molecules."

Cairns-Smith, A.G. (1985) *Seven Clues to the Origin of Life.* Cambridge University Press

What this chapter is about

The genomic information system is shared by all living things. It is worth learning a little of the history of the discovery of this genomic system because it serves to illustrate some of the important principles of heredity. Life relies upon the continuity of genetic information. This information is encoded in DNA, and copied from generation to generation. The practical upshot of these principles of heredity is that the DNA found in every living cell contains all the genetic information needed to construct the organism, as well as providing biologists with a wealth of information about biological history and evolutionary processes.

Key concepts

Evolutionary biology: heredity

Molecular evolution: DNA structure

Techniques: DNA extraction

Unity of life

All life on Earth has a common ancestor. If you trace your family tree back far enough, you will find you are related to the rats in your attic, the fly on your window, the mould on your bread, the rice in your cupboard, even the bacterium in your gut (Figure 2.1). We know this because every organism on Earth uses the same basic system to carry the information that it needs to grow and reproduce. This genomic system consists of information stored in DNA, which is then transcribed into RNA, and then translated into proteins. This system is so intricate, with so many complex interlocking parts, that we can be sure that it was not invented separately in different biological lineages. So we are all descended from an ancient, simple life form that carried the same fundamental genomic system that we share today with all of the Earth's biodiversity.

Not only is the genomic system of DNA, RNA, and proteins shared by lifeforms as different as rats, flies, mould, rice, and bacteria, but some of the actual information is also shared. In a sense this is not surprising, because all living things must share the instructions for constructing the genomic system itself. For example, all organisms must be able to convert the genomic information coded in DNA into RNA messages that provide the instructions needed to make a protein. One of the things they need in order to do this is a working copy of the enzyme RNA polymerase, which makes an RNA message from a DNA gene sequence. So all organisms must have a gene that codes for RNA polymerase.

In fact, eukaryotes (such as fungi, plants, and animals) have several different types of RNA polymerase

Figure 2.1 Rice (*Oryza sativa*) is one of the most important crop plants in the world. It is often said that humans are, genetically speaking, 99% similar to chimpanzees. Perhaps more remarkably, we share about half our genes in common with plants. Remember that next time you eat a bowl of rice.

Reproduced by permission of Oliver Spalt, 2002.

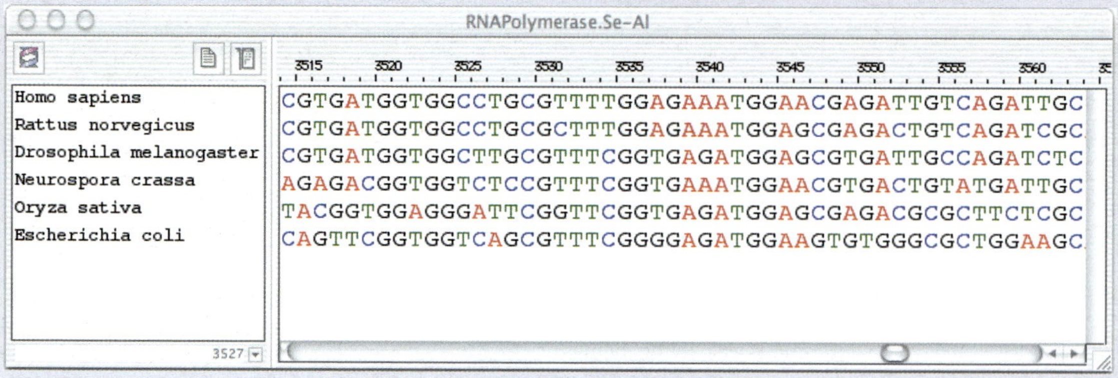

Figure 2.2 Part of the DNA sequence of the gene for the beta-subunit of the RNA polymerase II enzyme, from a human (*Homo sapiens*), a rat (*Rattus norvegicus*), a fly (*Drosophila melanogaster*), a mould (*Neurospora crassa*), rice (*Oryza sativa*), and a bacterium (*Escherichia coli*). This section represents approximately 1% of the entire sequence of the gene. This is a screen-shot of the freely available alignment editor program Se-Al, which can be downloaded from *http://tree.bio.ed.ac.uk/software/seal/*. See **TechBox 4.4** for further information about DNA sequence alignment.

All Se-Al screenshots in this book are reproduced by kind permission of Andrew Rambaut.

enzymes, and each one is constructed from many different subunits, in some cases combining a dozen or more different proteins. For simplicity, let's concentrate on a gene for a single subunit of one of these RNA polymerase enzymes, the beta-subunit of RNA polymerase II. **Figure 2.2** shows part of the DNA sequence of this gene. Each horizontal line of letters represents the sequence of bases in the version of the RNA polymerase gene found in a particular species. In this alignment, we can compare the RNA polymerase II beta sequence from humans, rats, fruit flies, bread mould, rice, and bacteria. Although the exact DNA sequence is slightly different in each of these organisms, each of the versions provides the necessary instructions for making a working copy of the RNA polymerase gene.

This is just a small part of one gene. The genomes of most organisms contain thousands of genes. Every time the body manufactures a new molecule to help it grow or move or respond to environmental change, particular genes must be located in the genome, unwound, transcribed, and translated. Then the newly manufactured molecules must be folded, combined, and transported to where they are needed. The expression of a single gene requires the co-ordinated action of dozens of enzymes, the manufacture of a great number of specialized molecular building blocks, and the co-ordination of a large number of tasks in time and space within the cell. Yet this complex process is being continuously performed by every single living cell to produce thousands of proteins and other molecules, all in the right place, at the right time, in the right amounts. The beauty and complexity of the genomic system never ceases to amaze me.

In order to truly appreciate the wonder of the natural world, you need to gain some insight into the workings of the molecular level of organization that underlie all of the functions of the living world. This is important for two reasons. Firstly, an appreciation of molecular biology is the best way to bring home the complexity and intricacy of organisms. Secondly, a grasp of the biochemical basis of heredity is essential to understanding evolution, as it is at this level that mutations occur, substitutions accumulate, causing lineages to change and diverge over time. We are going to briefly consider the history of the discovery of the genomic system in order to review some of the key principles of heredity. These ideas are not just of historical interest, nor are they only relevant to those interested in genetics itself. These principles of heredity explain why we can use DNA as an information source in evolutionary biology and ecology.

Principles of heredity

Our knowledge of the genomic system is surprisingly recent. The basic principle of heredity – that offspring tend to resemble their parents – has long been observed by human societies. But the exact mechanics of inheritance were subtle and unknown. It had always been recognized that animals and plants tend to arise from parents of the same species, but it was commonly believed that in certain circumstances living beings could arise spontaneously, such as flies being generated from rotting meat, wasps arising from galls on plants, or bacteria forming *de novo* in chicken broth. Spontaneous generation was finally put to rest 150 years ago, when conclusive experiments in which potential parent organisms were carefully excluded from sterilized material ultimately convinced scientists of the importance of genetic continuity. Only living organisms contained the necessary information to make another organism; life cannot arise without a copy of this genetic information. But what was the material basis of genetic continuity? Did reproductive cells contain tiny preformed creatures that grew into new adults? If so, then how were traits from both mother and father inherited? And how could an organism such as a sponge reproduce by budding, where a small piece of its body could be induced to grow into a new individual? Somehow, cells must be able to transmit information to create a new individual.

One early theory of heredity, published by Pierre Louis Moreau de Maupertuis in 1745 in his natural history tract *Vénus Physique*, held that particles corresponding to all parts of the body were provided by the parents, and used to build the developing offspring. These particles could be altered to give rise to new hereditary types, and might even undergo isolation in different parts of the world to produce new species. The idea that all parts of the parent's body contribute information to the offspring is referred to as pangenesis. Theories of pangenesis have a long history, going back to the philosophers of ancient Greece, but one of the most famous proponents of pangenesis was the father of evolutionary biology, Charles Darwin.

Darwin's theory of evolution is discussed in more detail in Chapter 5

Darwin recognized the central role of heredity in evolutionary theory, devoting whole volumes to recording and interpreting observations of inheritance in the natural world (largely in domesticated animals). Darwin knew that variations arose continually in natural populations, and that many variations could be inherited, but he could only guess at the mechanism. Critics of natural selection pointed out that if the characteristics of the parents were blended to create their offspring, then any favourable new variant would be diluted each generation and eventually lost.

The lack of a clear mechanism for inheritance had been, in many ways, a stumbling block for the development of evolutionary theory. To fill this gap, Darwin developed his theory of pangenesis, speculating that all the body's cells produced particles, called gemmules, which carried information. Gemmules collected in the reproductive cells prior to fertilization, ensuring the offspring inherited all the information needed to make a functioning body. Because gemmules formed in the adult body, Darwin's theory of pangenesis specifically allowed for the inheritance of acquired characteristics – modifications of the body during an individual's lifetime could be inherited by its offspring. The theory of pangenesis was politely ignored by some of Darwin's contemporaries, and criticized by others. One strong critic was Francis Galton, one of the founders of modern statistics and leader of the early eugenics movement. Galton showed that blood transfusions did not appear to move hereditary information from one individual to another, as would be expected if gemmules were carried in the blood. Even 20 years after the publication of Darwin's classic work on evolution, *The Origin of Species*, the material basis of heredity was still unknown, despite being the focus of much study.

The Weismannian barrier

At the present time there is hardly any question in biology of more importance than this of the nature and causes of variability, and the reader will find in the present work an able discussion on the whole subject. . . . Whoever compares the discussions in this volume with those published twenty years ago on any branch of Natural History, will see how wide and rich a field for study has been opened up through the principle of Evolution; and such fields, without the light shed on them by this principle, would for long or for ever have remained barren.

Charles Darwin, Foreword to *The Study of Heredity* by August Weismann (1880)

August Weismann transformed evolutionary biology by arguing forcefully against the inheritance of acquired characteristics. His arguments were largely made on theoretical grounds, by considering the implications of patterns of inheritance for the process of evolution. For example, how could the non-reproductive castes of social insects evolve if there was no way that a sterile worker could pass its bodily modifications, having no offspring of its own? How could mutations arise in organs that must be fully formed before use, therefore had no opportunity for acquiring new characteristics by use and disuse? And more importantly, how could information about the state of adult organs be translated into a form of inheritable instructions? (**Heroes 2**: August Weismann).

Weismann also argued on empirical grounds: despite widespread belief, there was simply no evidence that acquired characteristics could be inherited. For example, it was clear that human societies that practised male circumcision over many generations did not give rise to offspring that no longer had foreskins. Weismann carried out experiments that demonstrated that bodily modifications acquired in an individual's lifetime were not passed to their offspring. One of his most famous experiments was to dock the tails of mice, then breed from the tailless individuals, and dock the tails of their offspring. He continued this process through 21 generation of mice, docking the tails each generation. But the tailless mice never produced tailless offspring. Although this experiment seems trivial, it is important to realize that this demonstration ran counter to the prevailing opinion of the times. Many animal breeders believed it was possible to produce a tailless breed in this way, by cutting the tails of individuals then breeding from them. But experiments such as this denied the universality of pangenesis, because the docked tail did not appear to contribute to the genetic information inherited by the offspring.

Thus Weismann argued persuasively for 'hard inheritance': genetic information was not added to throughout life, as the body grew and changed, but was set immutable from conception. The reproductive cells (germline) were not influenced by changes in other cells of the body (soma) and so heritable information was passed from one generation to the next largely unchanged. Weismann defined two key principles of heredity. Firstly, he suggested that the germline was effectively isolated from the soma. This mean that changes made to the body during a lifetime could not be passed to the gametes. Secondly, Weismann considered that the germline represented an unbroken chain of information passed from parents to offspring, and so on down through the generations. These two principles combine to give us our modern view of heredity: that genetic information is passed from generation to generation, essentially unaffected by changes to the body.

Although he made detailed studies of developmental biology, Weissman's conceptual advances were primarily theoretical, made by considering the implications of various models of heritability for evolution. At the same time, largely unknown to those in the scientific community debating heredity and evolution, breeding experiments were being carried out that would shed light on the nature of the immortal germline. These experiments would eventually be used to counter Darwin's critics, by demonstrating that hereditary information did not blend and dilute down the generations, but was passed on in discrete units that could be carried over many generations.

HEROES OF THE GENETIC REVOLUTION

2

August Weismann

"I have gradually become aware, that, after Darwin, Weismann was the greatest evolutionary biologist of the nineteenth century. Further, the problems he was concerned with are often the same problems that concern us today"

Maynard Smith, J. (1989) *Weismann and Modern Biology*. Oxford Surveys in Evolutionary Biology, Volume 6, pages 1–12

NAME

August Freidrich Leopold Weismann

BORN

17 January 1834, Frankfurt am Main, Germany

DIED

5 November 1914, Freiburg, Germany

KEY PUBLICATIONS (English translations)

Studies in the Theory of Descent (1882) translated by R. Meldola. London, Simpson Low, Marston, Searle and Rivington.

The Germ-Plasm: a theory of heredity (1893) translated by W. N. Parker and H. Ronnfeldt, New York, Charles Scribener's Sons.

FURTHER INFORMATION

Facsimiles of Weismann's *Essays on Heredity* and *The Germ-Plasm* are freely available at *www.esp.org/books/chrono-lst.html*

Figure Hero 2
August Weismann.

Image from Conklin, E.G. (Oct–Dec., 1915) August Weismann. *Proceedings of the American Philosophical Society*, Volume 54, No. 220., pages iii–xii.

August Weismann was the first to be called a 'neo-Darwinian' (not intended as a compliment then, and, regrettably, often used in the same vein today). Like Darwin, much of his work was prescient, and it is surprising how many key ideas in modern evolutionary biology can be found in Weismann's work, such as the role of sexual reproduction in generating variation, and a discussion of the cellular causes of ageing. In particular, Weismann's careful observations of cell division led to the recognition of the role of chromosomes in heredity. Many of Weismann's books, like Darwin's, surprise modern readers with their vast catalogues of observations about the natural world. In Weismann's case, his special interest was in the coloration of caterpillars and butterflies. While these intimate studies of butterflies may seem whimsical, they provided Weismann with abundant raw material for understanding developmental biology and genetics.

One of Weismann's most important contributions was that he convincingly demonstrated that, counter to the prevailing viewpoint at the time, changes to the body during an individual's lifetime are not a source of heritable variation. Because he was working in the late 1800s when the molecular basis of heredity was unknown, Weismann's central theories are framed in terms of cell lines: the germline cells (which form sperm and eggs) are isolated from the somatic cells (all other cells in the body). We can now interpret Weismann's principles in terms of DNA: the information in DNA is passed on exactly as it was inherited (bar the occasional mutation), because information about the state of the body is not recorded in an individual's DNA during their lifetime.

The following extract from a contemporary review of Weismann's work give some sense of the impact of his ideas, and the controversy and excitement surrounding the problem of heredity:

> 'In spite of the difficulties involved in acceptance of Weismann's view, however, it has been enthusiastically accepted in England by the younger Darwinian school. . . . The old school of Lamarck seemed dead; even the ideas of Herbert Spencer and of Darwin himself as to "use and disuse" began to be looked upon as antiquated and unphilosophical. . . . At the present moment a reaction has set in; the battle is raging fiercely. . . . Alike in Germany and in England, criticism and doubt as to Weismann's premises are beginning to take place of the paean of exultation. . . . What is wanted now is some decisive experimental settlement of the question. Can it be shown that in any case a capacity or habit acquired beyond a doubt during the life-time of the individual is transmissible to the off-spring? If that can be proved, Weismannism falls at once to the ground, and we revert to the primitive Darwinian and Spencerian problem.'
>
> Allen, G. (1890) The new theory of heredity. *Review of Reviews*, Volume 1, pages 537–538

Discovery of the gene

'What [was] called for was a theory of heredity by which inheritance would be essentially discrete, discontinuous, and ensured by units that could be transmitted from generation to generation without losing their somatogenenic qualities. Such is the gene.'

Jacques Monod (1974) On the molecular theory of evolution. In *Problems of Scientific Revolution*, D. Harre (Ed.), Oxford University Press.

Gregor Mendel was a researcher and teacher at a monastery in Moravia (now in the Czech Republic), which had a thriving research programme in many aspects of natural science. For nearly a decade in the mid-1800s, Mendel conducted large-scale experiments in plant breeding. He systematically crossed 34 different pure-bred strains of peas and recorded the characteristics of over 10,000 individual plants. Through the pioneering application of statistical analysis to the problem of heredity, he was able to show that the heritable features of these pea strains were preserved down the generations. When Mendel crossed two varieties of peas, he found that offspring did not have a simple blend of their parents' characteristics. Instead of being intermediate between the two types, the offspring resembled one parent or the other. For example, a tall pea crossed with a short pea produced tall offspring, not offspring of medium height (**Figure 2.3**). But when these tall offspring were crossed together, they produced both tall and short plants. The variation in the parents' generation was not lost, because it reappeared in subsequent generations.

Furthermore, the proportion of offspring of each parental type varied in predictable frequencies. When tall and short plants were crossed, they produced all tall offspring. But when these tall offspring were crossed with each other, their offspring (the 'grandchildren' of the original tall by short cross) varied in height in predictable ratios: three tall offspring to every one short. Mendel had discovered that heritability was governed by discrete factors that were copied and combined down the generations, and did not disappear through interbreeding. He was therefore the first person to describe the action of the hereditary units, which he called 'factors' or 'elements', but are now known as genes.

But what were these inherited factors? Early geneticists and evolutionary biologists studied the behaviour of genes in great detail. They described patterns of inheritance, how different genes combined to produce particular traits, and how these heritable traits varied within populations. Yet they did not actually know what genes were made of, or where in the cell they were located. Chromosomes seemed a good candidate for the genetic material. Their ordered behaviour at cell division, with a copy of each chromosome going to each daughter cell, matched Mendel's description of the segregation and assortment of genetic factors. But chromosomes are made of both proteins and DNA. Which of these two types of molecules held the hereditary information?

DNA had been discovered in the 1860s by Friedrich Miescher. First, he collected cells, such as white blood cells taken from pus on bandages collected from a hospital. Then he used a number of protocols to lyse the cells and separate the cellular contents (see **TechBox 2.4**, p. 51). When he isolated the central nuclei of the cells, he found them to be full of a phosphorous-rich material. He called this substance nuclein. Nuclein was found in every cell that Miescher tested, but it appeared to be inert: that is, it was non-reactive and didn't appear to have any special metabolic role. Miescher initially concluded it might simply be a way of storing phosphorous in cells, though later he began to suspect it had some kind of role in fertilization.

As knowledge of nuclein was refined over the next 60 years, it was renamed deoxyribonucleic acid (DNA). It was shown that DNA was found in chromosomes, and that it was made up of phosphates, sugars, and bases linked together in long chains. But relatively few scientists were interested in DNA, since it did not seem to do anything exciting. DNA was always in the same inert form, did not appear to do anything other than lie around in chromosomes, and had only four different units (the nucleotide bases). Proteins, on the other hand, existed in huge variety, did much of the important work in a cell, and were made up of over 20 different units (amino acids). Many scientists thought proteins were the obvious choice for storing the vast amount of information need to make even the simplest cell. But the problem with the protein theory of heredity

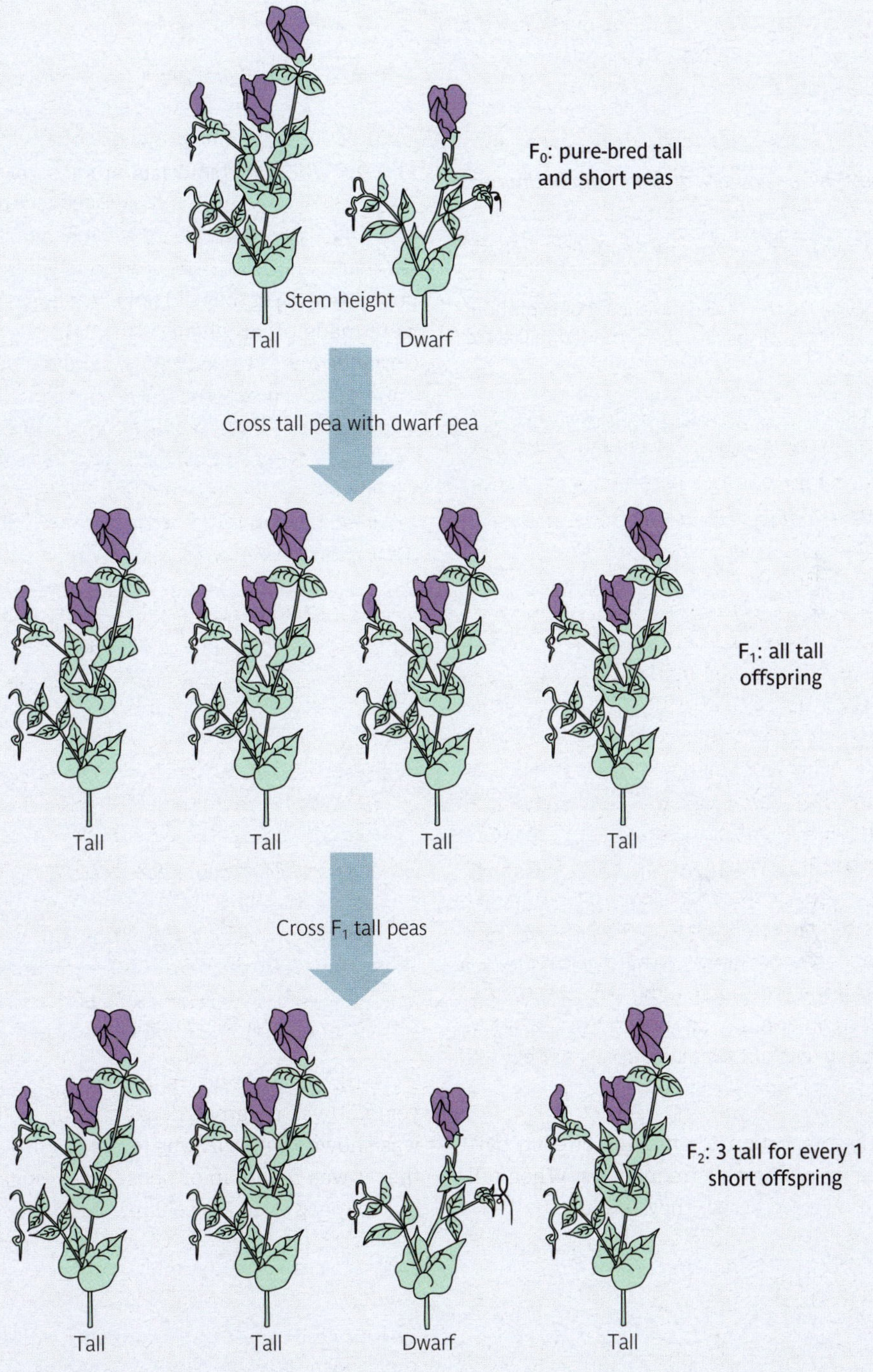

Figure 2.3 Gregor Mendel crossed distinct pure-bred lines of peas and showed that, for certain traits, the first-generation (F_1) offspring were not intermediate between the two parental types, but all resembled one parent. In the case illustrated here, crossing tall and dwarf varieties produced all tall offspring, no dwarfs. But the genetic information from the two parents was not lost. When Mendel crossed the first-generation (F_1) offspring with each other, the second-generation (F_2) offspring included both tall and dwarf plants. Mendel identified seven traits that varied discretely in this way, including wrinkly versus smooth seeds, yellow versus green peas, and purple versus white flowers. For more information see MendelWeb (*www.mendelweb.org*).

was not how information could be stored, but how it could be copied and passed to offspring. Could proteins be copied? And would a cell need to inherit a copy of every essential protein from its parent?

DNA as genetic material

In the 1940s several experiments had suggested that it was DNA, not proteins, that carried genetic information. For example, Oswald Avery and colleagues showed that genetic information could be passed from one strain of the bacterium *Pnuemococcus* to another. They used a series of experiments to show that the 'transforming factor' (genetic information) was preserved even when enzymes were used to remove all proteins, sugars, and RNAs. But if DNA was removed from the solution, then it could not transform cells. They concluded that it was the DNA that carried information from one cell to another. However, these experiments did not convince the majority of scientists working on the molecular basis of heredity, most of whom continued to concentrate on proteins.

Conclusive proof was provided by Alfred Hershey and Martha Chase in 1952. Their 'blender experiment' (named for their innovative use of kitchen equipment) showed that DNA was responsible for genetic continuity. By attaching different radioactive labels to the protein and DNA in viruses, they could demonstrate that it was the DNA, not the protein, that transmitted the information needed to make a new virus (**TechBox 2.1**). But it was the publication of a single, one-page scientific article in the journal *Nature* the following year that put DNA at the heart of modern biology. This paper, in which James Watson and Francis Crick described the molecular structure of DNA, concludes with a sentence of elegant understatement: 'It has not escaped our notice that the specific pairing we have postulated immediately suggests a possible copying mechanism for the genetic material.' This is the statement that launched the genetic revolution (see **Heroes 4**: Francis Crick).

Structure of DNA

It had been known since the 1920s that DNA was made up of regular patterns of three kinds of molecular subunits: phosphates, sugars, and bases. But how were they connected together? Watson and Crick had seen the outstanding X-ray diffraction pictures of DNA taken by Rosalind Franklin. These pictures suggested that DNA was a helix, with long chains of linked phosphate and sugar molecules twisted around each other in a regular pattern. Importantly, Watson and Crick combined this insight with an earlier observation made by Erwin Chargaff that the four types of bases of DNA were curiously evenly mixed. The number of adenine bases was always equal to the number of thymines, and the number of guanines was the same as the number of cytosines. Watson and Crick realized that Chargaff's pairing rule – A matches T, G matches C – was the key to the structure of DNA.

Watson and Crick constructed a large model, cutting shapes from tin plate to represent the four nucleotide bases (A, T, G, and C). This model, with its the elegant spiral staircase of two intertwined strands of phosphates and sugars, connected by rungs of paired bases, is familiar to many biologists as the star of one of the most widely used publicity photographs in the history of biology (**Figure 2.4**, p. 40). The complementary pairing of bases between the double strands of the helix, A with T and G with C, meant that each strand was an exact complement of the other. One strand could act as a template for the other, providing a means of copying information. The answer was so obviously right that Francis Crick is said to have announced in their local pub that evening: 'We have just uncovered the secret of life!' (**TechBox 2.2**).

From DNA to RNA to protein

The template-copying mechanism identified by Watson and Crick is the key to understanding not only the replication of DNA, but also the way that the information in DNA is used to build and operate living cells. The genome, made of DNA, is often described as a blueprint. It holds the instructions for making a cell, but it is not directly involved in construction. Instead, the genetic

TECHBOX
2.1

Hershey–Chase blender experiment

KEYWORDS

viruses

bacteriophage

radioactive labels

inheritance

DNA

proteins

FURTHER INFORMATION

The Race for DNA is an account of the discovery of the structure of DNA, told through contemporary documents and interviews. *http://osulibrary.orst.edu/specialcollections/coll/pauling/dna/narrative/page1.html*

RELATED TECHBOXES

TB 2.2: DNA structure

TB 2.4: DNA extraction

RELATED CASE STUDIES

CS 4.1: Glorious mud (using blenders to isolate DNA)

CS 3.1: Viruses within (viral genomes)

"When asked what his idea of happiness would be, [Hershey] replied, 'to have an experiment that works, and do it over and over again'."

Hodgkin, J. (2001) Hershey and his heaven. *Nature Cell Biology*, Volume 3, page E77

The experiment that put DNA at the centre of molecular genetics was carried out by Alfred Hershey and Martha Chase in 1952. They used bacteriophage to test whether it was proteins or nucleic acids that carried hereditary information. Bacteriophage (otherwise known as 'phage) are viruses that parasitize bacteria. Viruses lack the necessary equipment to replicate their own genomes, so to reproduce they must parasitize the molecular machinery

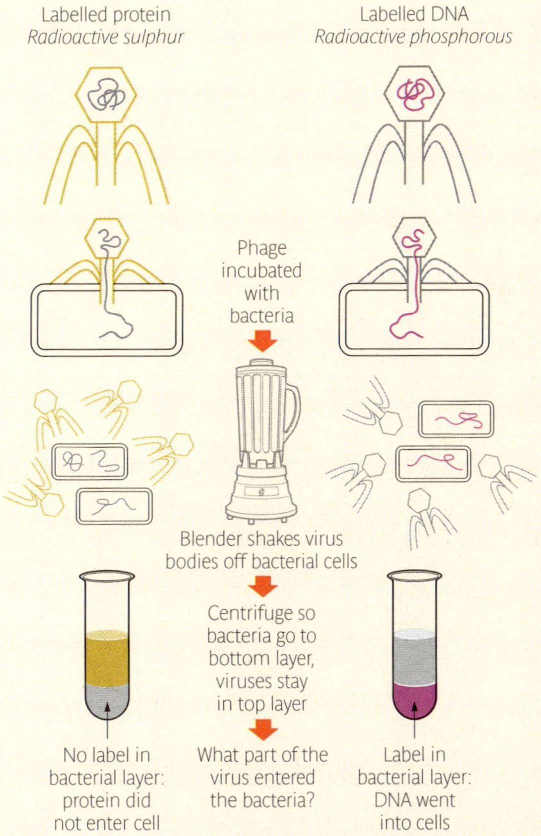

Figure TB2.1 The Hershey–Chase blender experiment.

of a living cell. A phage attaches to the host cell wall and injects its genome into the cell. The bacterial cell then makes copies of the virus genome, and the information in the virus genes is used to make the proteins that form the viral coat (i.e. the 'body' of the virus). The viral genomes are then packaged into the protein coats to form infectious virus particles.

Hershey and Chase labelled viral proteins with radioactive sulphur, and labelled viral DNA with radioactive phosphorous. They allowed these radioactive phages to infect bacteria, so that their genomes would be injected into the bacterial cells. Then they mixed the infected bacterial cells up in a Waring blender, an iconic 1950s domestic appliance. The blender separated the virus coats, which stayed outside the cell, from the genetic material which was injected into the bacteria. The radioactive labels allowed Hershey and Chase to show that the viral protein did not enter the bacterial cells, but the viral DNA did. DNA, not protein, was therefore the genetic material that carried the instructions for making new virus particles.

The effect of the Hershey–Chase experiment was to immediately convince both the leading scientists of the day (such as Linus Pauling) and less established researchers (such as James Watson and Francis Crick) that DNA was the key to understanding heredity. So the race began to discover the structure of DNA.

information stored in the DNA is expressed through the actions of RNA and protein molecules.

To illustrate this process of the conversion of genetic information from one form to another, let's consider the gene shown in **Figure 2.2**, p. 32. In humans, the gene that codes for the beta-subunit of RNA polymerase II (given the acronym *RPB2*) is found on the short arm of chromosome 4. It takes 3525 bases of DNA to specify the amino sequence needed to make this protein. When the cell needs to make an RNA polymerase II enzyme, the *RPB2* gene must be located. Then the DNA containing the gene is unwound from the chromosome, and the two strands of the double helix are unzipped to

Figure 2.4 Francis Crick (right) and James Watson with the tin-plate model they built in 1953 (with the help of workshop technicians at Cambridge) to demonstrate their proposed helical DNA structure with complementary base-pairing. You can find out more about the construction of this model by listening to *A Twist of Life*, an entertaining radio programme narrated by Steve Jones, including interviews with many of the leading figures in the race for DNA. *www.bbc.co.uk/radio4/science/atwisttolife.shtml*.

Reproduced by permission of A. Barrington Brown/SPL.

TECHBOX 2.2 DNA structure

KEYWORDS

nucleotide
base
purine
pyrimidine
ribose
sugar
phosphate
5′, 3′
double helix
phosphodiester bond

FURTHER INFORMATION

Any good biochemistry text will explain DNA structure in more detail. For example, a very clear account can be found in Elliott, W.H. and Elliott, D.C. (2005) *Biochemistry and Molecular Biology*, Oxford University Press.

RELATED TECHBOXES

TB 4.1: DNA replication
TB 1.2: DNA sequencing

RELATED CASE STUDY

CS 4.1: Glorious mud (DNA hybridization)

There are three basic subunits in DNA: bases, sugar, and phosphate.

Bases: The ring-shaped bases are the most charismatic of the DNA subunits. They come in four types which fit together in pairs. This pairing forms the basis of the information-carrying capacity of DNA. The bases are rings of oxygen, hydrogen, nitrogen, and carbon molecules. Two of the bases, the pyrimidines, are single rings.

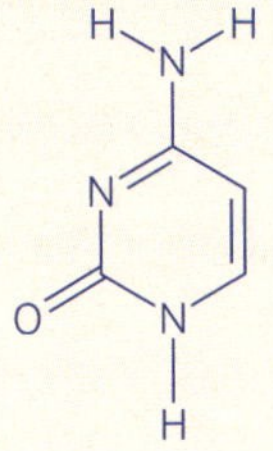

Cytosine (C)

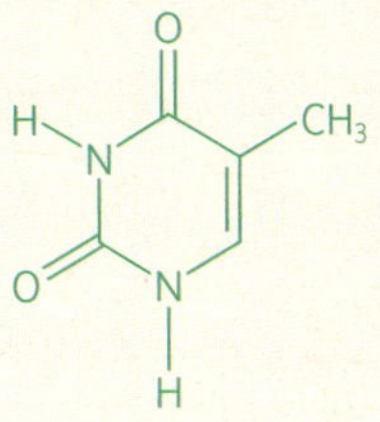

Thymine (T)

The other two bases, the purines, are made of double rings.

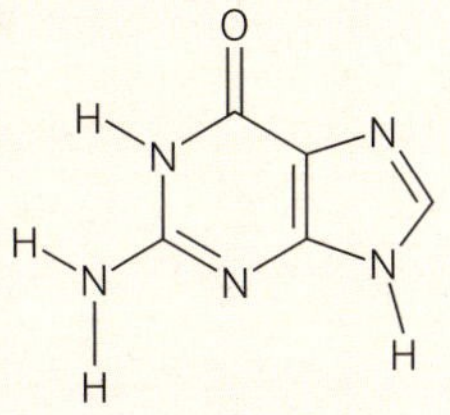

Guanine (G)

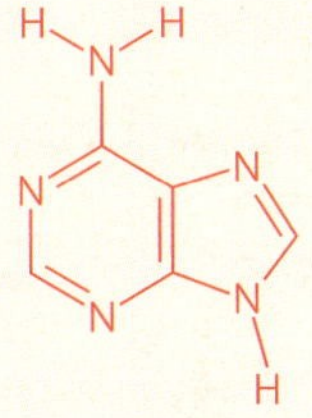

Adenine (A)

Sugars: Each base is joined to a sugar molecule, a 5-carbon (pentose) ring. In RNA, the sugar is ribose. In DNA, the sugar is a very similar molecule called deoxyribose (the 'deoxy' means that one hydroxyl (OH) group is missing from this form of ribose). Why do RNA and DNA have slightly different sugars? Deoxyribose makes DNA more chemically stable than RNA. It seems likely that RNA represents an earlier form of information storage, and DNA is the new improved version.

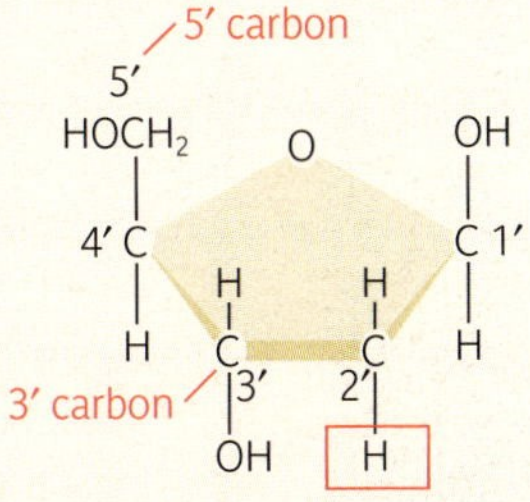

2-Deoxyribose

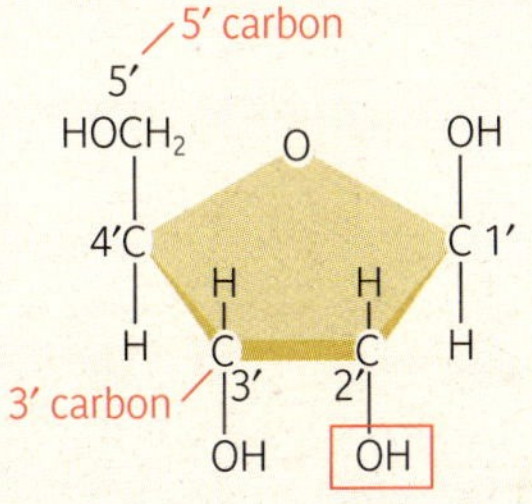

Ribose

Phosphate: Phosphate molecules provide the structural 'glue' that holds the DNA backbone together, because they form phosphodiester bonds (strong covalent bonds) which link the phosphate and sugar molecules.

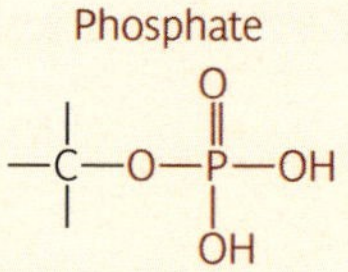

Putting it all together: The combination of base + sugar + phosphate is a nucleotide, the basic structural unit of DNA.

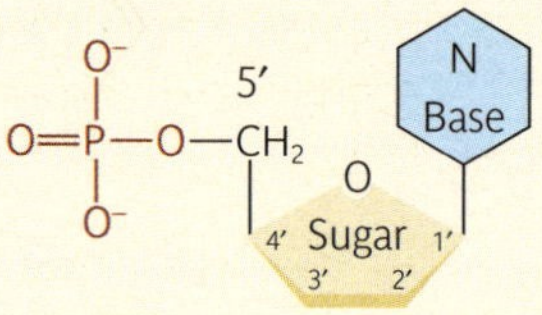

Nucleotides can form spontaneously under certain conditions, but linking them together into a DNA strand takes energy and specialized equipment (in the form of enzymes). A DNA polynucleotide strand is built by creating a phosphodiester bond that links the 3′ carbon on the sugar of the growing chain with the phosphate attached to the 5′ carbon of an incoming nucleotide. So the backbone of a polynucleotide strand is made of linked sugar–phosphate–sugar–phosphate, with one base joined to each sugar molecule.

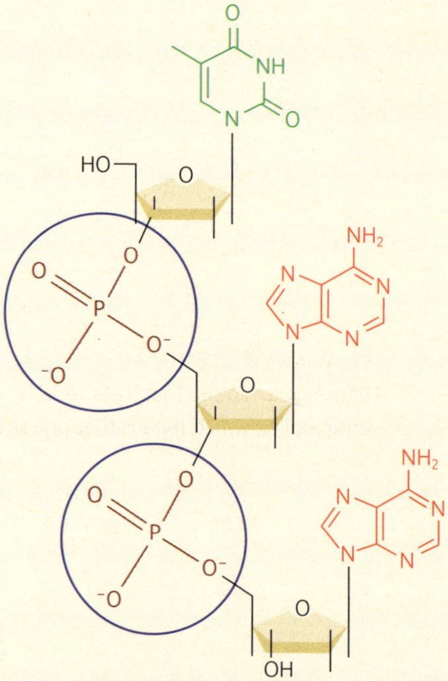

Base pairing: We have considered how the subunits of DNA – bases, sugars, and phosphates – link together to form a linear polynucleotide. Now we can consider how two strands fit together to make the famous double helix. If two polynucleotide strands face each other, the sugar–phosphate backbone runs down each side, and the bases stick into the middle, like the steps of a spiral staircase. Complementary pairs of bases can spontaneously form hydrogen bonds: three bonds between a C and a G, two between an A and a T. Each pair consists of one double-ring purine and one single-ring pyrimidine, so the complementary base pairs maintain an even 'step' width between the two sugar–phosphate strands.

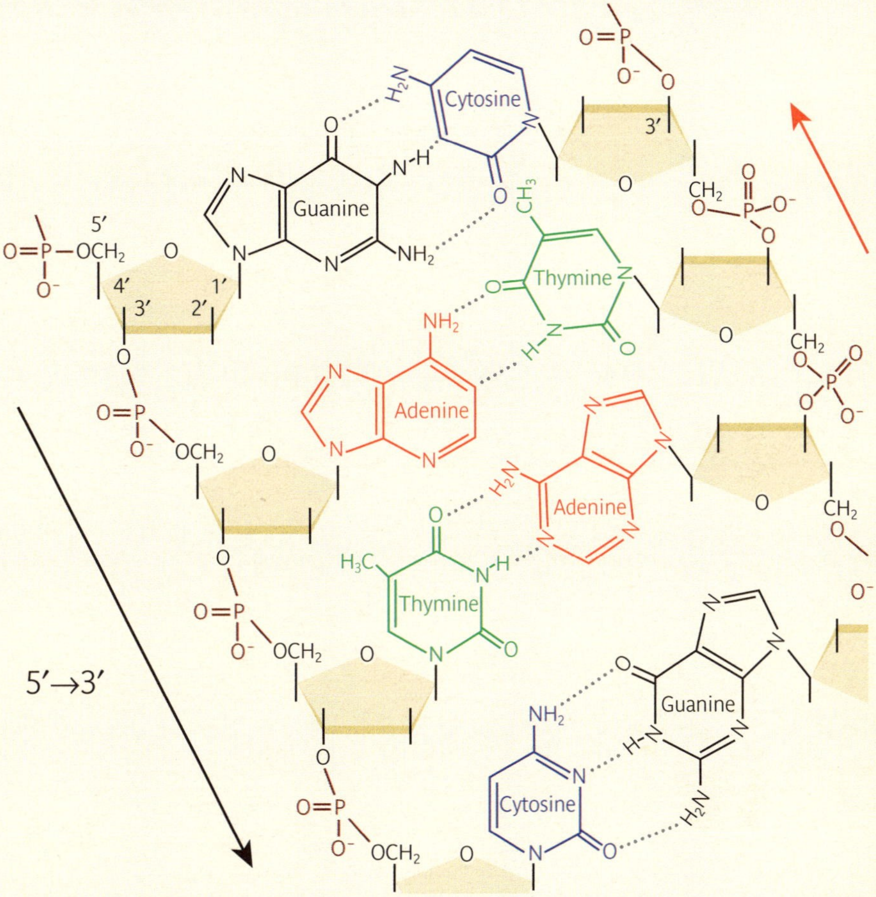

What does 5′ to 3′ mean? If you look closely at the double strand of DNA in the above figure, you can see that the strands are not mirror images of each other. They run antiparallel, which means that one strand is upside down with respect to the other.

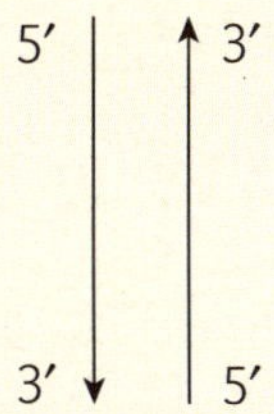

This may be hard to see at first, but you can use the numbering on the sugar rings to spot the difference. Each ribose has five carbon atoms. The carbon attached to the base is 1′ (pronounced 'one-prime'). Counting around the ring, the 5′ (five-prime) carbon is the one attached to the phosphate group of the nucleotide.

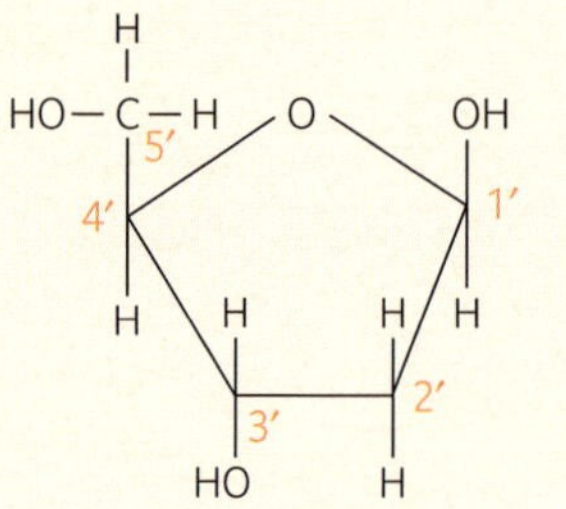

The 3′ (three-prime) carbon is the one that forms the phosphodiester bond to link to the phosphate group on the neighbouring nucleotide.

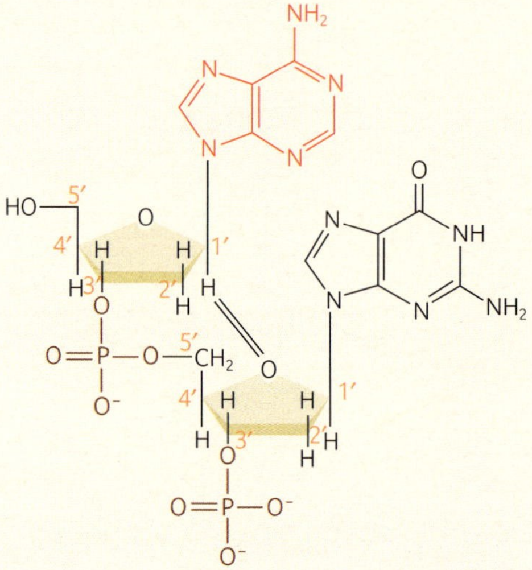

So, looking at the double-stranded diagram on p. 43, if you follow the series of connections in the two polynucleotide strands from top to bottom, the left-hand DNA strand runs 5′ to 3′ (phosphate connected to 5′ of sugar, which is connected at 3′ to the next phosphate, and so on), but its matching sister strand runs 3′ to 5′ (phosphate connected to 3′ of sugar, which is connected at 5′ to the next phosphate). By convention, when the sequence of bases in DNA is written down, it is usually given from the 5′ end and moving to the 3′ end (of course, if you have the base sequence on one strand you can work out the other strand using the base-pairing rules). So the sequence of bases in this short section of DNA would be written 'GATC'. 5′ to 3′ will also be important when we look at DNA replication (see Chapter 4).

Why is DNA a helix? Nucleotide bases attract each other along the edges, forming hydrogen bonds between the base on one strand and its matching base on the other strand. So why doesn't DNA form a simple ladder, with sugar–phosphate uprights and straight base-pair rungs? The flat faces of the bases are hydrophobic, so they repel water. If DNA was a simple ladder, the gaps between the 'rungs' would leave the bases exposed to water molecules, making the whole structure unstable. But if the bases are stacked not directly on top of each other but offset slightly and rotated, base pairs can fit snugly on top of each other and minimize the destabilizing effect of water molecules. Because each base-pair 'rung' turns at 32° from the previous pair, the double-stranded DNA molecule makes a complete turn every ten base pairs.

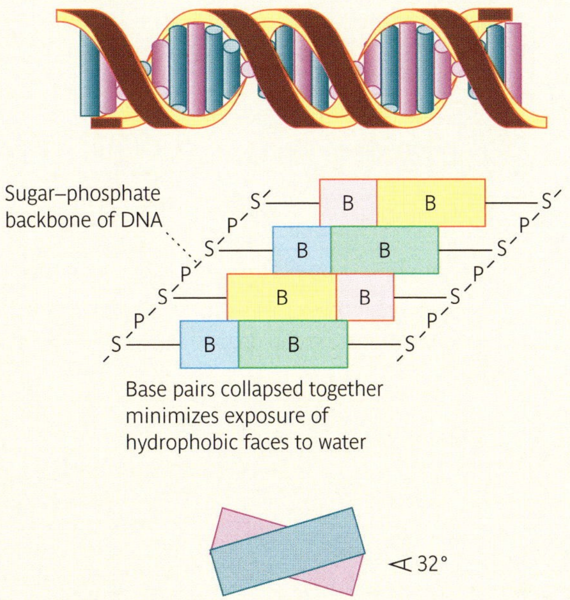

Figures from Elliott, W.H. and Elliott, D.C. (2005) *Biochemistry and Molecular Biology*, Oxford University Press.

expose the base sequence of the gene. An existing RNA polymerase enzyme then uses the DNA template to make an RNA copy of the gene.

RNA is more or less the same as DNA but there are a number of differences. RNA is single-stranded (not double-stranded like DNA), uses a different sugar molecule in its backbone (ribose not deoxyribose), and one of the four bases is slightly different (instead of thymine (T) it uses uracil (U)). The RNA copy of the gene is made by matching each base on the exposed DNA strand to its complementary RNA base. Where there is an A in the gene (DNA), it is matched by a U in the message (RNA), a T in the gene is matched by an A in the message, a G with a C, and a C with a G. So the DNA sequence of the human *RPB2* gene given in **Figure 2.2** begins 'CGTGATGGT. . . .', but its complementary RNA would read 'GCACUACCA. . . .' (**Figure 2.5**, p. 46).

Complementary base pairing is also covered in the DNA replication chapter (Chapter 4)

DNA is stuck in the nucleus. But RNA can move from the nucleus to the cytoplasm. Because this RNA strand is made by complementary base pairing it contains the same information as the DNA strand it was copied from. This RNA strand is known as messenger RNA (mRNA) because it can take the information from the nucleus to the cytoplasm where it can be used to build useful things.

When messenger RNA leaves the nucleus, it is taken to a ribosome. Ribosomes are the workbenches of the genomic system, where information from the nucleus, transported in messenger RNA, is used to construct a protein. Here, complementary base pairing is used again to translate the information in the messenger RNA into the amino acid sequence of the protein product. The sequence of bases in the messenger RNA is matched to bases on transfer RNAs, each of which brings a specific amino acid to the ribosome.

There is a host of transfer RNAs (tRNAs) in the cytoplasm. Each tRNA has a particular three-base recognition

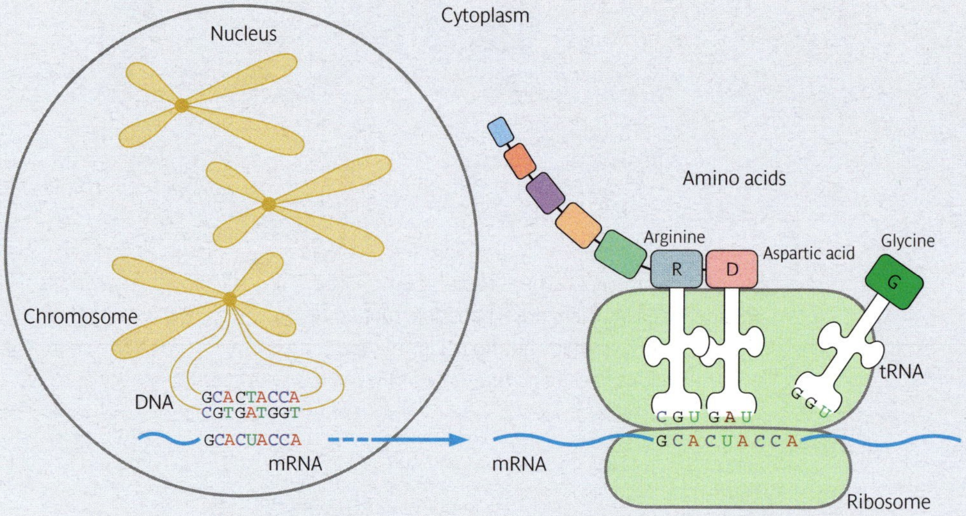

Figure 2.5 Simplified diagram illustrating the way that complementary base pairing is used to transfer information from the DNA in the nucleus to the messenger RNA (mRNA), which then moves to the cytoplasm, where transfer RNAs (tRNAs) complementary three-base sequences match the mRNA to bring the right sequence of amino acids to make the protein product.

sequence and carries a specific amino acid. So the bases CGT in the *RPB2* gene are transcribed to GCA in the mRNA, which is matched by a tRNA with the recognition sequence CGU, and carries the amino acid arginine. Similarly, the next three letters in the gene are GAT, matching CUA in the mRNA and a tRNA with the sequence GAU, which carries the amino acid aspartic acid to join the protein. In this way the base sequence in the gene (DNA) determines the base sequence in the message (mRNA) that matches the recognition sequence of a particular tRNA which determines the sequence of amino acids in the protein. Figure 2.6 shows the DNA sequence given in Figure 2.1 translated into amino acids (one amino acid for every three bases of DNA; **TechBox 2.3**).

The amino acids are attached to the growing peptide chain in the order specified by the gene. When all of the bases in the message have been 'read', the ribosome falls away from the message, releasing the chain of amino acids. The forces of attraction and repulsion between the amino acids in the chain cause parts of the sequence to spontaneously fold into energetically stable helices and sheets, which then twist around each other to form a

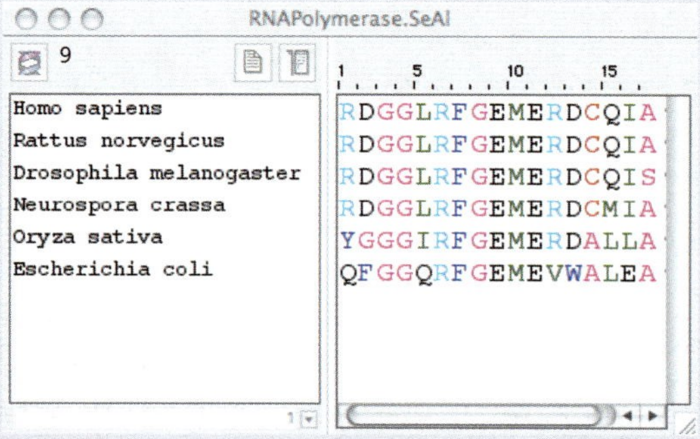

Figure 2.6 Part of the amino acid sequence of the beta subunit of RNA polymerase II, from five different species. The amino acid sequence shown here was translated from the DNA sequence shown in Figure 2.2. Each letter stands for a single amino acid, so R is arginine, D is aspartic acid, G is glycine, and so on (see **TechBox 2.3**).

TECHBOX 2.3 Genetic code

KEYWORDS

amino acid
translation
protein
gene
codon
redundancy
stop codon

FURTHER INFORMATION

You can read the intricacies of the amino acid notation rules at *www.chem.qmul.ac.uk/iupac/AminoAcid/A2021.html*

A list of genetic codes can be found on the NCBI website: *www.ncbi.nlm.nih.gov/Taxonomy/Utils/wprintgc.cgi*

RELATED TECHBOXES

TB 6.3: Multiple alignment

TB 2.2: DNA structure

RELATED CASE STUDIES

CS 3.1: Viruses within (identifying protein coding sequences)

CS 5.1: Sweet and sour (identifying protein-coding sequences)

There are 20 different amino acids commonly found in proteins. But DNA only has four different nucleotides with which to specify all of the amino acids. So the nucleotide sequence in a protein-coding gene is read in triplets: three-letter 'words' that each specify a particular amino acid. These triplets of nucleotides are called codons. With the four-letter alphabet of DNA, there are 64 possible three-base codons, which is more than the number of amino acids, so many amino acids are represented by more than one codon. Having multiple codons specifying the same amino acid is known as the redundancy of the genetic code. This redundancy leads to some interesting patterns of sequence evolution (see Chapter 8).

There are a number of possible 'start' codons which specify the beginning of a protein (e.g. ATG, which also codes for the amino acid methionine). There are also several stop codons. The tRNAs that match stop codons do not carry an amino acid, so when a stop tRNA attaches to the messenger RNA, amino acid chain elongation stops and protein synthesis is finished. A mutation that causes a stop codon to occur in the middle of a gene will give rise to a truncated protein product which will probably be non-functional **Figure TB2.3b** gives all 64 possible codons with the corresponding amino acid for the 'universal genetic code'. This is the code used in the majority of genomes. However, there are minor variants of this code (for example, mitochondrial genomes use a slightly different code; **Figure TB2.3a**).

Amino acids can be represented by their name (e.g. isoleucine), a three-letter abbreviation (Ile), or a single letter (I). The three-letter abbreviations are written with a capital and two lower-case letters: e.g. Thr for threonine or Asn for aspargine. When an international convention for symbols for the genetic code was established in 1968, there was an attempt to make the single-letter symbols of amino acids memorable using mnemonic associations. Some of the single letter codes are obvious, such as C for cysteine or M for methionine. Where more than one amino acid shared the same starting letter, it was given to the most commonly used amino acids: e.g. A for alanine but R for arginine. Other letter assignments are phonetic, like F for phenylalanine. And then the assignments start getting tenuous: for example, W for tryptophan because it's a big letter for a big double-ring molecule. U and O weren't assigned because they can be confused with V and zero, and J was left out because its not used in some languages.

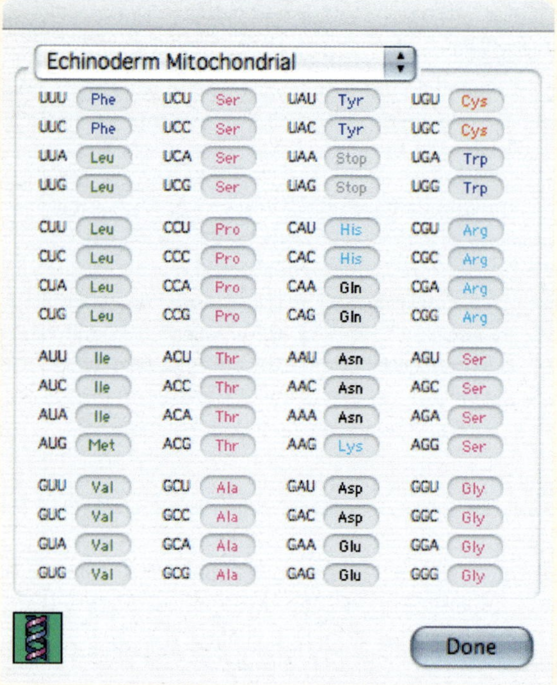

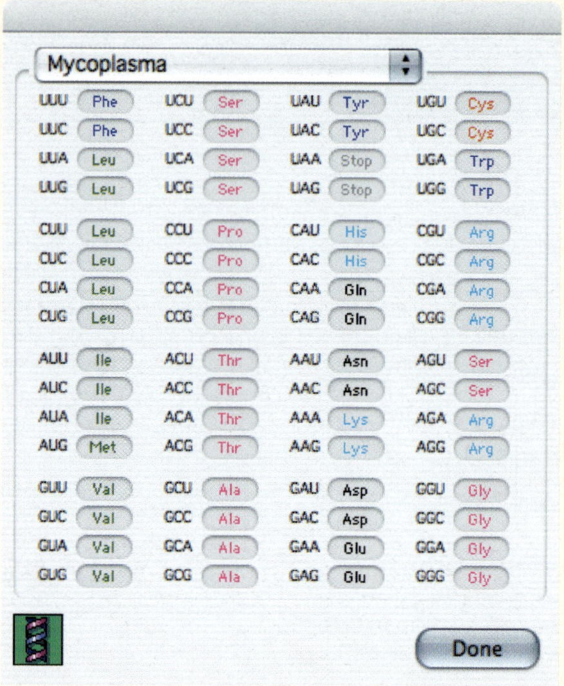

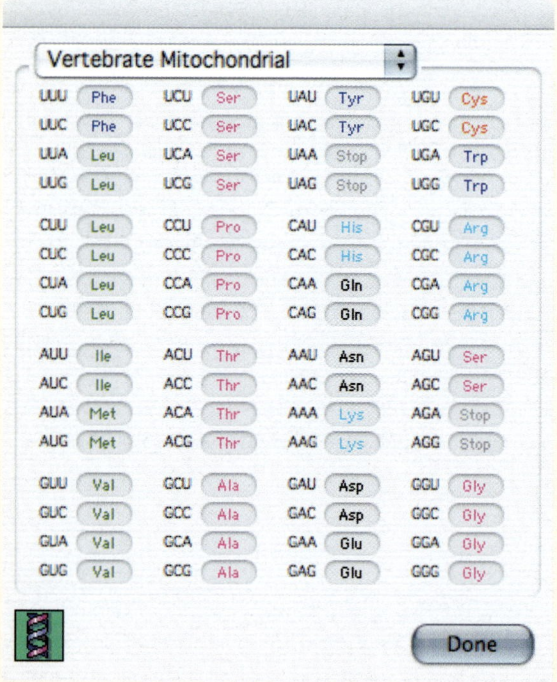

Figure TB2.3a Spot the difference: three examples of variations on the genetic code, shown here as screenshots from the alignment editor Se-Al. This program allows you to select any of 14 different genetic codes then toggle between the nucleotide sequence and amino acid sequence of a gene. As an aside, mycoplasmas belong to a charmingly named class of bacteria, the mollicutes, which contain the smallest and simplest single-celled organisms. Mycoplasmas have the smallest genomes of any non-virus organism, with as few as 580,000 bases (only twice as large as the longest human gene). DNA analysis suggests that mollicutes are secondarily simple, having lost many characteristics such as a cell wall as they adapted to a parasitic lifestyle.

1st	2nd: T	C	A	G	3rd
T	TTT F Phe Phenylanaline	TCT S Ser Serine	TAT Y Tyr Tyrosine	TGT C Cys Cysteine	T
	TTC F Phe	TCC S Ser	TAC Y Tyr	TGC C Cys	C
	TTA L Leu Leucine	TCA S Ser	TAA – *Stop*	TGA – *Stop*	A
	TTG L Leu	TCG S Ser	TAG – *Stop*	TGG W Trp Tryptophan	G
C	CTT L Leu	CCT P Pro Proline	CAT H His Histidine	CGT R Arg Arginine	T
	CTC L Leu	CCC P Pro	CAC H His	CGC R Arg	C
	CTA L Leu	CCA P Pro	CAA Q Gln Glutamine	CGA R Arg	A
	CTG L Leu	CCG P Pro	CAG Q Gln	CGG R Arg	G
A	ATT I Ile Isoleucine	ACT T Thr Threonine	AAT N Asn Asparagine	AGT S Ser Serine	T
	ATC I Ile	ACC T Thr	AAC N Asn	AGC S Ser	C
	ATA I Ile	ACA T Thr	AAA K Lys Lysine	AGA R Arg Arginine	A
	ATG M Met Methionine	ACG T Thr	AAG K Lys	AGG R Arg	G
G	GTT V Val Valine	GCT A Ala Alanine	GAT D Asp Aspartic acid	GGT G Gly Glycine	T
	GTC V Val	GCC A Ala	GAC D Asp	GGC G Gly	C
	GTA V Val	GCA A Ala	GAA E Glu Glutamic acid	GGA G Gly	A
	GTG V Val	GCG A Ala	GCG E Glu	GGG G Gly	G

Figure TB2.3b The universal genetic code. To translate a codon, find the first letter of the codon in the leftmost column, then read across to the correct column for the second letter of the codon, then read down to the correct line for the third letter. The colours of the one-letter amino acid codes represent the chemical properties of the amino acids: small non-polar (G, A, S, T: orange); hydrophobic (C, V, I, L, P, F, Y, M, W; green); polar (N, Q, H; magenta); negatively charged (D, E; red); and positively charged (K, R; blue). This is only one of many possible amino acid colouring schemes (for other schemes see *www.bioinformatics.nl/~berndb/aacolour.html*). Note that although this version of the code is referred to as 'universal', there are can be slight variations between genomes (**Figure TB2.3a**).

stable, three-dimensional structure (sometimes this folding process requires chaperone proteins to get the peptide to the right conformation). Finally, this completed protein combines with other protein and RNA subunits to make the working enzyme. This complex series of biochemical events – transcription of genes into messenger RNA, translation of mRNA to protein by transfer RNAs – occurs continuously in every cell. The genomic system is the most remarkably complex yet wonderfully effective organization, and it is the bedrock of all living processes.

The central dogma

'The central dogma was put forward at a period when much of what we now know in molecular genetics was not established. All we had to work on were certain fragmentary experimental results, themselves often uncertain and confused, and a boundless optimism that the basic concepts involved were rather simple and probably much the same in all living things.'

Crick, F.H.C. (1970) Central dogma of molecular biology. *Nature*, Volume 227, pages 561–563

This flow of information from DNA to RNA, and from mRNA to proteins, was described by Francis Crick (somewhat messianically) as the 'central dogma of molecular biology'. The simplest statement of the central dogma is: DNA copies DNA, DNA is transcribed into RNA, RNA is translated into protein (**Figure 2.7**).

Figure 2.7

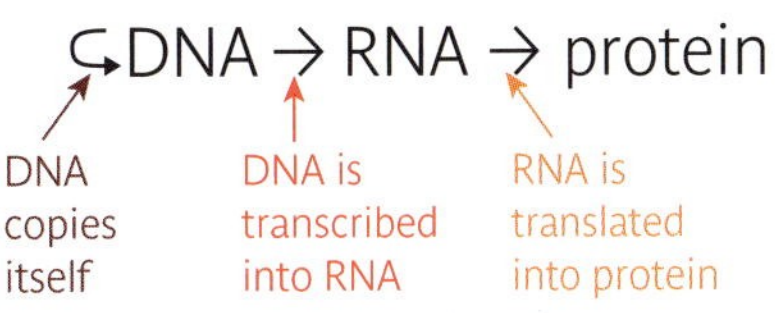
DNA → RNA → protein
DNA copies itself
DNA is transcribed into RNA
RNA is translated into protein

As with so many things in biology, the more we learn about the genome and cell function, the more complicated the picture gets. For example, a viral enzyme, reverse transcriptase, can use an RNA template to make a complementary DNA molecule. We can modify the dogma to include this pathway (**Figure 2.8**).

Figure 2.8

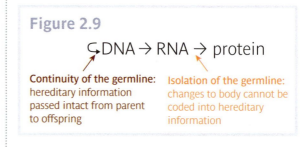

Reverse transcriptase can make DNA from RNA

↻DNA ⇆ RNA → protein

One of the important corollaries of the central dogma is this: the information in DNA and RNA is effectively interchangeable – DNA can be used to make a complementary RNA strand and vice versa (a fact exploited by many molecular technologies) – but the same is not true for the sequence of amino acids in a protein. Proteins don't have template copying. The sequence of amino acids in one protein cannot cause the formation of an identical copy of that sequence. Even prions, infectious proteins such as the one responsible for 'mad cow disease' (bovine spongiform encephalitis, or BSE) are thought to work by changing the conformation of existing prion proteins, rather than creating new copies of themselves. There is no known biochemical mechanism for translating the sequence of amino acids in a protein into the nucleotide code of a gene.

So the central dogma is a molecular statement of Weismann's barrier: information flows from the germline (DNA and RNA) to the soma (protein) but not back the other way (Figure 2.9). If the DNA sequence of a gene is changed, it may result in the formation of a protein with a novel sequence, and if that novel protein is advantageous, then carriers of that gene might be positively selected. But if a change is made directly to a protein, for example if the wrong amino acid is inserted as the protein is being constructed, then even if that change is advantageous, it cannot be coded back into the gene, so it is unlikely to passed on down the generations.

Figure 2.9

↻DNA → RNA → protein

Continuity of the germline: hereditary information passed intact from parent to offspring

Isolation of the germline: changes to body cannot be coded into hereditary information

Natural selection of genetic variants is discussed at more length in Chapter 5

Ubiquity of DNA

The template copying of DNA explains Weismann's principle of continuity: genetic information is copied from generation to generation because DNA can be faithfully replicated by complementary base pairing. The central dogma of molecular biology explains Weismann's theory of the isolation of the germline: changes to the body cannot be written back into an individual's DNA. We now know that the germline cells do not have to be physically isolated to preserve Weismann's barrier. The hereditary information is present in all cells, but the one-way flow of information from DNA and RNA to proteins means that the information in the genome is fundamentally unaffected by changes to the body.

The continuity and isolation of the genome make DNA a particularly handy molecule for biologists. If pangenesis was true, then each cell of the body could only provide information about the particular tissue it was drawn from. In that case, we could only read the complete set of genetic information for an organism if we obtained sperm or eggs or an undifferentiated zygote, where all the hereditary information was collected together. As it is, every cell (with a few exceptions) contains an entire copy of the genome. This means that we can collect nearly any biological tissue and extract DNA from it. Incidentally, this fact also makes cloning possible, because each cell carries the instructions for the construction of the whole organism (see **TechBox 6.4**).

How to get DNA samples

Typically, biologists working with DNA data will sample tissue from living individuals; for example, a seed, a blood sample, or, for the more unfortunate study animal, a piece of liver. Some types of specimen are easier to extract DNA from than others (see **TechBox 2.4**). Generally, the more biological material, and the fresher

TECHBOX 2.4 DNA extraction

KEYWORDS

lysis

purification

nuclease

proteinase

DNAase

centrifuge

FURTHER INFORMATION

A number of websites provide simple instructions for extracting DNA at home or in the laboratory: for example *http://gslc.genetics.utah.edu/units/activities/extraction/*.

RELATED TECHBOXES

TB 4.2: DNA amplification

TB 6.4: Cloning and conservation

RELATED CASE STUDIES

CS 2.1: On the origin of faeces (DNA surveys)

CS 3.1: Glorious mud (DNA hybridization)

DNA is present in most biological samples. But before DNA can be sequenced, it must be isolated and purified. For some samples, DNA extraction is routine and reliable. For others, successful DNA extraction is an art that requires endless patience and a great deal of tinkering in the laboratory. The details of DNA-extraction techniques will vary from lab to lab, and different procedures will work best for particular samples. For this reason, there are a very large number of extraction protocols (which are essentially laboratory recipes) available on the internet, in scientific journals, and in laboratory manuals. However, all extraction protocols follow the same basic steps of cell lysis (to free the DNA), nuclease inactivation (to prevent DNA breakdown), and purification (to remove non-DNA molecules).

1. **Cell lysis:** First, sample tissues have to be broken up. For example, a leaf may be frozen in liquid nitrogen then pounded in a mortar and pestle, or a piece of liver might be pulverized in a blender. Then the crushed material is spun or strained to remove extraneous material, leaving just disassociated cells. A chemical (such as proteinase K) is added to the cells to burst the cell walls and release the DNA. Once cell lysis is complete, the digested material can be spun in a centrifuge, so that the cellular debris sinks to the bottom of the sample, permitting a liquid containing the DNA to be pipetted off the top. While it is possible to isolate organelle DNA (from mitochondria or chloroplasts) from nuclear DNA, it is more usual for all cellular DNA to be mixed together.

2. **Nuclease inactivation:** Enzymes that degrade DNA (DNases) are present in most biological samples. Various chemicals must be added to stop these enzymes from destroying the DNA in the sample, such as SDS (sodium dodecyl sulphate), EDTA (ethylenediaminetetra-acetate), and proteinase K.

3. **DNA purification:** The DNA solution contains other biological molecules such as lipids, polysaccharides, and proteins, which need to be removed. This is usually done by phenol–chloroform extraction or by running the solution through a column to separate the DNA from the other components. The DNA can then be precipitated out of solution using ethanol. The condensed DNA is usually extracted by a quick spin in the centrifuge which leaves the DNA in a pellet at the bottom of the test-tube. More dramatically, a glass rod or stick can be used to spool long strands of DNA out of the solution (**Figure TB2.4**).

DNA extraction at home: It is possible to carry out all of these steps using commonly available materials, for example using household detergent to lyse cells, contact lens cleaning solution to inactivate nucleases, and rubbing alcohol to purify the DNA. With a bit of experimentation you can produce clearly visible strands of DNA.

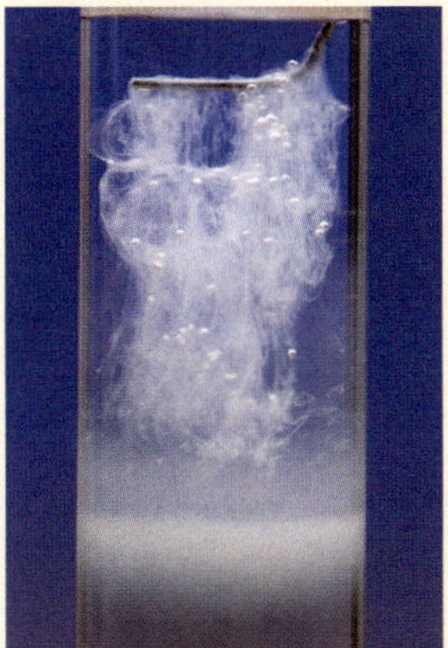

Figure TB2.4 It is possible to produce visible white strands of DNA from a biological sample, as shown here in this photo, where DNA is being spooled out of the solution.

it is, the easier it is to extract DNA. However, usually only a small sample is needed, such as a few grams of tissue or a few millilitres of blood. Sometimes DNA sampling is destructive, resulting in the demise of the sampled individual. For example, an entire beetle may be ground up for DNA extraction. Other organisms may survive being partially sampled, such as taking a leaf from a plant.

The harm done to study organisms by DNA sampling is an important ethical consideration. For example, it has been shown that the practice of toe-clipping amphibians – a common means of marking and taking tissue samples from captured individuals – can reduce their probability of survival. The distress caused to animals by being captured and handled to take a blood sample, or the destruction to habitat created by the search for elusive organisms, should not be underestimated. For many species, non-invasive methods of DNA sample collection are becoming widely advocated, such as using hair-traps, collecting faeces, or finding cast-off skin or feathers. Although these techniques may provide poorer-quality samples than destructive sampling, they can sometimes provide a practical way of collecting DNA and may even provide a rich source of data on individuals' movement and behaviour (see **Case Study 2.1**).

DNA extraction

DNA extraction is generally best performed on fresh tissue, because DNA, like most biomolecules, degrades over time. DNA degradation can be reduced by preserving samples, particularly by freezing or immersion in ethanol. But although fresh material is the easiest to work with, DNA samples have been successfully taken from Egyptian mummies, frozen mammoths, dehydrated penguins, pickled thylacines, carved whale teeth, preserved food, ancient timber, and 1000-year-old marine sediments. Even the last meal of Ötzi, the 'iceman' whose 5000-year-old frozen body was found in the European alps, has been determined through DNA analysis of his intestinal contents. Museum and herbarium specimens are proving particularly valuable for DNA analysis, although understandably many curators are not terribly keen on having pieces taken out of rare specimens. However, the inevitable decay of DNA means that there is little chance of recovering DNA from very old biological specimens.

DNA from preserved specimens (ancient DNA) is discussed in Case Studies 2.2 and 5.2

DNA extraction involves three basic steps: breaking up the cells to release the DNA, halting the action of enzymes that would destroy the DNA, then separating the DNA from the other cellular components. However, biological samples vary widely in their structure and contents, so most biologists find they have to fiddle with DNA-extraction protocols to get the best results for their samples. If a serious amount of fiddling in the lab is needed to make a particular sample yield useable DNA, then DNA extraction can seem like more of an art than a science, but the degree of elation when DNA is successfully extracted is often proportional to the amount of time spent trying to get it to work (**TechBox 2.4**).

CASE STUDY 2.1

On the origin of faeces: using DNA from scats to survey endangered lynxes

KEYWORDS

DNA extraction
species identification
conservation
contamination
primers
controls
PCR

RELATED TECHBOXES

TB 2.4: DNA extraction
TB 4.3: Primer design

RELATED CASE STUDIES

CS 1.2: DNA surveillance (monitoring endangered species)
CS 6.2: Keeping the pieces (DNA and conservation)

Pires, A.E. and Fernandes, M.L. (2003) Last lynxes in Portugal? Molecular approaches in a pre-extinction scenario. *Conservation Genetics*, Volume 4, pages 525–532

> *“DNA methods on scats, or ‘molecular scatology’, are a relatively new means of gleaning information from animals that may be otherwise difficult to survey.”* [1]

Background

When a species becomes critically endangered, accurate estimates of the number of surviving individuals are crucial to developing effective conservation strategies. But, naturally, the more endangered an animal gets, the harder it is to spot. Indirect methods of surveying are often the only feasible approach. The presence of endangered animals can in some cases be established by finding scats (faeces). Scats can often be identified to species by characteristic shape and size, but this visual identification is not always reliable[1,2].

Aim

The Iberian lynx (*Lynx pardinus*), the most endangered cat species in the world, is close to extinction in the wild. There may be as few as 100 individuals left in the wild, in several isolated populations. In Portugal, the Iberian lynx are now so scarce that they are almost never seen. This study aimed to determine whether there are enough individuals remaining in the wild in Portugal to form a viable population.

Methods

Potential lynx habitats were surveyed between 1997 and 2000, and a total of 104 putative lynx scats were collected. The tip of each scat was removed and placed in a plastic vial containing silica isolated with porous tissue, and the rest of the scat was frozen in plastic bags. Faeces are a relatively poor source of DNA, so choosing the best DNA-extraction method is critical to success. The authors tested a number of DNA-extraction methods (**TechBox 2.4**) and found that commercial DNA-extraction kits gave the best results. Following standard procedures for these kits (with some modifications), faecal material taken from the outside of the scats (the part most likely to have picked up cells on its way out of the lynx) was lysed, incubated with proteinase K, washed, and purified. Extractions were performed with blank controls to test for contamination. They used lynx-specific primers (sequences from mitochondrial genes known to match only lynx DNA, not other species: **TechBox 4.3**) in a PCR analysis to amplify any target DNA from the scats (**TechBox 4.2**). The PCR was also performed on scats from captive lynx to provide positive controls (to prove that the methods can amplify lynx DNA when it is present). Dog and wildcat scats were used to provide negative controls (to test whether the methods could give a false result in the absence of lynx DNA).

Figure CS2.1a The Iberian Lynx (*Lynx pardinus*) has the dubious distinction of being the most critically endangered cat species. Lynx populations were reduced by trophy hunting, trapping, and habitat loss. Development and road building have contributed to the fragmentation of the remaining lynxes into very small, isolated populations. With somewhere between 30 and 300 breeding females left in the wild, much hope is being placed in captive breeding programs to build up viable populations of Iberian lynx. Iberian lynx news, including latest counts of wild lynxs and announcements of cubs born in captivity, can be found at *www.iberianature.com/material/iberianlynxnews.htm*.

Results

Of 104 scat samples, 95 were found to contain DNA. However, only two of these could be shown to be lynx DNA. This result provides the first conclusive proof that there have been lynx in the area in 1997. However, with only two positive samples, it is difficult to tell whether these individuals were visitors, or the last members of a population, or whether this survey method underestimates lynx numbers. Some samples showed evidence of cross-contamination, highlighting the importance of checks for contamination.

Conclusions

The authors conclude that lynx have virtually disappeared from Portugal. If they are present, they are in such low numbers that they would be unable to maintain viable populations.

Limitations

The problem of negative results – concluding absence of lynx due to failure to amplify lynx DNA – is an issue for remote surveying using DNA-poor samples such as faeces. Here, the authors used control samples, taken from captive lynx faeces, to demonstrate that their technique can detect lynx DNA from scats. Methods using multiple primers may be able to give rapid results and more protection against false negatives. For example, Dalén and

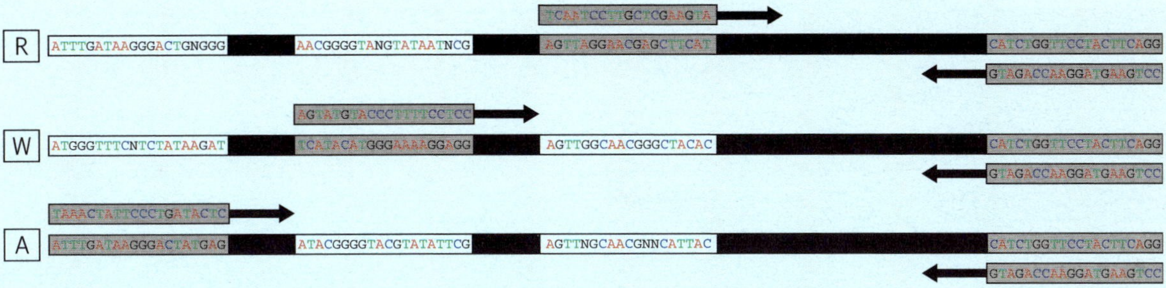

Figure CS2.1b Illustration of the concept of RCP-PCR. One of the three species-specific primers will react with the general primer. The resulting fragment size depends on whether red fox (R), wolverine (W), or arctic fox (A) DNA is present in the extract. Intraspecific variable sites in the template are shown as (N). From Dalén *et al.* (2004)[3].

coworkers developed a *rapid classificatory protocol PCR* (RCP-PCR), where several different primers are added to the same sample: a universal primer that binds to all mammal species, and a number of species-specific primers, producing different-size fragments characteristic of each species (**Figure CS2.1b**)[3].

Future work

'Molecular scatology' has the potential not only to survey the presence or absence of particular species in a given area, but also to provide a picture of the activity of species in space and time. For example, DNA sampling from scats has been used to show how spatial overlap between arctic and red foxes changes between seasons[4]. DNA analysis of faeces can identify both predator and prey, revealing the species that made up the carnivores' most recent meals[5], and even allowing individual differences in prey-preference to be studied[6].

References

1. Davison, A., Birks, J.D.S., Brookes, R.C., Braithwaite, T.C. and Messenger, J.E. (2002) On the origin of faeces: morphological versus molecular methods for surveying rare carnivores from their scats. *Journal of Zoology*, Volume 257, pages 141–143.
2. Pires, A.E. and Fernandes, M.L. (2003) Last lynxes in Portugal? Molecular approaches in a pre-extinction scenario. *Conservation Genetics*, Volume 4, pages 525–532.
3. Dalén, L., Götherström, A. and Angerbjörn, A. (2004) Identifying species from pieces of faeces. *Conservation Genetics*, Volume 5, pages 109–111.
4. Dalén, L., Elmhagen, B. and Angerbjörn, A. (2004) DNA analysis on fox faeces and competition induced niche shifts. *Molecular Ecology*, Volume 13, pages 2389–2392.
5. Deagle, B.E. *et al.* (2005) Molecular scatology as a tool to study diet: analysis of prey DNA in scats from captive Steller sea lions. *Molecular Ecology*, Volume 14, pages 1831–1842.
6. Fedriani, J.M. and Kohn, M.H. (2001) Genotyping faeces links individuals to their diet. *Ecology Letters*, Volume 4, pages 477–483.

CASE STUDY 2.2

More moa: using ancient DNA to catalogue diversity of extinct giant birds

KEYWORDS

ancient DNA
DNA extraction
museum specimens
genotyping
phylogeny
control region
molecular clock

RELATED TECHBOXES

TB 7.3: Bayesian inference
TB 8.4: Molecular dating

RELATED CASE STUDIES

CS 6.1: Barcoding nematodes (DNA taxonomy)
CS 8.1: Same but different (defining species using DNA)

Baker, A.J., Huynen, L.J., Haddrath, O., Millar, C.D. and Lambert, D.M. (2005) Reconstructing the tempo and mode of evolution in an extinct clade of birds with ancient DNA: The giant moas of New Zealand. *Proceedings of the National Academy of Sciences USA*, Volume 102, pages 8257–8262

“. . . *Ancient DNA methods provide powerful tools for inferring the number of lineages, as well as the tempo and mode of evolution of entire extinct groups of animals.*” [1]

Background

Found only in New Zealand, moa were a morphologically diverse family of birds belonging to the order Struthioniformes, which includes ostriches and emus and their allies (**Figure CS2.2a**). Currently six moa genera are recognized, ranging in size from the Giant Moa *Dinornis giganteus* (up to 250 kg, twice the size of an ostrich) to the relatively small Coastal Moa *Euryapteryx curtus* (20 kg, the size of a largish turkey). Moa went extinct not long after human settlement of New Zealand, less than 1000 years ago. However, moa bones are abundant, both in natural collections such as caves and swamps, and in middens (prehistoric rubbish dumps).

Aim

Assignment of fossil remains to species is usually dependent on morphological similarity. Yet members of the same species may be morphologically very distinct, for example juveniles and adults, or males and females. Conversely, distinct non-interbreeding species may appear very similar from skeletal evidence alone. By sequencing DNA from a very large number of moa specimens, these biologists hoped to determine how many distinct lineages of moa had existed, and to explore reasons for the diversification of this endemic New Zealand group.

Methods

Moa DNA samples were obtained by taking bone cores or shavings from 125 museum specimens. DNA was extracted from between 0.1 and 0.5 g of sampled bone using EDTA with proteinase K (**TechBox 2.4**). Samples were purified by phenol–chloroform extraction, then extracted DNA was amplified using PCR (**TechBox 4.2**). They checked for the presence of multiple bands that might indicate nuclear copies of mitochondrial genes. DNA was sequenced along both strands. DNA was amplified in laboratories in Canada and New Zealand. Identical sequences were obtained from specimens sequenced in both laboratories. Baker *et al.*[1] used 658 nucleotides of the mitochondrial control region to genetically type their specimens, because it has a rapid rate of molecular evolution, so is expected to differ between different species or populations. For a subsample of specimens, they used a longer alignment of 2184 nucleotides of mitochondrial genes to estimate a phylogeny and molecular dates. They used a Bayesian phylogenetic method (**TechBox 7.3**)

Figure CS2.2a Moa – large flightless birds from New Zealand – went extinct long before cameras were invented. This scene was posed in 1899 using a museum model in the botanic gardens being 'hunted' by medical students. One of the students (on the left) was Sir Peter Buck (Te Rangi Hiroa), who, amongst a great many other achievements, was the first Maori to qualify as a doctor. For more information on Te Rangi Hiroa see *www.nzedge.com/heroes/buck.html#FIRST*.

Image courtesy of Alexander Turnbull Library, New Zealand. Kehoe, E.L. Mock Moa Hunt – Photograph taken by Guy. 1899. Ref no:PACol-1308.

to estimate a phylogeny of the DNA samples, then used the genetic distance between samples (represented by branch lengths on the phylogeny in **Figure CS2.2b**), and the Bayesian probabilities on groupings, to judge which lineages represented distinct populations or species (**TechBox 6.2**).

Results

Baker *et al.*[1] identified 14 distinct lineages of moa. Nine of these are currently recognized as species. The other five may be either geographically separate populations or newly recognized species. The phylogeny showed that some sequences did not group with the rest of the sequences for that species. For example, a sequence from a sample which had been labelled as a stout-legged moa *Euryapteryx geranoides* (yellow squares in **Figure CS2.2b**) is more similar to sequences from the heavy-footed moa *Pachyornis elephantopus* (dark green triangles). The authors suggest that as many as a third of the museum specimens had been incorrectly assigned to species.

Conclusions

While other ratite lineages are relatively species-poor, the authors of this study suggest that moa diversity was as high as other classic island endemic radiations, such as Darwin's

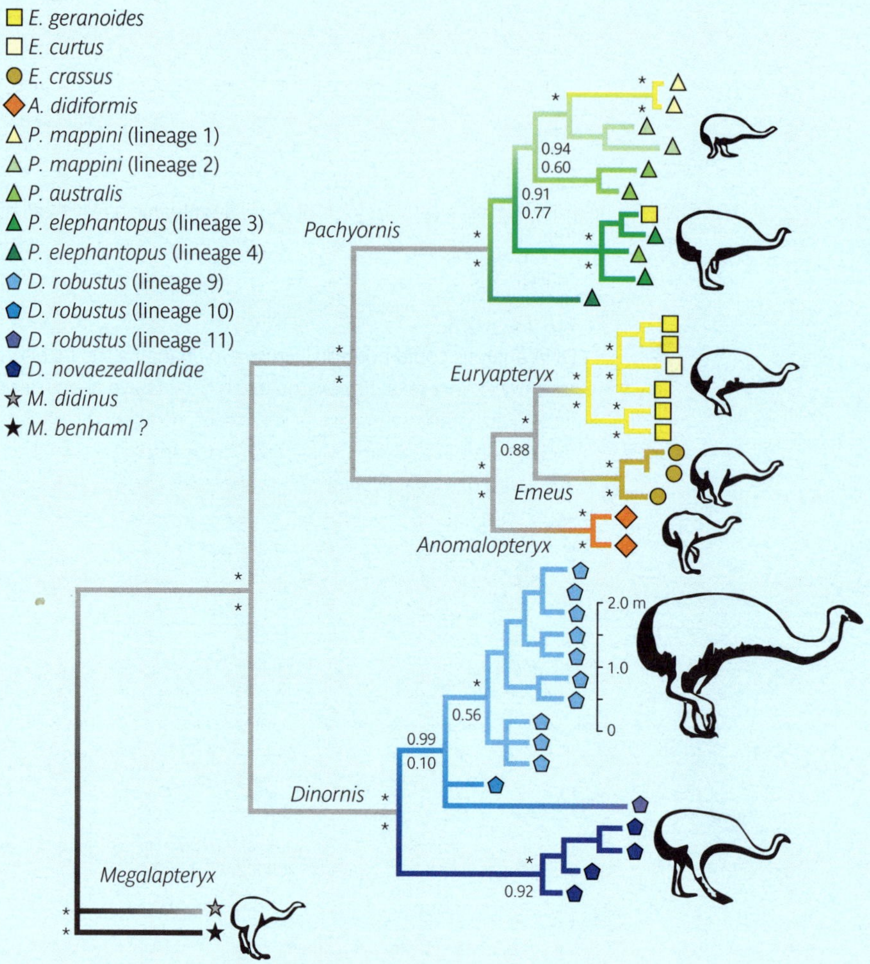

Figure CS2.2b Phylogeny (evolutionary tree) of moa, inferred from DNA extracted from moa bones. Moa bones are found in abundance in middens (prehistoric kitchen rubbish), suggesting moa were eaten by early colonists of New Zealand. A study that compared the species compositions of natural collections of bones in New Zealand (e.g. from animals that died from falling into caves) with those found in middens concluded that animals that were more often hunted and eaten were more likely to have gone extinct. From Baker A.J. *et al*. (2005) Reconstructing the tempo and mode of evolution in an extinct clade of birds with ancient DNA: The giant moas of New Zealand. *Proceedings of the National Academy of Sciences USA*, Volume 102, pages 8257–8262.

finches. On the basis of molecular dates, they propose that much of the diversification of moa lineages was relatively recent, possibly driven by geographic reshaping of New Zealand, 4–10 million years ago. Populations became isolated and ecologically specialized as mountains rose and islands separated (see Chapter 8 for a discussion of estimating evolutionary time from DNA sequence data).

Limitations

Genetic difference is a continuous scale, and species definitions are a matter of opinion: what one biologist considers a separate species, another may consider a regional subtype. Some biologists dispute that molecular data are a valid way of identifying species, as morphologically identical populations may show consistent molecular differences (see Chapter 6 for a discussion of the use of molecular data to define species). Identification of misassigned specimens is a potentially valuable use of DNA data, but must be conducted with due recognition that DNA taxonomy is also prone to errors, through contamination, sequencing errors, or incorrect phylogenetic inference.

Future work

DNA analysis could potentially result in changing the taxonomic assignment of many recently extinct taxa. For example, DNA data have been used to show that bones previously attributed to two co-occurring species of moa are actually from sexually dimorphic males and females of the same species (with the females twice the size of the males)[2]. Where abundant data are available, DNA analysis can provide a picture of the population biology of an extinct species. For example, DNA analysis has suggested that moa were present in much greater numbers than previously suspected[3]. The accuracy and precision of such estimates will be improved by increased sampling: more genes, longer sequences, more individuals, and broader geographic coverage.

References

1. Baker, A.J., Huynen, L.J., Haddrath, O., Millar, C.D. and Lambert, D.M. (2005) Reconstructing the tempo and mode of evolution in an extinct clade of birds with ancient DNA: The giant moas of New Zealand. *Proceedings of the National Academy of Sciences USA*, Volume 102, pages 8257–8262.
2. Bunce, M. *et al*. (2003) Extreme reversed sexual size dimorphism in the extinct New Zealand moa Dinornis. *Nature*, Volume 425, pages 172–175.
3. Gemmel, N.J., Schwartz, M.K. and Roberston, B.C. (2004) Moa were many. *Proceedings of the Royal Society London B: Biology Letters Supplement,* Volume 271, pages S430–S432.

Conclusions

The mechanisms of inheritance have been revealed in an astonishingly short period of time. From the point of view of someone interested in the role of DNA in evolutionary biology, there have been a number of notable leaps forward in understanding the nature of genetic information (Figure 2.10). It is convenient to think of these advances as connected to particular scientists, but of course, like all science, there were a great many other scientists who contributed to each of these fields.

Firstly, Charles Darwin noted that there was a great deal of heritable variation in natural populations, and the natural selection of successful variants provided a plausible mechanism for the evolution of adaptations. Without knowing about genes, Darwin clearly illustrated the role of heredity in the process of evolution by descent with modification. Secondly, August Weismann reasoned that the information needed to make an organism was passed from one generation to the next, essentially unaffected by bodily changes acquired during an individual's lifetime. Thirdly, Gregor Mendel showed through breeding experiments that this

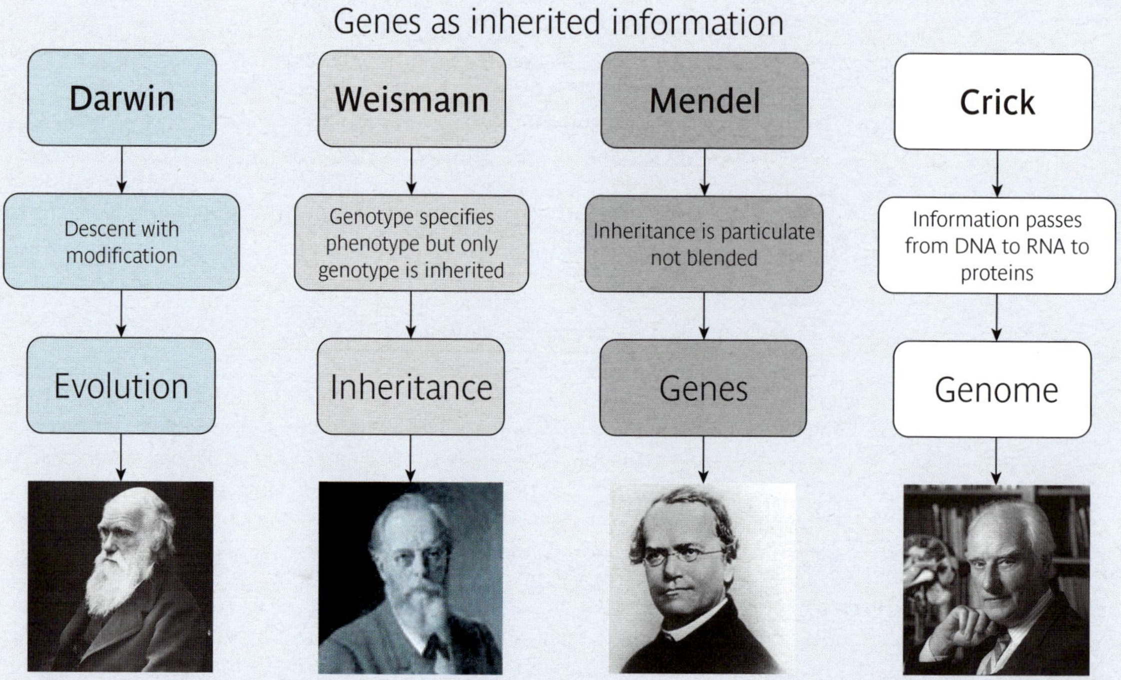

Figure 2.10 Some of the conceptual leaps forward in the development of the modern theory of heredity and evolution. Although a single name is listed beside each advance, this is shorthand for a great number of scientists. For example, the evolution of life on Earth had been studied for a hundred years before Darwin: in particular, much progress had been made in the interpretation of the stratigraphic record which revealed an apparent increase in complexity and diversity over long timescales. Similarly, Francis Crick is used to represent the advances in molecular genetics because of the key role he played in uncovering the structure of DNA, deciphering the genetic code, and describing the flow of information from the genome to the construction and maintenance of organisms, but of course there were a great many other scientists who contributed to this field (including Crick's coauthor James Watson).

Weismann photo from Conklin, E.G. (Oct-Dec., 1915). August Weismann. *Proceedings of the American Philosophical Society*, Volume 54, No. 220, pages iii–xii.

variation was particulate: inheritance was controlled by genetic factors that were not blended but passed intact down the generations as discrete units of information. Fourthly, Francis Crick and others showed that complementary pairing of bases explains how DNA stores information, how it can be faithfully copied from one generation to the next, and how the genetic information is used to construct RNA molecules and proteins. This provided a molecular explanation for the observations of Darwin, Weismann, and Mendel: evolution acts on genes (information coded in DNA), which are then expressed in the body (through the action of RNA and proteins).

These scientists are four among many who have shown that understanding inheritance is the key to understanding evolution. But DNA is also incidentally a source of information about evolution. Genomes would not carry this information if every individual had exactly the same DNA sequence. It is because DNA sequences differ between individuals, between species, and between evolutionary lineages that we can use DNA to understand evolutionary history and processes, as well as illuminating gene function itself. In the next chapter, we will consider the process of mutation, whereby the genome of an individual is permanently changed. Mutation creates heritable variation, which is the raw material of evolution. It also makes each one of us unique, a fact that is increasingly exploited in biology and medicine.

Further information

A very readable account of the development of evolutionary biology can be found in:

Young, D. (2007) *The Discovery of Evolution*, 2nd edn. Cambridge University Press.

The development of ideas about heredity is outlined in:

Cobb, M. (2006) Heredity before genetics: a review. *Nature Reviews Genetics,* Volume 7, pages 953–958.

There are a number of excellent online resources covering the development of molecular genetics. It is worth exploring the excellent *DNA Interactive Timeline*, an animated history of genetics which includes many of the less-well known players:

www.dnai.org/timeline/index.html

Any biochemistry book will cover DNA and protein structure. The following provides a very accessible, biology-friendly account:

Crowe, J., Bradshaw, T. and Monk, P. (2006) *Chemistry for the Biosciences: the essential concepts*. Oxford University Press.

Animations of cellular processes such as DNA replication, transcription, and translation are available at:

www.johnkyrk.com

3 We are all mutants

Or: how do I identify individuals?

"An ideal situation would be if the organism were to respond to the challenge of the changing environment by producing only beneficial mutations where and when needed. But nature has not been kind enough to endow its creations with such a providential ability."

Theodosius Dobzhansky (1951) *Genetics and the Origin of Species*, 3rd edition. Columbia University Press

What this chapter is about

A mutation is any heritable change to the genome, which will be passed on to the next generation when the genome is copied. Mutation is essential for evolution: without it, all genomes would be identical. Mutation is common enough that, in most cases, we can expect each individual's genome to have a unique DNA sequence. Yet, because mutations are inherited, we can also expect the genomes of individuals to be similar to those of close relatives. So by comparing DNA sequences, we can distinguish individuals, identify family groups, and study the heritability of traits across generations. In this chapter we will review some of the ways of generating heritable differences between genomes, such as chromosomal rearrangements, transposable elements, changing repeat number, and altering the nucleotide sequence of DNA. The development of new and faster ways of revealing which mutations individuals carry has wide-ranging implications for both research and wider society.

Key concepts

Evolutionary biology: variation within populations

Molecular evolution: mutation

Techniques: genotyping

Mutants

Would you like to be able to walk through walls, lift objects using only the power of your mind, or have the capacity to heal any wound instantly (and therefore be able to get into an awful lot of fights without risking lasting damage)? These are just some of the attributes of the X-Men, superhero-like characters first introduced in Marvel Comics in 1963, and popularized in a string of movies beginning in 2000 (**Figure 3.1**). The source of the X-Men's incredible powers were not alien birth (as for Superman), gamma radiation (the Incredible Hulk), or the bite from an irradiated spider (Spiderman). Instead, the X-men owed their supernatural abilities to mutation. Each of the X-Men carried a mutation that had the dramatic effect of conferring unnatural abilities. These mutants were vilified by the general public, yet considered to represent the evolutionary future of humanity (which is good news, if you want your descendants to be able to control the weather with their psychic powers).

Figure 3.1 A mutant with a mission (from the movie *X-Men 2*): does the evolutionary future of humanity hold the development of potentially useful new appendages, or a gradual decline into obesity and short-sightedness?

In fact, we are all mutants. It is estimated that each human embryo begins life with as many as a hundred new mutations. Each of these mutations is a change that occurred in the genome of one of the parents during the production of sperm or eggs. If that egg or sperm forms an embryo, the mutation will be copied into every new cell added to the embryo as it grows into an adult. And if sperm or eggs from that adult go on to form another embryo, then the mutation will be perpetuated to a new generation. Alas, these mutations do not tend to confer super powers. Many mutations do not have any noticeable effect, especially those that occur in parts of the genome that do not seem to code for anything. Mutations that occur in functional sequences are often harmful, decreasing the chances of survival of that individual and its descendants. Only very rarely will a mutation produce any beneficial changes.

The word 'mutation' predates the genetic revolution, and can have different meanings in different contexts. For the purposes of this book, we will refer to a mutation as being any permanent, heritable change to the genetic material. Using this definition, we will call a change to the genome a mutation if it can be passed on to any copies made of that genome (usually when DNA is replicated before cells divide). Changes to the genome may arise from DNA damage or errors made in copying

DNA. Most of these changes will be corrected by one of the multitude of DNA repair pathways that monitor and repair the genome. But there are two ways that a change to the genome can become permanent, such that the original DNA sequence is no longer recoverable. Firstly, while DNA repair is on the whole very efficient, the occasional change to the sequence goes unrepaired. Secondly, sometimes the process of repair itself results in a change in the DNA sequence. If the mutation happens to occur in a cell that will give rise to a new individual, then that mutation can be copied to a new generation.

See Chapter 2 for the differing evolutionary fates of body and germ line cells

In order to understand how genomes evolve, we need to become familiar with the way genomes can be changed. In this chapter, we will look at the kind of accidents that can befall DNA, which shuffle and change the information in the genome, from large-scale genome rearrangements to the change of identity of a single nucleotide. We are going to start by looking at chromosomal rearrangements, which can have severe effects on individual development, but may also play an important role in speciation in some lineages. Then we will look at the action of transposable elements, which can jump from one place to another, moving DNA sequences around the genome and disrupting normal gene function. We will then consider duplications and deletions of DNA, from the scale of whole genes down to simple repeat sequences. Then we will consider how DNA damage can create the smallest and most fundamental kind of mutation, the exchange of one nucleotide in the DNA sequence for another. All of these types of mutation leave a mosaic of changes in the genome. It is these changes that allow us to distinguish individual genomes, trace relationships, and uncover the evolutionary history of species.

Rearranging chromosomes

Mutations can involve large-scale changes to the genome, such as the loss or duplication of whole chromosomes. Loss of all or part of a chromosome is usually lethal, because it might delete information essential to an organism's development or functioning. Less obviously, creating an extra copy of all or part of a chromosome can be just as damaging, probably due to the disturbing effect of having extra doses of the duplicated genes' products. There are some chromosomal losses and gains that can be tolerated, though with potentially severe consequences. People with Down's syndrome have an additional copy of part or all of chromosome 21, a condition that gives rise to a range of characteristics including a wide 'sandal gap' between the first and second toes, and, for most (but not all) affected people, some degree of mental retardation. However, not all organisms are intolerant of chromosome doubling. Many plant breeds, both natural and artificially created, have one or more extra sets of chromosomes.

Mutations may involve less dramatic chromosomal rearrangements. A DNA sequence can become inverted by being cut out and reinserted in the genome back-to-front (Figure 3.2). A large-scale inversion may retain all the genetic information in the DNA sequence, so an individual carrying an inversion may be completely normal. But inversions can cause problems during cell division. We will briefly review the behaviour of chromosomes at cell division, so that we can understand how some chromosomal mutations occur. However, this is not the place to provide a summary of cell division. If you are unfamiliar with the way cells

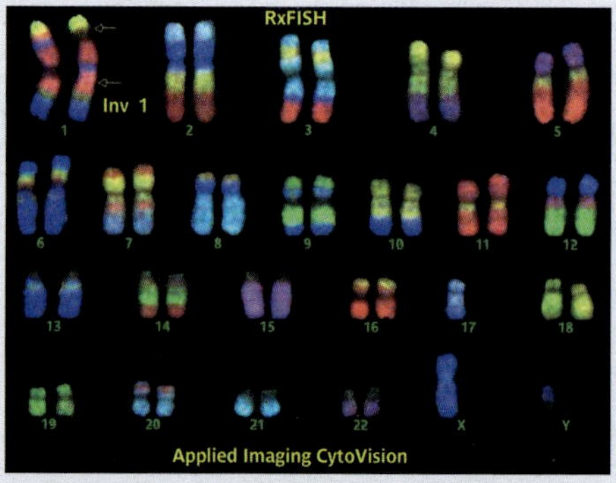

Figure 3.2 A karyotype is a picture of all of the chromosomes from a cell, neatly arranged in homologous (matching) pairs. In this karyotype, one copy of chromosome 1 has an inversion. If we describe the banding patterns using letters for each colour band (Y for yellow, R for red, P for purple, B for black, and L for blue) then we can see that one version of chromosome 1 has the banding pattern YRPRBRL, but the other has YBRPRL. The RPRB segment in one is reversed to give BRPR in the other.

Image: Cytopix.com

copy and divide their chromosomes, it's a good idea to pick up an introductory biology text at this point and have a look at meiosis and mitosis.

Your body cells are diploid, just like the body cells of most animals and plants (and some fungi), which means every cell contains two copies of each chromosome. These duplicate chromosomes contain the same genes, so their DNA sequences will be very similar, but not necessarily identical. In most diploid organisms, one copy of each chromosome came from each parent through the process of sexual reproduction, where the genomes of two individuals are fused together to make a new individual. Clearly, sexual organisms can't keep doubling their genomes every time they reproduce. They avoid an ever-increasing number of genome copies by undergoing reduction/ division (meiosis). The diploid germ cells that will give rise to gametes (such as sperm and eggs) go through two rounds of cell division: reduction (meiosis I) where the two copies of the genome are split to make two haploid daughter cells, each with only one copy of each chromosome pair, then division (meiosis II) when each haploid cell divides again to produce a total of four haploid gametes. So when two gametes fuse to form an embryo, each contributes one copy of each chromosome to make a new diploid individual.

When diploid germ cells divide to make haploid gametes, the pairs of matched chromosomes line up with each other, so that one chromosome from each pair can go into each gamete. When chromosomes pair, they form chiasmata, cross-over points where DNA is exchanged between chromosome arms (**Figure 3.3**). Crossing over is a reasonably common occurrence, resulting in the reshuffling of sequences between sister chromosomes. This shuffling is an important source of genetic variation in sexually reproducing populations.

When a chromosome with an inversion lines up with its non-inverted sister chromosome, it has to loop around so that the gene order matches up. This inversion loop is not a problem for normal meiosis. But if chiasmata form within the loop, the sister chromosomes will

Figure 3.3

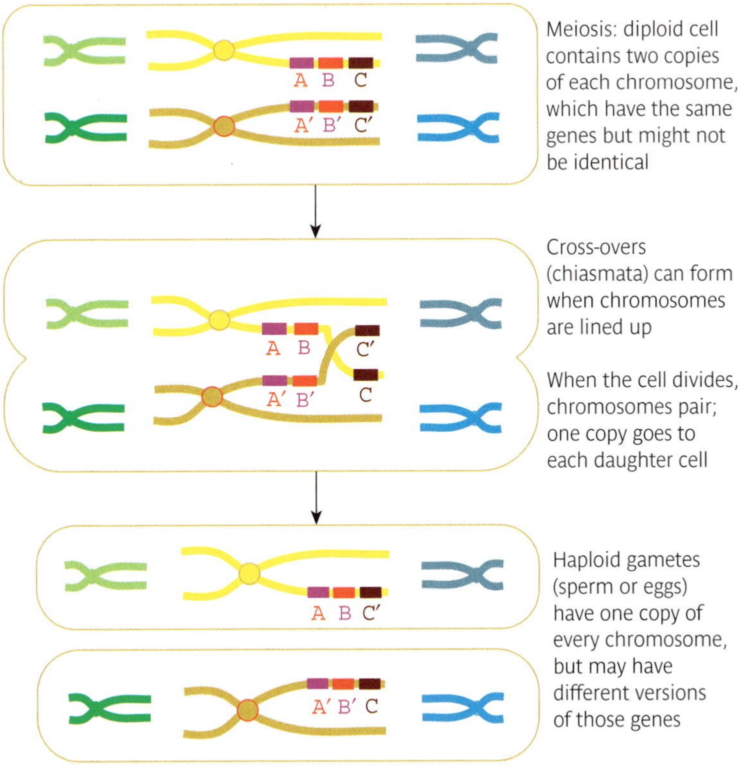
A B C
A′ B′ C′
Meiosis: diploid cell contains two copies of each chromosome, which have the same genes but might not be identical
A B C′
A′ B′ C
Cross-overs (chiasmata) can form when chromosomes are lined up
When the cell divides, chromosomes pair; one copy goes to each daughter cell
A B C′
A′ B′ C
Haploid gametes (sperm or eggs) have one copy of every chromosome, but may have different versions of those genes

exchange non-equal parts, creating unbalanced chromosomes, one with two centromeres and one with none.

Figure 3.4

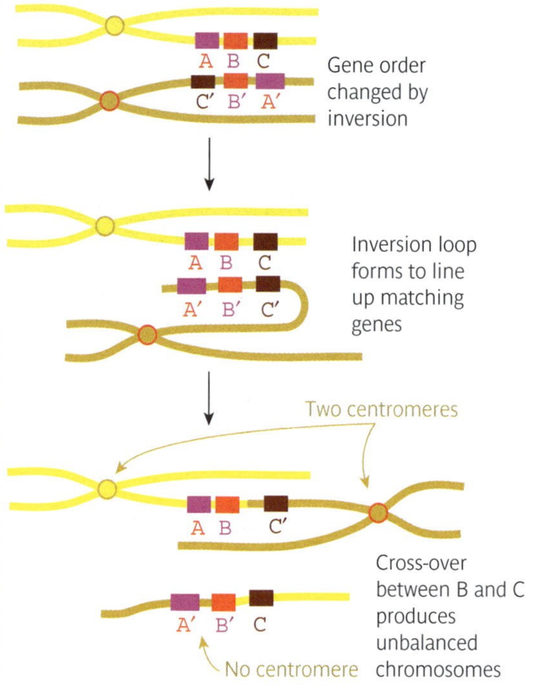
A B C
C′ B′ A′
Gene order changed by inversion
A B C
A′ B′ C′
Inversion loop forms to line up matching genes
Two centromeres
A B C′
Cross-over between B and C produces unbalanced chromosomes
A′ B′ C
No centromere

Since it is the centromere that guides the movement of the chromosomes at cell division, chromosomes with two centromeres might be pulled in two directions and broken, and chromosomes with no centromere will wander undirected. Therefore, these unbalanced chromosomes are likely to get lost at cell division, creating gametes without a full chromosome complement. Such gametes are unlikely to be able to form viable offspring. The practical result of all this is that individuals that carry one inverted and one normal chromosome are likely to have reduced fertility.

Because of the reduced reproductive success of individuals with mixed chromosomes, those that only mate with others with the same chromosome type will have a greater chance of producing viable offspring. Thus inversions can lead to the separation of a population into two non-interbreeding species. For example, researchers at the University of Ouagadougou examined the karyotype (chromosome complement) of over 4000 *Anopheles funestus* mosquitoes from villages around a swamp in Burkina Faso. *Anopheles funestus* mosquitoes are vectors for malaria, a disease that kills more than a million people every year (**Figure 3.5**). The researchers found two distinct types of *Anopheles funestus* mosquitoes, physically identical but each with a characteristic pattern of chromosomal inversions. Importantly, they found relatively few hybrids that carried both types of inversions. From this observation they inferred that although the two types of *Anopheles funestus* occur in the same places and are indistinguishable to humans, they don't tend to interbreed successfully. This implies that the two chromosomal strains of *Anopheles funestus* are in the process of evolutionary divergence, perhaps ultimately leading to the formation of two distinct species.

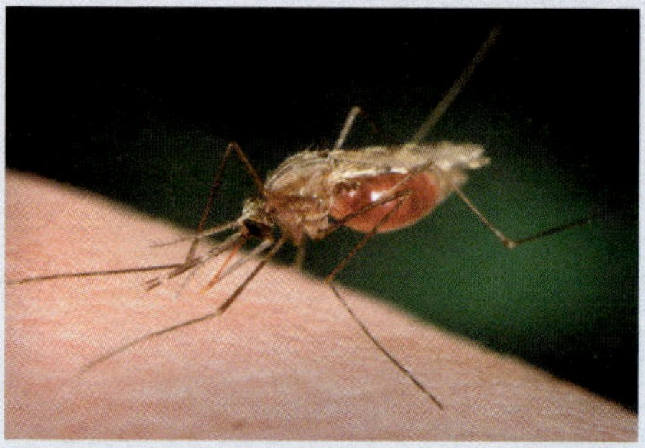

Figure 3.5 Malaria derives its name from the early hypothesis that it was carried by bad air arising from swamps, but we now know it is carried by another factor that arises from swamps: mosquitoes. There are dozens of different species of *Anopheles* mosquitoes that can transmit the four different forms of *Plasmodium* parasite that cause malaria. *Anopheles funestus*, shown here, and *Anopheles gambiae* are the main vectors for malaria in Africa, so are indirectly responsible for the death of nearly one million people every year.

Photograph: James Gathany.

Chapter 6 includes a discussion of the process of speciation

Transposable elements

Chromosomal rearrangements can move large chunks of DNA around the genome. Transposable elements are another mechanism for moving sequences around the genome, albeit on a finer scale. Originally described as 'jumping genes' (see **Heroes 3**), these mobile genetic

HEROES OF THE GENETIC REVOLUTION

3

Barbara McClintock

"I was entranced at the very first lecture I went to. It was zoology, and I was just completely entranced. I was doing now what I really wanted to do, and I never lost that joy all the way through college."

B. McClintock, quoted in Keller, E.F. (1983) *A Feeling for the Organism*

NAME

Barbara McClintock

BORN

16 June 1902, Hartford, Connecticut, USA

DIED

2 September 1992, Huntington, New York, USA

KEY PUBLICATIONS

Creighton, H.B. and McClintock, B. (1931) A correlation of cytological and genetical crossing-over in *Zea mays*. *Proceedings of the National Academy of Sciences*, Volume 17, pages 492–497.

McClintock, B. (1953) Induction of instability at selected loci in maize. *Genetics*, Volume 38, pages 579–599.

FURTHER INFORMATION

Keller, E.F. (1983) *A Feeling for the Organism*. W.H. Freeman and Company, New York

Figure Hero 3
Barbara McClintock.
Reproduced by permission of Cold Springs Harbor Laboratory Archives.

Barbara McClintock's contributions to genetics were profound, from her technical advances that allowed her to provide the first physical evidence of chromosomal crossing over to her insight into the existence of mobile genetic elements. And yet she almost missed out on a career in science, due to family resistance to her gaining a college education (in addition to being a financial burden on the family, her mother apparently feared a degree would make McClintock unmarriageable). However, her father acquiesced and McClintock graduated from Cornell's College of Agriculture in 1923. One of her lecturers recognized McClintock's promise and encouraged her to undertake postgraduate study in genetics.

The 1920s and 1930s were a time of great excitement in genetics. Many of the foundations of the field were being built, particularly in the *Drosophila* labs, such as those of Thomas Hunt Morgan at CalTech. With her background in plant breeding, McClintock developed maize as an alternative experimental system. She and her research group spent time in the maize fields, making experimental crosses and observing the progeny. At the time genetics consisted of inferring genes from the observable results of breeding experiments, rather than by direct observation of the genetic material. McClintock brought about a revolution in plant genetics by combining these breeding experiments with cytogenetics (observations of chromosomes). With her talent for fine-scale microscope work and her endless patience, she developed ways of staining and preparing maize chromosomes so that they could be recognized and followed across generations of maize. By identifying specific maize chromosomes and their banding patterns, and using this physical geography of the chromosomes to observe the results of crosses, McClintock played a key role in connecting genetics theory to physical observations of the behaviour of chromosomes, long before the biochemical basis of heredity was known. Most significantly, she and her student Harriet Creighton were the first to witness recombination between chromosomes, which had been predicted on the basis of breeding experiments but never before directly observed. Furthermore, her observation of ring chromosomes led her to predict the presence of specialized ends of chromosomes, telomeres, that regulated chromosomal behaviour at cell division.

In the 1940s, McClintock's ground-breaking work in cytogenetics was recognized by a number of honours, including election to the National Academy of Sciences of the USA and the presidency of the Genetics Society. She moved to Cold Spring Harbor laboratories where she began the work she is now most famous for: the recognition and description of unstable patterns of inheritance in maize. Through a combination of recording colour patterns on corn cobs over several generations and observation of minute changes in chromosomal structure, she described the action of genetic loci that appeared to move around the genome, generating mosaic patterns in the phenotype as they switched genes on in some cells, and off in others. Although McClintock published her evidence for 'jumping genes' and spoke about it at many meetings, the work was not well received, and eventually she gave up trying to convince people of her heretical hypothesis. However, in the 1960s and 1970s, McClintock's observations of genetic regulation (switching genes on and off) and mobile elements (genetic loci transposing across the genome) were vindicated by independent research by other scientists. She was retrospectively recognized as the discoverer of transposition, for which she received a Nobel Prize in Physiology and Medicine in 1983.

McClintock's work was characterized by an intimacy with her study system, such that she frequently 'knew' what the genetic mechanism was, before she had the observational evidence to prove it. This intuitive approach to her research inspired some students and colleagues, but irritated others. However, her intuitions were often right, and the combination of her 'feeling for the organism' and her innovative techniques revealed many important aspects of chromosomal evolution, once controversial but now taken for granted in genetics.

> 'I found that the more I worked with [chromosomes] . . . the bigger and bigger they got, and when I was working with them I wasn't outside, I was down there. I was part of the system. . . . It surprised me because I actually felt as if I was right down there and these were my friends.'
>
> B. McClintock, quoted in Keller, E.F. (1983) *A Feeling for the Organism*

elements are now better understood as genomic viruses that contain sequences that cause them to be copied and propagated through the genome. The simplest transposable elements are short sequences (in some cases less than 1000 nucleotides) that can move by simple transposition, by being cut out of one location and reinserted in another. The most sophisticated transposable elements are very similar to some 'free-living' viruses, with genes encoding enzymes responsible for copying the virus genome, which can then propagate through the host genome, and, in some cases, move between individuals (**Case Study 3.1**). The evolutionary success of these genomic parasites is evident by their sheer numbers. Based on rough estimates from genomic sequences, it seems that DNA derived from transposable elements makes up a much larger proportion of your genome than the genes needed to construct a human being (you may feel this raises an interesting philosophical question about exactly who you are, given that your genome could be considered to be more virus than human).

Transposable elements make a substantial contribution to the mutation rate in many species, causing mutations in several different ways. Firstly, if a transposable element inserts into a working gene, it will almost certainly disrupt the genetic information, causing a 'knock-out' mutation (see **TechBox 3.1**, p. 73). Secondly, they can disrupt normal gene expression (**Heroes 3**). Many transposable elements contain strong regulatory signals that attract the cellular machinery that makes RNA transcripts. The presence of these strong signals can cause overexpression of genes in the vicinity. This may be how some genomic viruses cause diseases such as leukaemia (see **Figure CS3.1b**, p. 72). Thirdly, transposable elements can move DNA around the genome: they might pick up pieces of the host genome when they are copied and then translocate them to another part of the genome (or even to the genome of a different individuals). Fourthly, transposable elements can be a source of novel DNA. Some sequences in the human genome were derived from viral sequences, such as syncytin, a gene active in the formation of the placenta which was derived from a virus envelope gene.

CASE STUDY 3.1

Viruses within: using bioinformatics to describe an endogenous retrovirus

KEYWORDS

BLAST
datamining
endogenous retrovirus
EST
transcription
phylogeny
ORF

RELATED TECHBOXES

TB 1.1: GenBank
TB 3.4: BLAST

RELATED CASE STUDIES

CS 5.1: Sweet and sour (characterizing genes through bioinformatics)
CS 8.2: Molecular detective (tracing paths of infection with bioinformatics)

Bromham, L., Clark, F. and McKee, J.J. (2001) Discovery of a novel murine type C retrovirus by datamining. *Journal of Virology*, Volume 75, pages 3053–3057

“*So, naturalists observe, a flea*
Has smaller fleas that on him prey;
And these have smaller still to bite 'em;
And so proceed ad infinitum”
Jonathan Swift (1733) *On Poetry: A Rhapsody*

Background

All viruses replicate by infecting a host cell and hijacking its cellular machinery to make more copies of the virus. Retroviruses do this by making a DNA copy of their RNA viral genome (using the enzyme reverse transcriptase: see Chapter 2, page 49), then inserting this viral DNA into the host's own genome. Using strong promoters and enhancers, the virus deceives the host's transcription machinery into making many RNA copies of the viral genome. These RNA copies of the virus genome can be packaged into new virus particles, or may insert into more places in the host genome (transposition; see **Heroes 3**). By and by, host genomes tend to get filled with retroviral DNA, which can have a number of effects. Viral DNA may simply bloat the genome with 'junk' sequences, or they may contribute useful sequences that are co-opted by the host genome. Alternatively these endogenous retroviruses may retain the ability to replicate as a virus, or transpose throughout the genome, causing mutations as they insert into or near important sequences[1].

Aim

Whole genome sequences provide a valuable opportunity to study the number, distribution, and activity of endogenous retroviruses (ERVs) embedded in the genome. Previous studies have used 'datamining' approaches, usually based on BLAST searches (**TechBox 3.4**, p. 93), to identify endogenous retroviral sequences in genomic sequences[2,3]. This study aimed to go beyond identification, using bioinformatics to gauge the potential activity of an endogenous retrovirus. This may provide an efficient way of identifying genomic viruses capable of creating mutations within the genome, and may even provide a way of judging the potential for cross-infection between individuals[4].

Methods

We used BLASTn (**TechBox 3.4**) to search Genbank (**TechBox 1.1**) for matches to two known retroviral genomes. All high-scoring matches greater than 2000 nucleotides in length, plus 5000 nucleotides on either side, were cut out and examined for genomic features such as intact viral genes, long terminal repeats (LTRs) that define the edges of the viral genome and are required for insertion into the host genome, and regulatory sequences that control transcription (**Figure CS3.1a**). These regulatory sequences include the TATA box (where the transcription machinery binds), primer-binding sites and polypurine signals (for initiation of reverse transcription), and polyadenylation signals and splice sites (for processing the RNA transcript)[3]. We also blasted any intact genomes we found against the expressed sequence tag (EST) database, which contains the sequences of mRNA transcripts collected from cells, and can therefore be used to identify whether a particular gene is expressed in a given tissue. We performed phylogenetic analysis (see Chapter 7) on candidate viral genomes to see how they were related to other known viruses.

Results

Of all the sequences we found showing similarity to our query sequences, we focused on one that appeared to be a novel viral genome embedded in a mouse genomic sequence. We called this sequence *Mus musculus* endogenous retrovirus (MmERV). Phylogenetic analysis showed that the protein-coding part of the MmERV genome is related to 'type-C' family of mammalian leukaemia-causing gammaretroviruses that can have both endogenous (passed on with the host genome) and exogenous (capable of infecting other individuals) forms. But MmERV seemed to be a new 'species', because its genome differed from that of its nearest known relative (the *Mus dunni* endogenous retrovirus, MDEV) at more than 10% of the sites in the protein-coding nucleotide sequence. The MmERV genome is recombinant: the middle part of the genome is from a gammaretrovirus, but the ends of the genome (the LTRs) are very similar to those from VL30 transposable elements. There are multiple copies of MmERV in the mouse genome, on several different chromosomes. The presence of matching sequences in ESTs from mouse heart, skin, and T-cells suggests that MmERV is actively transcribed in many different mouse tissues. The genome analysis revealed a number of features suggesting that MmERV is both endogenous (inherited) and exogenous (infectious). MmERV was apparently present in cellular RNA from multiple mouse strains, suggesting it is widespread in mice, and not simply an infection of a single laboratory population. The MmERV genome appears to have all the features it needs to replicate and be packaged into viral particles, and is structurally similar to both the MDEV virus, which is able to produce virions (infectious particles), and VL30 elements, which can move between individuals. Curiously, MmERV transcripts contains two to six copies of a 70-base repeat sequence that contains the polyadenylation sequence, a possible case of multimerization of regulatory elements (making extra copies to enhance transcription). These poly-A repeats are also found on non-viral RNA transcripts in mouse cells, possibly indicating that they are being used to regulate the expression of host genes as well as viral genes (a possible case of co-option of viral sequences into the host genome).

Conclusions

Identification of new endogenous retroviruses is now a commonplace part of analysis of whole genome sequences. This study shows that BLAST analyses can be used not just to identify viral sequences within the genome, but also to gauge the likely activity of those viruses. This may provide an efficient first step for identifying genomic viruses that present a risk of cross-species infection, for example through xenotransplantation (using animal organs in human patients: the two major donor species, pigs and baboons, both carry genomic viruses that, under lab conditions, can replicate and infect human cells[1]). This case also shows that you can carry out genomics research with no more than a computer, an internet connection, and a great deal of enthusiasm and patience. There are so much data becoming publicly available that you may not have to search very long before you find something rather interesting.

```
        |U3→
1       TGAAGAATAAAAGTTACTGAACTCTTCCTCACCCCAGAACTCGTTCCCAGAACACTCCTG
61      AACTCTTCACCCTAGAATGCATTCCTGAACTCCTCACCCTAGAGTTCGAACCCTCCCAAC
121     TAAAGACTGTTCCAAGAACATTTTTGAGATAAGGGCCTCCTGGAACAACCTCAGAATGAA
181     CCGGGTACATTGCCAAATAATAGGACATGACCCCTTAGTTACGTAGAATCCCTTGGCAGA
241     ACCCCTTGTCCCTTGGCAGAACCCCTTAGTTATGTAAACTTGTACTTTCCCTACCCCGCT

                           promoter         ←U3|R→
301     CTCCCCCCTTGAGTTTTCCTATATAAGCCTGTGAAAAATTTGGCTGGTCGTCGATTCTCC

                            |          pol motif
361     TCTACACCACTAGGTGTATGAGTTTCGACCCCAGAGCTCTGGTCTATGTGCTTTCATGCT

               polyA signal   ←R|U5→  |
421     GCTGCTTTATTAAATCTTGCCTTCAACATTTTGAGTTCGGTCTCAGTGTCTTCTTGGGTC

                                                  ←U5|primer binding
481     CGCGGCTGTCCCGAGGCTTGAGTGAGGGTCTCCCTTCGGGGGTCTTTCATTTGGGGGCTC

        site                                     splice donor sie
541     GTCCGGGATCAGTGCGACCACCCAGAGACCCTAGACCCACTTTAAGGTAAGATTCTTTGA
601     CCTGTCTTGGTTTGGTGTCTCTGTTCTGTTTCTAAGTTTGGTGCGATCGCAGTTTCGGTT
661     TTGCGGACGCTCAGTGAGACTGCGCTCCGAGAGGGAACGCGGGGTGGATAAGGATAGACG
721     TGTCCAGGTGTCCGCCGTCCGTTCGCCCTGGGAGACGTCCCAGGAGGAACAGGGGAGGAC

                                                       [  Rpt1
781     CAGGGACGCCTGGTGGACCCCTTTGGAGGCCAAGAGACCATCTGGGGTTGCGAGATCGTG

                            ]    [  Rpt2                          ]
841     GGTTCGAGTCCCACCTCGTGCCTTGTTGCGAGATCGTGGGTTCGAGTCCCACCTCGTGCA

                                                   |glyco-gag→
901     GAGGGTCTCAATCGGCCGGCCTTAGAAAAGCCATCTGATTCTCTGAGTTGCTTGTGGTCG
                                                  LeuSerCysLeuTrpSer

961     ACGCGAAGTCGCCGCCGCTTTTGGTTTCTTTTTTGTCTTAGTCTCGTGTTCGCTCTTGTT
        ThrArgSerArgArgArgPheTrpPheLeuPheCysLeuSerLeuValPheAlaLeuVal

                            |gag→
1021    GTGTCTACTATTGTTCTGGAAATGGGACAATCTGTGTCCACTCCCCTTTCTCTAACTCTG
        ValSerThrIleValLeuGluMetGlyGlnSerValSerThrProLeuSerLeuThrLeu

1081    GAGCATTGGAAGGAGGTGCGGGTCAGAGCTCACAACCAGTCGGTGGAGGTCAGAAAGGGT
```

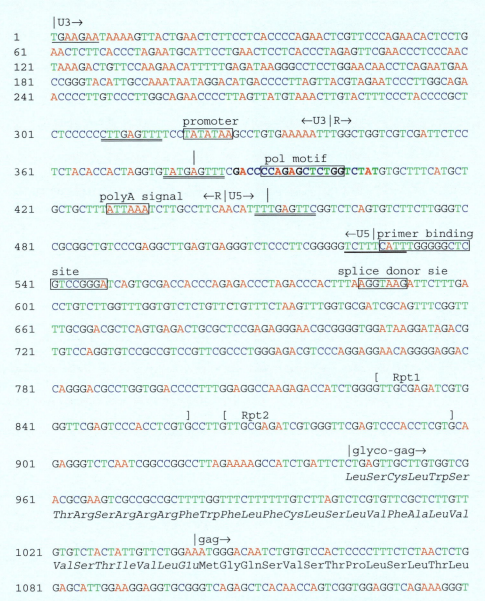

Figure CS3.1a The first kilobase or so of the MmERV genome, with annotation of regulatory features and genes. From Bromham, L., Clark, F. and McKee, J.J. (2001) Discovery of a novel murine type C retrovirus by datamining. *Journal of Virology*, Volume 75, pages 3053–3057.

Limitations

This kind of analysis is best applied to non-redundant whole genome sequences, but the number of available whole genomes is rapidly increasing. Bioinformatic analysis of virus genomes is powerful and fast, but it can never replace labwork. The MmERV genome appears to be replication-competent, but only experiments could demonstrate that it could replicate in living cells or infect whole animals.

Figure CS3.1b Though regarded as icons of cuddliness, koalas are plagued by a number of communicable diseases. Many koala populations have been significantly reduced by the sexually transmitted disease *Chlamydia*, and koalas are prone to a contagious form of leukaemia that is apparently caused by a type-C mammalian retrovirus, similar to the one described here, that appears to be both endogenous (inherited) and exogenous (infectious; see Hanger *et al*. 2000 in the Bibliography at the end of the book).

Photographer: Erin Silversmith.

Future work

Characterizing the likely activity and infectiousness of genomic viruses is important for gauging the risk of cross-species infections, for example the transfer of retroviruses between cell cultures of different species used in medical research. It may also help to predict the safety of gene vectors. For example, a close relative of MmERV, the gibbon–ape leukaemia virus (GALV), is being developed as a gene-therapy vector, even though it can apparently cross-infect distantly related species[1] (**Figure CS3.1b**).

References

1. Bromham, L. (2002) The human zoo: endogenous retroviruses in the human genome. *Trends in Ecology and Evolution*, Volume 17, pages 91–97.
2. McCarthy, E.M. and McDonald, J.F. (2004) Long terminal repeat retrotransposons of *Mus musculus*. *Genome Biology*, Volume 5, page R14.
3. Tristem, M. (2000) Identification and characterization of novel human endogenous retrovirus families by phylogenetic screening of the Human Genome Mapping Project database. *Journal of Virology*, Volume 74, pages 3715–3730.
4. Martin, J., Herniou, E., Cook, J., Waugh O'Neill, E. and Tristem, M. (1999) Interclass transmission and phyletic host tracking in murine leukemia virus-related retroviruses. *Journal of Virology*, Volume 73, pages 2442–2449.

TECHBOX 3.1 Gene names

KEYWORDS

protein

nomenclature

knock-out

gene family

homologue

paralogue

FURTHER INFORMATION

There are many species-specific guidelines for naming genes, for example:
human (*www.gene.ucl.ac.uk/nomenclature*)
mouse (*www.informatics.jax.org/mgihome/nomen*)
yeast (*www.yeastgenome.org/gene_guidelines*)
and even the slime-mould *Dictyostelium* (*http://dictybase.org*).

RELATED TECHBOXES

TB 1.1: GenBank

TB 6.1: Taxonomy

RELATED CASE STUDIES

CS 3.1: Viruses within (identifying sequences through homology)

CS 5.1: Sweet and sour (gene duplication)

Genes, like species, should have unique identifiers (**TechBox 6.1**). Technically, a gene should have a full-length formal name and a more commonly used abbreviation (usually written in italics). If the gene has a protein product, the protein is usually given a related name (no italics, sometimes written in capitals). For example, the *Huntington disease* gene is commonly referred to as *HD*, or by the name of the protein product, huntingtin, the function of which is still unknown (**Figure 3.12**, p. 76). Gene names may reflect relationships to other genes, either within the same genome (paralogues) or in different species (homologues; see **Case Study 5.1**). For example, the gene human period 2 (abbreviation *hPer2*) is one of three related genes (paralogues) in humans (*hPer1*, *hPer2*, and *hPer3*) that produce protein products related to the PERIOD family of proteins that control circadian rhythms in many animals. The mouse homologue of this gene is named *mPer2*.

However, while attempts are being made to standardize gene nomenclature, naming rules tend to be specific to particular organisms. There is, as yet, no equivalent of Linnaean nomenclature for genes, and you will find that different authors have different ways of recording gene names. Assigning unique and informative names to genes is in many ways even more difficult than species taxonomy. Whereas species taxonomy has a clear hierarchical structure, which aims to reflect evolutionary relationships, gene names are more haphazard. Related genes in different species may be given entirely different names. As an example, we can look at a very charismatic developmental gene, the human version of which is known as *paired box gene 6* (*Pax6*). This gene plays a key role in the induction of eye formation in a wide range of animals (it also has a number of other roles, particularly associated with the development of the central nervous system). One of the active sites of the Pax6 protein, the homeodomain, is so well conserved between animals that the homeodomain from a mouse can be used to trigger eye formation in fruit flies; and not just in the usual place, but anywhere the gene is expressed (resulting in rather disturbing flies with eye tissue on their wings, legs, and abdomen). But although these animals clearly have the same homologous gene which they all inherited from a common ancestor, this gene is known by different names in different organisms. In *Drosophila* (fruit flies), there are two copies (paralogues) of this gene, called *eyeless* (*ey*) and *twin of eyeless* (*toy*). In the roundworm *Caenorhabditis elegans* it is known as *Variable ABnormal morphology* (*vab-3*). In the mosquito *Anopheles gambiae*, it has the unlovely (and presumably temporary) name of *ENSANGP00000017427*. Furthermore, the *Pax6* gene may be sometimes referred to by its disease phenotypes, such as *Aniridia* (in humans) and *Small eye* (in mouse). Yet clearly it is the same gene and, astoundingly, does the same basic job in all of these divergent animals.

The changing fashions of gene names reflect technical progress in genetics. Genes were originally identified through breeding experiments: mutants with noticeable phenotypic effects were identified and bred to isolate the genetic factor responsible for the change. Genes identified through knock-outs (a mutation that destroys the normal functioning of the gene) are typically named for what happens when they don't function, rather than what they do when they work. For example, ordinary *Drosophila* have red eyes, but individuals in which the *white* gene (*w*) has been knocked out develop white eyes. So a working copy of the *white* gene is responsible for making eyes red. Confusing.

Figure TB3.1 The caterpillar on the right is plump and healthy. The caterpillar on the left has gone floppy and is about to die, thanks to a bacterial toxin, Mcf, produced by the gene *makes caterpillars floppy* (*mcf*). The bacteria do not directly infect caterpillars, but are carried in the guts of nematodes that do.

Reproduced by permission of Richard ffrench-Constant from Daborn, P.J. *et al.* (2002) *Proceedings of the National Academy of Sciences USA*, Volume 99, pages 10742–10747.

In the 1980s and 1990s, gene discovery rates began to increase massively. There was a palpable air of excitement surrounding gene discovery, and bright young things were attracted to work in genetics labs. Geneticists kept themselves amused dreaming up gene names with references to popular culture, one of the most famous being the *sonic hedgehog* gene (*SHH*) named after the indomitable hero of a computer game (*SHH* is a homologue of the *hedgehog* gene, *hh*), the knock-out of which causes spines to form on *Drosophila* embryos. Postgenomic gene discovery has lost a little of the enthusiasm of earlier days, and high-throughput sequencing has automated gene identification: new genes tend to get assigned a systematic identifier (such as *ENSANGP00000017427*) until someone finds the time, resources, and inclination to work out what the gene actually does.

But a quick flick through genetics journals reveals the odd quirky gene name. My personal favourite is the gene that makes caterpillars floppy (*mcf*; **Figure TB3.1**)[1]. Curiously, this is not a caterpillar gene at all, but a gene from a bacterium, *Photorabdus luminescens*, which lives as an essential symbiont in the guts of some nematodes (such as *Heterorhabditis bacteriophora*). So how does it make caterpillars floppy? The nematode makes its living as an endoparasite, burrowing into insect larvae and digesting their tissues. The bacteria help the worm by releasing the Mcf toxin which causes the larva to lose body turgor (i.e. go floppy) and die. *Photorabdus* also helps the nematode by releasing antibacterial and antifungal agents to prevent the larva being invaded by other competing pathogens. So *Photorabdus luminescens* is simultaneously a symbiont (of nematodes) and a parasite (of insects). This is important because *Heterorhabditis* worms are used as a biological control agent of insect pests, but it is the bacterial gene *mcf* that provides the lethal weapon. So the Mcf toxin might be developed as an insecticide, and the *mcf* gene could potentially be inserted into plants to confer defence against caterpillars (like the *Bt* gene from *Bacillus thuringiensis*; see Chapter 4).

Reference

1. Daborn, P.J. *et al.* (2002) A single *Photorhabdus* gene, makes caterpillars floppy (*mcf*), allows *Escherichia coli* to persist within and kill insects. *Proceedings of the National Academy of Sciences USA*, Volume 99, pages 10742–10747.

Duplications and deletions

We have seen how regions of the genome can be lost, doubled, or turned around by chromosomal rearrangements, or moved about by transposable elements. Now we will consider finer-scale rearrangements that act on smaller genomic regions. Duplications and deletions can occur at any scale, from the loss of a single nucleotide to the doubling of the whole genome, but in this section we will consider the gain or loss of whole genes, which can provide fuel for genomic evolution. Duplicate genes can be created when a copy of a gene is inserted elsewhere in the genome. If the new copy is inserted next to the existing gene, the genes are said to be tandem (side-by-side) repeats. Once created, tandem repeats are prone to increase in number through a process called unequal crossing over. When chromosomes pair at meiosis, tandem repeat sequences may be sufficiently similar that they can pair with neighbouring sequences. If recombination occurs between these misaligned repeats, one chromosome may end up with more repeats, and the other less (Figure 3.6).

Tandem gene duplication can generate gene families: sets of related sequences with similar functions. Gene duplication can provide the raw material for evolutionary change by creating a 'spare copy' of essential sequences. If one copy of the gene can keep producing the original gene product, the spare gene copy may be able to change without jeopardizing the function of the original gene. Many duplicate genes simply decay, acquiring changes that destroy the sense of the coding information (these non-functional copies are usually referred to as pseudogenes; see Figures 5.25 and 5.28). But in some cases duplicate genes can evolve new functions, or become specialized to perform just one of the functions of the original gene (see Case Study 5.1).

Figure 3.6

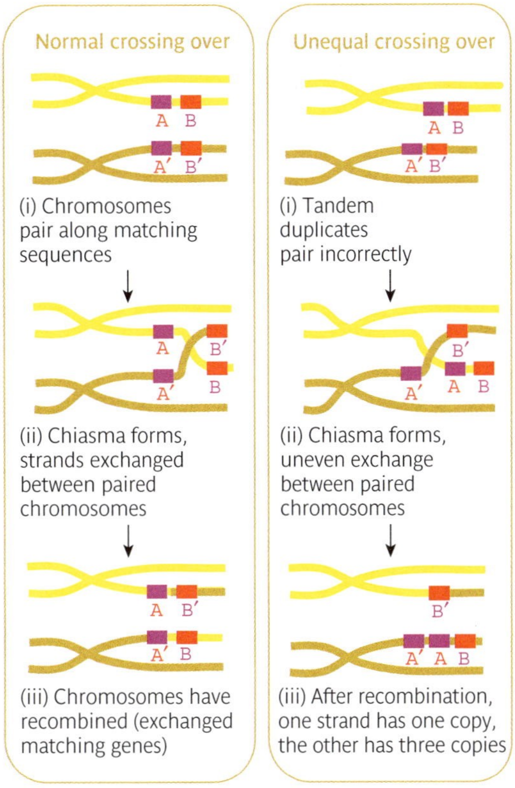
Normal crossing over
A B
A′ B′
(i) Chromosomes pair along matching sequences
A B′
A′ B
(ii) Chiasma forms, strands exchanged between paired chromosomes
A B′
A′ B
(iii) Chromosomes have recombined (exchanged matching genes)
Unequal crossing over
A B
A′ B′
(i) Tandem duplicates pair incorrectly
B′
A′ A B
(ii) Chiasma forms, uneven exchange between paired chromosomes
B′
A′ A B
(iii) After recombination, one strand has one copy, the other has three copies

Repeat sequences

> *If a replication fork is a train speeding along a track of double-stranded DNA, the trinucleotide repeats are a hairpin curve in the track. Experiments have demonstrated that the train can become derailed at the hairpin curve, resulting in significant damage to the track. Repair of the track often results in contractions and expansions of track length.*

Lenzmeier, B.A. and Freudenreich, C.H. (2003) Trinucleotide repeat instability: a hairpin curve at the crossroads of replication, recombination, and repair. *Cytogenetic and Genome Research*, Volume 100, pages 7–24

Not all duplications and deletions are of whole genes. In fact, some short sequences of DNA are particularly prone to duplication and deletion. Most eukaryotic genomes contain large numbers of short tandem repeats (where the same short sequence of nucleotides is repeated over and over again). These regions of highly repetitive DNA were originally called satellite DNA, because they formed a distinct band (called a satellite band) when DNA was separated into different fractions by centrifuging it through a density gradient. These runs of tandem repeats are thought to primarily be generated by slippage during replication (though unequal crossing over may also generate variable numbers of repeats; see Figure 3.6).

Slippage is a by-product of the universality of complementary base pairing (see Chapter 2). So for example, the repeat run CACACACACACA pairs with the complementary sequence GTGTGTGTGTGT (Figure 3.7).

Figure 3.7

```
——CACACACACACA——
——GTGTGTGTGTGT——
```

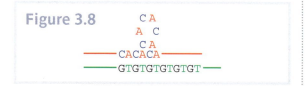

But any CA repeat can pair with any GT. If the sixth CA was to pair with the third GT, the intervening repeats would loop out.

Figure 3.8

```
          C A
         A   C
          C A
——CACACA——
——GTGTGTGTGTGT——
```

This loop can be removed by the genome's repair machinery (which can detect such distortions of the DNA helix), causing a reduction in the number of repeats.

Figure 3.9

```
——CACACA——————
——GTGTGTGTGTGT——
```

Again, genome-repair machinery will detect the distortion due to the unequal number of GT repeats and may delete the extra copies in order to correct the mismatch between the strands.

Figure 3.10

```
——CACACA——
——GTGTGT——
```

Alternatively, misalignment of repeat sequences can cause expansion of the number of repeats, if the newly synthesized strand loops out.

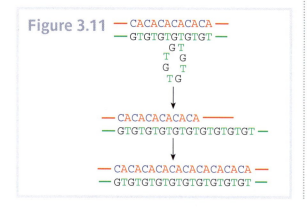

Figure 3.11

```
—CACACACACACA—
—GTGTGTGTGTGT—
        G T
       T   G
       G   T
        T G
         ↓
—CACACACACACA———
—GTGTGTGTGTGTGTGTGTGT—
         ↓
—CACACACACACACACACACA—
—GTGTGTGTGTGTGTGTGTGT—
```

Figure 3.12 A woman living in a stilt village on Lake Maracaibo in the early nineteenth century unwittingly gave rise to one of the most informative human genetic studies in the world. This woman (who died of Huntington's disease) gave rise to ten generations of descendants, encompassing around 15,000 people, many of whom carry disease-causing alleles of the *HD* gene. These kindreds are now studied by the US–Venezuela Collaborative Research Project (see Wexler *et al*. 2004). Although the Huntington's disease gene was one of the first human disease genes to be mapped (in 1983), it took a decade to be sequenced, a heroic undertaking in the pregenomic era.

Reproduced courtesy of Randy Trahan; Photograph: Geoffery Charles.

The more repeats there are, the more chance for slippage to occur, and the greater the likelihood of increasing or decreasing the number of repeats. So parts of the genome with many repeat sequences are unstable, prone to mutations that change the repeat copy number.

This instability can cause havoc in the genome. For example, Huntington's disease is a neurodegenerative disorder that causes movement abnormalities and loss of cognitive functions. Huntington's disease is a dominant genetic disorder, which means a single copy of the disease-causing allele is sufficient to cause the disease. So a person carrying the Huntington's disease allele has a 50% chance off passing the disease allele to each offspring. And any child that receives the disease allele will definitely get the disease, though they may not fall sick until after they have had their own children, each of which also has a 50% chance of inheriting the Huntington's disease allele (**Figure 3.12**).

Recessive traits, where two copies of an allele are needed to cause a disease, are discussed in Chapter 5

But Huntington's disease can show patterns of inheritance not expected from a normal Mendelian trait. Although most carriers don't develop symptoms until after the age of 40, some individuals develop the disease earlier (but only if they inherited the disease from their father, not their mother). Furthermore, in some cases the age of onset gets earlier with each subsequent generation, so that the son of a man with early-onset Huntington's disease may have even earlier disease onset.

These mysteries of inheritance were solved when the *Huntington disease* gene (*HD*) was sequenced in 1993. The beginning of the gene had multiple repeats of CAG, which codes for the amino acid glutamate (see **TechBox 2.3**). But the number of CAG repeats varied between individuals. Most people have between 9 and 35 CAG repeats. People with more than 40 CAG repeats in the *HD* gene develop Huntington's disease. The age of onset is correlated with the number of repeats: the more repeats, the earlier disease begins. Furthermore, once the number of repeats goes above 40, the frequency of changes to the number of repeats also increases, a process known as 'dynamic mutation'. Expanding repeats are mutations that increase their own rate of mutation.

There are at least nine human diseases that, like Huntington's disease, are known to be caused by unstable repeats. For example, Fragile X syndrome, the most common heritable form of moderate mental retardation, is caused by a massive increase in the number of copies of a CGG repeat in the upstream regulatory region of gene called *FMR1*. But many tandem repeats, especially those occurring in non-coding regions of the genome, do not have noticeable effects on their carriers.

Before we go on, we need some terminology to describe the different regions of tandem repeats in the genome. A particular place in the genome is usually referred to as a 'locus'. A run of repeats occurring in a given place in the genome is a termed a 'repeat locus'. Being a Latin word, the plural of locus is 'loci'. Repeat loci are known by many different names, including variable number of tandem repeats (VNTRs), simple sequence repeats (SSRs), and short tandem repeats (STRs). If the sequence that is repeated is between 10 and 100 nucleotides, it is often called a minisatellite. If the sequence that is repeated is only a few nucleotides long, such as CACACACACA, the repeat region is usually referred to as a microsatellite.

Thanks to dynamic mutation, these repeat regions provide an extremely useful tool for distinguishing individuals in a population. Because the mutation rate of tandem repeats may be orders of magnitude faster than in the rest of the genome, these sequences are so prone to change that two individuals are unlikely to share exactly the same number of repeats in all of their repeat loci. Counting the number of repeats at many different loci can often provide a unique identifier for individuals, often referred to as a DNA fingerprint.

Comparing repeat sequences

If you wanted to compare repeat sequences between individuals, you could sequence the entire repeat sequence then simply count the number of repeats. But there is a faster (and cheaper) way of comparing repeat number, which takes advantage of the fact that increasing or decreasing the number of repeats at a locus changes the length of the repeat region. To compare the size of repeat regions between individuals, you need to identify non-repeated DNA sequences (flanking regions) on either side of the run of repeats.

Figure 3.13

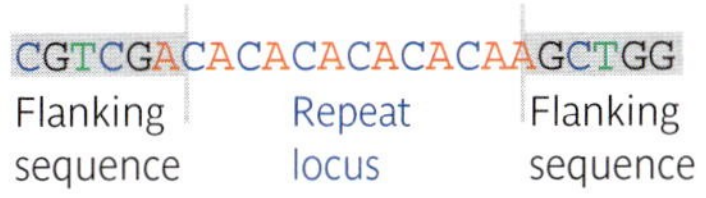

If the flanking regions have a lower rate of change than the repeat region, they may stay the same even when the number of repeats between the flanking regions change. You can then use these conserved flanking sequences to produce fragments of DNA that start and end with the same sequence, but have variable numbers of repeats in between.

Figure 3.14

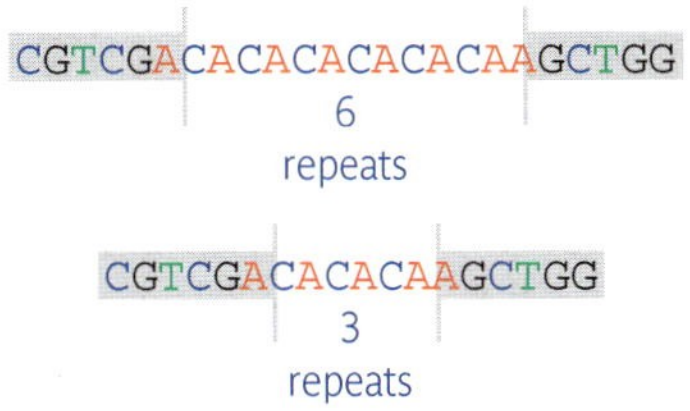

There are two basic approaches to producing DNA fragments containing repeat sequences. One approach is use restriction enzymes, which are endonucleases that cut DNA at particular recognition sequences. So, for example, the repeat locus illustrated in Figure 3.13 can be isolated by cutting the DNA with the restriction enzyme *AluI*, which cuts the sequence AGCT between the G and the C, thus dividing the DNA into different size fragments that all begin and end with the recognition sequence.

Figure 3.15

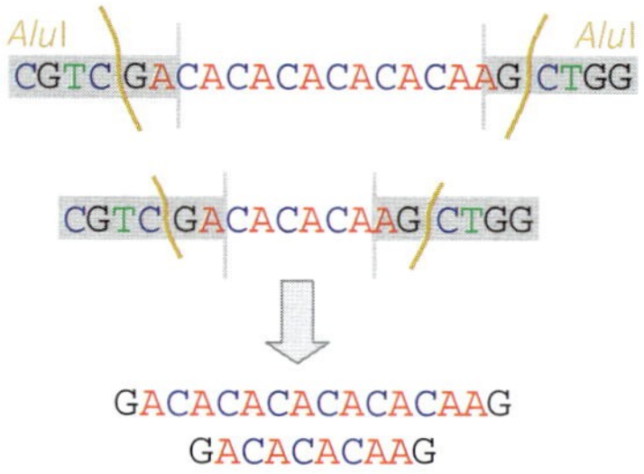

The restriction enzyme approach requires a relatively large DNA sample to produce enough fragments. The second approach is to design primers that stick to the flanking regions (see **TechBox 4.3**), then use PCR to produce vast numbers of copies of the fragment containing the repeat. This is the most commonly used approach, and it can be applied to a wider variety of DNA samples because it does not require large amounts of DNA to start with.

Chapter 4 explains DNA amplification by the polymerase chain reaction (PCR)

Once you have your fragments, you can separate them by size using electrophoresis, which is the movement of a charged substance in an electric field. DNA has a negative charge, thanks to the phosphate ions in the backbone of the helix (see **TechBox 2.2**). So if you subject DNA molecules to an electric current, they will tend to move towards the positive electrodes. But if your DNA samples are in a porous medium, such as a gel, then they will meet some resistance as they move towards the positive electrode. The bigger the fragment, the more it will resist movement through the medium (imagine using the same force to drag a basketball and a golfball through a bog: the larger ball will move more slowly as it encounters more resistance). So the shorter DNA fragments will move more rapidly than the longer fragments, and therefore get further down the gel (Figure 3.16). The gel can then be stained to reveal bands of DNA of different fragment lengths.

Figure 3.16

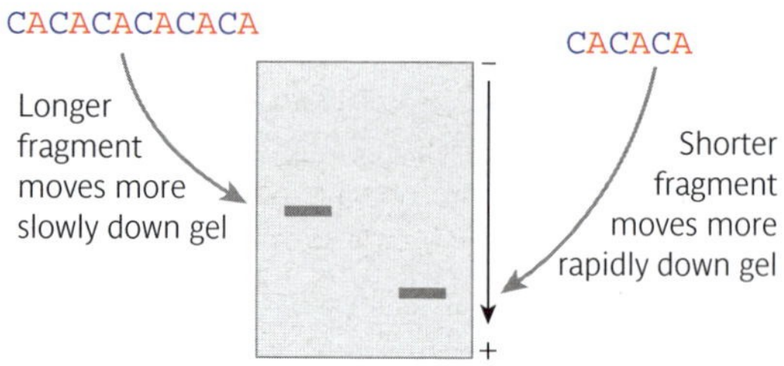

An alternative method is capillary electrophoresis, in which DNA fragments are drawn through thin glass tubes containing a porous medium. The movement of the fragments is detected using fluorescent dyes (added during PCR amplification of the fragments).

Using DNA fingerprinting to identify individuals

> ***"I took one look, thought 'what a complicated mess', then suddenly realized we had patterns. There was a level of individual specificity that was light years beyond anything that had been seen before. It was a 'eureka!' moment. Standing in front of this picture in the darkroom, my life took a complete turn. We could immediately see the potential for forensic investigations and paternity, and my wife pointed out that very evening that it could be used to resolve immigration disputes by clarifying family relationships."***
>
> Alec Jeffreys (originator of DNA fingerprinting) 2004, http://genome.wellcome.ac.uk

The mutational instability of short tandem repeats, and the relative simplicity of comparing repeat regions, makes them ideal for identifying individuals. The number of repeats changes so frequently that even closely related individuals are likely to have different numbers of repeats in at least some places in the genome. The first step is to identify repeat loci that tend to vary in the number of repeats between individuals. A particular number of repeats at a given locus is considered to be an allele, a discrete heritable variant. Individuals may be characterized by a specific combination of alleles

Locus	D3S1358	vWA	FGA	D8S1179	D21S11	D18S51	D5S818
Genotype	15, 18	16, 16	19, 24	12, 13	29, 31	12, 13	11, 13
Frequency	8.2%	4.4%	1.7%	9.9%	2.3%	4.3%	13%

Locus	D13S317	D7S820	D16S539	THO1	TPOX	CSF1PO	AMEL
Genotype	11, 11	10, 10	11, 11	9, 9.3	8, 8	11, 11	X Y
Frequency	1.2%	6.3%	9.5%	9.6%	3.52%	7.2%	(Male)

Figure 3.17 This is the CODIS profile of Bob Blackett, who is apparently a forensic DNA analyst. At each of 13 loci, Bob has two alleles, one on each chromosome. The 14th box records his sex-chromosome status (see **Figure 7.58**). Each allele is given a number. The frequency of this allele in the reference population is given below. These frequencies are sometimes used to calculate how likely a random match would be; that is, the probability that two people could have exactly the same pattern of alleles. However, these probability calculations depend on identifying an appropriate reference population, since the frequencies vary between countries, and ethnic groups within countries (see **Figure TB3.3**, p. 92). You can find out more about Bob's DNA at: *www.biology.arizona.edu/human_bio/activities/blackett2/overview.html*.

Reproduced by permission of Bob Blackett/ Rick Hallick, University of Arizona.

across many repeat loci. If you score enough repeat loci, you should be able to produce a 'DNA fingerprint': a pattern of alleles that is statistically unlikely to be found in more than one individual.

DNA fingerprints have been embraced by the legal system as a way of determining whether a particular individual is associated with a biological sample found at a crime scene (such as hair, blood, or semen). The United States Federal Bureau of Investigation keeps a database called CODIS (Combined DNA Index System) of DNA samples recovered from crime scenes, from convicted offenders, and from people who have been arrested. A CODIS profile scores the alleles present at 13 particular repeat loci (**Figure 3.17**). CODIS is currently the largest database of DNA fingerprints in the world, containing more than four and a half million profiles. However, the United Kingdom's National DNA Database contains a larger proportion of its source population, with DNA fingerprints from around 5% of the UK population (i.e. more than three million people). Anyone arrested in the UK can have their DNA fingerprint permanently stored in the database, whether they are subsequently convicted or not. DNA databases such as these have been used to identify suspects in crimes that would otherwise have been unsolved. But retaining DNA fingerprints on a database raises important ethical issues (**TechBox 3.2**). For example, concerns have been raised over the strong racial skew of the UK National DNA Database, and the large number of samples from children, many of whom have never been formally cautioned or charged.

Using microsatellites to uncover relationships

Although tandem repeats have a very high rate of change, they still carry historical signal. The repeat loci in your parents' genomes were copied when they made the gametes that came together to make you. Some of these repeat sequences might have increased or decreased in length, but on the whole your DNA fingerprint will be more similar to your parents than it is to random members of the population. Therefore, analysis of repeats can reveal family relationships. Indeed, one of the first applications of DNA fingerprinting was in an immigration dispute in 1985, where genetic data were used to prove that a child that had been refused British residency was indeed the son of a UK citizen.

More broadly, DNA fingerprinting can be used to understand social dynamics and population structure. In particular, analysis of repeat regions has provided a way of assessing mate choice and paternity in wild populations, often revealing relationships that would not have been evident from behavioural studies alone. For example, the alpine marmot (*Marmota marmota*; **Figure 3.18**, p. 86) is one of the few mammal species considered to be monogamous (forming exclusive breeding

TECHBOX
3.2

Biobanking

KEYWORDS

haplotype

association studies

disease

ethics

LOD score

diabetes

informative pedigree

FURTHER INFORMATION

An internet search will reveal the URLs of many biobanks, including the two mentioned here: deCODE (*www.decode.com*) and the Genographic Project (*www3.nationalgeographic.com/genographic*).

RELATED TECHBOXES

TB 1.1: GenBank

TB 3.3: SNPs

RELATED CASE STUDIES

CS 3.2: A modern Icelandic saga (biobanking in action)

CS 4.2: Peopling the Pacific (tracing human migration)

A biobank is any large collection of biological samples, usually linked to information about the individuals from which the samples were taken. Large-scale collection of biological samples from a population is not a new concept. For example, many countries have been storing heel-prick blood samples taken from all new-born babies, along with individual information, for over three decades. The largest collections of human biological samples are those used in forensic identification of individuals (see page 79), but the term 'biobank' is nowadays most commonly applied to samples collected for research into human health. These biobanks essentially act as large-scale extensions of traditional family association studies.

The classic way to identify genes that cause human diseases is to find an 'informative pedigree' (a family that shows a clear pattern of inheritance of the disease), then try to establish which parts of the genome are shared by affected family members, but not those unaffected. The gene for Huntington's disease, an early success story of genetic mapping, is a good example of a disease suited to association studies: it is caused by a single gene, has high penetrance (all people with the disease allele will develop the disease), is genetically dominant (a single copy of the disease allele will give rise to the disease, so half of the offspring of an affected person are also expected to be affected by the disease), and with large and well-described pedigrees (particularly because Huntington's disease is usually expressed later in life after carriers have had children; **Figure 3.12**). Association studies have revealed the genetic basis of many single-gene disorders with clearly defined pedigrees. But they are limited in power for complex diseases that may be influenced by many factors, both genetic and environmental. For example, heart disease has a strong genetic component (it runs in families) but it is clearly influenced by many different genes and environmental factors, so has incomplete penetrance (e.g. even in the presence of inherited risk factors, the chance of developing heart disease is influenced diet and exercise).

Instead of identifying informative pedigrees from particular families, biobanks take a large number of samples from a particular population and perform statistical analyses to identify genomic markers that are significantly associated with specific diseases. This approach does not work well for rare diseases: even in a large sample, there will be too few affected people to allow association with genetic markers to be detected. Instead, biobanks focus on 'common alleles', those present in 5% or more of the population. Therefore, biobanks work best for identifying complex, multifactorial traits, influenced by variation in many genes, rather than for searching for single mutations that cause rare diseases. These complex diseases are often described by the phrase 'common disease, common variant': it may be the combination of many different factors, both genetic and environmental, that leads some people to be affected by a certain disease.

Biobanks target multifactorial diseases by connecting individual genetic information with health data, then looking for a statistically significant association between certain genotypes and particular health outcomes. For example, the deCODE Icelandic biobank (see **Case Study 3.2**, p. 83) published a study identifying a genetic risk factor for Type 2 diabetes. They compared the frequency of genetic markers (SNPs; see **TechBox 3.3**, p. 90) in 1185 people affected by Type 2 diabetes and 931 unaffected people. They found that a particular haplotype was over-represented in the affected sample (and a different haplotype that was under-represented in people with diabetes)[1]. However, the frequencies illustrate how these alleles play only a small part of the risk of developing diabetes: only one-third of the affected patients carried the at-risk haplotype, and one-quarter of the unaffected people also carried the at-risk haplotype. Type 2 diabetes clearly has many common genetic variants that may contribute to disease risk, but it also has strong environmental risk factors, particularly related to diet. Type 2 diabetes has increased rapidly in the developed world within a single generation, so this increase cannot be due to genetic factors alone.

Many countries are now establishing biobanks, often partly or wholly government funded, because they are considered a valuable research tool in medical genetics. Indeed, the OECD (Organization for Economic Co-operation and Development) specifically provides funding for countries to establish biobanks. Not all biobanks are formed for medical research. For example, the privately funded Genographic project is currently establishing a global database of DNA samples to trace human migration patterns by selling kits that allow individuals to take a cheek swab and post it to a laboratory for sequencing. But the establishment of biobanks has not been without controversy. In particular, there have been protests from activists representing various indigenous populations who fear that, by donating their DNA to a biobank, individuals may be giving away a valuable resource: the genetic material of their particular population, which may contain medically useful variations. To some extent, this issue of shared genetic material affects all donors to biobanks, because any individual is likely to share genomic variants with their relatives, who may not have consented to the donation.

Some of the ethical issues raised by large-scale biobanks are[2]:

Consent: The rapid pace of advance in the field of genomics poses problems for the informed consent of donors, when the use of the sample may change as technology and scientific knowledge develop. Therefore, it is not possible to explain to a donor the uses to which their sample will be put (such explanation would normally be considered essential for informed consent). The problem of informed consent for future use is also raised by existing collections of biological samples. For example, most people born in Australia in the last three decades have had a blood sample taken and stored (potentially indefinitely) on Guthrie cards. These cards have been accessed by medical researchers[3] and the police[4] without patient knowledge or consent. Should a donor have the right, at any point in the future, to request information on the use of their sample, or to demand the samples be destroyed? Or should 'informed consent' be assumed to cover all unknown future uses?

Profit: Some biobanks are either wholly or partly owned by pharmacogenomics companies, who are investing in the future rights to marketable medicines arising from research on the biobank. There may be tension between the potential public health benefits of biobank discoveries and private company profits. A compromise strategy might be profit-sharing, but who should benefit: the whole community or only those who donated to the biobank? The ownership and accessibility of the data impacts on this debate, as public benefit is best

Figure TB3.2 Dr Robert Guthrie, who developed the heel-prick blood test now widely used to test new-born babies for potentially debilitating illness, is unlikely to have foreseen that his test would lead to criminal convictions due to DNA testing of stored blood samples.

served by researchers having open access to biobank data, but commercial gain relies on at least some degree of non-disclosure or data protection.

Anonymity: Biobanks differ in the degree to which samples and information can be identified to specific individuals. Some are encrypted, so that only a limited number of people can connect individuals to their samples and data. Others are anonymized, so that individual identity is stripped from the data and samples (in which case, it would be no longer possible for the individual to exert rights over their biobank entry). Concern has been raised about the risk of identifying individuals from biobank data, which might then make aspects of medical records or genetic risk factors available to employers or health insurance companies.

References

1. Grant, S.F.A., *et al*. (2006) Variant of transcription factor 7-like 2 (TCF7L2) gene confers risk of type 2 diabetes. *Nature Genetics*, Volume 38, pages 320–323.
2. Cambon-Thomsen, A. (2004) The social and ethical issues of post-genomic human biobanks. *Nature Reviews Genetics*, Volume 8, pages 866–873.
3. Barnes, G., Srivastava, A., Carlin, J. and Francis, I. (2003) Delta-F508 cystic fibrosis mutations is not linked to intussesception: Implications for rotavirus vaccine. *Journal of Paediatric and Child Health*, Volume 39, pages 516–517.
4. Phillips, G. (2003) Guthrie cards. Transcript of Catalyst, ABC television www.abc.net.au/catalyst/stories/s867619.htm.

CASE STUDY **3.2**

A modern Icelandic saga: biobanking identifies a schizophrenia gene

KEYWORDS

SNPs

genome-wide scan

haplotype

genealogy

LOD score

association studies

FURTHER INFORMATION

The database of health and genetic data is currently only available to Icelandic researchers and through licences granted by deCODE (*www.decode.is*). The genealogical database of Icelanders, *www.islendingabok.is*, is accessible to Icelanders; however, Engilbert Sigurdsson provides his user name and instructions for entering the database in the appendix to his 2004 article[2].

RELATED TECHBOXES

TB 3.2: Biobanking

TB 3.3: SNPs

RELATED CASE STUDIES

CS 5.2: Time flies (detecting substitutions associated with traits)

CS 4.2: Peopling the Pacific (human population history)

Stefansson, K. *et al*. (2002) Neuregulin 1 and susceptibility to schizophrenia. *American Journal of Human Genetics*, Volume 71, pages 877–892

> *Some eleven centuries after the Age of Settlement, a singularly goal-orientated descendent of the Vikings, Kari Stefansson, returned to Iceland in 1996 having spent two decades in the U.S . . . His plan was to make use of the extensive available genealogical and census data on the Icelandic population, the relative homogeneity of the Icelandic gene pool and the general willingness of the Icelandic population to participate in research projects.* [1]

Background

The *Landnáma* – or *Book of Settlements* – is a twelfth century document detailing the history, lands, and descent of the Icelandic people, written by Ari the Learned. A modern version – the *Íslendingabók* (*Book of Icelanders*) – containing the genealogy of half the Icelanders who have ever lived, is now being created by a biotechnology company called deCODE Genetics Inc[1]. An act of parliament gave deCODE the rights to Iceland's database of health information and blood samples from all Icelanders (except those who choose to opt out of the project). When combined with genealogical information from the *Íslendingabók*, this database may provide a powerful tool for tracing genetic factors that influence complex diseases, particularly heart disease and mental health. One of their first success stories was the identification of a gene apparently associated with risk of schizophrenia.

Aim

Schizophrenia shows evidence of heritability, so researchers have been hunting for genes associated with schizophrenia for decades. The slow progress of this search has been attributed, at least in part, to the complexity of the condition and the likelihood that many different genes contribute to the disease[1]. Association studies, where informative pedigrees are used to identify regions of the genome that segregate with the disease, identified a number of candidate chromosomal regions on nine different chromosomes[2]. In particular, several independent studies identified a locus on the p-arm of chromosome 8 that seemed to be associated with risk of developing schizophrenia, but it was not known which of the many genes in this region was involved with the disease. The deCODE project used data from Icelandic families to fine-scale map the region of chromosome 8 implicated in schizophrenia in order to identify candidate genes that may influence the development of the disease.

Methods

This study used a sample of over 400 patients with schizophrenia plus unaffected family members in a 'hypothesis-free genome-wide scan'; that is, they looked for any genetic

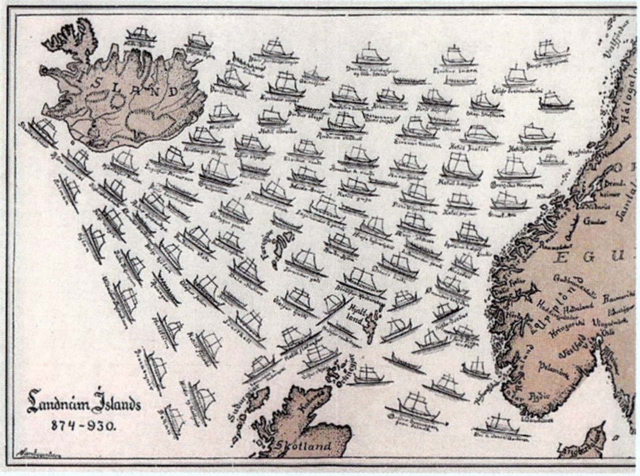

Figure CS3.2 The well-documented history of the Icelanders, and their relatively genetically homogeneous population, has made them an excellent population for the hunt for genes involved in common human diseases.

markers significantly associated with the disease, rather than targeting suspected candidate genes. They used the genealogical database (the *Íslendingabók*) to construct large pedigrees for the affected people. Then they conducted a genome-wide scan using microsatellite markers to identify regions of the genome linked to the inheritance of schizophrenia. This analysis rests on the assumption that both patients and unaffected people will have an equal chance of carrying any haplotype that is not associated with schizophrenia, but that a higher-than-expected representation of a haplotype in the sample of patients indicates genetic linkage between those genetic markers and some gene that influences schizophrenia. When they had identified a promising region of the genome, they used a genetic library (BAC clones) to identify many more closely spaced genetic markers along the region, to give an average of one marker every 75,000 bases. These fine-scale markers were used to search for any haplotypes shared across families affected by schizophrenia.

Results

The highest LOD score (a measure of linkage disequilibrium: the non-randomness of association between genetic markers and disease) was for a region of chromosome 8. The fine-scale mapping of this region narrowed the target region down to a 600 kb region containing the gene *neuregulin-1* (*NRG-1*), involved in cell-to-cell signalling, and, more specifically, in the development of neurones. They then genotyped all of the patients and unaffected controls for 58 SNPs along this locus. None of these SNPs was identified as being specifically associated with schizophrenia in these patients, but they found a core haplotype consisting of seven linked genetic markers (two microsatellite loci and five SNPs) that was present in 15.4% of the people with schizophrenia, but only 7.5% of the control (unaffected) people. The identification of *NRG-1* as a candidate gene in schizophrenia has also been supported by more detailed experiments using animal models. Mice that are heterozygous for defective *NRG-1* express less neuregulin than normal mice, and they have reduced numbers of particular neurotransmitter receptors, which matches observations from the brains of schizophrenic patients. These *NRG-1*-deficient mice show behavioural abnormalities that can be reversed with the antipsychotic drug clozapine, used to treat schizophrenia in humans.

Conclusions

The haplotype identified by this study is significantly associated with schizophrenia. However, this study did not prove that *NRG-1* is causally linked to schizophrenia, nor can this haplotype be used to reliably predict who will develop schizophrenia, because some of the people carrying the haplotype do not suffer from schizophrenia, and nearly 85% of schizophrenia suffers do not have this haplotype. But the identification of *NRG-1* as a candidate gene in schizophrenia points to the potential involvement of specific neutrotransmitter pathways, which will now receive more research attention.

Limitations

Although all records are encrypted to prevent the identification of individuals, the access to personal data granted to deCODE has stirred controversy, on commercial, political, and ethical grounds. One of the scientific criticisms of the deCODE project was that such a geographically restricted and genetically homogenous population may reveal little about heritable disease in other populations and, from a commercial point of view, there is little incentive to produce new diagnostic tests or treatments that can only be marketed to 300,000 Icelanders. But like the HapMap project (see **TechBox 3.3**, p. 90), the aim of the Icelandic genomic project is to identify genetic markers common to all human populations. This finding has now been replicated in other human populations in Scotland, China, Ireland, and the Netherlands. Because genome-wide scans survey a very large amount of data, there is a risk of false positives (obtaining a significant result for a spurious association)[3], so any gene identified by a hypothesis-free scan requires further independent investigation.

Future work

The precise role of *NRG-1* in schizophrenia remains to be determined, which will take a great deal more laboratory work and medical investigation. There are many more candidate genes for schizophrenia to be investigated, in addition to the influence of environmental factors. But deCODE's success suggests that biobanks may provide an effective way to identify candidate genes involved in complex diseases.

References

1. Sigurdsson, E. (2004) Genomics and genealogy provide an Icelandic springboard into the human gene pool. *Journal of Mental Health*, Volume 13, pages 21–27.
2. Harrison, P.J. and Owen, M.J. (2003) Genes for schizophrenia? Recent findings and their pathophysiological implications. *Lancet*, Volume 361, pages 417–419.
3. Hirschhorn, J.N. and Daly, M.J. (2005) Genome-wide association studies for common diseases and complex traits. *Nature Reviews Genetics*, Volume 6, pages 95–108.

relationships between one male and one female). Alpine marmots live communally in family groups, consisting of a breeding pair, plus non-breeding subordinates and juveniles. But, using DNA extracted from hair samples, analysis of six microsatellite loci revealed that one-fifth of the marmot juveniles were not fathered by the resident male in the family group, and a third of the litters contained the offspring of two different fathers.

Because repeat alleles are inherited, close relatives will tend to have similar alleles. This means that a small, interbreeding population will tend to share a relatively small number of alleles, which will be distinct from other such populations. So repeat sequences such as microsatellites can, in many cases, be used to tell which population an individual is from. This is becoming increasingly useful as a way of monitoring the trade in endangered species. For example, DNA can be extracted

Figure 3.18 The alpine marmot (*Marmota marmota*). Females are not above mating with their own sons.

Photographer: François Trazzi, 2004.

from timber to check not only the species of tree it was taken from, but also the geographic region of origin. Genetic analysis will play an increasingly important role in policing the illegal logging of tropical hardwoods, to make sure that commercially available timber has not been harvested from endangered populations. DNA fingerprinting can also improve traceability and quality control of timber. The global increase in wine sales has led to a growth of the cooperage (barrel-making) industry, and subsequent shortages of suitable oak timber. DNA fingerprinting can be used to check the provenance of oak timber, to ensure the barrels are made from the optimum type of timber for maturing wines.

The effect of population size on genetic diversity is discussed in Chapter 5

Microsatellites are handy because they evolve so rapidly that they can differ between members of a population, and even closely related populations can have distinctive sets of microsatellite alleles. However, this rapid rate of change can make microsatellite loci difficult to amplify. If you want to find a particular gene in a given species, you can design a primer based on the gene sequence from a related species (see **TechBox 4.3**). Some microsatellites are located in well-conserved flanking regions, so a primer used in one species may amplify the repeat region in another. But, in general, this approach doesn't work: a microsatellite locus known from one species is likely to have changed so much over time that it is unrecognizable in a different species. In this case, researchers must identify appropriate repeat regions by screening a genomic library of small fragments of DNA. The library is then screened with a probe that will hybridize to a particular repeat sequence. This approach can be rather hit-and-miss, as it can take many attempts to find appropriate microsatellite loci, but new improved protocols continue to be published.

Point mutations

Insertion, deletion, and rearrangement of DNA sequences are major drivers of genomic evolution. But as this book is primarily concerned with the comparisons of DNA sequences between species, the mutations we will most commonly encounter are point mutations, which create single nucleotide changes to the DNA sequence. Broadly speaking, changes to the nucleotide sequence of DNA come about through two processes, copy errors and damage. However, there is no clear division between these processes: copy error can arise through incorrectly repaired damage encountered during replication, and damage may occur more often when DNA is exposed during replication. In the next chapter, when we look at DNA replication, we will consider the mutations caused by copy errors, which form a substantial proportion of the mutations accumulated each generation. But in this chapter, we will focus on point mutations caused by DNA damage.

There are many physical and chemical factors that can damage DNA and cause change to the information in the genome. When people worry about mutagenic effects of radiation or chemicals, it is this incidental DNA damage they fear. But many mutagens are common features of the environment. So although damage to the genome may be increased by exposure to hazards such as radiation, DNA damage occurs all the time, in all cell types. In fact, there is a good chance that, in the time it takes you to read this paragraph, at least one of your cells will sustain damage to its genome.

Damage to DNA can be disastrous, because it can destroy the information needed to make and maintain living cells. Because of the potentially dire consequences of DNA damage, cells have a variety of mechanisms for fixing damage. But, however efficient they may be, no repair mechanisms are perfect. Sometimes, damage that changes the DNA sequence persists unrepaired, so that the altered sequence is copied along with the genome and passed to the cell's descendants. And sometimes the repair process itself causes a permanent change to the DNA sequence. Once the sequence is altered so that the original sequence can no longer be recovered, that change is a mutation that will be included in any copies made from that genome.

There are so many types of mutation, and such a diversity and complexity of repair systems, that we could not hope to cover them all. Instead, we will look at a single mutagen that causes a particular type of DNA damage, and give a brief summary of the strategies a cell employs to deal with this damage. The mutagen we will consider is ultraviolet light (a component of sunlight), which can cause the formation of thymine dimers (**Figure 3.19**, p. 88). Thymine dimers are formed when two adjacent Ts become linked together, causing a distortion in the DNA helix. Formation of thymine dimers is potentially ruinous for a dividing cell, for two reasons. Firstly, when two Ts on one strand pair with each other, they can no longer pair with the opposite strand. Since DNA carries information by complementary base pairing, any bases that don't pair represent a loss of information. If information is lost from a critical sequence, it may prevent the cell from making something important. Secondly, and more seriously, a thymine dimer can block the movement of a polymerase enzyme along the DNA helix, halting DNA replication. If a cell cannot replicate its DNA then it can't divide. And a cell that can't divide is a cell with no evolutionary future. So there has been strong selection pressure for cells to evolve mechanisms for detecting and repairing UV-induced thymine dimers. We will follow a typical bacterial repair pathway, but similar responses can be found in most organisms.

There are several different repair pathways that deal with thymine dimers (**Figure 3.19**). Firstly, there is a specific enzyme that detects thymine dimers, and cuts the bonds between the neighbouring Ts (this enzyme gets its activation energy from visible light). This pathway reverses the damage caused by the thymine dimer, restoring the sequence to two normal Ts. If the dimer is not removed, it will attract the attention of enzymes that scan the genome for distortions in the helix, and trigger a general excision repair pathway. The damaged DNA strand is cut out, then a special DNA polymerase makes a new strand, and a ligase enzyme glues the newly made strand to the helix. If a thymine dimer is not detected and repaired by the excision repair pathway, then when that DNA is next copied, it will cause the polymerase enzyme to stall. To avoid halting DNA replication altogether, the cell may initiate a DNA repair pathway that bypasses the blockage caused by the thymine dimer, and continue replication after the block. This leaves a gap in the new DNA strand, which may be filled by a postreplication repair pathway that uses part of the other strand of the replication fork to replace the missing DNA sequence. If this strategy does not work, then some organisms have a last-ditch option: the SOS pathway. SOS is a complex series of responses to an increasingly desperate situation: if the cell cannot fix the dimer, replication will cease and the cell will die. In a race against time, the cell tries a number of strategies to overcome the block in replication. The SOS response is error-prone: it may fix the damage, but in doing so it will probably change the DNA sequence. Inducing SOS repair is likely to lead to many mutations, each of which may be just as disastrous as the original damage itself. This is why the SOS response is the final attempt of a damaged cell to divide or die.

This is just one example of a specific type of DNA damage (thymine dimers) caused by a particular mutagen (UV light), to which the cell has a sophisticated response (enzymes that reverse damage, failing that excision repair, and failing that postreplication repair, and finally the last-chance SOS response). There are many other kinds of damage, brought about by a range of different mutagens, and each will trigger particular

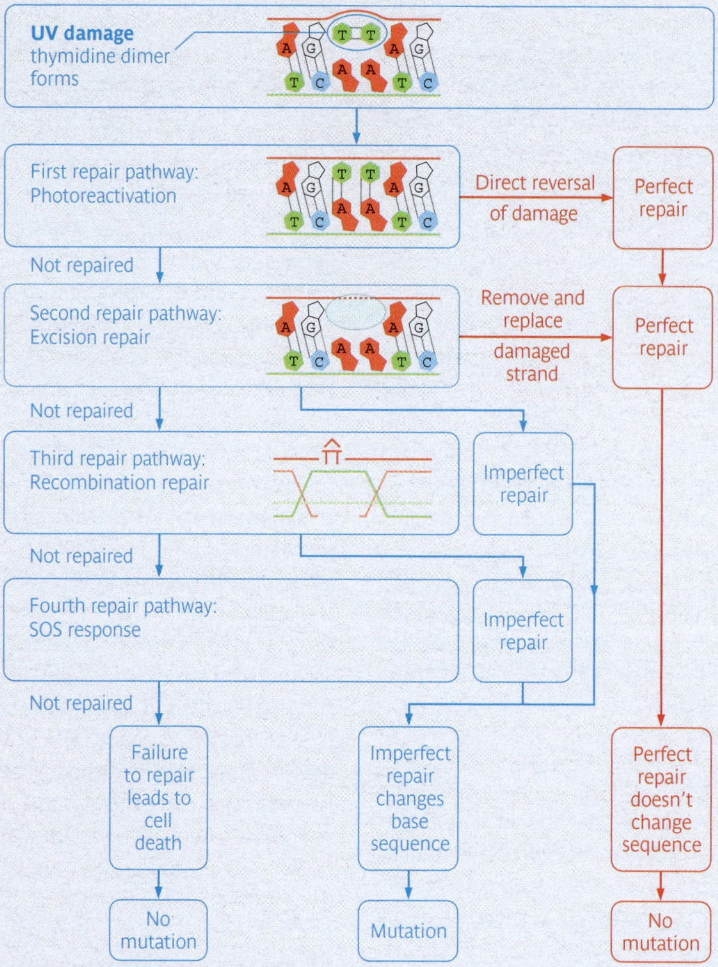

Figure 3.19 A stylized repair pathway for a thymine dimer. This flowchart describes four possible responses to UV damage, rather than representing the precise sequence of events in any given organism. Damage may be directly reversed, removed and replaced with excision repair, replaced with DNA from another strand by recombination repair, or bypassed using the SOS response (this pathway has been described in bacteria, but it has been suggested that vertebrate genomes may have some of the elements of this pathway too). Imperfect repair creates a change in the base sequence (mutation), but perfect repair does not (sequence is returned to its original state) nor does failure to repair (damaged DNA can't be copied to next generation).

repair pathways. It would not be practical to list them all here. If you are interested, then you can learn about specific mutagens, types of DNA damage, and particular repair pathways in any general genetics or biochemistry textbook. For the purposes of this book, it is enough to know that DNA damage occurs frequently, and it is usually repaired. But occasionally damage is imperfectly repaired, resulting in a change to the DNA sequence. From an evolutionary perspective, DNA damage only becomes a mutation when imperfectly repaired changes become a permanent part of the DNA sequence, so that it will be passed on to all offspring.

Single nucleotide polymorphisms (SNPs)

Although most errors are corrected, and most damage repaired, the occasional change slips through and becomes a heritable change to the DNA sequence. This has the dual effect of making every genome unique, yet making related genomes similar. This is, of course, why children resemble their parents. This is good, when your children inherit your cute nose, keen eyesight, or charming smile. But it is not so good when your children inherit your flat feet, attractiveness to mosquitoes, or propensity for heart attacks. We have already seen that we can use mutations to identify individuals, track family relationships, characterize populations, and record movement of individuals in space and time (e.g. **Case Study 1.2**). We can also use the patterns of inheritance of mutations to try to determine why some sequence variants have undesirable effects. The first step is to genotype the individuals in the population we are studying, by using a set of genetic markers that we expect to differ between individuals. Until recently, genotyping has commonly relied on characterizing repeat loci (DNA fingerprinting), but increasingly researchers are turning to methods of identifying single nucleotides that differ between individuals.

Sites in the genome where the nucleotide sequence differs between some members of a population are often referred to as single nucleotide polymorphisms, or SNPs ('snips'; see **TechBox 3.3**, p. 90). 'Single nucleotide' refers to the fact that these are (usually) one-base changes to the DNA sequence that arise from point mutations (or single-base insertions or deletions). 'Polymorphism' means that these sequence variations are carried by some, but not all, members of a population: that is, that there are multiple (poly) different variants (morphs) of the sequence in the same population. Technically, a particular variant must be present in at least 5% of individuals in a given population to be considered a polymorphism. But, in practice, people often refer to any locus that is not the same in all individuals as polymorphic. This is because it is difficult to accurately judge the frequency of sequence variants where sampling is limited to a relatively small number of individuals, as is often the case in molecular ecology or medical genetics. For example, imagine you have sampled 10 individuals, and one of them has a unique sequence variant found in no other individuals in your sample (such a variant is referred to as a singleton). In this case, you would be unable to tell whether that allele was present in only one individual or in 10% of the population (or more or less).

We will consider polymorphism in more detail in Chapter 5

If each human carries around 100 new mutations, and if around one in 1500 nucleotide positions in the human genome is polymorphic, then how is it possible to produce one universal human genome sequence? The answer is that it's not possible. Published human genome sequences are either a single person's genome, or a consensus sequence for a sample of people, representing the most likely nucleotides to be found at any given position in the DNA sequence. So it is extremely unlikely that any given individual would have the exact sequence reported by the human genome sequencing projects. But any given human would have a high probability of having the same sequence at the majority of sites in the genome. Often in evolutionary biology, it is most helpful for us to use the consensus sequence, representing the average sequence for a population. However, for some genetic studies, it is the polymorphic sites we wish to study. In particular, SNPs are very useful for identifying genes that cause particular inherited traits, such as genetic diseases.

The bioinformatic tools for finding similar sequences in a database are covered in TechBox 3.4, p. 93

There are two ways that SNPs can be used to reveal the genetic basis of a trait. Firstly, the SNP may actually cause a particular trait by disrupting an important sequence. So, for example, two SNPs in the human period 2 (*hPer2*) gene in humans have been found to be associated with being a 'morning person' (that is, being annoyingly chirpy early in the mornings, but socially challenged due to an inability to stay up late at parties). When researchers studied a particular family with a known pattern of inheritance for an extreme form of morningness called Advanced Sleep Phase Syndrome (ASPS), they found that all of the family members with ASPS had one particular SNP in the coding region of the gene, causing a glycine to be included in the protein

TECHBOX 3.3 SNPs

KEYWORDS

HapMap

genetic diversity

haplotypes

heteroduplex

ethics

microarray

marker-based breeding

FURTHER INFORMATION

An introduction to the methods for analysing SNP data can be found in:
Gibson, G. and Muse, S.V. (2004) *A Primer of Genome Science*, 2nd edn. Sinauer Associates.

RELATED TECHBOXES

TB 3.2: Biobanking

TB 4.2: DNA amplification

RELATED CASE STUDIES

CS 3.2: A modern Icelandic saga (genome-wide scans)

CS 4.2: Peopling the Pacific (human genetic diversity)

Any two individuals from a given species will have identical DNA sequences across most of the genome. But, with the possible exception of clones (including some identical twins; **TechBox 6.4**), any two individuals will have occasional differences in their DNA sequences. Identifying the sites that differ between individuals is important for studying the inheritance of genetically determined traits, and for studying population dynamics.

Detecting SNPs

There are a very large number of methods for detecting positions in the genomes where the sequence can differ between individuals (single nucleotide polymorphisms, SNPs), and new methods are being constantly devised. It would be impractical to provide a list of methods here, but it is possible to consider a few of the general approaches to finding SNPs and gauging their diversity.

The most obvious method is to sequence the whole genome of as many individuals as possible and directly compare their DNA sequences, to identify sites where their sequences differ. But this approach is time-consuming and expensive if you wish to sequence many individuals from a species with a large genome. Sequencing only part of the genome is a hit-and-miss strategy, as you may fail to find informative polymorphic sites. A faster way to detect SNPs by direct sequencing is to mix the DNA from several different individuals together. When the mixed DNA sample is sequenced, polymorphic sites will show up as a site with a mixed signal, because there is more than one possible base at a given position. However, it may be difficult to tell the difference between a true polymorphic site and a sequencing error, so it is usually necessary to resequence the locus of interest once it has been detected by this method.

There are several alternative approaches to direct sequencing of the regions containing the SNP. If you know what your target sequence is, you can design an array of single-stranded DNA sequences that differ by just one nucleotide. If your sample pairs with one of these sequences, you will know what the base sequence of your sample is without actually having to sequence it. So you take the PCR product from your DNA sample, label it with some kind of reporter, and hybridize it to the array of sequences so that your sample sticks to the matching sequence on the array. One way to do this is to position the sequences on a chip (referred to a microarray) which can carry tens of thousands of alternative versions of the sequence, corresponding to all possible variants for a decent stretch of genomic DNA (say, 50 kb). By labelling your PCR product with a fluorescent marker you will be able to tell to which sequence on the array it hybridizes. Alternatively, the array can be attached to tiny beads only micrometres in diameter, each of which carries one specific sequence and is coded using varying amounts of two coloured dyes. After the sample DNA is hybridized to the beads, they are passed through a flow cytometer, which is able to separate the beads by colour and indicate the amount of DNA hybridized to each kind.

Alternatively, we can take advantage of the fact that mismatched nucleotides cause a distortion in the DNA helix. When DNA from different individuals is hybridized together (termed heteroduplex DNA), the bases cannot pair at points in the sequence where they differ, so there will be a 'bubble' which may destabilize the surrounding helix. There are several ways of detecting these mismatches in heteroduplex DNA, most of which rely on mismatched DNA having a lower melting temperature or retarded movement down a gradient (such as a gel or a chromatography column) due to the distorted shape of the molecule[1].

Analysing SNPs

Relatively few SNPs occur within genes, and even fewer play a causal role in determining differences between individuals. Instead, the aim of SNP analysis is to identify regions of the genome that contain alleles associated with the trait of interest. If a particular SNP is located near the gene of interest, then it will tend to be inherited along with the gene. If we are lucky, then a causal difference in the gene (e.g. a diseases allele) will just happen to be connected to an identifiable SNP nearby. Then if we find someone has that SNP, there is a good chance they also have the allele we want to study. It is this incidental association between the SNP and the trait that allows the position of the gene to be deduced. However, researchers will generally not find a perfect link between SNP and trait, such that having the SNP guarantees you have the disease. Instead, most studies report a statistically significant association between a SNP haplotype and a disease (see **TechBox 3.2**, p. 80).

Results from the HapMap project (see below) suggest that the human genome is divided into blocks, defined by recombination hotspots. Between these recombination sites, variation tends to be inherited together as a haplotype, but the association between the blocks may be altered by recombination. This means that one good informative SNP, termed a haplotype tagging SNP (htSNP or tagSNP), should be sufficient to represent all of the alleles in the same block. A genome-wide coverage of tagSNPs allows the genome to be scanned for SNPs or sets of SNPs that 'segregate with' (tend always to be found in connection with) the trait you are interested in. There is then a good chance that the SNP is located near a sequence that influences that trait.

Once an informative SNP is located, there may be a dozen or more genes nearby (see **Case Study 3.2**, p. 83). The next step can be to do a more targeted sequencing programme of a candidate gene located near the SNP, from a large sample of both patients and unaffected people (this strategy is often referred to as 'medical resequencing')[2]. However, even when the gene itself is unknown, that SNP may serve as an informative marker. SNPs have been widely used in 'marker-based breeding' to develop crops resistant to disease. Once an informative marker has been identified, very large numbers of seedlings can be genotyped, and only those that have inherited the marker grown up, saving a lot of time and resources in plant breeding.

SNP databases

The number, diversity, and size of SNP databases is growing, and it would be pointless to provide a list here of even the major ones. Instead, I will briefly describe a flagship SNP database, a human haplotype map known as HapMap[3]. Unlike biobanks (**TechBox 3.2**, p. 80), no data on individuals were collected with the samples, so the data from HapMap cannot in itself be used to identify disease genes[4]. The goal of the HapMap project was to provide a comprehensive and useful set of genomic markers that any researchers can use to screen genetic data. HapMap data are freely available in a number of formats, and has a useful website with information, analysis tools, and tutorials (www.hapmap.org).

The HapMap international consortium aimed to identify common haplotypes found in all human populations by analysing DNA from 269 people from four distinct populations: Japanese from Tokyo, Yoruba from Ibadan in Nigeria, Han Chinese from Beijing, and people

Figure TB3.3 A Yoruba man and his daughter. Yoruba people from Ibadan in Nigeria are one of four source populations for the human HapMap database. But the people in this photograph illustrate the complicated nature of human population genetics: the man is from Lagos, not Ibadan, and he now lives in London. His daughter has an Australian mother who herself has one Dutch and one Sicilian parent. Which set of SNPs would be an appropriate reference for either of these people? By using four disparate human populations, the HapMap project hopes to identify haplotypes common to all populations.

Reproduced by permission of photographer, Marcel Cardillo.

of European descent from Utah, USA. But it is also clear that human populations differ in the presence or frequency of many haplotypes. This means that while many SNPs appear to be informative across many ethnic groups, others are specific to particular populations (SNPs found only in one population often go by the somewhat dubious name of 'private alleles').

SNP databases are being developed with the hope that they will dramatically reduce the time and effort required to identify genes that contribute to disease, not only in humans but in livestock, companion animals, and crops. However the human databases generate complex ethical issues. As these databases expand and multiply, it may become necessary to rigorously protect the identity of individuals. Even though no information about donors is given in HapMap, it seems possible that relatively small amounts of genetic information can be used to identify an individual on a database, then gain the entire publicly available genomic information for that individual[5]. Because of the complex issues surrounding public use of genetic data, the HapMap project involved substantial input from ethicists at every stage of its planning and enactment[4].

References

1. Gibson, G. and Muse, S.V. (2004) *A Primer Of Genome Science*, 2nd edn, Sinauer Associates.
2. Gibbs, R. (2005) Deeper into the genome. *Nature*, Volume 437, page 1233.
3. The International HapMap Consortium (2005) A haplotype map of the human genome. *Nature*, Volume 437, pages 1299–1320.
4. The International HapMap Consortium (2005) Integrating ethics and science in the International HapMap Project. *Nature Reviews Genetics*, Volume 5, pages 467–475.
5. Lin, Z., Owen, A.B. and Altman, R.B. (2004) Genomic research and human subject privacy. *Science*, Volume 305, page 183.

TECHBOX
3.4

BLAST

KEYWORDS

GenBank

alignment

database

homology

identity

FURTHER INFORMATION

There are many online tutorials (see links from NCBI webpage). Korf, I., Yandell, M. and Bedell, J. (2003) *BLAST*. O'Reilly & Associates covers practical and theoretical aspects of blast searching from the group up. It is available as a hard-copy book or online, *http://proquest.safaribooksonline.com/0596002998*

RELATED TECHBOXES

TB 1.1: GenBank

TB 6.3: Multiple sequence alignment

RELATED CASE STUDIES

CS 3.1: Viruses within (identifying genes in whole genome sequences)

CS 5.1: Sweet and sour (identifying genes in whole genome sequences)

The bioinformatics industry is built on BLAST. Technically, BLAST stands for Basic Local Alignment Search Tool, though the name was originally intended as a play on the name of an earlier search method called FASTA[1] (so BLAST is an example of a common phenomenon in bioinformatics programming: think of acronym first then come up with an name that fits it). The scientific paper describing the BLAST algorithm was one of the most highly cited papers in the 1990s, now with over 22,000 citations in the scientific literature[2] (and a follow-up paper with improved methods has over 19,000 citations[3]). BLAST has now been modified, extended, and diversified into a great variety of flavours by a large number of scientists and programmers.

BLAST is used when you wish to take a particular sequence and identify the closest possible match to it in a sequence database. Essentially, this is done by making a large number of pairwise alignments: the query sequence is aligned against a database of sequences, and the highest-scoring alignments are reported (see **TechBox 6.3**). The main problem with this approach is that the databases are huge: it would be impossible to provide an exhaustive alignment search against the millions of DNA sequences currently in GenBank in any reasonable timeframe. The breakthrough with BLAST was its speed: it is a heuristic search, which means it doesn't look at all possible alignments so it can't be guaranteed to find the absolute best match. But in practice, BLAST performs well and returns results almost immediately.

Rather than trying to align the whole query sequence to the database in one go, BLAST works by initially matching short segments of the query sequence against the database (hence the 'local alignment' in the acronym). Imagine that you lined up all of the sequences in GenBank. Then you cut your query sequence up into short fragments (say, 11 bases long) and you slid each fragment along the billion nucleotides of sequence in GenBank until you found a match between the fragment and a sequence somewhere in the database. Then, each time you got a hit, you tried to extend that match by seeing if the query sequence on either side of the 11bp fragment also matches to that sequence in the database. If you can find matches either side of the query fragment, then you have increased the chances that you have found an informative match. You can then use an alignment algorithm to score the match of the whole query sequence to that particular database sequence. Repeat the process until you have made a thorough search of the database, then rank all of the good matches by their alignment scores.

When you blast you can choose whether you want to match a DNA sequence to the nucleotide database (e.g. BLASTn), an amino acid sequence to the protein database (e.g. BLASTp), or even to match a protein sequence to the nucleotide database (e.g. tBLASTn). Search strategies are becoming more sophisticated. For example, PSI-BLAST uses an iterative procedure to refine its search: after the first round of BLAST, positions in the matched pairs are scored according to how conserved the nucleotide sequence is across all pairs. These scores are used to generate a profile that informs a second round of BLAST searching. This iterative procedure that favours conserved positions makes PSI-BLAST good for finding divergent members of protein families.

Figure TB3.4 The quagga is an extinct subspecies of zebra. Like other zebras, the quagga could produce occasional hybrids with horses (zebra hybrids are given the sci-fi-like name of zebroids). A breeding programme was set up to resurrect the quagga by selectively breeding Plains Zebras for quagga-like coat patterning. Of course, this creates a simulcrum of the lost subspecies, but does not retrieve the quagga genome. The quagga was the subject of one of the first ancient DNA studies (see **TechBox 7.1**).

BLAST hits are reported with E-values that reflect the probability that the observed match is due to chance alone and therefore does not reflect a meaningful association between the sequences. It is similar to the *P* value reported for many statistical tests: the lower the E-value, the more likely it is that the sequences match due to their evolutionary relationship, rather than similarity by chance. The probability of a random match decreases with increasing length of the query sequence. A four-base query sequence, say AGTC, will perfectly match to very many sequences, because there is a fairly high chance of those four bases occurring together by chance. But any particular 25 base sequence has much less chance of occurring in unrelated sequences: for example, the nucleotide sequence CGTAGGGGTCAACATAATTTTCTTC matches exactly to only one sequence in the whole of GenBank (the cytochrome oxidase gene of a quagga, *Equus quagga*: **Figure TB3.4**).

People running large numbers of searches using vast quantities of data often put a lot of time and effort into fine-tuning BLAST (or other search algorithms) to optimize speed or performance for particular types of searches. As with most bioinformatics programs, you could just use the default parameters, opening the browser window, pasting your sequence into the box, and hitting 'Blast!' without worrying about the many different options on the page. But if you take the time to learn about the options, and compare the output of searches using different parameters, you may get some interesting results (and possibly get hooked on blasting, which keeps many bioinformaticians amused for hours).

References

1. Altschul, S.F. (2003) Foreword. In *BLAST*, eds Korf, I., Yandell, M. and Bedell, J., O'Reilly & Associates.
2. Altschul, S.F., Gish, W., Miller, W., Myers, E.W. and Lipman, D.J. (1990) Basic local alignment search tool. *Journal of Molecular Biology*, Volume 215, pages 403–410.
3. Altschul, S.F. *et al*. (1997) Gapped BLAST and PSI-BLAST: a new generation of protein database search programs. *Nucleic Acids Research*, Volume 25, pages 3389–3402.

where a serine should be. Since that SNP was not found in unaffected family members, it is a strong candidate for causing ASPS. But other cases of genetic causality are less clear cut. Another SNP found to be significantly associated with morningness occurs in the upstream regulatory regions of the *hPer2* gene, rather than in the protein-coding part of the gene. However, of the 210 people included in that particular study, 14% of morning people had the upstream SNP, compared to only 3% of night-owls, and 6% of intermediate ('normal') people. So while this single nucleotide change seems to be associated with differences in daily rhythms, it would be problematic to say that it 'caused' morningness, as most morning people did not have the SNP, and some evening people did (see **Case Study 3.2**, p. 83).

Secondly, and far more commonly, a SNP may be located in a non-coding part of the genome, but close enough to the gene of interest that it is usually inherited along with a particular version (allele) of the gene. In this case, the SNP itself may have no effect on its carrier, but it can serve as a marker that indicates the presence of the disease-causing allele. SNPs are particularly useful as genomic markers because they are so abundant. On average, there is at least one SNPs in every thousand bases of DNA in the human genome, so there is a high likelihood of finding one near each gene. It is because of the usefulness of SNPs that there are many international consortia developing databases of SNPs for humans and other species (see **TechBoxes 3.2** and **3.3**, pp. 80 and 90).

Mutation rate

The genome is constantly bombarded by mutagenic agents that cause DNA damage, and every time the genome is replicated there is a chance of acquiring copy errors. How frequently do mutations occur? We have defined mutation as a permanent change to the genome that will be included in any copies made of that genome. To measure the mutation rate, we need to observe the accumulation of changes in the genome over many generations. In some short-lived organisms that can be easily kept in laboratory conditions, it is possible to measure the mutation rate directly. One classic way of measuring mutation rate is to grow bacteria on a medium containing a food source that they cannot readily metabolize, then wait for a mutation to occur that allows its lucky carrier, and all its descendants, to utilize the new food source, giving rise to a growing bacterial colony (**Figure 3.20**, p. 96). In fact, bacterial cultures are routinely used to judge whether particular chemicals are mutagenic (and therefore pose a danger to human health), by looking for an increase in the rate at which bacteria produce offspring that can metabolize something their parents can't.

But our ability to measure mutation rates directly is limited in two ways. Firstly, the direct approach cannot be applied to the majority of organisms because they cannot be grown in the lab or do not have sufficiently short generation times to allow experiments to be conducted in a reasonable timeframe. Secondly, direct measurement of mutation rates by looking for novel traits in bacteria can only detect a very small fraction of mutations that occur in the genome: those that have a measurable and beneficial effect, allowing the bacteria to thrive in the specific test environment. For these

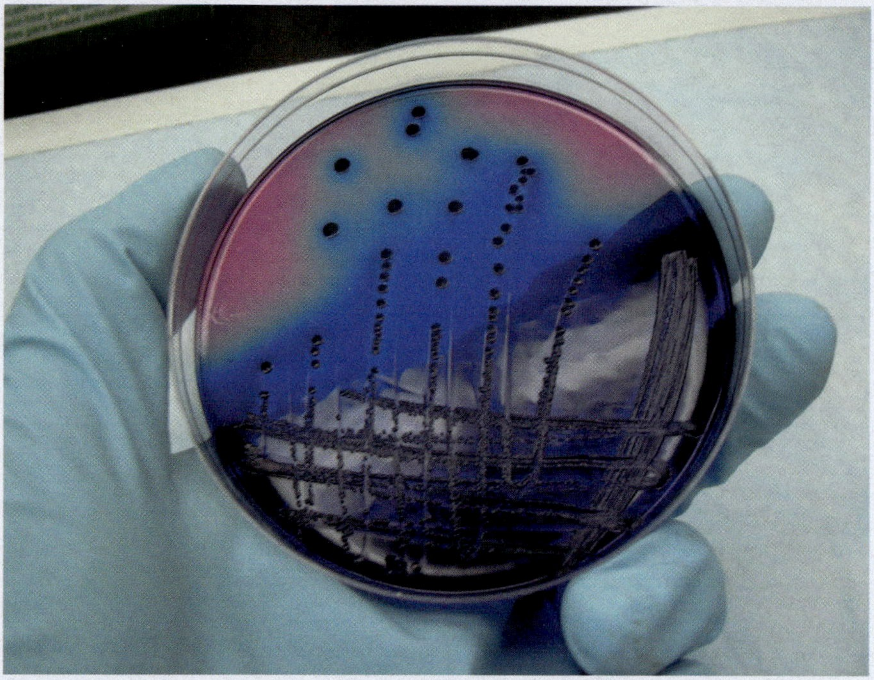

Figure 3.20 The bacterium *Escherichia coli* is the workhorse of mutation studies. The potentially mutagenic effects of environmental agents, such as new commercially produced chemicals, are often investigated by measuring the rate of mutation reversal in *E. coli* colonies when exposed to that agent.

Reproduced by courtesy of Stephen Hill.

reasons, most estimates of mutation rate are made by comparing the genomes from populations or species that have been separated over evolutionary time for hundreds of thousands of generations.

Why are useful mutations so rare? We have discussed how organized and complex the genomic information system is (Chapter 2). So when considering mutation we can think about what happens to an organized, complex, information-rich system when it is randomly altered. Here's an experiment to try at home: put on a blindfold, open up your computer and make a random change to its internal wiring. If I was to ask you to place a bet on one of three possible outcomes of this random rewiring – the computer will still work, the computer will no longer work, or the computer will run better than ever – which would you choose? I would guess that, unless you are particularly lucky (or well-practised at blind computer repairs), you stand a very good chance of ruining your computer with a random change. The same basic principle applies to genetic information. Genes (and their associated regulatory elements) are highly organized and finely tuned to produce the right product in the right place at the right time. Just like a random change to a computer, a random change will almost certainly wreck a gene. So the majority of mutations in DNA sequences that hold important information will be deleterious: they will decrease the chance of that organism surviving. Mutations in non-important sequences are unlikely to be deleterious, because the organism is unaffected by such changes, but by the same token they are also unlikely to be beneficial.

The disastrous nature of most mutations, and the sophistication of the DNA repair mechanisms in even the simplest cells, raises an important question: if natural selection against deleterious mutations drives the evolution of such a complex and finely tuned repair system, then why doesn't that selection pressure result in a perfect repair system that always returned a damaged sequence to its original state, without reducing replication time? After all, mutation is costly, because many mutations reduce that individual's chances of having great-great-grandchildren. Shouldn't all organisms have the lowest possible mutation rate? To explore this paradox, let's consider two possible strategies for perfecting the repair system: reduction of total mutation rate, and directing mutations towards useful changes.

Chapter 8 covers some of the factors that influence mutation rate in different species

Mutation rates are a compromise

Reduction in the rate of mutations is costly in terms of time and energy. Any increase in repair efficiency must be paid for out of a cell's total metabolic budget: it takes energy and resources to make and operate repair enzymes. Furthermore, repair takes time: copying accurately is slower than copying with less regard for precision. No genome has unlimited resources. No genome can take an infinite amount of time to copy. Therefore, at some point, the amount of resources used or time taken for repairs will outweigh the benefit gained from higher accuracy.

It may be helpful to consider the parable of the snail and the parasite. Snails have shells to protect their soft, vulnerable bodies from predator attack. A shell that is too thin will not save its owner from predation, so alleles for soft-shelledness are less likely to be passed on to the next generation. You may think that selection will tend to favour the thickest shell that the snail can carry. Yet this is clearly not the case, because snails infected with a certain parasite grow thicker shells than normal. Many parasites influence their host's morphology or behaviour in order to enhance their own chance of transmission, so it seems possible that the thicker shells are a results of the parasite's machivellian developmental manipulation. Why would a parasite make its snail host grow a thicker shell? Because if the snail gets eaten, so does the parasite. The parasite benefits from more robust predation-protection just as much as the snail does. So if the snail is capable of growing a thicker shell when infected, then why does it not always do so? Surely natural selection should maximize protection from predators to increase the snail's chances of survival? A snail that spends all of its metabolic resources on growing a shell may have little left over for investing in reproduction. Natural selection cannot favour snails that survive predation but do not reproduce, because, by definition, there will be no snail offspring to inherit the trait. The economy of nature must balance the costs and benefits of predation protection in the currency of reproductive output.

Economic considerations must also be applied to DNA repair. In the next chapter, we will look at the mechanisms employed to reduce mutations created by mistakes in DNA replication. One of these strategies is proofreading: the polymerase enzyme that copies DNA can check the accuracy of newly added bases and move 'backward' to delete them if they are incorrect. So any polymerase must balance 'forward' replication activity that adds nucleotides to the new DNA strand, with 'backward' proofreading activity that removes incorrect bases (Figure 4.12). Mutations in the polymerase gene can change the balance between replication and proofreading, favouring copy accuracy at the expense of replication speed, or favouring speed over accuracy. For example, bacteria put in an experimental environment that favours fast reproduction can evolve a higher mutation rate, because individuals with a 'mutator' polymerase that copies more quickly by proofreading less can have a reproductive advantage over individuals with a 'antimutator' polymerase that copies more slowly by proofreading more. In this particular case, it appears that the benefit of fast replication outweighs the cost of acquiring more mutations. It is also possible that selection for rapid adaptation could alter the balance between the accuracy and speed of DNA replication. For example, it has been suggested that bacteria with a mutator polymerase are more likely to evolve antibiotic resistance. Because they copy fast with lots of errors, mutator bacteria have a greater chance of accidentally producing a mutation that confers resistance. However, although mutation is essential for adaptation, it carries a high cost. Mutators may, in certain circumstances, produce more 'winners' (advantageous mutations), but it will be at the expense of producing an awful lot of 'losers' (deleterious mutations).

Is mutation random?

A random process is one that is not directed at a particular outcome. Another way to state this is that a random process is unpredictable: we may be able to describe the average result of a series of random events, but we cannot predict the exact result of any one event. The classic example is tossing a coin. Overall, we expect as many tails as heads, but on any given coin toss, we cannot predict whether we will get a head or a tail. You will often read in evolution textbooks that mutation is random. It is important to think about this and ask 'random with respect to what?'.

The influence of chance events on molecular evolution will be discussed again in Chapter 5

The types of changes generated by mutation are not entirely random, because some mutations are more likely to occur than others. There are three important sources of bias (non-randomness) in mutations. Firstly, a genome can only step from where it is now. This arises as a logical consequence of the continuity of genetic information. The genome of the offspring is derived from that of its parents. Species evolve from existing species and are not created *de novo*. By definition, a mutation can create a genome that is one accidental change away from its parent genome. So the set of possible mutations for a genome is defined by the set of sequences that are one change (such as a chromosomal inversion, a genome duplication, a single nucleotide change) away from the genome as it is now.

Secondly, some genomic loci are more likely to mutate than others. We have already seen how repeat sequences are prone to increases in copy number. The *Huntington disease* gene is more likely to mutate than many other human genes because it contains an unstable repeat region. Point mutations are also more likely to occur at certain places in the genome, referred to as mutational 'hotspots'. When studies are made of the frequency of mutations along a particular gene, it is often found that a handful of sites account for most of the mutations. The positional bias in mutation can occur at a very fine scale. In many eukaryote genomes, cytosines (C) are far more likely to mutate if they are next to a guanine (G). This is because cytosines tend to be methylated when they occur in a CG dinucleotide, and methylated Cs are prone to spontaneous deamination, where the amide group (NH_2) attached to the base is replaced with an oxygen molecule. When methylated cytosine is deaminated, it turns into thymine (T). So C is more likely to mutate to T when it is sitting next to a G. Similarly, a thymine is more likely to undergo UV-induced mutation if it is next to another T than if it is next to a G, because the TT dinucleotide can form a thymine dimer (Figure 3.19, p. 88).

Thirdly, some kinds of mutation are more likely to occur than others. For example, deletions are typically more common than insertions. Some base changes are more likely to occur than others. A pyrimidine (a one-ring base, T or C) is more likely to mutate to the other pyrimidine than it is to be exchanged for a purine (two-ringed base, A or G). Similarly, a purine will more often be mutated to the other purine than to a pyrimidine. A change from one purine to the other purine, or from one pyrimidine to the other, is referred to as a transition. Transitions are usually far more frequent than transversions (changes from a purine to pyrimidine or vice versa). This may be partly due to the many mutation pathways that create specific transitions (e.g. deamination of C produces T).

Purines and pyrimidines are explained in TechBox 2.2

So the types and locations of mutations are not entirely random: some mutations are more likely to occur than others. However, there is one very important sense in which mutations are considered to be random: mutations are random with respect to fitness. Another way of saying this is that mutation is not directed towards a particular outcome. As a general rule, mutations should be considered to be accidents. They arise without being designed for a purpose, in the same sense that accidentally breaking your coffee cup is not 'for' anything, it just happens.

Fitness is discussed in TechBox 5.1

If beneficial mutations contribute to adaptation, and deleterious mutations result in a reduced chance of leaving descendants, then surely there must be massive selection pressure to only produce good mutations? Wouldn't any genome that could decrease the number of deleterious mutations be selected for? Possibly so. In the late 1980s, there was a flurry of excitement when experiments were reported that appeared to show that *Escherichia coli* bacteria could preferentially generate mutations that would rescue them from a life-threatening situation. *E. coli* can normally use lactose as an energy source by means of the enzyme lactase, but these experiments involved growing lactase-deficient *E. coli* (Lac^-) on a lactose-only medium. These Lac^- bacteria would eventually starve if left on a lactose medium that they couldn't metabolize. After some time, colonies of bacteria arose that could grow on lactose, thanks to a reversal mutation in the lactase gene. The lactase-enabling mutations appeared to occur at a higher rate than expected by chance, leading to the hypothesis that the bacteria could somehow direct the process of genetic change to just the right genes that would rescue them from starvation. Because it challenged a long-held tenet of evolutionary biology – that mutation is random with respect to fitness – the argument over directed mutation was often heated. Indeed, the debate got so acrimonious that the mere mention of the phrase 'directed mutation' was enough to get blood boiling. One researcher suggested that, to allow the debate to proceed in a civilized fashion, the phrase 'directed mutation' should be replaced by something neutral, like 'Fred'. So for a while the debate continued concerning the role of 'Fred' in the molecular evolution of bacteria.

But the debate has become much quieter and, as yet, no concrete evidence has been put forward that beneficial mutations are more likely to happen than expected from random permutations of existing DNA sequences. Note that, although random mutation is often considered a pillar of evolutionary theory, it need not necessarily be true for natural selection to work. In fact, selection would be more efficient if mutation was biased toward useful changes. But, as Theodosius Dobzhansky (one of the great founding fathers of genetics) noted in the quote at the beginning of this chapter, unfortunately for real organisms, it does appear to be the case that mutation is random with respect to fitness.

Conclusions

On the one hand, mutation is rare. A barrage of DNA repair pathways ensure that most damage to DNA is corrected before the genome is replicated, and that most errors made in DNA copying are corrected (see Chapter 4). On the other hand, mutation is very common, for although mutation rates are very low, genomes are constantly subject to damage. Every nucleotide of the genome represents an opportunity for mutation to occur. The size of the genome, and its constant replication and activity, makes mutations inevitable and common. Not all of these mutations end up in successful offspring, but some do. Studies suggest that there at least 100 new mutations in the genome of every person. Some sites, particularly repeat sequences, are sufficiently mutable that we can expect to find a different set of mutations in every individual in a population. In this way, mutation makes each one of us unique.

Without mutation there could be no evolution, because each genome would be exactly the same as all other genomes. Yet, although mutation is necessary for evolution, in most cases the results of mutation are either disastrous or unnoticeable. Very rarely does mutation produce a desirable feature. All your children will be mutants, but it is, alas, unlikely that these mutations will confer X-men like superpowers on them. However, mutation does incidentally create something that is very useful for biologists: it creates a genome that is a storehouse of information. Mutations make genomes unique, so that we can use genetic markers to identify individuals. Because these mutations are inherited, related individuals are more similar than random members of a population, so we can use DNA to describe family relationships, and identify members of interbreeding populations. Conversely, we can use known family relationships to find genetic markers that are associated with important traits, such as inherited disease risk, and use those markers to track down the genes responsible. In order to understand how to track descent with DNA, we need to look at how mutations are copied from the genome they occur in then passed on to that genome's descendants. So the next topic we need to cover is perhaps the most important topic in the whole of biology: DNA replication.

Further information

A popular account of the processes and consequences of mutation is given in:

Leroi, A.M. (2004) *Mutants: on genetic varieties and the human body*. HarperCollins.

Mark Ridley's reader in evolution (not to be confused with his popular textbook of the same name) is a fabulous collection of writings by leading evolutionary biologists on a wide range of topics, including several papers on mutation:

Ridley, M., ed. (2004) *Evolution*. Oxford Readers Series, Oxford University Press.

This is a fantastic book on molecular evolution, including patterns and consequences of mutation:

Lynch, M. (2007) *The Origins of Genome Architecture*. Sinauer Associates.

Endless copies

4

Or: how do I amplify DNA?

"If the account of heredity . . . is correct, it follows that the whole pageant of evolution since pre-Cambrian times – ammonites, dinosaurs, pterodactyls, mammoths, and man himself – is merely a reflection of changed sequences of bases in nucleic acid molecules. What is transmitted from one generation to another is not the form and substance of a pterodactyl or a mammoth, but primarily the capacity to synthesize particular proteins. The development of specific form is a consequence of this capacity, and the capacity itself depends on the self-replication properties of DNA."

John Maynard Smith (1958) *The Theory of Evolution*. Penguin

What this chapter is about

The evolution of life depends on hereditary information being copied from one generation to the next. A basic grasp of DNA replication is therefore important to anyone wishing to understand evolution. Furthermore, familiarity with the processes of DNA replication is the key to understanding many molecular techniques. In particular, DNA amplification (making millions of copies of a DNA sequence in the laboratory) relies upon the domestication of the DNA-copying processes that occur in living cells. Understanding DNA replication is also central to appreciating the nature of biological information stored in DNA. DNA replication creates a nested hierarchy of differences between genomes that reveals the relationships between organisms and the processes of evolution.

Key concepts

Evolutionary biology: descent creates a hierarchy of similarities

Molecular evolution: DNA replication

Techniques: DNA amplification

Blame your father

In the previous chapter, we learned that we are all mutants. Each of us begins life with dozens of new mutations, most of which occurred in our parents' bodies, in the cells that divided again and again to make sperm or eggs. You might assume that, as you inherit half your DNA from your mother and half from your father, these new mutations had an equal chance of coming from either parent. But this is not the case. Most of the new mutations in your genome came from your father, not your mother.

As we saw in Chapter 3, some mutations are caused by damage to the DNA, biochemical accidents that muddle the order or identity of the nucleotide sequence, causing a permanent change to the information in the genome. But a large fraction of the mutations that occur every generation arise not from incidental damage due to mutagens, but to mistakes made when copying DNA. DNA replication is astoundingly accurate, but it is not perfect. The error rate varies between species (and even between enzymes within a species). In mammals, the DNA copy error rate is typically around one in a billion – that is, for every billion bases of DNA copied, there will usually only be one mistake. But, given that the human genome is more than three billion nucleotides long, this is still enough to create several new mutations every time the genome is copied. And the human genome is copied many times per individual generation, as cells divide and diverge to form the germ cells that will carry a copy of the genome to a new generation.

DNA is copied almost (but not entirely) without error, and most (but not all) mutations are repaired. So your DNA is almost, but not exactly, the same as your parents'. When you copy your DNA to give to your own offspring, new mutations will be added to the ones you collected, and more still when your children copy their genomes to produce your grandchildren, and so on. Every time DNA is copied, there is a chance for copy errors to occur. So the more times the genome is copied, the more mutations should accumulate. And this is why you can blame your father for most of the new mutations you have (since advantageous mutations are rare, I am assuming most of these mutations have sadly not resulted in improvements for which you wish to thank your father).

For an explanation of why most mutations are deleterious, see Chapter 3

'Male-driven evolution' arises from the different ways that sperm and eggs are produced. In mammals, like yourself, eggs are formed by a process of symmetrical cell divisions. A germ cell divides, then each of the daughter cells divides again, and so on until all the eggs (and their accompanying cells) are produced (**Figure 4.1**). The egg cells are all produced before sexual maturity is reached. This means that each egg is formed from approximately the same number of cell divisions. But sperm cells are produced by an asymmetrical pattern of cell divisions. When each germ cell divides in two, one daughter cell goes on to produce sperm cells, and one goes back into the pool of germ cells to divide again another day, when once again it will divide to produce one cell fated for sperm production and another to go back into the pool of germ cells. Unlike eggs, sperm cells are produced continuously in adult males, so the pool of germ cells is constantly dividing and redividing.

Figure 4.1

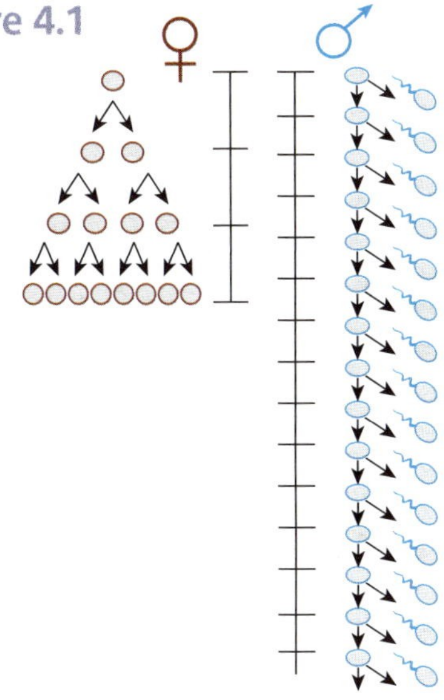

The pattern of sperm cell formation has two important consequences. Firstly, it takes far more cell divisions to produce sperm than eggs. This means that there are more opportunities for copy errors to occur in the manufacture of sperm, simply because the genome is copied more times to produce each sperm cell than it is to produce each egg (**Figure 4.2**). Secondly, the more sperm are produced, the more cell divisions the germ cell line undergoes. This means that, as a male ages, he accumulates ever more mutations in his germ cell line, and these mutations are copied to the newly formed sperm cells. The older the father, the more mutations his sperm are likely to carry.

Figure 4.2

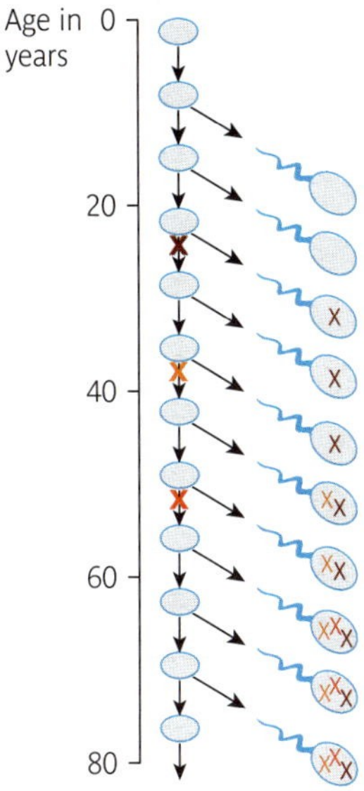

The example of male-driven evolution is not included to make older fathers feel nervous, but to illustrate an important point. The fact that the mutation rate is measurably higher in males than in females demonstrates that DNA copy errors are a major component of the mutations that accumulate every generation. The more the genome is copied, the more chances there are for copy errors to accumulate (**Figure 4.3**, p. 104). And the nature of DNA replication is such that once a mutation occurs, through damage or copy errors, it will then be faithfully copied and distributed to offspring produced from that genome.

In this chapter we are going to start with an overview of DNA replication, then see how the process of DNA replication has been domesticated to provide a means of copying DNA in the laboratory. Then we will take a closer look at the mechanisms that make DNA replication so phenomenally accurate, in order to understand how the occasional copy error slips through the net to create a new mutation. Finally we are going to step back and see the bigger picture: how the process of copying DNA, generation after generation, leads to the accumulation of a hierarchy of genetic differences between individuals, populations, and species. It is this hierarchy of changes that makes DNA such a useful tool for biologists.

Figure 4.3 Male (left) and female of the common eider (*Somateria mollissima*), which breeds in the Artic, keep its chicks warm by lining its nest with soft down (which can also be used to keep humans warm when collected and made into eiderdown quilts). In birds, as in mammals, most heritable mutations occur in males, because it takes far more cell generations to make sperm than to make eggs. Male-driven evolution has been demonstrated in mammals by showing that sequences on Y chromosomes (present for all of their time in males) have faster rates of molecular evolution than sequences on autosomes (present for equal time in males and females), and autosomes have faster rates than sequences on the X chromosome (present for two-thirds of their time in females). However, since mammalian females have two copies of the X chromosome (XX), but males have only one (XY), sequence evolution on the X chromosomes may be slow due to the removal of recessive mutations expressed in hemizygous males. The detection of male-driven evolution in birds proves that it is not simply an artefact of slower rates of molecular evolution in the mammalian X chromosome, because in birds, the situation is reversed: females are hemizygous (WZ) but males are homozygous (ZZ) for the sex chromosomes. Sure enough, Z chromosomes (present for more time in males) have a faster rate of molecular evolution than autosomes, which are faster than W chromosomes (present for all their time in females).

Template reproduction

The discovery that DNA is the hereditary material came about when scientists demonstrated that DNA, not proteins, carries genetic information from one individual to another. Proteins had been considered the natural candidate for the hereditary material because they were made of 20 different amino acids, giving a generous 'alphabet' for information storage. But information storage alone cannot provide a basis for heredity. The information must be able to be copied from one generation to the next, and there was no obvious way to copy the information in proteins. When James Watson and Francis Crick revealed the structure of DNA, by fitting together models of the molecular subunits of ribose, phosphates, and bases, they knew they had found 'the secret of life' because the complementary arrangement of bases provided a way of copying information (**Heroes 4**).

Watson and Crick's discovery of the structure of DNA is covered in Chapter 2

The base-pairing rules of DNA (C ≡ G, A = T) provide a mechanism for the transfer of genetic information through template copying. Any strand of nucleic acid (say, GATCC) can be used as a template to make a complementary strand (CTAGG). Then if you repeated the process, using that complementary strand as a template to make a new strand, you would recover the sequence of the original strand (GATCC). If you repeated the process again and again, using a nucleic acid strand as a template to make a complementary strand, then using the complementary strand to make another copy, you would keep producing the same nucleotide sequence (**Figure 4.4**, p. 106). So the

HEROES OF THE GENETIC REVOLUTION

4

Francis Crick

"I will always remember Francis for his extraordinarily focused intelligence. . . . Being with him for two years in a small room in Cambridge was truly a privilege."

James Watson (2004) http://news.bbc.co.uk/1/hi/sci/tech/3937475.stm

NAME

Francis Harry Compton Crick

BORN

8 June 1916, Weston Favell, Northamptonshire, UK

DIED

28 July 2004, San Diego, USA

KEY PUBLICATIONS

Watson, J.D. and Crick, F.H.C. (1953) Molecular structure of nucleic acids. *Nature*, Volume 171, pages 737–738.

Crick, F.H.C., Barnett, L., Brenner, S. and Watts-Tobin, R.J. (1961) General nature of the genetic code for proteins. *Nature*, Volume 192, pages 1227–1232.

FURTHER INFORMATION

The Wellcome Trust holds the archive of Crick's papers, accessible from *http://genome.wellcome.ac.uk*

Figure Hero 4
Francis Crick.
Photographer: Marc Lieberman.

Although Crick is best known as one of the co-discoverers of the double-helix structure of DNA, for which he shared a Nobel Prize in 1962 with James Watson and Maurice Wilkins, his contributions to the development of molecular genetics were legion. He brought his training in physics to understanding biological systems, using X-ray crystallography to explore the structure of proteins. Crick's commitment to modelling biochemical structures, drawing together information from other scientists' experiments, provided the key to unlocking the secret of DNA. In their famous *Nature* paper of 1953, Francis Crick and James Watson demonstrated that DNA consisted of a double helix with interconnecting pairs of complementary bases. Rarely has a single scientific paper changed a whole field so rapidly. It was instantly clear that Watson and Crick had shown how DNA could carry genetic information through complementary base pairing. In a second *Nature* paper, published the following month, they showed how the double-helix structure provided a means of self-replication by template copying.

Crick took this discovery forward, proposing an 'adaptor hypothesis' to explain how the DNA code could be translated into proteins (later proved correct by the discovery of transfer RNA molecules). Working with other scientists at the Cavendish Laboratory in Cambridge, Crick conducted genetic experiments on viruses to demonstrate that the code was formed of triplets of nucleotides, and that the code was degenerate (many different triplets could code for the same amino acid). His pioneering work in molecular genetics led Crick to deduce the 'central dogma' of molecular biology – one-way transfer of information from nucleic acid to protein – which remains one of the key principles of modern evolutionary biology ('dogma' was intended to indicate central importance, rather than unchallengability).

Crick wrote on a wide variety of topics, from selfish DNA to directed panspermia (the hypothesis that life on Earth has an extraterrestrial origin). He devoted much of his research energy into understanding consciousness, exploring topics such as the function of dream sleep and the neural activity underlying visual awareness.

> "I think what led me into biological research was really because I felt there was a mystery which I thought [had] to be explained scientifically. And one of these areas was the borderline between the living and the nonliving and the other one was the problem of how the brain works – the problems of consciousness. Of course nowadays we call those areas molecular biology and neurobiology, but those terms weren't known at that time."
>
> Francis Crick (1989) www.accessexcellence.org/AE/AEC/CC/crick.html

information coded in the original strand is preserved through endless copies.

Figure 4.4

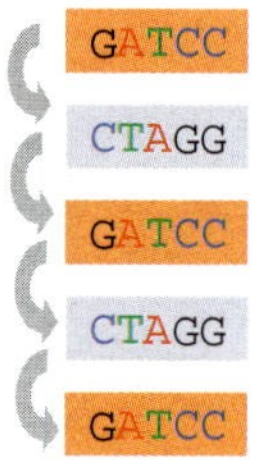

This process of template reproduction is the key to the evolution of life on Earth because it is the single known means of self-replication, the fundamental property of living systems. All life on Earth uses the same nucleic acid replication system: there are no known alternatives. If you want to understand life, and its evolution, the best place to start is to become familiar with the wonders of DNA replication. There is no better system for understanding why life exists and evolves.

Understanding the template reproduction of DNA is also the key to understanding the practical use of molecular data in biology, for two reasons. Firstly, most molecular genetic techniques are based upon the ability to make myriad copies of DNA sequences in the lab, which is achieved through the 'domestication' of the processes of DNA replication. Secondly, template copying is the key to understanding the evolutionary information stored in DNA. As DNA is copied from parent to offspring to grandchildren and great-grandchildren, it accumulates sequence changes that are inherited and copied down the generations. DNA replication creates a hierarchy of genetic differences: more closely related individuals have more similar genomes, more distantly related individuals have more differences between their genomes. Many molecular techniques in biology exploit this simple fact.

DNA replication

DNA replication involves the co-ordinated action of a vast number of enzymes, regulatory factors, and molecular building blocks. The exact details differ between species, but the basic processes of DNA replication are common to all organisms.

The entire genome must be copied every time a cell divides. So, prior to cell division, complexes of replication proteins form, which then bind to specific sites in the genome. Bacterial genomes, being small and circular, commonly have a single origin of replication. But the much larger eukaryote genomes, with multiple, linear chromosomes, must initiate replication at numerous sites throughout the genome, to allow the entire genome to be copied in a reasonable time (which, depending on how rapidly the cells are dividing, may range from a few minutes to a few hours). There can be many thousands of replication origins in a large genome. The co-ordination of replication origins is very important, to make sure the entire genome is copied before cell division, and no region is copied more than once (**TechBox 4.1**).

At the site of replication initiation, the DNA double helix is melted apart to form a replication bubble (**Figure 4.5**).

Figure 4.5

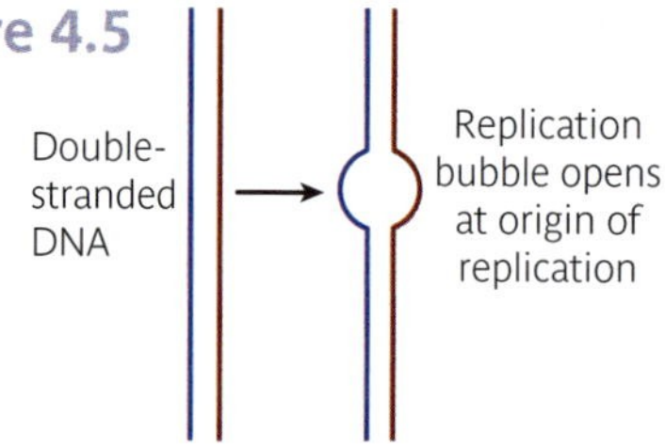

At the site of the replication bubble, the DNA helix is unwound by a suite of enzymes, separating the two complementary strands of the DNA helix to expose the sequence of nucleotide bases (**Figure 4.6**). This unzipped section of the DNA helix is referred to as the replication fork.

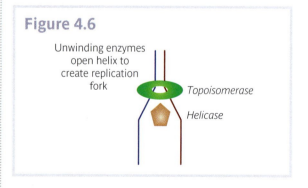

Figure 4.6

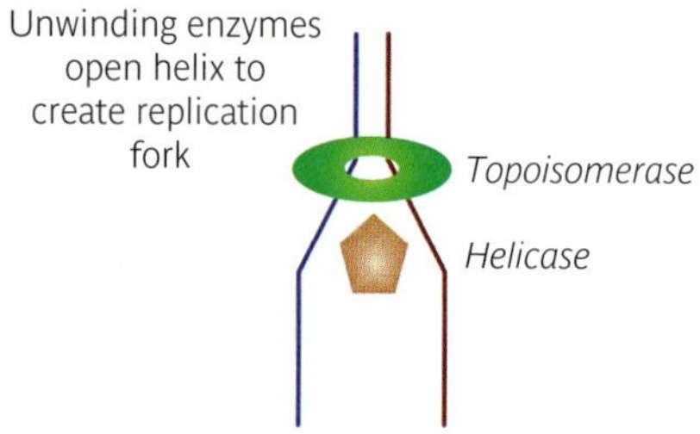

Each of the exposed strands forms a template for the construction of a new complementary strand. An enzyme with primase activity makes short RNA polynucleotides along the exposed single-stranded DNA. These RNA primers provide a starting block for DNA polymerase.

TECHBOX 4.1

DNA replication

KEYWORDS

primer
helicase
polymerase
melt
unwind
topoisomerase
replication fork
Okazaki fragment
5′→3′
primase

RELATED TECHBOXES

TB 2.2: DNA structure
TB 3.2: DNA amplification

RELATED CASE STUDIES

CS 4.2: Peopling the Pacific (copying processes)
CS 4.1: Glorious mud (DNA hybridization)

The details of DNA replication differ between organisms (most notably between prokaryotes and eukaryotes). But in all organisms, DNA replication involves the formation of a replication complex made of a large number of enzymes, which split the two strands of DNA and construct two new complementary strands to create two double-stranded helices from one. The basic series of events is:

1. Melt: strands are separated at the origin of replication
2. Unwind: replication forks expose single strands to act as templates
3. Prime: synthesis of a new strand begins with a short RNA strand
4. Grow: DNA polymerase adds nucleotides to form a complementary strand.

Since different species have different enzymes, here we will use enzyme names (ending in -ase, such as polymerase) to represent particular activities, which might be performed by different proteins in different organisms.

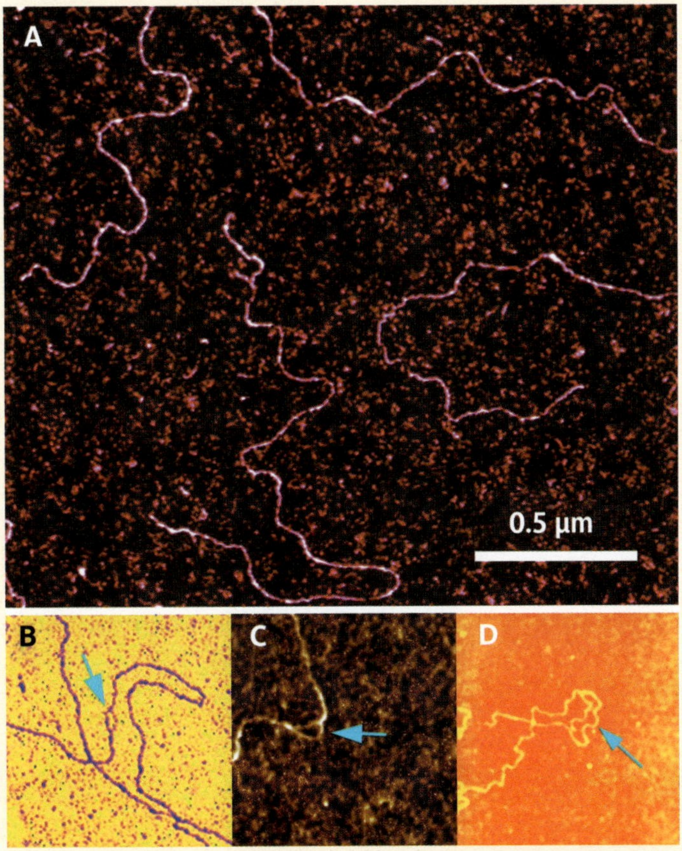

Figure TB4.1a Electron micrographs of replication bubbles in DNA from budding yeast.
Reproduced courtesy of Professor Zhifeng Shao, University of Virginia.

Melt: DNA replication occurs by template reproduction: the base sequence on one strand is used to create a new complementary strand. So to replicate DNA, the first step is to separate the two strands of the double helix so that each can act as a template for the production of a new strand. Replication origins are sequences in the genome (typically less than 250 bases in bacteria, but often much longer in eukaryotes) recognized by the enzymes that can locally destabilize the DNA helix to break the hydrogen bonds between base pairs. Then helicase 'melts' the double helix, separating base pairs to open up the two DNA strands to form a replication bubble (**Figure TB4.1a**).

Unwind: If you take a two-stranded rope or string, fixed at one end, and pull the strands apart, you will soon find that the rope snarls up and the strands cannot be separated. The same thing would happen to the DNA double helix if it were not for the unwinding enzymes (topoisomerases) that prevent the formation of positive supercoils as the strands are separated. Topoisomerases cut one or both DNA strands ahead of the replication fork, allow the strands to unwind, then rejoin the strands. In circular chromosomes, such as those found in mitochondria or bacteria, two replication forks move in opposite directions around the chromosome, until they meet on the other side of the chromosome, having made an entire copy of the genome.

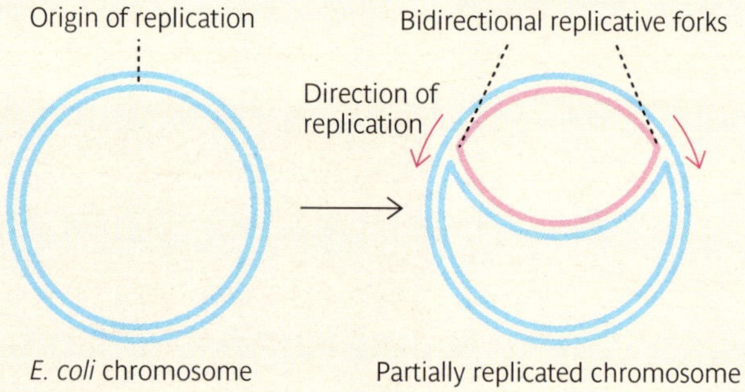

Figure TB4.1b

Linear eukaryotic chromosomes split at a numerous locations, then the replication forks move in both directions along the chromosome until they meet up.

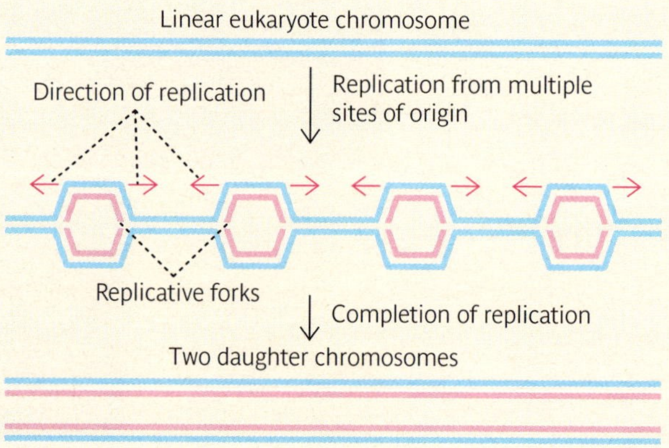

Figure TB4.1c

Prime and grow: DNA polymerase, the enzyme responsible for making new DNA strands, works by adding nucleotides to an existing polynucleotide chain: it forms a phosphodiester bond between the 3′ carbon of the sugar of the last nucleotide in the chain to the phosphate of the incoming nucleotide (**TechBox 2.2**). This mode of chain extension introduces two important limitations. If DNA polymerase adds nucleotides to the 3′ end of an existing nucleotide chain, then how does it get started? And how does it copy both the 5′→3′ and 3′→5′ strands?

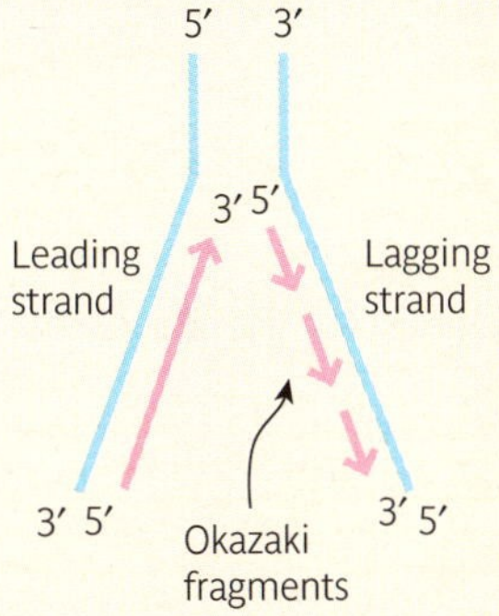

Figure TB4.1d

Unlike DNA polymerase, RNA polymerase is able to start from scratch, joining two nucleotides together to start a chain. So primase begins the DNA replication process by making a short RNA molecule (approximately 10 nucleotides long), matching nucleotides to the open replication fork. This short RNA strand acts as a primer to which DNA polymerase can then add more nucleotides. The RNA primer is later excised from the newly synthesized DNA strand.

Primase also plays a role in solving the problem of the directionality of DNA polymerase. Each DNA molecule consists of two antiparallel strands: the 'leading' strand, which runs 3′→5′, and the 'lagging' strand, which runs 5′→3′. DNA polymerase can only make a new strand in the 5′→3′ direction, by adding nucleotides to the 3′ carbon atom of the sugar of the previous base. So it can easily make a complementary (antiparallel) strand to the leading strand, moving along the exposed strand towards the replication fork, making a 5′→3′ strand to match the 3′→5′ leading strand. But DNA polymerase can't synthesize in the 3′→5′ direction to make the complementary strand to the lagging strand, because it can't add nucleotides to the 5′ end of the new chain. Instead, as the DNA unwinds, primase repeatedly places short RNA primers along the lagging strand, between 1000 and 2000 bases apart. DNA polymerase then uses these primers to make a series of short fragments, running in the 5′→3′ direction along the exposed strand, moving away from the replication fork. Then nuclease enzymes remove the RNA primers and the Okazaki fragments are joined together by ligase.

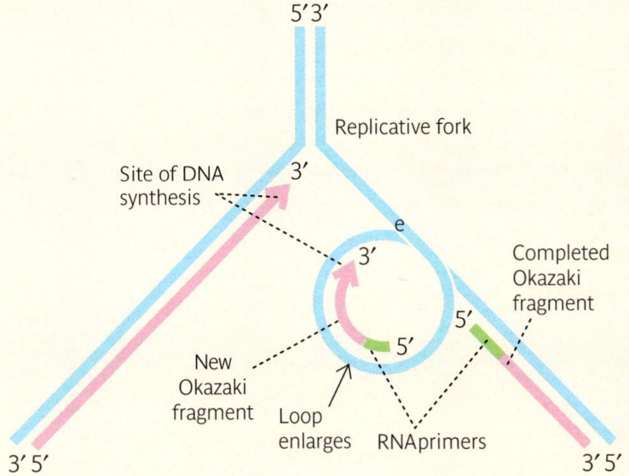

Figure TB4.1e Replication forks are usually drawn 'open-ended', such as in this diagram (with a handle and tines like an item of cutlery). In reality, DNA replication is usually bidirectional: rather than splitting at the end of a strand, a double helix is opened in the middle, and two replication forks move away from each other.

Starting at the end of the primer, DNA polymerase 'grows' a new complementary strand by adding nucleotides that pair with those on the existing strand, and binding them together along their phosphate–sugar backbone. DNA polymerase can only work in one direction (see **TechBox 4.1**), so it moves forward along one strand, creating continuous polynucleotide to match the leading strand, and backwards along the other strand, creating a series of short polynucleotides (Okazaki fragments) that are then glued together to make a continuous strand to match the lagging strand (**Figure 4.7**).

TechBox 4.1 explains why the leading strand is copied in one piece but the lagging strand is copied in many short fragments.

Figure 4.7

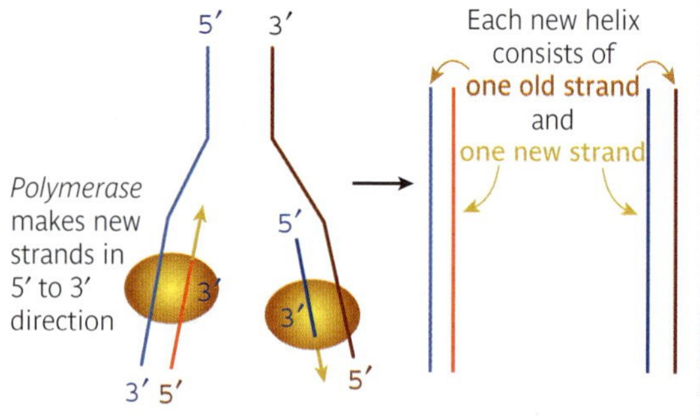
5′
3′
Each new helix consists of
one old strand
and
one new strand
Polymerase makes new strands in 5′ to 3′ direction
5′
3′
3′
3′ 5′
5′

The replication fork moves along the helix. Unwinding enzymes continue to untangle the DNA ahead of the replication fork, and the replication complex extends the newly created strands to match the unzipped DNA. The end result is two identical, double-stranded DNA molecules, each consisting of one strand of the original molecule and one newly synthesized strand.

Domesticating DNA replication

DNA replication is occurring constantly, in every cell every time it divides. Copying DNA is also an essential step in most molecular genetic techniques, which often rely on the ability to successfully amplify DNA from natural samples. For example, DNA sequencing requires a very large number of copies of the same DNA sequence (**TechBox 1.2**). But DNA amplification was not, strictly speaking, entirely a human invention (despite resulting in a particularly lucrative patent). Instead, it relies on the domestication of natural processes of DNA replication.

DNA replication *in vivo* (in living cells) is a very complex process. The replication machinery, consisting of dozens of enzymes all working together, must be formed at the right time, in the right places, to allow the entire genome to be copied in a co-ordinated fashion before cell division. Enzymes and cofactors must come together at the sites of replication origin to form the machinery that will melt open the double-stranded DNA to form replication bubbles, unwind the DNA helix to expose replication forks, make short DNA primers along the exposed strands, then grow new DNA strands to match both the leading and lagging strands, creating two double helices from one. This is far too complex a process to replicate in the laboratory.

DNA replication *in vitro* (in the laboratory) must deal with several important differences from the well-organized living cell. Firstly, rather than dealing with DNA that is packaged into chromosomes, the process of extracting DNA from biological samples (**TechBox 2.4**) usually results in the genome being chopped into pieces and mixed up. Secondly, it would be extremely tricky to get all of the correct enzymes working together *in vitro* with the right conditions and cofactors to recreate the complex replication machinery found in the living cell. So the DNA replication method adopted by a living cell – starting at defined origins of replication and moving replication forks along the double helix until the whole genome is copied – is not practical (or indeed possible) in a test tube.

Lab-based DNA amplification overcomes these problems by being much, much simpler. Instead of using dozens of enzymes working in concert, the most common techniques rely on essentially three factors to copy DNA: heat, primers, and polymerase. Instead of enzyme complexes creating replication forks at defined sites of replication, lab-based methods commonly take the brutally effective approach of heating the sample to break the hydrogen bonds between base pairs, melting the entire helix so that all of the DNA in the sample is converted to single strands. This bypasses the need to have enzyme complexes to initiate and maintain replication forks: instead, all DNA in the sample is instantly exposed as a template (for alternatives that don't use heat, see **TechBox 4.2**). Living cells begin DNA

TECHBOX
4.2

DNA amplification

KEYWORDS

primer

nucleotides

dNTP

polymerase

PCR

thermal cycling

strand-displacement amplification (SDA)

helicase-dependent amplification (HDA)

isothermal

FURTHER INFORMATION

Examples of protocols and guides to primer design can be found at Protocol-Online *http://www.protocol-online.org/prot/Molecular_Biology/PCR/*

RELATED TECHBOXES

TB 4.1: DNA replication

TB 1.2: DNA sequencing

RELATED CASE STUDIES

CS 2.1: On the origin of faeces (amplifying DNA from suboptimal samples)

CS 2.2: More moa (amplifying DNA from museum specimens)

> *“the truly astonishing thing about PCR is precisely that it wasn’t designed to solve a problem; once it existed, problems began to emerge to which it could be applied”*
>
> Stephen Scharf, quoted by P. Rabinow (1998) on http://sunsite3.berkeley.edu/PCR/whatisPCR.html

Because DNA amplification is a domesticated version of DNA replication, it is important to read through the DNA replication box first (**TechBox 4.1**). DNA amplification will be explained with respect to the processes that occur in normal DNA replication, because it follows the same basic pattern of melt/unwind, prime, and grow.

DNA amplification, like many lab procedures, is like following a recipe, with a set of ingredients and a series of ‘cooking’ instructions. The ingredients for most amplification methods are:

Primer: oligonucleotides (short strings of nucleotides) that are complementary to a part of the target sequence in the sample DNA, so they can provide a starting point for DNA polymerase (**TechBox 4.3**).

Nucleotides: the raw material for DNA synthesis is provided in the form of deoxyribonucleotides (dNTPs), each consisting of a phosphate plus sugar plus base. Four dNTPs need to be provided, one for each of the four ‘letters’ of DNA (i.e. dATP, dCTP, dGTP, and dTTP) (**TechBox 2.2**).

Polymerase: the enzyme that builds new DNA strands by matching nucleotides to the single-stranded DNA in solution.

Mg^{2+}: magnesium is an essential cofactor for polymerase function.

Buffer: provides the correct chemical environment for the DNA amplification reactions.

The cooking instructions for the most common technique for DNA amplification, polymerase chain reaction (PCR), are essentially multiple repeats of the following cycle (often up to 30 cycles):

1. Melt (heat to separate double helices into single strands)
2. Prime (cool so that the primers bind to the target sequences in the DNA)
3. Grow (polymerase adds nucleotides to make a complementary strand of DNA)
4. Melt again (so that all of the DNA in the sample, both original and newly copied, can be used as templates for the next round of prime and grow).

If you compare this to the DNA replication box you will see that ‘unwind’ is missing. That’s because in the standard DNA amplification procedures, heat is used to denature all DNA in the sample into single strands, so there are no helices to be unwound. However, there are new DNA amplification methods that use unwinding enzymes rather than heat to separate the strands: see below.

1. **Melt/unwind:** The most common DNA-amplification methods use heat to denature all the DNA in the sample. Because melting temperature depends on the strength of bonds between the DNA helices, it may vary between samples. For example, DNA with a high proportion of Gs and Cs generally requires higher melting temperatures, because there are three hydrogen bonds between each GC pair but only two for each AT pair. It may be necessary to adjust the temperatures to get a PCR reaction to work well for your samples.

Not all DNA amplification techniques require temperature cycling to expose single-stranded DNA templates. For example, strand-displacement amplification (SDA) uses a single round of heating with enzymes and primers to begin a reaction that then continues at low temperature: endonucleases cut completed strands which are then displaced by the polymerization of a new strand. Helicase-dependent amplification (HDA) avoids heating altogether by using a helicase enzyme to unwind DNA at low temperature (37°C). By circumventing the need for a thermal cycling machine, single-temperature (isothermal) amplification techniques may allow the development of mobile DNA-amplification kits that could be taken into the field or on doctors' rounds.

2. **Prime:** There are two roles to priming. One is to provide a starting block for polynucleotide synthesis, because DNA polymerase cannot begin a new chain from scratch (see **TechBox 4.1**, p. 107). The second is to specify the DNA sequence you wish to amplify. The success of a PCR reaction depends on designing short DNA sequences (each around 18–30 bases long) that will stick to the right sequence at the right time with the right strength (**TechBox 4.3**, p. 115). In fact, PCR usually uses two primers, one which defines the beginning of the sequence to be copied, and one which defines the end. This requires some knowledge of the sequence you wish to amplify. However, primer walking can be used to extend amplification into unknown territory: the PCR product from one reaction can be used to design primers to sequence an adjacent unknown sequence. Not all DNA amplification techniques use pairs of linear primers. For example, padlock probes have their recognition sequence split into two connected by a linking sequence, so when both ends hybridize to the target sequence the probe forms a circle. These circularized probes can be labelled for gene detection, or used to prime rolling-circle amplification.

3. **Grow:** Many different 'domesticated' polymerases are used to copy DNA sequences *in vitro*. All of them work by adding nucleotides to the 3′ end of a nucleotide chain (hence the need for a primer), matching the sequence on a template strand (hence the need for heat or enzymes to expose single-stranded DNA). The most common is the *Taq* DNA polymerase (originally isolated from the hot-springs bacterium *Thermophilus aquaticus*; **Figure 4.9**, p. 120), but other thermostable polymerases can survive thermal cycling, such as *Pfu* and *Pwo* (from the archaebacteria *Pyrococcus furiosus* and *Pyrococcus woesii*). Some amplification technique use multiple polymerases. Long-range PCR techniques amplify much longer sequences than standard PCR by combining the strengths of two polymerases, such as *Taq* (efficient 5′→3′ polymerase activity) and *Pwo* (3′→5′ proof reading abilities). Isothermal DNA amplification techniques can make use of a wider range of polymerases, because they do not need to survive temperature cycling.

4. **Melt again:** This cycle – melt, prime, grow – is repeated again and again, often around 30 times. The number of copies of the target sequence increases exponentially, as each round copies the original DNA plus the copies made previously. Temperature cycling is usually done by placing DNA plus reagents in small lidded test tubes (sometimes referred to as Eppendorf tubes), which are then placed in a thermocycler, a machine that is programmed to heat or cool the test tubes to particular temperatures for defined periods of time.

synthesis by using primase to create short RNA polynucleotides along the exposed replication fork. The need for primase is circumvented *in vitro* by the addition of short DNA sequences (primers; **TechBox 4.3**, p. 115) that bind to specific sequences on the exposed single-stranded DNA, providing a starting point for DNA polymerase. And instead of the co-ordinated set of polymerase enzymes and cofactors found in most living cells, generally only one robust DNA polymerase is used in the lab to do all of the DNA synthesis.

Polymerase chain reaction (PCR)

"The thing that was the 'Aha!' the 'Eureka!' thing about PCR wasn't just putting those [things] together . . . the remarkable part is that you will pull out a little piece of DNA from its context, and that's what you will get amplified. That was the thing that said, 'My God, you could use this to isolate a fragment of DNA from a complex piece of DNA, from its context. That was what I think of as the genius thing. . . . In a sense, I put together elements that were already there. . . . You can't make up new elements, usually. The new element, if any, it was the combination, the way they were used. . . ."

Kary Mullis, one of the inventors of PCR (quoted by P. Rabinow (1998) on http://sunsite3.berkeley.edu/PCR/whatisPCR.html)

Rarely has an invention had such a rapid impact on science and the wider culture as the development of the polymerase chain reaction (PCR). Within a few years of the invention of PCR, it had been used to detect infectious agents in samples of blood and water, to identify missing persons, to settle paternity disputes, and as a forensic tool that resulted in both convictions and acquittals in criminal courts.

The PCR protocol is essentially a recipe describing the chemicals and conditions that can be employed to make multiple copies of a DNA sequence in a test tube (**Figure 4.8**). It works by cycling the DNA sample through a series of heating and cooling cycles, with the raw materials and enzymes needed for DNA synthesis. Through the process of template reproduction, the number of copies of the target DNA sequence increases exponentially. In the first round, complementary copies are made of the original DNA, then in the next round copies are made of the original DNA and the first set of copies, and so on. Three main technical breakthroughs were needed to make PCR work: polymerase, primers, and temperature cycling.

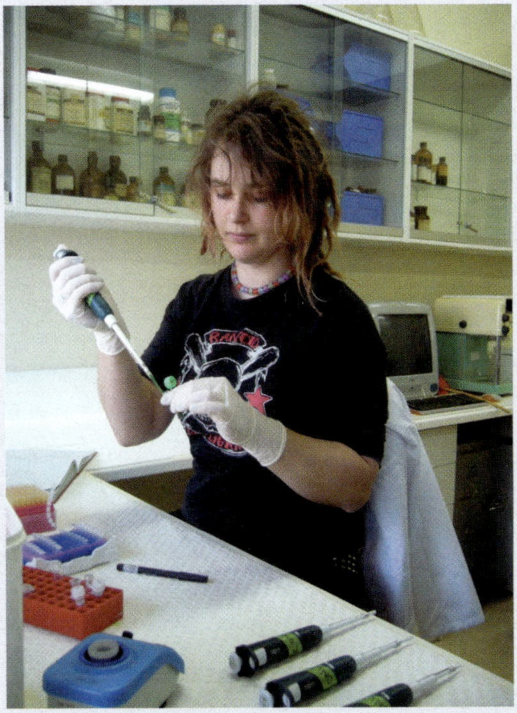

Figure 4.8 Preparing a PCR reaction (here, DNA from the Five-toed Earless Skink, (*Hemiergis initialis*) is being amplified in order to investigate why some five-toed skinks have five toes, while others have four or three).
Photograph: Lindell Bromham.

Primers: starting blocks for DNA synthesis

One of the first technical breakthroughs that allowed the development of DNA amplification was the ability to make short DNA molecules with a specific base sequence. DNA polymerase cannot start from scratch, but can only add nucleotides to the end of an existing polynucleotide chain. In living systems, an enzyme with primase activity manufactures short RNA polynucleotides that match the sequence of bases on the exposed DNA of the replication fork. But in the lab, researchers circumvent the need for primase by adding ready-made primers. Primers are short sequences of DNA that stick to complementary DNA sequences in the sample, providing starting blocks for polymerase. Vast numbers of copies of the primer are added to the extracted DNA. These primers sticks to the DNA in the sample, providing a free 3′ OH to which new bases are attached.

In addition to providing a starting block for DNA polymerase to act upon, primers also solve the problem of where to begin DNA synthesis. *In vivo*, the replication enzyme complex recognizes specific sequences in the genome that act as replication initiation sites. But *in vitro*, this approach won't work for two reasons. Firstly, the genome is chopped up. So, instead of providing co-ordinated starting sites to copy the whole genome, replication initiation sequences will be found on a number of disassociated fragments. If polymerase began at these recognition sequences it may not get very far before it fell off the end of the fragment, and fragments that did not contain an initiation sequences would not get copied at all. Secondly, most researchers wish to amplify a specific part of the genome. For example, they may wish to sequence the same gene from many different species. But the sequence to be amplified might not be near a replication initiation site. Happily this problem can be solved by adding primer sequences that stick to the exact sequence of DNA you wish to copy (**TechBox 4.3**).

Because complementary base pairing is a universal feature of DNA, if you make a short polynucleotide chain of a particular sequence and add it to a sample of single-stranded DNA, then it will stick to any complementary sequence throughout the DNA in the sample. So if you know (or can guess) a short base sequence from the region of DNA you wish to amplify, you can design a primer that will stick to that sequence (and to no other sequence in the genome). Adding large numbers of copies of this primer sequence to your sample will make the polymerase enzyme start exactly where you want it to, and copy only your target DNA, even if that sequence represents less than a ten-thousandth of a per cent of the DNA in the sample.

Temperature cycling: from double-stranded to single-stranded and back again

Primers work by complementary base pairing, so they can only stick to exposed nucleotides on single-stranded DNA. In living systems, the double-stranded DNA helix is gently prised apart by the complexes of enzymes that create and maintain the replication fork. But most lab techniques employ a much less sophisticated means of exposing single-stranded DNA templates. The entire sample is heated to the temperature at which the hydrogen bonds holding the nucleotide 'rungs' of the helix melt away (typically between 94 and 96°C; see **TechBox 4.2**, p. 111). Heating is a pretty brutal way to treat a delicate and complex molecule. DNA loses its higher-order structure when heated: the supercoils unravel, the helix unwinds, and the base pairs separate. All that is left are single strands of polynucleotides. This loss of structure is referred to as 'denaturing', and it's an appropriate term because it's exactly what should not happen in a natural system.

When the sample is cooled, the single-stranded DNA begins to anneal, sticking back together along the exposed nucleotide sequences to form double helices once more (**TechBox 4.2**). The degree of matching between two strands of DNA will determine how strongly they stick together, because the hydrogen bonds linking the bases form between matched base pairs. The more bases match, the more bonds form between the two strands, so two perfectly matched strands will be more strongly stuck together than an imperfect match (see **Case Study 4.1**).

The universality of complementary base pairing means that, as the single-stranded DNA is cooled, any complementary DNA sequences will anneal together when the sample is cooled again. As primers are added in excess and outnumber the source DNA, it's more likely that a target sequence will be matched up with a primer than with a complementary strand of source DNA. So when the DNA is cooled, most copies of your target sequence will be bound to a primer. If a primer sticks to its exact complement, it is more likely to stay stuck as the sample is heated again to raise it to the temperate needed for polymerase to extend the sequences (typically around 72°C). DNA polymerase can now start at the primer sequences and make the rest of the matching strand. Now you repeat the cycle of heating and cooling

TECHBOX 4.3

Primer design

KEYWORDS

PCR

melting temperature, T_m

annealing temperature

GC content

degenerate primers

nested PCR

FURTHER INFORMATION

There is a range of primer design software available, both licensed and freeware. For example, Primer3 (*http://frodo.wi.mit.edu/cgi-bin/primer3/primer3_www.cgi*)

RELATED TECHBOXES

TB 1.2: DNA sequencing

TB 4.2: DNA amplification

RELATED CASE STUDIES

CS 1.2: DNA surveillance (species identification using PCR)

CS 2.1: On the origin of faeces (DNA surveys using PCR)

Careful primer design is essential for successful DNA amplification, because it is one of the main ways that PCR can be optimized to a particular task (that is, to amplify a specific sequence in the context of a particular sample). Typically, primers are designed in pairs: a forward primer defines the beginning of the sequence to be amplified, and a reverse primer defines the end.

The goal of primer design is to choose a short nucleotide sequence, usually between 18 and 30 bases long, that will stick only to your target sequence (and nowhere else) when your DNA sample is cooled and reheated. You can work out the chance of any sequence occurring at random in the genome. If all bases are equally frequent, then we would expect a particular sequence of 4 bases, say ACTG, to occur, on average, every 4^4 bases (or once every 256 bases). So we would expect to find hundreds of thousands of instances of this four-letter sequence in the human genome. But any given 17-base sequence has a probability of occurring only once in every 4^{17} bases (over 17 billion), so it is less likely to occur twice in the genome simply by chance (see **TechBox 3.4**).

The temperature at which the primer will stick to the target DNA is determined primarily by primer sequence and length. The simplest formula for estimating the melting temperature (T_m) of a primer is the Wallace formula: add together 4°C for every G or C and 2°C for every A or T. This is because GC pairs are held together by three hydrogen bonds, so they require more energy to melt than AT pairs, which are held together by only two bonds (**TechBox 2.2**). However, primer-design programs will usually give a more accurate indication of T_m, by including the influence of other factors such as salt concentration. Accurate estimation of T_m is important: if the T_m is lower than the annealing temperature used in a PCR reaction, the primer will stick to many non-target sequences. If it is too high, it may fail to anneal to the right sequence. The two members of the primer pair should have similar melting temperatures.

Most laboratories design their primers to suit the particular amplification task, then order them from a company that specializes in manufacturing oligonucleotides (short DNA segments). There are three main considerations when designing primers:

Sequence: A primer should match a sequence that is unique to the target DNA (that is, a sequence that is not repeated elsewhere in the genome), so that it will bind by base pairing to the target DNA and provide a free 3′ end for polymerase to work from. For primer pairs, one will be complementary to the beginning of the sequence, and the other will be the reverse complement of the end of the sequence. It is also worth avoiding repeat sequences, runs of Cs and Gs, or any self-complementary sequences that could bind with another part of the primer sequence.

Length: The shorter the primer sequence, the more chance that it will match a random sequence in the genome. But the longer the primer, the higher the annealing temperature, which means that the primer may begin annealing at the wrong point in the temperature cycle (e.g. the primers may bind indiscriminately during the elongation phase, resulting in non-target sequences being amplified).

GC content: around 30–60% of the primer sequence should be Gs or Cs. This is to ensure that the primer anneals and melts at an appropriate temperature.

By following these principles, it is possible to design primers manually. However, it is increasingly common to use computer programs that calculate the T_m and identify any problems with potential primer sequences. In addition, there are lots of tricks and quirks in primer design to increase success rates, so it is worth reading protocols and talking to people who are experienced with your particular sequence, species, or sample to get hints on the best primer-design strategies.

Degenerate primers: Sometimes it is useful to add many different versions of a primer, each one differing at one or two positions, so that they can stick to a range of similar sequences. This may be necessary when designing primers from a protein sequence: because of the degeneracy of the genetic code (see **TechBox 2.3**), a particular amino acid sequence could be coded for by many different nucleotide sequences. So you may need to include primers corresponding to each of the possible nucleotide sequences that could code for the amino acid sequences of the gene you wish to amplify. Similarly, if you don't know the exact sequence of the gene you are looking for, but you do know what the corresponding sequence is in a related species, then you might design degenerate primers that cover alternative versions of the DNA sequence. But, of course, the less specific your primers are, the more chance there is that they will stick to a non-target sequence in the genome and amplify the wrong piece of DNA.

Nested primers: Different primers can be used one after the other to increase the specificity of the PCR. After PCR is performed with one set of primers, the amplified DNA is then put through a PCR reaction with a different set of primers, corresponding to a sequence just inside the previous primer pair. This reduces the probability of non-target amplification, because of the low probability that a randomly amplified sequence will match both primer pairs (see **Case Study 2.1**).

to create more and more copies of the sequence to which the primers stick.

This cycle of heating and cooling can be done manually by anybody with several water baths (one at each temperature), a reliable stopwatch, and sufficient patience. However, PCR has been made more efficient by the development of thermal cyclers (commonly referred to as 'PCR machines') which do all the tedious heating and cooling for you. But the real breakthrough for automating DNA amplification was in the taming of the polymerase enzyme.

Taq polymerase: Molecule of the Year 1989

Denaturing DNA would spell disaster for a living cell, because without its higher-order structure DNA could no longer be reliably maintained, copied, and transcribed. Heating also denatures most proteins, including those involved in DNA synthesis. When PCR was first developed in the mid-1980s, back when samples had to be manually heated and cooled, fresh polymerase had to be added each cycle to replace the denatured enzymes ruined by the last round of heating. Discovery of a DNA polymerase enzyme that could survive thermal cycling was an important breakthrough in the development of biotechnology.

There are a great many variants of the DNA polymerase enzyme in the living world: not only does the enzyme vary slightly between species, but some species have many different kinds of polymerase, which perform different DNA replication tasks in the cell. For example, mammalian cells have a dozen or more different kinds of polymerase, functioning in both DNA synthesis and

CASE STUDY
4.1

Glorious mud: using DNA-reassociation kinetics to quantify biodiversity of soil bacteria

KEYWORDS

Cot curve
denature
DNA hybridization
metagenomics
conservation

FURTHER INFORMATION

For a beginners guide to metagenomics see *www.bioteach.ubc.ca/Biodiversity/Metagenomics/*

RELATED TECHBOXES

TB 2.4: DNA extraction
TB 6.2: What is a species?

RELATED CASE STUDIES

CS 6.1: Barcoding nematodes (DNA taxonomy)
CS 6.2: Keeping the pieces (DNA and conservation)

Sandaa, R.-A., Torsvik, V., Enger, O., Daae, F.L. and Castberg, T. D. H. (1999) Analysis of bacterial communities in heavy metal-contaminated soils at different levels of resolution. *FEMS Microbial Ecology*, Volume 30, pages 237–251

> “*Our results demonstrate long-term effects of heavy metals on total bacterial communities in contaminated samples from field sites. Compared to noncontaminated soil, a pronounced reduction of bacterial diversity as well as changes in bacterial community structure were obtained even in the presence of low metal concentrations below the upper legal limits set by the European Union.*” [1]

Background

For many people, the word ‘biodiversity’ brings to mind a rainforest bursting with tree species, vines, beetles, butterflies, and exotic mushrooms. But perhaps the biggest slice of the world’s biodiversity lies beneath our feet: a single gram of soil may contain many thousands of species of soil bacteria. Furthermore, the biodiversity of soil microbes may have a much greater practical impact on our lives than the more charismatic rainforest dwellers, due to their effect on soil fertility. One of the problems with conserving soil biodiversity is that it is difficult to quantify. We need to be able to measure the diversity of bacteria if we are to monitor potentially damaging changes to soil microbe communities. Bacterial identification has traditionally relied upon culturing cells in the laboratory and describing their appearance, activity, and biochemistry. But the vast majority of soil bacterial species cannot be successfully cultured.

Aim

Genomic difference has been put forward as a universal species definition for bacteria, traditionally measured using DNA-reassociation kinetics[2]. When denatured (single-stranded) DNA is cooled, the rate of reassociation into double strands will be partly determined by the similarity of nucleotide sequences of the strands in the sample (this procedure is also referred to as DNA hybridization). When DNA from different genomes is mixed, the rate of reassociation will be determined by the amount of difference between the genomes: the more the genomes in the sample differ, the longer it takes for all of the DNA to return to the double-stranded state. This can be plotted as percentage reassociation as a function of concentration (C_0) and time (t), commonly known as a Cot curve. As well as measuring the genomic difference between two species of bacteria, DNA-reassociation studies provide a way of estimating the genetic diversity of a mixed microbial sample. Researchers in Norway used DNA-reassociation studies to test whether contamination with heavy metals reduced the biodiversity of soil bacteria.

Methods

The authors of this study collected soil samples from experimental field sites, some of which had been treated with sludge contaminated with heavy metals. To extract bacteria, they homogenized soil samples in a Waring blender (see **TechBox 2.1**) with detergent and nuclease inhibitor. The bacteria were then concentrated by centrifugation and washed. The cells were lysed to release DNA, which was purified through a column (**TechBox 2.4**). Because large fragments of DNA are desirable for hybridization, DNA was sheared in a French press, which uses pressure to force DNA through a small hole. The diversity of DNA in the sample was estimated using DNA-reassociation kinetics, measured using a UV spectrophotometer.

Results

This study showed that, although the total numbers of bacteria remained the same in both contaminated and uncontaminated soils, bacterial diversity decreased dramatically in soils treated with heavy metals. The DNA-reassociation kinetics suggested that the uncontaminated soil samples contained approximately 16,000 different kinds of bacterial genome for every gram of soil, but soil contaminated with heavy metals had as few as 2000 different kinds of genome per gram. However, this estimate is based on the assumption that all species of bacteria in the sample are equally abundant. This is unlikely to be true, since most ecological communities contain some very abundant species, and many rare species. Reanalysing the data using standard models of species abundance, Gans and coworkers[3] estimated that there were at least 830,000 different genomes per gram of uncontaminated soil. They suggested that diversity was reduced by 99.9% in heavily contaminated soil.

Conclusions

While the actual number of different species of soil bacteria may be difficult to estimate, comparison of Cot curves shows that there is much less genetic diversity in the samples from contaminated soil.

Limitations

DNA reassociation is a blunt tool, and its use in species classification has been criticized: bacteria that differ in ecology and morphology may have very similar genomes, and strains

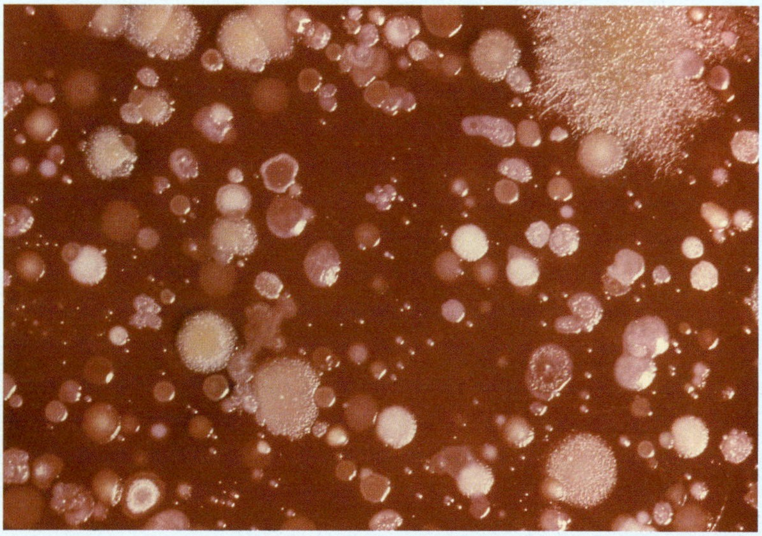

Figure CS4.1 Going underground: a culture of naturally occurring soil bacteria, including *Bacillus anthracis* (see also **Figure 4.20**). There is far more biodiversity beneath the ground than above it. But measuring the biodiversity of soil bacteria is difficult as the vast majority of bacteria found in soils cannot be cultured in a laboratory. DNA-based techniques avoid this problem by extracting bacterial genomic DNA directly from soil samples.

Courtesy of CDC/ Dr James Feely.

with many genomic differences may be functionally similar. It is not clear how measures of genomic diversity in a sample are related to meaningful definitions of species (**TechBox 6.2**). Furthermore, varying assumptions made in the estimation of number of different genomes from Cot curves can give dramatically different answers (in this case, an order of magnitude difference).

Future work

Metagenomics, habitat-based genomic analysis, takes analysis of soil bacteria to a new level[4]. Rather than simply measuring the difference in the DNA of a mixed microbial sample, the DNA is cloned into libraries which can be used for expression studies. For example, metagenomic studies of soil bacteria have revealed that there is a wealth of previously undescribed antibiotic proteins coded in the genomes of soil bacteria.

References

1. Sandaa, R.-A., Torsvik, V., Enger, O., Daae, F.L. and Castberg, T. D. H. (1999) Analysis of bacterial communities in heavy metal-contaminated soils at different levels of resolution. *FEMS Microbial Ecology*, Volume 30, pages 237–251.
2. Konstantinidis, K.T. and Tiedje, J.M. (2005) Genomic insights that advance the species definition for prokaryotes. *Proceedings of the National Academy of Sciences USA*, Volume 102, pages 2567–2572.
3. Gans, J., Wolinsky, M. and Dunbar, J. (2005) Computational improvements reveal great bacterial diversity and high metal toxicity in soil. *Science*, Volume 309, pages 1387–1390.
4. Steele, H.L. and Streit, W.R. (2005) Metagenomics: advances in ecology and biotechnology. *FEMS Microbiology Letters*, Volume 247, pages 105–111.

repair. Lab-based DNA amplification requires an enzyme that can work in unusual conditions. The logical place to look for a polymerase enzyme able to survive repeated heating was in an 'extremophile' organism living at temperatures that would cook ordinary organisms.

Thermophilus aquaticus lives in hot springs, at temperatures that denature most genomes (**Figure 4.9**, p. 120). Not surprisingly, its version of DNA polymerase is able to operate at high temperatures. *T. aquaticus*' DNA polymerase, now referred to as *Taq*, can persist in the PCR sample during the heating phase, then begin copying when the sample is cooled. *Taq* is not the only polymerase enzyme that can be used in PCR, but it is currently the most common. DNA replication *in vitro* using *Taq* differs from DNA replication in living systems in a number of important ways. Firstly, *Taq* can only copy relatively short DNA sequences, typically between 600 and 2000 bases. Secondly, unlike most polymerase enzymes, *Taq* has no proof reading activity, so it has an inherent error rate of around one mistake for every ten- to hundred-thousand bases copied. Given that a PCR reaction involves making millions of copies of a DNA sequence, this error rate is not trivial. Mistakes can be identified through multiple coverage, so most genome projects sequence each part of the genome many times over. New PCR techniques are being developed to take advantage of other thermostable polymerases that can proof read (**TechBox 4.2**). This should reduce the PCR error rate.

Feral DNA replication: the problem of contamination

The great strength of DNA amplification – that it can take a tiny amount of target DNA and make millions of copies of it – is also one of its dangers. There is DNA everywhere: on your hands, in your lunch, on your lab bench. So it is possible that your sample contains DNA from more than one source. DNA from all sources is structurally the same, and it is all copied by the same mechanism. Therefore, a PCR reaction will amplify any sequence that your primer sticks to, whether or not it was the sequence you wanted. You may be attempting to PCR DNA sequences from an endangered fern species, but instead you could end up with millions of copies of your own DNA from a skin cell that brushed

Figure 4.9 *Thermophilus aquaticus*, an 'extremophile' bacterium first found in hot springs at Yellowstone National Park, USA, is the original source of the *Taq* polymerase now used to amplify DNA in the lab. Because of its key role in the development of molecular genetics, *Taq* was awarded the inaugural 'Molecule of the Year' award in 1989. *Taq* earned the patent holders approximately $200 million (US dollars), but the patent has been the subject of lengthy legal battles.
Photograph: Gregor Wenzel.

into your test tube, or the DNA from a different fern specimen stored in the same container, or the DNA from fern spores blown in through the lab window. Contamination is a particularly serious problem when dealing with very small samples, or samples with degraded DNA, such as museum specimens. In these cases, contaminating DNA can overwhelm the target DNA, so the PCR product can consist entirely of non-target sequences. The problem of contamination can be countered by scrupulous attention to lab hygiene, by careful primer design, and through careful consideration of the resulting DNA sequences.

In some cases, contamination can be detected by the surprising similarity of your amplified DNA to something other than your intended sample. You may not know what the exact DNA sequence of a gene is in your target species, but you will have some idea of which other species you expect it to be related to. For example, in the mid-1990s two separate laboratories made the surprising claim that they had sequenced DNA isolated from dinosaur bones and eggs. Two lines of argument were used to refute these claims. Firstly, there is currently doubt that DNA can survive in a sample longer than a million years without being totally degraded, so there is virtually no hope of finding 80 million-year-old dinosaur DNA. Secondly, comparing the dinosaur-derived sequences to the DNA of living species revealed some unexpected relationships. Dinosaur origins and evolution may be debated, but most people would expect dinosaur sequences to be most similar to birds and other reptiles. But careful analysis of the supposed dinosaur sequences showed that, in one case, they were most similar to sequences from plants and fungi, and in the other case, most similar to human sequences. It is easier to believe that these sequences were actually contaminants than it is to rethink dinosaurs as giant mushrooms, or to imagine that the dinosaur genome underwent an improbable degree of convergence with the human genome.

Contamination will be less obvious when the target DNA is closely related to the contaminant. In fact, one of the most pervasive sources of contamination is from the PCR product of a previous reaction. A drop of liquid from a previous PCR reaction can contain many thousands of copies of a sequence, overwhelming the target DNA in your sample. Good lab practice will reduce the risk of contamination, such as conducting DNA extraction and pre- and post-PCR preparation in different parts of the lab, not sharing lab equipment, and generally avoiding activity that will spread aerosols of PCR product. In addition, a number of checks for contamination can be performed. Blank controls, with no sample material, should be included in PCR procedures: a positive PCR result for a blank control demonstrates that the PCR product came from contamination (**Case Study 2.1**).

Faithful but not perfect

We have now considered how DNA is replicated by using the two strands of the helix as templates to make complementary strands to form two new helices. We have also seen how the process of DNA replication has been dramatically simplified to provide innumerable copies of a DNA sequence for molecular genetic analysis. Now we are going to have a more detailed look at the process of DNA replication in order to understand how the combination of replication and mutation turns the genome into a phenomenally useful source of historical information.

Copy errors are an important source of mutation (heritable changes to the DNA sequence). Some of the mistakes made in copying essential genetic information will be disastrous. But some mutations are not lethal, nor even harmful. And, luckily for biologists, it is these non-lethal mutations that make DNA the most useful tool we have for uncovering evolutionary patterns and processes. First, we will consider how the process of replication itself creates mutations in the genome. Then, we will look at how the process of copying the genome, mutations and all, generates hierarchies of similarity between individuals, populations, and lineages.

Copy errors

There are three basic error-checking steps of DNA replication that detect and remove copy errors: base selection, proof reading, and mismatch repair. It is rare for a copy error to pass through all three stages without being corrected. The neat geometry of the DNA helix plays an important role in the detection of errors. As Watson and Crick discovered, correct pairs (A with T, G with C) fit snugly in the DNA helix. Incorrect pairs do not. Distortions to the helix caused by damage or mispairing underlies each of the three error-checking stages of replication.

Complementary base pairing is discussed in TechBox 2.2

This is particularly important when considering the interaction between the polymerase enzyme and the DNA helix. Think of polymerase as being like a hand, with the fingers and thumb wrapped around the helix (**Figure 4.10**, p. 122). The hand needs to be able to close its grip around the helix in order to engage the polymerase active site. A distorted helix prevents the hand closing, so that the polymerase function cannot be employed, but the exonuclease site may be engaged.

The first step of error correction is base selection. If a nucleotide to be added to the new strand pairs correctly with the base on the template strand, it will activate the polymerase site (**Figure 4.11**, p. 122). The polymerase activity will then bind the incoming nucleotide to the sugar–phosphate backbone of the growing strand. But if the incoming nucleotide does not pair correctly, it is unlikely to engage the polymerase site, and so will not get joined to the new strand. Only one in a million base-incorporation errors are likely to get through this first checking mechanism.

The second step is proof reading. Most DNA polymerase enzymes have two main active sites, each of which does

Figure 4.10

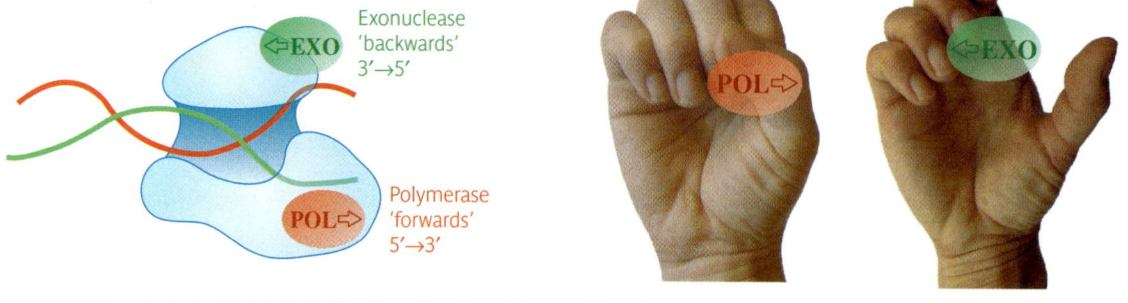
EXO
Exonuclease 'backwards' 3′→5′
POL
Polymerase 'forwards' 5′→3′
POL
EXO

a different job. The polymerase site takes DNA synthesis 'forward' (5′→3′), adding nucleotides to the end of a growing strand. The exonuclease site takes the enzyme 'backwards' (3′→5′), removing nucleotides from the end of the growing strand. If an incorrect base is added to the new DNA strand, it will not pair tightly with the template strand (**Figure 4.12**, p. 123). The unpaired new strand is free to flap out of the polymerase active site, and into the exonuclease site. The exonuclease activity chews up the most recently added nucleotides, erasing the incorrect base. Efficiency of proof reading ranges from 0% of errors corrected to 99.99%, depending on the particular polymerase enzyme. Even so, the occasional error will make it past the proof reading stage.

The importance of 5′ to 3′ is explained in **TechBox 2.2** *and* **TechBox 4.1**

The third stage of copy-error correction is postreplication mismatch repair. If an incorrect nucleotide has managed to avoid correction by base selection and proof reading, it may be incorporated into the new DNA strand. But the wrong nucleotide will not pair correctly with the mismatched base on the template strand. This causes the DNA helix to bulge out. Mismatch repair enzymes scan the DNA for bulges that indicate an incorrect base. Once found, the base can be chopped out using an endonuclease, which removes nucleotides from the middle of a DNA strand. The excised strand is then replaced with the correctly matching bases (**Figure 4.12**, p. 123).

If a mismatch is detected in the double-stranded DNA helix, how does the repair machinery know which of the mismatched bases is correct, and which is the error? DNA strands are methylated, with a methyl group (CH_3) added to cytosines (specifically, those that occur in CG dinucleotides). But a new strand is not methylated immediately. Therefore, when a mismatch is detected, the base on the unmethylated strand is

Figure 4.11

1. Base selection

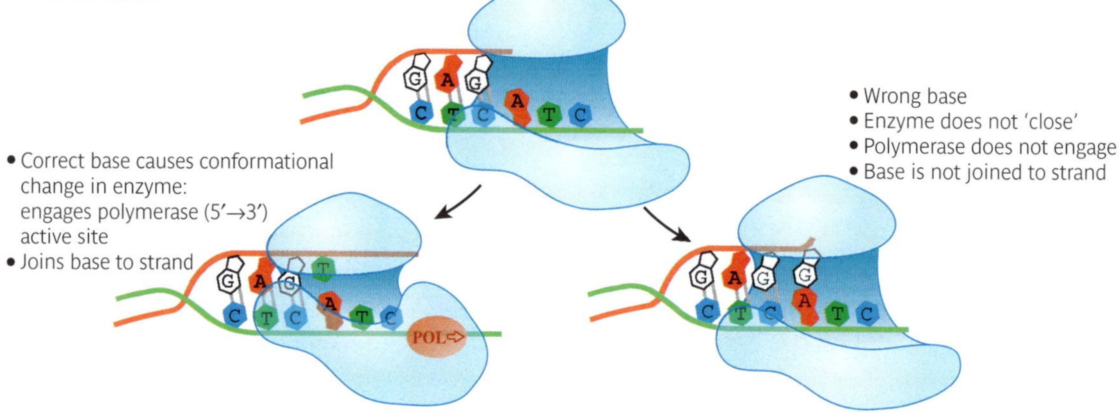
G A G
C T C A T C
• Correct base causes conformational change in enzyme: engages polymerase (5′→3′) active site
• Joins base to strand
POL
• Wrong base
• Enzyme does not 'close'
• Polymerase does not engage
• Base is not joined to strand

Figure 4.12

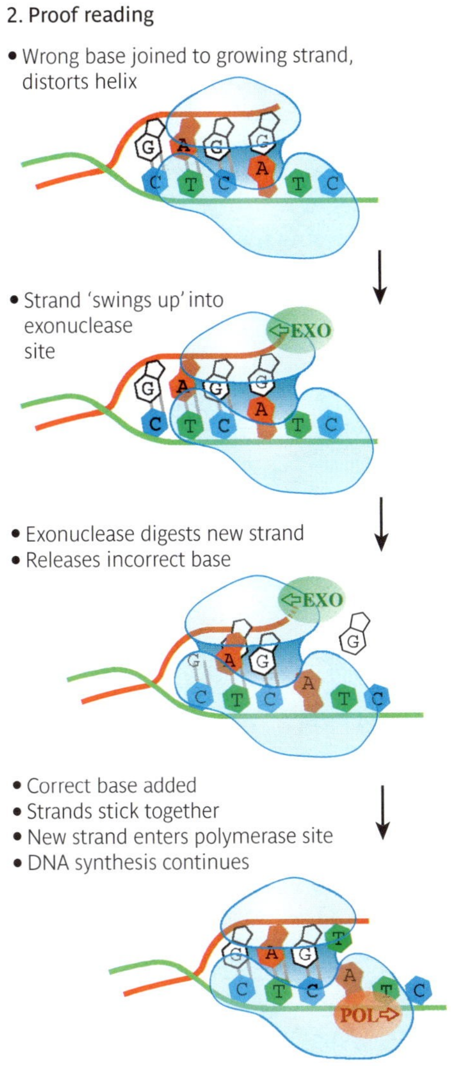

2. Proof reading
• Wrong base joined to growing strand, distorts helix
• Strand 'swings up' into exonuclease site
EXO
• Exonuclease digests new strand
• Releases incorrect base
EXO
• Correct base added
• Strands stick together
• New strand enters polymerase site
• DNA synthesis continues
POL

Figure 4.13

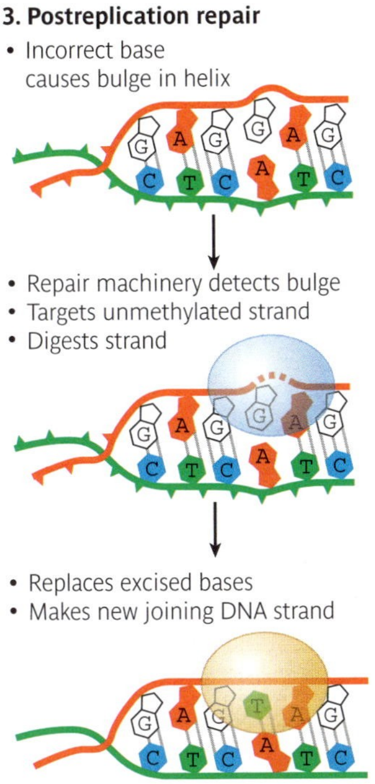

bound to be the copy error. Once the new strand is methylated, the repair machinery could not tell which of the mismatched nucleotides was correct, and would have an equal chance of excising and repairing the wrong strand, thus permanently changing the base sequence (Figure 4.13).

If it remains uncorrected, the mismatch could persist until the next replication. When the mismatch is unzipped and those strands are used as templates, each base will be paired with its correct partner. In the example shown here, an AG mismatch has persisted after replication (Figure 4.14). When this DNA is copied, the strand with the A will be paired with a T, but the strand with the G would be paired with a C. In this way, a single DNA molecule gives rise to two daughter molecules with different base sequences: AGGAG and AGTAG. This is now a permanent change that will be inherited by every copy made from these DNA sequences.

Figure 4.14

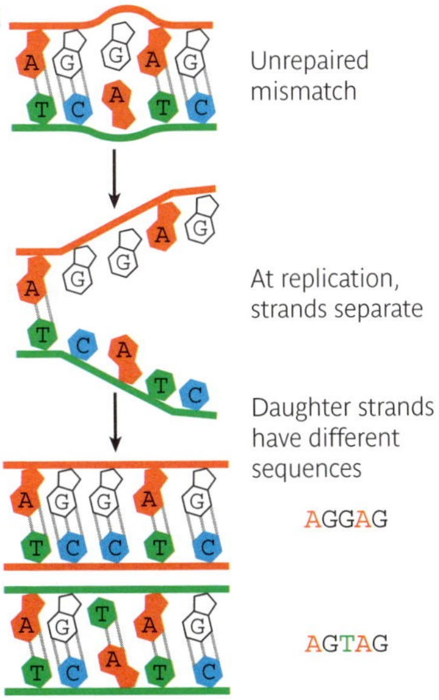

History in the genome

DNA replication is the basis of life on Earth. It is also the basis of bioinformatics, because the process of copying creates an array of genomes, each of which is more similar to close relatives, and increasingly dissimilar to ever more distant relatives. In other words, copying creates a hierarchy of similarities between genomes. Due to the process of descent with modification, this hierarchy spans all levels of biological organization, from individuals to populations to lineages to distantly related kingdoms. We saw this in Chapter 1: DNA from the mystery blob could be tracked back through time, from the relationship of a single whale to its parents, grandparents, to the past sperm whale populations, to the diversification of the whales, the radiation of mammals, the origin of the animal kingdom, and potentially way back further still. In order to look more carefully at how DNA replication leaves a hierarchy of changes that we can read at all levels of biological organization, we are going to revisit an example introduced in Chapter 2: the gene that codes for a subunit of the transcription machinery, RNA polymerase II beta.

Same but different: how do important sequences change?

All organisms must have the information needed to make all the essential components of the genomic information system, including the enzyme RNA polymerase. And yet we saw in **Figure 2.2** that the DNA sequence for the RNA polymerase II beta subunit gene is not identical in all organisms. If they all require the same enzyme to do the same job, then why not have exactly the same gene?

To picture how such an important gene could change over evolutionary time, we can invent a fable about a recipe which is handed down through the generations of a family. Imagine that someone, let's call her Mrs Smith, was famed for her excellent sponge pudding. The pudding was so much admired by all who partook of it that each of Mrs Smith's sons and daughters took a copy of the recipe with them when they left home, so they too could create the much-loved dessert. Their children, too, would copy the recipe upon leaving home, and so would their grandchildren, and great grandchildren, and so on down the generations. Occasional mistakes were made in copying, some of which ruined the recipe so that it was no longer used and copied. But if the copy mistake did not prevent the cooking of a delicious sponge pudding, it was not noticed, and so some little mistakes in the recipe would be copied again by the next generation of pudding-makers. Eventually, after many generations, all 75 of Mrs Smith's great-great-great-grandchildren have their own copy of the recipe, yet their recipes are all slightly different – a spelling mistake here, a change of ingredients there. But every single recipe produces a fine sponge pudding.

The same process of minor changes being perpetuated down chains of copies occurs with any process that involves sequential copies made with low error rate. Because each copy inherits the mistakes made in the previous copy, the copying process produces a hierarchy of similarities. Each copy is more similar to those closest to it in the chain of copies, so the more closely related two copies are, the more similar they are likely to be. This principle has also been used to reconstruct the history of languages, manuscripts, and cultural artefacts (**TechBox 4.4**, p. 126; **Case Study 4.2**, p. 127).

In Chapter 6 we will see how the hierarchy of similarities is revealed by sequence alignment

Much the same process of copying with occasional errors has occurred in the evolution of the RNA polymerase II beta gene. All organisms must have a functioning copy of this gene, which they inherit from their parents. Very rarely, mistakes are made in copying this gene; such as an A inserted instead of a T, or a G where there should be a C. Most mistakes will simply ruin the gene; which can then only make a faulty copy of the enzyme, reducing the chance that an organism carrying that mutation will reproduce and pass the mutation on to the next generation. But occasionally, a copy mistake is made that does not ruin the enzyme, which can still function even though it is slightly different from its parent copy. If the individual with this mutation survives and reproduces, that gene may be copied to their offspring, mutation and all. The mutation will then become an inherited difference, present in descendants of that lineage. Note that this mutation did not have to make the enzyme better, it simply didn't make it much worse, so that it didn't reduce its chances of ending up in successful progeny.

Sequences show a nested hierarchy of similarities

When you examine the sequences for RNA polymerase II beta gene given in **Figure 2.2**, you notice that, although they are all different, some sequences are more different than others. The two most similar sequences are from the two most closely related species – the two mammals, rat (*Rattus*) and human (*Homo*). Their sequences are almost identical, differing only at four places in the sequence (the rat has a C instead of a T at position 3531, a G instead of an A at 3546, and a C instead of a T at positions 3553 and 3561; **Figure 4.15**).

The two mammal sequences are also similar to the other animal sequence, from the fruit fly *Drosophila*. Here we can see that the human and fly sequences differ at nine places in the sequence (**Figure 4.16**).

When we compare sequences from the animal and fungus kingdoms, we see more sequence differences. The human and bread mould (*Neurospora*) sequences have 11 differences between them. The plant sequence is even more different from both the fungus or the animals: there are 20 differences between human and rice. The most distantly related organism to us, the bacterium, is also the most different, with 23 sequence differences between human and *E. coli*. Yet all of these

Figure 4.15

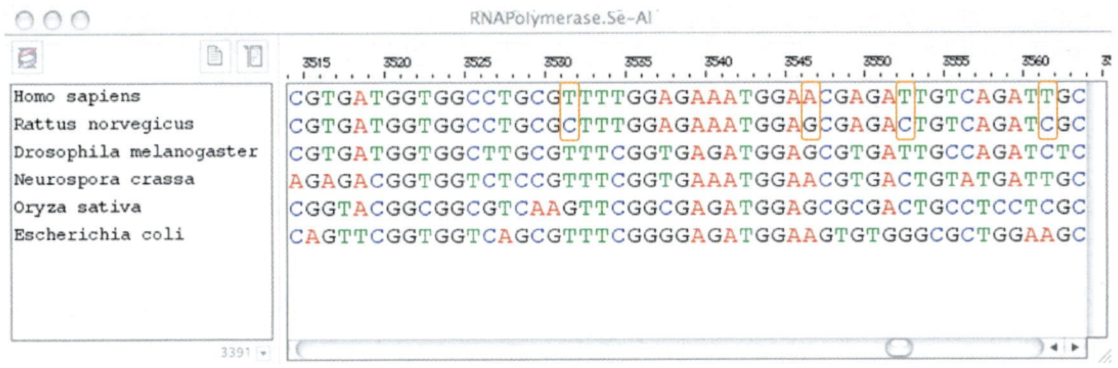

Figure 4.16

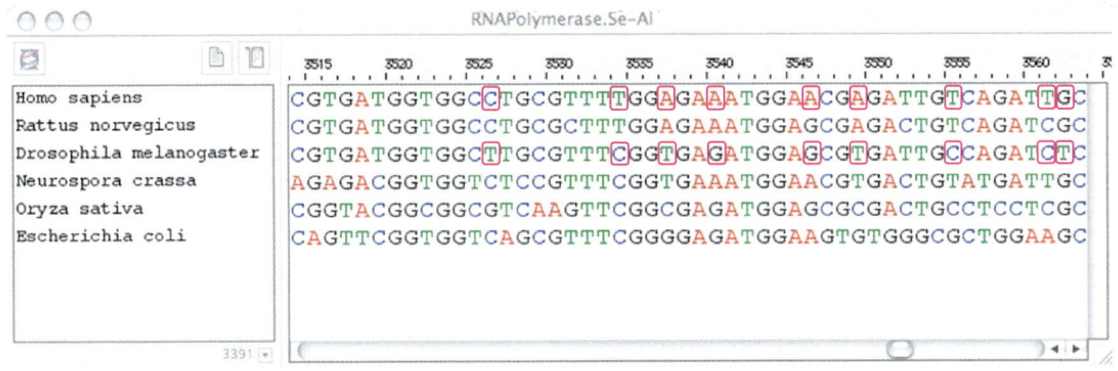

TECHBOX 4.4

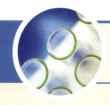

Copy errors reveal history

KEYWORDS

language

learning

manuscripts

phylogeny

FURTHER INFORMATION

You can find a range of studies that have taken a phylogenetic approach to cultural evolution in: Mace, R., Holden, C. J. and Shennan, S. (2005) *The Evolution of Cultural Diversity: a phylogenetic approach*. UCL Press, London.

A summary of the approach can be found in the following article: Mace, R. and Holden, C.J. (2005) A phylogenetic approach to cultural evolution. *Trends in Ecology and Evolution*, Volume 20, pages 116–121.

RELATED TECHBOXES

TB 7.1: Distance methods

TB 7.2: Maximum likelihood

RELATED CASE STUDIES

CS 4.2: Peopling the Pacific (language trees)

CS 7.1: Up the river (tracing disease origins)

For any process of sequential copying – where each copy is used as a template for the next copy – comparison of shared copy errors can be used to reconstruct the copying process. Much of this book concerns the application of this line of thinking to DNA, given that each organism's DNA was copied from its parents, which was copied from their parents, and so on. But the same idea can be applied to reconstructing the history of any copied information. Techniques developed for the analysis of DNA data are now being applied to understanding language evolution, reconstructing the history of ancient texts, and studying the development of cultural artefacts (see **Case Study 4.2**).

Before the advent of printing technology, books were copied individually by scribes. A mistake made in copying (or an 'improvement' added by the scribe) would be perpetuated in any copies made from the new manuscript. For example, Geoffrey Chaucer's *Canterbury Tales*, a series of stories told by pilgrims on their way to Canterbury, was written in the late fourteenth century. The original version of the Canterbury Tales has disappeared, but there are around 80 surviving copies. By comparing the similarities and differences of the surviving copies, it is possible to group the manuscripts into a nested hierarchy. A research group including specialists in both literature and evolutionary biology compared 850 lines from one of the tales, *The Wife of Bath's Prologue*, from 58 surviving fifteenth-century manuscripts[1]. They constructed nested hierarchies of shared differences in spelling, word usage, and punctuation, and used this information to draw a diagram indicating the relationships between the manuscripts. They found five distinct groups of related manuscripts, each representing a chain of copies. Importantly, their research also identified a group of key texts that appear to be much closer to the ancestral text, suggesting that these are the manuscripts that will give the most clues to Chaucer's original version.

In fact, any culturally learned behaviour can be interpreted as the result of a copying process with occasional errors. So it is sometimes possible to reconstruct the history of technologies and crafts by reconstructing hierarchies of similarity. For example, the nomadic Turkmen people (from Turkmenistan and neighbouring regions of Iran and Afghanistan) used woven textiles to make both practical and ceremonial artefacts, such as saddle bags and wedding decorations. Carpet weaving is a difficult skill to learn, particularly

Figure TB4.4 Geoffrey Chaucer himself is depicted as a pilgrim in the Ellesmere Manuscript of the *Canterbury Tales*. The Ellesmere Manuscript is one of the oldest surviving copies of the *Tales*, and may have been created not long after Chaucer's death. An electronic version of the *Prologue to the Canterbury Tales*, both in Middle English and modern English, is available at *http://pages.towson.edu/duncan/chaucer/index.htm*. You can download the *Canterbury Tales* from Project Gutenberg (though, alas, a plain text version) from *www.gutenberg.org/etext/2383*.

as Turkmen carpets incorporate intricate designs, so Turkmen girls learned to weave from their mothers over a long period of time. Particular carpet designs were passed from mother to daughter, with the occasional modification. An analysis of Turkmen textiles produced from the eighteenth to twentieth centuries reveals a hierarchy of similarities that result from a pattern of descent with modification, with particular designs copied down generations, just as genes are[2].

References

1. Howe, C.J., Barbrook, A.C., Spencer, M., Robinson, P., Bordalejo, B. and Mooney, L.R. (2001) Manuscript evolution. *Endeavour*, Volume 25, pages 121–126.
2. Tehrani, J. and Collard, M. (2002) Investigating cultural evolution through biological phylogenetic analyses of Turkmen textiles. *Journal of Anthropological Archaeology*, Volume 21, pages 443–463.

CASE STUDY **4.2**

Peopling the Pacific: combining genetic, language, and archaeological data to trace human oceanic settlement

KEYWORDS

linguistics
migration
tree-like data
mitochondria
Y chromosome
cultural evolution

RELATED TECHBOXES

TB 4.4: Copy errors
TB 8.4: Molecular dating

RELATED CASE STUDIES

CS 7.2: Non-tree-like flowers (horizontal gene transfer)
CS 8.2: Molecular detective (tracing infection)

Hurles, M.E., Matisoo-Smith, E., Gray, R.D. and Penny, D. (2003) Untangling Oceanic settlement: the edge of the knowable. *Trends in Ecology and Evolution*, Volume 18, pages 531–539

> *"During major migrations to previously uninhabited lands, we might expect the simultaneous transmission of genes, language and culture. By contrast, we might expect the decoupling of biology, culture and language during other periods. Thus, correspondence between the different forms of data needs to be evaluated rather than assumed."*[1]

Background

The people of Oceania arose from one of the world's greatest ocean-travelling civilizations. Melanesian people have inhabited the islands of Near Oceania (such as Papua New Guinea) for at least 50,000 years. But around 3300 years ago, the first Polynesian people (with the distinctive Lapita culture and artefacts), settled the coastal areas of these islands. Within 200 years, there were Lapita settlements throughout the Pacific, on previously uninhabited islands such as Vanuatu, Fiji, and Samoa. Migrating in ocean-going canoes, Polynesian people reached even the remotest islands of Oceania, such as Hawaii, Rapanui (Easter Island), and Aotearoa (New Zealand). But the precise pattern of Oceanic settlement has been debated: was there a fast, eastward migration of Polynesian people, creating a distinct series of related populations (the 'express train' model), or has Pacific history involved much more movement and exchange between established populations (the 'entangled bank')? Colonists arriving in a new land bring with them their genes, culture, and language, so each of these data sources can show the historical signal of descent with modification. However, it is also possible for the historical signal to be erased, for example when genes mix by

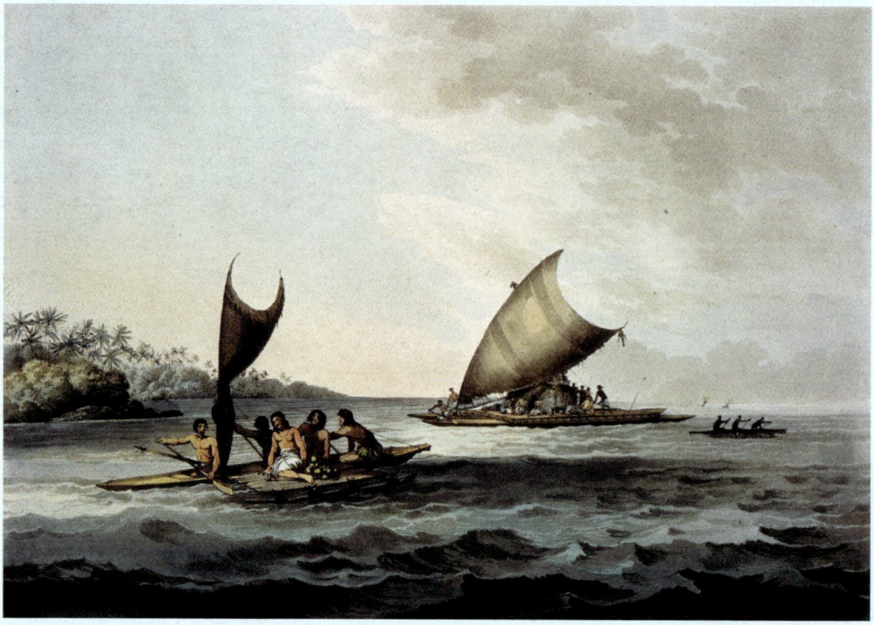

Figure CS4.2a The Polynesians, who travelled between the distant islands of Oceania, are one of history's greatest sea-faring cultures. Some early anthropologists believed that the colonization of the Pacific had been primarily due to chance, with the occasional lucky canoe chancing upon an uninhabited island. But DNA analysis is concordant with the oral history preserved by many Polynesian cultures, which tells of accomplished navigators able to cross vast distances between Oceanic islands in a clearly directed migration, and, in some cases, to return to their island of origin. A website dedicated to Captain Cook's observation of the transit of Venus in New Zealand in 1769 has a good section on oceanic voyages in traditional Polynesian wakas *http://transitofvenus.auckland.ac.nz/wakavoyaging/index.html*.

migration and intermarriage, languages mix through trade or conquest, and cultures mix with the borrowing of technology.

Aim

The researchers wished to compare the express train and entangled bank models of oceanic settlement, by testing the predictions they make about genes, cultures, and languages. If the express train model is the best description of Oceanic settlement, then populations might show a nested hierarchy of similarities, each founding population receiving 'copies' of the genes, language, and culture of its parent population (**TechBox 4.4**). Because a nested hierarchy can be drawn as a phylogenetic tree, such data are described as 'tree-like' (see Chapter 7). If the entangled bank model is a better description of oceanic settlement, then there should be no clear hierarchical pattern, so the data would not be 'tree-like'. The researchers aimed to trace migration through the genetic history of the people, their language, and the commensal animals that travelled in their canoes with them.

Figure CS4.2b (*opposite*) Map of Oceania showing the similarity between languages (a) and genes (b), and the dates of colonization estimated from the archaeological record (c). Shading indicates the approximate coastline during the last glacial maximum. In (b) the distribution of the predominant paternal (blue) and maternal (red) Polynesian lineages among modern populations is shown.

From Hurles, M.E., Matisoo-Smith, E., Gray, R.D. and Penny, D. (2003) Untangling Oceanic settlement: the edge of the knowable. *Trends in Ecology and Evolution*, Volume 18, pages 531–539.

(a)
(b)
(c)
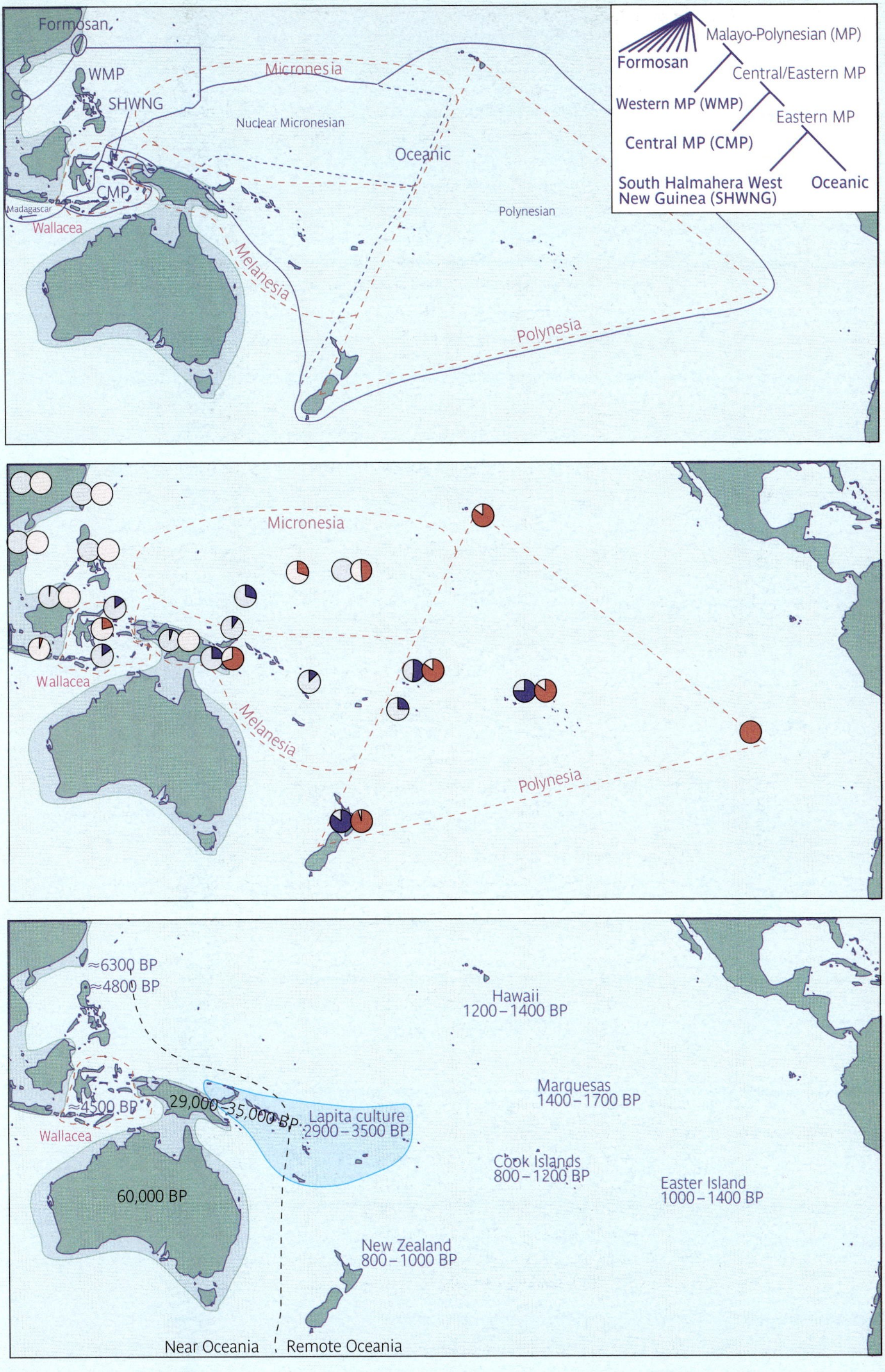
Formosan
WMP
SHWNG
CMP
Madagascar
Wallacea
Micronesia
Nuclear Micronesian
Oceanic
Polynesian
Melanesia
Polynesia
Formosan
Malayo-Polynesian (MP)
Central/Eastern MP
Western MP (WMP)
Eastern MP
Central MP (CMP)
South Halmahera West New Guinea (SHWNG)
Oceanic
Micronesia
Wallacea
Melanesia
Polynesia
≈6300 BP
≈4800 BP
Hawaii
1200 – 1400 BP
≈4500 BP
29,000–35,000 BP
Lapita culture
2900 – 3500 BP
Marquesas
1400 – 1700 BP
Wallacea
Cook Islands
800 – 1200 BP
Easter Island
1000 – 1400 BP
60,000 BP
New Zealand
800 – 1000 BP
Near Oceania
Remote Oceania

Methods

A cross-disciplinary team reviewed analyses based on DNA sequences, linguistic information, and archaeological studies of Polynesian cultures. Over 5000 shared words for over 200 Austronesian languages were coded as a matrix, scoring the presence of cognate words in different languages. For example, the ocean-going canoes used by Polynesian people are known as waqa in Fiji, wa'a in Hawaii, and waka in Aotearoa. DNA sequences have been obtained from Polynesian people from across Oceania, including both mitochondrial DNA (passed on by females) and Y-chromosome sequences (passed on by males). Genetic data from the Pacific rat (*Rattus exulans*) were also analysed. These rats were often carried in ocean-going canoes as a food source by Polynesians as they migrated across the Pacific.

Results

Analysis of the language data show a clearly structured hierarchy, suggesting an eastward chain of colonization, from continental Asia to Remote Oceania. The pattern closely matches the archaeological evidence of the spread of Polynesian people. The mitochondrial data also suggest eastward expansion, with a hierarchy of changes from Asia to Micronesia and Melanesia, to the islands of Remote Oceania. Both mitochondrial and Y-chromosome data suggest that there was little genetic mixture between Melanesians and Polynesian people as the 'express train' of Polynesian colonization passed through. However, genetic data from both contemporary and archaeological specimens of the Pacific rat suggest that there was ongoing exchange between remote oceanic settlements.

Conclusions

The combined data support an eastward migration of Lapita people, as predicted by the express train model of Pacific settlement. This conclusion is also supported by a decrease in both genetic and linguistic diversity from west to east, as new populations were founded by a series of small 'sub-samples' from previous populations (as few as 100 individuals may have initially colonized some islands; see **TechBox 5.3** for the effect of population size on genetic diversity). However, genetic data from humans and rats also suggest that there was postsettlement contact throughout the history of oceanic people, a testament to their remarkable navigational skills that allowed the Polynesians to travel repeatedly between remote islands.

Limitations

Although the different data sources all show a common pattern of eastward migration, with decreasing diversity along the chain of oceanic colonization, there is some disagreement between the datasets. For example, Y-chromosome and mitochondrial data give different estimates of the beginning of the migration of the oceanic people out of Asia (see Chapter 8 for discussion of molecular dates). This discrepancy may have arisen from assumptions made in the analysis, or it could indicate a complex cultural history (e.g. males and females could show quite different migration patterns).

Future work

Sex-specific demographic patterns could be explored by comparing Y-chromosome, mitochondrial, and autosomal DNA sequences. The developing technique of 'glottochronology' (using linguistic datasets to estimate divergence times of languages) may allow the timing of the movement of people across the Pacific to be estimated from language data alone, and compared to genetic and anthropological dates.

Reference

1. Hurles, M.E., Matisoo-Smith, E., Gray, R.D. and Penny, D. (2003) Untangling Oceanic settlement: the edge of the knowable. *Trends in Ecology and Evolution*, Volume 18, pages 531–539.

sequences contribute to the manufacture of a functional RNA polymerase II beta subunit.

This hierarchy of similarity can also been seen in the amino acid sequence for this section of the RNA polymerase II beta protein subunit. If we translate the DNA sequences in **Figure 4.16**, using the series of three-base 'words' (codons), we get the amino acid sequences that we first met in Chapter 2 (**Figure 2.6**). Looking at these amino acid sequences (reproduced in **Figure 4.17**), we can see that (for this small section of the gene) although the three animals (human, rat, and fly) have different DNA sequences, they all translate to the same amino acid sequence. This is because not all changes in the DNA sequence cause a change in the amino acid sequence, due to the redundancy of the genetic code (see **TechBox 2.3**, p. 115).

Figure 4.17

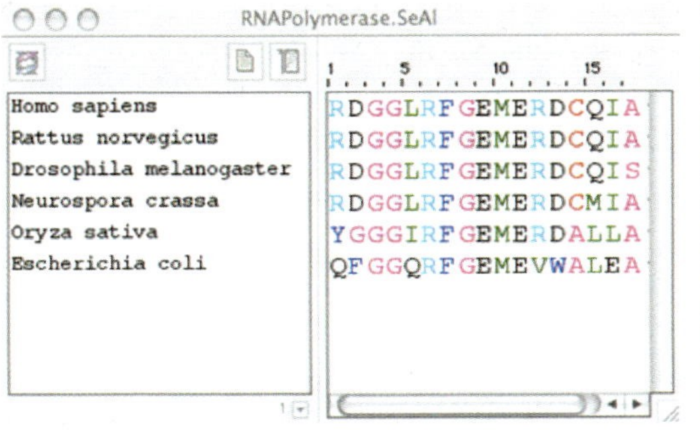

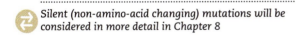

Silent (non-amino-acid changing) mutations will be considered in more detail in Chapter 8

Looking at the nucleotide alignment above (**Figure 4.15**), read along the scale bar at the top to locate position 3529. Where the human has CGT, the rat has CGC. But both of these codons specify the amino acid arginine (R). So although the DNA sequence of the gene changed, the amino acid sequence remained the same: both versions of the sequence make exactly the same protein (**Figure 4.17**). Since this change from T to C in the rat lineage didn't make any difference to the functioning of the RNA polymerase enzyme, it continued to be copied down the rat generations. Similarly, in **Figure 4.16**, you can see at position 3532, TTT (human and rat) and TTC (fly) both code for phenylalanine (F), so when the C mutated to T in a distant mammalian ancestor, it continued to be copied and was eventually inherited by both the rat and human lineages (**Figure 4.16**).

Similarly, the fungus sequence differs from the human sequence by 11 nucleotides, but differs in only one amino acid (M instead of Q at amino acid position 1186: **Figure 4.17**). Rice differs from the animal sequences at six amino acid positions, and the bacterium differs from the animal sequences at eight positions. All of the sequences, from bacteria to humans, have the central amino acid motif RFGEME (arginine–phenylalanine–glycine–glutamine–methionine–glutamine). The universality of this motif suggests two things – firstly, that this sequence is so important to the function of the enzyme that any change to it is likely to be disastrous and, secondly, that this sequence was present in the last common ancestor of animals, plants, fungi, and bacteria.

See ***TechBox 2.3*** *for an explanation of how triplets of DNA nucleotides code for particular amino acids in the protein sequence*

Hierarchies reveal history

Comparing sequences for the same gene across different species gives us an insight into how DNA sequences evolve. By noting that the RFGEME motif is conserved across all these organisms, we can predict that it must have such an important and specific function that it cannot be changed. A quick search of a protein database, such as PFam (www.sanger.ac.uk/Software/Pfam), reveals that this motif is part of domain 7 of the RNA polymerase II beta protein. Domain 7 interacts with another domain of the enzyme to form the clamp that locks the polymerase enzyme onto the DNA strand. So this amino acid sequence has a critical role to play in the formation of the transcription machinery.

Sequence comparisons also provide information about the relationships between species. The action of copying DNA every generation, occasionally passing on a mutation, leaves a historical record in the genes. Shared inherited changes reveal lines of descent. We can draw the shared sites of this short amino acid sequence as a nested hierarchy. All of the animals have the same sequence, which differs at one position from fungi. Plants share 10 of these amino acids with fungi and animals, and the bacterium shares only eight amino acids with the other species (**Figure 4.18**). This pattern is unsurprising, given that human, rat, and fly are all more closely related to each other than each is to fungi, and that animals and fungi are more closely related to each other than either is to plants. Bacteria are the most distantly related. So this hierarchy of shared changes reflects the evolutionary relationships between the species. Indeed, we should expect that most heritable characters will show a comparable pattern of shared similarities.

Figure 4.18

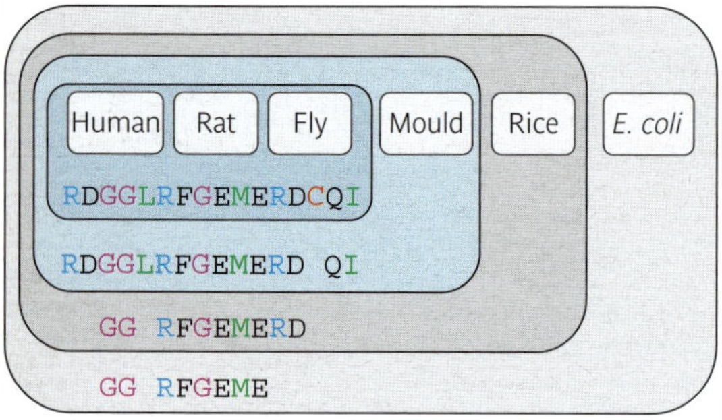

If we choose a different gene and compare it between species we are likely to find a similar pattern. For example, you can compare the gene for threonyl-tRNA synthetase between these species. This gene codes for another key component of the genomic system. Threonyl-tRNA synthetase is an enzyme that binds the amino acid threonine to the right transfer RNA (tRNA) molecule, which will then carry threonine to the ribosome so that it can be incorporated into a newly synthesized protein molecule (see **Figure 2.5**). You would find that the protein sequences of the human and rat versions of threonyl-tRNA synthetase were most similar (96% of amino acids the same), then the fruit fly (75%), then the mould (56%), then rice (55%; you can get this information from the Homologene database at www.ncbi.nlm.nih.gov). Less than 50% of this sequence is identical between all of the species including the bacterium *E. coli*.

We will see why the percentage divergence does not increase linearly with evolutionary time in Chapter 8

In fact, we would expect to see a similar pattern of a nested hierarchy of similarities for other heritable traits. The two mammals are very similar in body organization, in which they differ from the fly. But all three animals share a great many features of cell structure and organization, which are not shared with fungi or plants. Plants, fungi, and animals all share features of the eukaryotic cell – such as the cell nucleus – that differ from bacterial cells. All six species – human, rat, fly, mould, rice, and *E. coli* – share features of the DNA–RNA–protein genomic system, common to all life on Earth.

The evolutionary relevance of homologies (characters shared by descent) will be discussed in Chapter 6

Hierarchies of similarity in *Bacillus* bacteria

This hierarchy of similarities can be seen at all levels of biological relationships. We have seen how the RNA polymerase II beta gene can be compared between the different kingdoms of life. Now let us use a different part of the gene sequence to look at the way the same gene can reveal relationships between closely related populations and species. **Figure 4.19** shows a section of

Figure 4.19

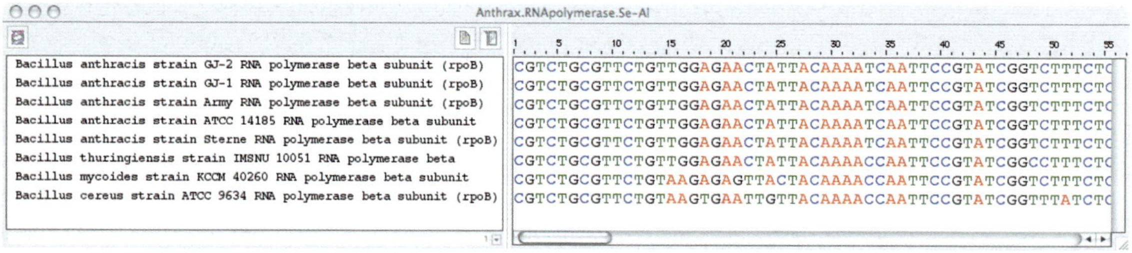

Figure 4.20 The beauty of bacteria: colonies of *Bacillus anthracis*.

CDC/ Courtesy of Larry Stauffer, Oregon State Public Health Laboratory.

RNA polymerase II beta gene from a number of different types of *Bacillus* bacteria, a group of spore-forming bacteria found in soil, water, and dust (**Figure 4.20**). Most *Bacillus* are harmless to humans. Some *Bacillus* are actually useful, such as *Bacillus thuringiensis*, the original source of the Bt insecticide now widely used in agriculture (see also **TechBox 3.1**). Bt insecticide can be sprayed on plants, or produced by crops genetically modified to carry the *Bt* gene which was isolated from *B. thuringiensis*. (Interestingly, Bt spray is allowed under organic farming regulations, but GM crops that produce their own Bt are not.)

One particular species of *Bacillus* is deadly to humans. If *Bacillus anthracis* enters the human body through a cut on the skin, it can cause nasty black ulcers to develop (cutaneous anthrax). If *Bacillus anthracis* enters the lungs or digestive tract, it can cause a potentially fatal illness (inhalation or gastrointestinal anthrax). Anthrax can be passed from cows and sheep to humans, but cannot be passed from human to human. Its ability to form tough spores makes anthrax an ideal biological warfare agent, because it can be dispersed in powder form, and can survive a wide variety of conditions, even low-level radiation (**Figure 4.21**, p. 134). The importance of anthrax to human health and agriculture, and its potential use in bioterrorism, makes reliable identification of suspected anthrax samples essential.

We could draw a nested hierarchy of genetic differences between this sequence from different *Bacillus* strains, just as we did between members of different kingdoms. **Figure 4.22** (p. 134) gives the percentage similarity between RNA polymerase II beta genes between various types of *Bacillus* bacteria.

We can see that, for this section of the RNA polymerase II beta gene, the nucleotide sequence is identical between different strains of *Bacillus anthracis*. But this sequence differs from other *Bacillus* species. This observation suggests that DNA sequencing could be used to rapidly identify the presence of anthrax in a suspected bioterrorist sample. You could extract DNA from the sample, and use PCR to amplify the RNA polymerase II beta gene, then sequence it. If the sequence generated was identical to the *B. anthracis* sequence, you could be fairly confident that the sample contained anthrax.

The use of DNA to identify species (sometimes called DNA barcoding) is discussed in Chapter 6

Hierarchies of similarities exist within species as well. Your genome will be more similar to your cousin's than it will be to an unrelated individual, because your genome and your cousin's were both copied from the same grandparents' genomes. The same principle is true for *Bacillus*. Strains that are recently derived from a

Figure 4.21 Robert Koch's work on *Bacillus* was pioneering not only in his use of micrographs to make images of bacteria, but also because his study of anthrax was, arguably, the first case where an infectious agent was proved beyond doubt to be the cause of a disease. Koch's work on this and other infectious diseases earned him a Nobel Prize in 1905, and his approach continues to be used today by those who follow 'Koch's postulates' to demonstrate disease causation: association (agent found in affected but not unaffected individuals), isolation (disease-causing organism is isolated from affected individuals), inoculation (isolated organism causes disease in healthy individuals), and re-isolation (to prove that the original organism can be recovered from inoculated individuals).

Figure 4.22

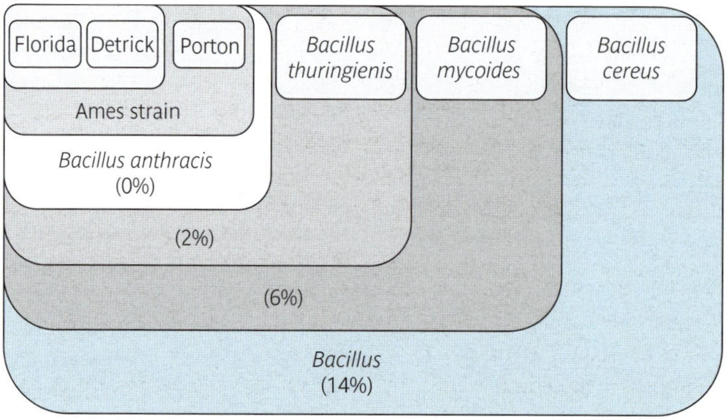

Approximate percentage difference between the sequences of the RNA polymerase II beta gene from three strains of *Bacillus anthracis* (one isolated from the Florida victim of the 2001 bioterrorist attacks, and two from military research facilities in the US (Fort Detrick) and the UK (Porton Down)), and three other species of *Bacillus*.

common stock will have more similar genomes than more distantly related strains. So although the RNA polymerase II beta gene may be identical in all *Bacillus anthracis*, there are other genomic differences that reveal the history of particular anthrax samples.

In 2001, when America was reeling from the shock of the attacks on the World Trade Centre and Pentagon, five people died and 17 more were infected by anthrax spores that had been sent through the post. The perpetrators of this bioterrorist attack have still not been identified, but scientists have explored the possibility of using DNA sequence data to trace the source of the outbreaks. They compared DNA sequences from *Bacillus anthracis* isolated from the first identified victim of the attacks, a journalist from Florida, with a whole genome sequence of a laboratory strain of anthrax from the UK military research facility Porton Down. Previous research has suggested that anthrax genomes are all nearly identical, making identification of different strains difficult. But by using whole or nearly complete genome sequences, this research team were able to identify a number of polymorphic sites that differed between strains, including single nucleotide differences (SNPs), repeat regions (VNTRs), and insertions and deletions (indels). Some of these changes may have no noticeable effect on phenotype, but several were predicted to change important genes, and thus might have an effect on the activity or infectivity of the Florida strain.

SNPs, repeat regions, and indels are introduced in Chapter 3

Furthermore, the genetic analysis of *Bacillus anthracis* strains demonstrated that these genetic differences can arise very rapidly, even in as little as a few years. This suggests that whole genome sequences may be used in future to identify the sources of outbreaks, even for a bacterial species previously considered to show little genetic variation. In this case, although the perpetrator of the attacks has never been identified, the investigation demonstrated that the Florida bioterrorist anthrax was most similar to the strain maintained and distributed by the US military research facility Fort Detrick.

The use of genetic differences to genotype individuals or identify strains is covered in Chapter 3

Conclusions

Understanding DNA replication is important for all biologists, not just biochemists or geneticists. Everyone who wants to understand the basis of life on Earth should familiarize themselves with the way that the information in DNA is copied by template reproduction. The astounding complexity of the DNA replication machinery ensures that the genome is copied accurately and efficiently at cell division. This process is considerably simplified in the laboratory, where the entire DNA sample is melted to expose single strands, primers are added to provide a starting point for a domesticated polymerase enzyme to work from, and the sample run through repeated cycles of melting and copying to produce millions of copies of the same DNA sequence.

No copying process is perfect. Errors in laboratory DNA amplification are to be avoided at all costs. But ironically it is errors in natural DNA replication that provide us with one of the most useful tools in biology. Errors in DNA replication can be passed on to the next generation, thus the chain of copies leaves a record of its history. An organism's history can therefore be reconstructed by considering the hierarchy of similarities between its genome and the genome of other individuals, strains, or species. We have seen here how comparing gene sequences can illuminate the relationships between the great kingdoms of life, or the origin of particular bacterial strain used in a bioterrorist attack. We can also see that the hierarchy is predictive: we can identify the putative dinosaur DNA as a human contaminant because we expect dinosaur sequences to be more similar to sequences from close relatives such as birds and reptiles, than to sequences from a more distantly related mammal. It is the hierarchy of DNA differences that allows us to read stories from DNA, revealing history, current patterns, and evolutionary processes.

Taking stock: what have we learned so far and where are we going next?

In the first chapter, we took an overview of the evolutionary information in DNA. We saw how it can be used to shed light on individuals, populations, and lineages. To understand how

DNA stores information, we took a brief look at DNA structure in Chapter 2. We saw how the complementary base pairing of nucleotides provides a means of transferring information between nucleic acids, providing a pathway for the genetic instructions to move from the genome to the sites of production of the working parts of a cell, as well as a mechanism for the transfer of information between generations. In Chapter 3 we saw how mutation changes the information in the genome, by altering the way genes function, and by creating inherited differences that will be passed on to descendants. In Chapter 4 we learned how DNA is copied almost, but not entirely, without error. The process of copying with occasional errors creates the historical information in DNA by creating a chain of increasingly dissimilar genomes.

Thus far, we have primarily focused on individual genomes, considering the way the genetic information is expressed, altered, and passed on to offspring. In the next four chapters we are going to move beyond the forces that affect individuals, and begin to take the long view. In Chapter 5 we will take the leap from individuals to populations in order to see how a mutation that occurs in a single individual can become a standard part of the genome of all members of a population. Then in Chapter 6 we move from the population to the species, as we learn how to compare the genetic differences between species and higher taxa. Chapter 7 leads us from the species to the lineage, by considering how the differences in DNA sequences can be used to reconstruct the evolutionary history of biological lineages. Finally, in Chapter 8, we will use our understanding of the processes of molecular evolution to explore the possibility of using DNA sequences to estimate the timescale of evolutionary change.

Further information

A useful review of male-biased mutation is given in:

Ellegren, H. (2007) Characteristics, causes and evolutionary consequences of male-biased mutation. *Proceedings of the Royal Society of London B*, Volume 274, pages 1–10.

Any good biochemistry or genetics text will explain DNA replication. The following is a very friendly but detailed account:

Elliott, W.H. and Elliott, D.C. (2005) *Biochemistry and Molecular Biology*, 3rd edn. Oxford University Press.

5

Descent with modification

Or: how do I detect natural selection?

"... can we doubt (remembering that many more individuals are born than can possibly survive) that individuals having any advantage, however slight, over others, would have the best chance of surviving and procreating their kind? On the other hand, we may feel sure that any variation in the least degree injurious would be rigidly destroyed. This preservation of favourable individual differences and variations, and the destruction of those which are injurious, I have called Natural Selection, or the Survival of the Fittest. Variations neither useful nor injurious would not be affected by natural selection, and would be left either a fluctuating element, as perhaps we see in certain polymorphic species, or would ultimately become fixed, owing to the nature of the organism and the nature of the conditions."

Darwin, C. (1872) *The Origin of Species by Means of Natural Selection, or the Preservation of Favoured Races in the Struggle for Life*, 6th edn. John Murray

What this chapter is about

Mutation creates differences among individuals in a population. Some of these mutations will be lost when their carriers fail to reproduce. But some mutations will increase in frequency with each passing generation until they replace all alternative alleles in the population. The process of molecular evolution is, at its most basic, the substitution of one base in the DNA sequence for another, so that all members of a population carry a copy of the same mutation. In this chapter, we will explore two ways that a mutation can become a substitution. Firstly, some mutations influence their own chance of being copied to the next generation, and these mutations can increase by positive selection until they replace all alternative alleles, or decrease by negative selection until they are lost from the population. Secondly, the frequency of alleles in a population can fluctuate randomly, drifting up and down, and these fluctuations occasionally result in the loss of all alleles but one. Both of these processes – selection and drift – are evident when we compare DNA sequences.

Key concepts

Evolutionary biology: selection and drift

Molecular evolution: substitution

Techniques: detecting selection

Variation

You will notice that quotes from Charles Darwin keep popping up in this book (**Figure 5.1**, p. 139). This is partly due to the unshakeable hero worship that I, along with many evolutionary biologists, have for Darwin. It is hard to think of anyone who has had a greater impact than Darwin on the way that humans understand themselves and the world around them. But more importantly, the frequency of quotes from Charles Darwin is a demonstration of the quality of Darwin's writing on evolutionary biology, and the extent of his remarkable insight into the workings of the natural world. In fact, why don't you put down this book and go and find a copy of *The Origin of Species* and read it now? Much has changed in evolutionary biology in the 150 years since Darwin wrote *The Origin of Species*, most importantly the discovery of the molecular basis of inheritance. Yet there is still no better book on evolution than the book that started the whole field, a book usually referred to, rather appropriately, as *The Origin*.

The Origin of Species is not Darwin's only great work, though it is his most famous and arguably his most important. Before he wrote *The Origin*, he wrote a popular account of his travels on the Beagle, a survey ship in which he circumnavigated the world. This journey gave Darwin many valuable opportunities to make

Figure 5.1 Charles Darwin, who often employed his children's help in making natural history observations around their home in Kent. For example, Darwin studied the path-following behaviour of bees for many years, describing the way bees stopped at particular sites (which he termed buzzing places): 'I was able to prove this by stationing five or six of my children each close to a buzzing place, and telling the one farthest away to shout out "here is a bee" as soon as one was buzzing around. The others followed this up, so that the same cry of "here is a bee" was passed on from child to child without interruption until the bees reached the buzzing place where I myself was standing.' A nice example of inclusive fitness.

Photograph: J. Cameron, 1869.

observations on geology and natural history. After *The Origin*, he tackled the thorny subject of human evolution in *The Descent of Man* (1871), which is notable for its exploration of sexual selection (the influence of mate choice on the evolution of morphological and behavioural traits). As long as you can cope with typically lengthy Victorian prose, Darwin's books are often entertaining as well as enlightening. This is particularly true of *The Expression of the Emotions in Man and Animals* (1872) which is a delight to read (and, incidentally, one of the first scientific publications to use photographs).

Darwin's curiosity about the natural world was boundless, as was his capacity for careful and thorough observation: any one of his books is a primer on how to be an effective scientist. For example, in *The Formation of the Vegetable Mould, Through the Action of Worms, with Observations on their Habits* (1881), Darwin reports his systematic investigations into the contribution of earthworms to the formation of soil through an exhaustive series of behavioural experiments and observations in the field (he concludes that 'worms have played a more important part in the history of the world than most persons would at first suppose' and that 'worms, although standing low in the scale of organization, possess some degree of intelligence'). But, unlike *The Origin* or *The Descent of Man*, the majority of Darwin's books – on domestication, formation of coral reefs and islands, orchids, climbing plants, carnivorous plants, and pollination biology – are rarely read nowadays.

One such work is Darwin's four-volume taxonomic treatise on barnacles, published between 1851 and 1854. This work represents 8 years' worth of careful dissections of barnacles and detailed examinations of fossils, collected by Darwin himself on the Beagle voyage or sent to him by his global network of correspondents. Darwin's barnacle work became such an all-consuming project that one of his children is said to have asked a friend 'where does your daddy do his barnacles?', on the assumption that every father spent their days absorbed in the dissection of tiny crustaceans. Darwin, whose close observations of the natural world continued to instil in him 'awe before the mystery of life', described the form of his 'beloved barnacles' in such rapturous terms that his children likened his words to an advertisement. However, by the end of the 8 years he grew understandably weary of the little creatures and declared 'I hate a Barnacle as no man ever did before, not even a Sailor in a slow-sailing ship'. Which leads us to an important and perplexing question: why would one of history's greatest thinkers devote such a large amount of time to dissecting barnacles and describing the minute differences between them? Shouldn't he have been working on something more important and interesting?

Thomas Henry Huxley, a friend of Darwin's and staunch promoter of evolutionary biology, wrote to Darwin's son Francis that 'in my opinion your sagacious father never did a wiser thing than when he devoted himself to the years of patient toil that the Cirripede [barnacle] book cost him'. Huxley considered that, while Darwin had gained field experience in geology and natural history from the Beagle voyage, he needed a solid basis in anatomy and development on which to build his theory of natural selection. Darwin's work on barnacles was critically important to the formation of evolutionary theory because it was a key demonstration of the ubiquity of variation in natural populations. Variation is the raw material of evolution. It is because individuals within a population are not all the same that evolution occurs.

Substitution

In Chapters 3 and 4, we saw how mutation is responsible for generating variation. Although DNA is copied with extraordinary accuracy and nearly all damage to DNA is repaired, the occasional mutation slips through the net and becomes a permanent change to an individual genome. Mutation occurs sufficiently frequently that, for many types of organisms, each individual has a unique genome. Furthermore, these mutations can be passed to the offspring, so that related individuals will share many of the same mutations. Mutation creates a wealth of heritable variation. Variation is the fuel of evolution. Some variants eventually become fixed in the population so that all members of the population now carry that mutation. In molecular evolutionary terms, this process is referred to as substitution: one variant has been substituted for another. In this chapter, we will look at how a mutation (a permanent change in a particular genome) becomes a substitution (when that mutation is now carried by all members of a population).

We should start by clarifying some of the terms we will be using in this chapter. I think the most useful definition of evolution is Darwin's phrase: descent with modification. 'Descent' emphasizes continuity of heritable information: that change over many generations is what we are concerned with, not the changes that happen to an individual in its lifetime. 'Modification' reminds us that it is change itself that is the stuff of evolution, which need not always be adaptation.

We also need to describe units of genetic variation, and the most flexible term we can adopt is 'allele'. An individual chromosome carries one possible allele (genetic variant) at any given locus (point in the genome). Mutation generates a new allele which can be inherited by its carrier's descendants. Each new allele starts as a mutation in a single individual, but through the process of descent can come to be carried by an increasing (or decreasing) proportion of individuals in the population with each passing generation. The basic process of molecular evolution is the substitution of one allele for another, which occurs when a single allele increases in frequency to the point where it replaces all other alleles at that locus. At this point, all members of a population carry a copy of the same allele, and we say the allele has reached fixation (or become fixed).

To understand the process of substitution, we need to think about the effect a particular mutation has on the organism that carries it, and the processes that can cause the frequency of a particular mutation to increase or decrease in the population. Specifically, we need to think about the roles of natural selection and chance events in determining allele frequencies.

Natural selection

Darwin was not the first to propose that the living world was the product of change. The debate about the history of the natural world had raged in both academic circles and in the general public for decades. But Darwin brought about a revolution in two ways. Firstly, he amassed an impressive catalogue of evidence in favour of transformation of species, from his observations of fancy pigeon breeds to his description of finding fossils of giant ground sloths near their modern day relatives in South America. Secondly, Darwin did what no-one had done before: he provided a plausible mechanism for the transformation of species.

Darwin's argument can be summarized as follows. Individuals in a population vary, in a myriad of ways, from the most trivial differences in minor features to critical alterations to important traits. These variations are often heritable, so that offspring tend to resemble their parents. Any heritable variation that increases the chance of an individual producing successful descendents will tend to be perpetuated down the generations, because successful reproducers will tend to produce offspring with the same characteristics that promoted their own reproductive success. Because they result in more successful offspring, these advantageous heritable variants will increase in frequency in the population at the expense of other less successful variants, eventually replacing all other variants. And that's natural selection. Simple, isn't it?

Natural selection is the name we give to the process that occurs when heritable variation influences the number of copies made of different variants. It can apply to any population of things that have heredity (they copy) and variation (not all copies are identical). In common with the rest of life on Earth, you are a template-copier, because you use your own genome to make a copy of your genetic information to give to your offspring. Mutations that change the template will be passed on to all copies made from that template. Any change to the template that increases the number of copies made from it will be present in more copies, naturally. So the world becomes full of copiers that are good at copying. This simple logic that explains so much of our world is just as mind-blowing today as it was one hundred and fifty years ago when *The Origin* was published.

Consider a heritable change that makes baby spiders feed on their mother's body. Matriphagy (mother-eating) might give those spiderlings a nutritional head start in life, and thereby increase the chances that they will survive to maturity and become mothers themselves (and get eaten by their own offspring who have also inherited the mother-eating tendency). This reproductive success could result in an increase in the number of mother-eating spiders next generation, each of which will also stand a good chance of having baby spiders that survive to eat their own mothers. By and by, within that particular spider population, spiders that eat their mothers may outnumber those that don't by virtue of their nutritional advantage in early life. Evidently, this has not happened to all spider species. But it has happened in several species in which all offspring now consume their mothers (**Figure 5.2**). Are mother-eating spiders 'better' than their non-cannibalistic relatives? That can only be a matter of personal opinion. But we can definitely say that matriphagy is, in this particular case, an adaptation, because it is a heritable variant that increases its own chance of transmission. This is a case of positive selection: a mutation that has a positive effect on its own chances of being copied is a mutation that will increase in frequency each generation.

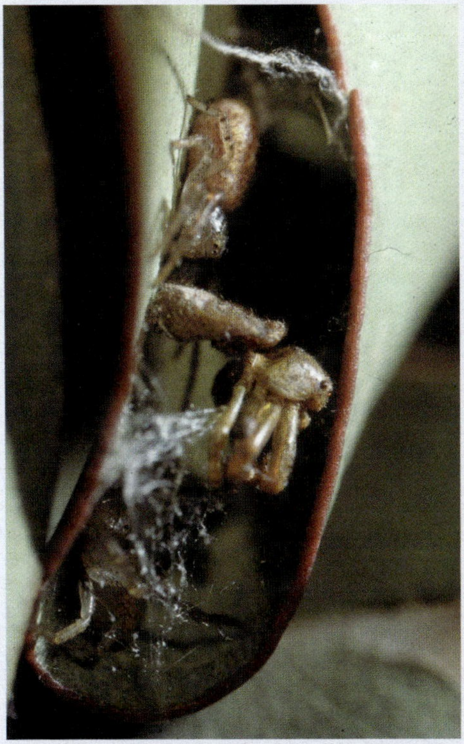

Figure 5.2 All spiders provide some degree of maternal care, such as constructing a safe environment for eggs to hatch or bringing prey items for spiderlings to eat. In the extreme case, the mother allows herself to be cannibalized by her own offspring. This *Diaea* female has been consumed by her young, who sucked her body fluid out from her elbows (Evans, T. *et al.* 1995). Obviously, species with suicidal maternal care produce only one brood of eggs per female, but the increased mass and dispersal ability of the offspring increases the reproductive success of the cannibalized female.
Photograph courtesy of Theo Evans.

The flipside of positive selection is negative selection: a mutation that reduces the chances of its carrier reproducing will decrease in the population with each passing generation. For example, when a female *Drosophila subobscura* (fruit fly) encounters a male, she tests his mettle in a dancing competition. The female moves rapidly side to side, and the male must follow her movements in order to keep facing her. If the female out-dances the male, she rejects him. If he keeps up with her fancy footwork, he is allowed to mate with her. So a male fruit fly may be possessed of all the faculties he needs to survive and flourish, he may be as fertile as any other fly, and devilishly handsome to boot. But if he can't dance, his genes are going nowhere. Any mutation that reduces dancing ability is less likely to find its way into the next generation of *Drosophila*. So with every passing generation any mutations that reduce dancing ability will decline in frequency, as the lousy dancers will tend to have fewer offspring than the good dancers.

Gradualism

The impact of Darwin's *The Origin of Species* was immediate and profound. Many scientists who read it were immediately convinced of its explanatory power. Others remained sceptical, or in some cases down-right hostile to the idea. On the whole, though, *The Origin* brought about a rapid change in the world view of the biological community. By 1872 when the last edition of *The Origin* was published, Darwin could state with some satisfaction that 'almost every naturalist admits the great principle of evolution'. But although *The Origin* marked a turning point for the acceptance of the transformation of species and the evolutionary history of the natural world, it took considerably longer for Darwin's proposed mechanism of change – natural selection – to be universally accepted as the primary agent of adaptive evolution. For example, Thomas Henry Huxley, that tenacious defender of Darwinism, was 'prepared to go to the stake if requisite' for Darwin's arguments about the geographic and geological distribution of species. However, he was less convinced by Darwin's insistence on gradual change.

Darwin based his theory of evolution by natural selection on the principle of gradualism: that descent with modification occurred by continuous accumulation of many small changes. Darwin followed the lead of the founding father of modern geology, Sir Charles Lyell, in explaining deep history in terms of processes that could be witnessed today. Lyell's central argument can be summarized as 'the present is the key to the past'. He explained major geological events, such as mountain building and change in the course of rivers, in terms of the gradual and continuous action of observable forces such as sedimentation, erosion, and uplift. Similarly, Darwin explained the massive changes in the biological world – the transformation of species over time – in terms of the accumulation of many small, heritable variations leading to the formation of distinct varieties and races, processes that any naturalist could witness working on a small scale today. It is important to note that hypothesis of gradualism does not concern the speed of change: Darwin recognized that evolutionary change may run faster or slower at different periods. But Darwin insisted that, for the most part, we can explain the large changes of the past by considering the cumulative effect of many tiny differences, continuously accumulated over thousands of generations.

We will explore the pace of evolutionary change in Chapter 8

But this gradualist explanation did not satisfy some scientists, including Thomas Henry Huxley. He was suspicious that there was no experimental evidence for the origin of new species, and remained convinced that there was a role for heritable changes of large effect in the process of adaptation and the formation of new kinds. It was not until after Darwin's lifetime that the power of natural selection to bring about evolutionary change was demonstrated by a combination of laboratory experiments, field studies, and mathematical models. Laboratory studies, such as those on the fruit fly *Drosophila*, showed that new mutations arise constantly, and that these mutations can rise in frequency under particular conditions or due to selective breeding. Field studies, such as those on predation of stripy-shelled snails in English hedgerows, showed that heritable variations can influence the chances of survival and reproduction of individuals in the wild (**Figure 5.3**). From the point of view of understanding patterns of molecular evolution, the most important advances came from setting population genetics within a mathematical framework.

Figure 5.3 Nice and stripy: the variation in banding patterns on the snail *Cepaea nemoralis* was initially interpreted as neutral variation, but classic studies in the 1950s showed that some banding patterns were more likely to end up as thrush food than others, suggesting a role for banding and colour in predator protection.

Photograph © Andrew Dunn, 1 May 2005.

The power of selection

Natural selection is a numbers game, a consequence of reproducing entities with heritable variations that result in differential numbers of copies. Yet while *The Origin* included a vast catalogue of observations and inferences, it contained no maths, not one single equation (and only one diagram, but more of that in Chapter 7). In the 1920s and 1930s, a triumvirate of scientists were instrumental in putting Darwin's theory of evolution by natural selection into a firm mathematical framework: R.A. Fisher (**Figure 5.29**, p. 182), founder of many of the statistical techniques biologists use today, who was also a keen promoter of eugenics; Sewall Wright (**Figure 5.4**), whose analysis of agricultural breeding programmes led to new analytical methods and to the enduring 'fitness landscape' model of adaptive evolution; and J.B.S. Haldane, a brilliant and outrageous polymath who illuminated the implications of selection and whose socialist beliefs led him to be a committed popularizer of biology (**Figure 5.5**).

The mathematical theory of natural selection is based in probabilities. That is, it describes the likely set of outcomes of a repeated process, rather than predicting with certainty the outcome of a particular case. The important thing about thinking in terms of probabilities is that you cannot predict the fate of a particular mutation, but you can make statements about the overall expected outcome of many mutations within a given population.

Let's think about the fate of a single mutation. A mutation arises in a single individual's genome. If this individual breeds, then that mutation has a chance of being copied to the next generation. If the individual does not reproduce, or all their offspring die, then that mutation will disappear. The loss of a new mutation might occur because the mutation itself reduces the chance of that individual successfully reproducing (negative selection). But many new mutations will be lost by chance

Figure 5.4 Sewall Wright (left) and John Maynard Smith. Wright denied ever using a guinea pig to clean a blackboard (see Crow 1988).

Photograph courtesy of Dr James Bull, University of Texas.

Figure 5.5 J.B.S. Haldane was described (by Peter Medawar) as 'in some ways the cleverest and in others the silliest man I have ever known'. Unlike R.A. Fisher and John Maynard Smith, J.B.S. Haldane's reading glasses did not prevent him from enlisting, and he admitted to quite enjoying the First World War, despite receiving several serious injuries while serving with the Black Watch (he later pondered whether he had not actually died in the war and imagined the rest of his 'rather outrageous' life). He was more of a danger to himself than the enemy was, at one point riding a bicycle across the front line, accurately predicting that the German solders would be too stunned to shoot him. His life-endangering activities continued in civilian life as he frequently used himself as a guinea-pig in experiments to test the limits of human physiology. If you read only one biography of a scientist, for entertainment's sake choose one about Haldane (e.g. Clark 1968).
Photograph courtesy of University of Texas.

events, when their carrier is hit by a bus, falls into a swamp, or fails to find sufficient water in a drought. Because every mutation must start at very low frequency, very many will be lost when their carrier fails to reproduce, regardless of the effect of the mutation on phenotype.

However, if a new mutation is not lost, and is copied to more individuals in the subsequent generations, then its fate is subject to two forces. One of these forces is random: the carriers of any given allele are subject to the same blind chance that affects us all, which can increase or decrease the frequency of an allele. The other force is deterministic: natural selection promotes the frequency of some alleles at the expense of others by virtue of their effect on the probability of survival and reproduction. We are not going to look at any of the details of the mathematical theory of natural selection here, suffice to say that it was demonstrated that, in a large population, a mutation with only a tiny relative advantage could be driven to fixation by natural selection.

In summary, mutations occur continuously, but most will be lost when their carriers fail to reproduce. Of those that are not lost, most will be either deleterious or harmless, but some will, by a lucky chance, confer slight benefits. We can expect some of these beneficial mutations to go to fixation by natural selection. The substitution of a mutation that makes an almost immeasurable difference to phenotype may seem a hopelessly inefficient means to evolutionary change. But natural selection is a numbers game that plays out in populations consisting of hundreds, thousands, or even millions of individual genomes, each of which has a small chance of producing a beneficial mutation, and each beneficial mutation has a small chance of going to fixation. And the ages of species are measured in hundreds of thousands of generations, so even a small rate of fixation would allow a species to accumulate a large net amount of change, even if each selected mutation had only a tiny effect.

Why are most mutations harmful? See Chapter 3

Novel mutations and standing variation

We can actually witness the process of substitution in organisms with rapid generation turnover and high selection pressures. Human immunodeficiency virus (HIV) has an average mutation rate of 0.2 mutations every time the genome is copied. This means that, on average, one in every five new HIV virions is likely carry a novel mutation. These mutations will normally be harmful to their carrier (reduce the virus' chance of replication) or occasionally harmless (have no effect on the chance of the virus replicating). But every now and then, one of these new mutations might increase the chance of the mutant virus replicating. For example, a mutation in the HIV genome that makes the viral particle unrecognizable to the immune system will allow its carrier to replicate freely, and therefore produce more descendants than viruses with alternative alleles that cause them to be recognized and suppressed by the immune system.

Large populations have more chance of beneficial mutations arising, simply because they contain more genomes, each of which has a chance of producing a new, potentially useful, mutation. Compare an infected human with 100 HIV virions in their body to a human with one million HIV virions. The large population of HIV will produce 10,000 times the number of new mutations per day than the small population will. So although the chance of any given mutation being beneficial remains the same, there is more chance that at least one of those million viruses will produce a mutant offspring that can outwit the immune system or resist the effect of drugs and go on to produce a whole new generation of HIV virions. If you had one million people with lottery tickets, your population would be more likely to include a winner than if you only had ten people with tickets.

But selection does not always need to wait for a new mutation to occur. The response to selection will often be to promote the frequency of variants already in the population: that is, selection commonly acts on standing variation. A change in the environment may give existing variants a new advantage (see **Case Study 5.2**). For example, in the famous case of Darwin's finches on the Galapagos islands, in a drought year the average beak size increased, because big-beaked birds could crack harder seeds, and could therefore eat a greater range of seeds than the smaller-beaked birds. The drought did not cause those big-beak variants to occur, they were already there, even in the good years. But in a drought year, they stood a better chance of reproducing than their small-beaked comrades, so there were relatively more copies of alleles contributing to large-beakedness in the generation following the drought. Strong directional selection can reduce the supply of standing variation in the population because, as selection pushes a single variant to fixation, all alternative versions of that trait are lost. Many artificial selection experiments report a decline in the response to selection over time, which may reflect a depletion of the available genetic variation in the population.

The influence of environment

When John Maynard Smith, one of the greatest evolutionary biologists after Darwin, joined the war effort in Britain in 1942, his poor eyesight prevented him from joining the army (**Figure 5.4**). He commented that, in his generation, bad eyesight was a selective advantage in that it prevented him from getting shot (instead he designed aircraft, a field that would lead him to evolutionary biology through consideration of flight in animals). The point of this story is that whether a trait is 'good' or 'bad' depends on the circumstances. Fitness is relative: the selective advantage of an allele must be considered relative to other alternative alleles in the population in the particular environment in which it arises (see **TechBox 5.1**, p. 155). The 'environment' of an allele has many aspects. We immediately think of the external conditions of the population – climate, predators, food sources – but environment also includes other members of the population and even the other mutations in the genome. And if the environment changes, the fitness of the mutation may change with it.

We can see the dramatic effect of changing environment on the fitness of an allele by considering the human metabolic disorder, phenylketonuria (PKU). PKU is caused by mutation in the *phenylalalanine hydroxylase* gene (*PAH*). There are over 400 different mutations in the *PAH* gene that cause PKU, but the phenotype is the same for all of them. A child born without a working copy of *PAH* cannot make the phenylalanine hydroxylase enzyme, and therefore cannot metabolize the amino acid phenylalanine, which builds up in their brain causing the gradual onset of mental retardation. Since a person affected with PKU is unlikely to reach

CASE STUDY 5.1

Sweet and sour: evolution of taste receptors in vertebrates through gene duplication and loss

KEYWORDS

BLAST

positive selection

deletion

pseudogene

paralogue

cDNA

intron

exon

ORF

RELATED TECHBOXES

TB 3.4: BLAST

TB 5.4: Detecting positive selection

RELATED CASE STUDIES

CS 3.1: Viruses within (BLAST)

CS 7.2: Non-tree-like flowers (horizontal gene transfer)

Shi, P. and Zhang, J. (2006) Contrasting modes of evolution between vertebrate sweet/umami receptor genes and bitter receptor genes. *Molecular Biology and Evolution*, Volume 23, pages 292–300

> *The past five years have brought the powerful combination of molecular biology, electrophysiology, genetics, and behaviour to bear on the problem of peripheral taste and revolutionized the taste field.* [1]

Background

Taste is an important sense for most vertebrates, allowing them to detect useful nutrients and avoid dangerous toxins. There are five basic tastes: sweet, sour, salty, bitter, and umami (a savoury taste, due to glutamates found in aged or fermented foods such as parmesan cheese and soy sauce). Detection of these tastes is discrete: particular cells can detect particular tastes[1]. Taste receptor cells use specific G-protein-coupled receptors (GPCRs) to detect various chemicals (a vast array of GPCRs are involved in other stimulus–response pathways, such as vision, smell, behaviour, and immune response). In particular, the T1R family of receptor proteins allow detection of sweet and umami tastes, and the T2R receptors allow detection of bitter tastes.

Aim

The authors aimed to reconstruct the evolutionary history of these taste receptor genes in vertebrates by identifying all the type 1 and type 2 taste receptor genes (*Tas1R* and *Tas2R*, respectively) in the nine complete (or near-complete) genomes that were available at the time: five mammals (mouse, rat, human, dog, and opossum), a bird (chicken), an amphibian (western clawed frog), and three fish (zebrafish, pufferfish, and fugu). They wished to contrast the pattern of evolution seen in the *Tas1R* genes, which are found in roughly the same numbers in all vertebrates, to *Tas2R* genes, which occur in much greater diversity.

Methods

The researchers used a five-stage strategy for locating taste-receptor genes in the nine vertebrate genomes. First, the researchers found the genomic location of putative *Tas1R* and *Tas2R* genes by blasting known gene sequences against the whole genomes. *Tas2R* genes have a relatively simple structure, consisting of a single, short coding sequence, which makes them ideal query sequences for BLAST searching (**TechBox 3.4**). But *Tas1R* genes have a more complex structure, with protein-coding sequences (exons) interspersed with non-protein-coding sequences (introns). Although the protein-coding exons tend to be conserved by negative selection to maintain protein function, the non-coding introns are probably selectively neutral and therefore evolve rapidly. Second, they aligned the putative

Tas1R genes they found against cDNAs, which are sequenced from messenger RNA transcripts. cDNA transcripts contain only exons, so this alignment allowed them to identify the protein-coding parts of the sequence. Third, they translated each sequence into amino acids and checked for the presence of seven transmembrane domains known to be functionally important for T1R and T2R proteins. Fourth, they checked that each putative gene sequence had a continuous sequence of translatable codons (an open reading frame, or ORF) and was therefore at least theoretically capable of producing a working protein. Any sequence with a disrupted ORF, or that was missing any of the seven trans-membrane domains, or was less than 200 nucleotides long, was considered to be a pseudogene (a non-functional copy of a gene). Finally, all putative *Tas1R* and *Tas2R* sequences were blasted against the whole GenBank database to make sure their closest matches were genes from the T1R and T2R families, not sequences for other GPCR transmembrane proteins. The researchers then aligned amino acid sequences (**TechBox 6.3**) and constructed phylogenetic trees (see Chapter 7). They tested for evidence of positive selection in the *T1R* genes by statistically comparing the fit of two alternative models of sequence evolution to their data. In one model, all codons had a ratio of non-synonymous to synonymous substitutions (*dN/dS*, otherwise known as ω) of one or less than one. In the alternative model, some codons had a *dN/dS* greater than one, which is considered to be evidence of positive selection (**TechBox 5.4**, p. 175).

Results

Comparison of the number of genes in each species shows that the two families of genes show different evolutionary histories. The sweet and umami receptors, T1R1, T1R2, and T1R3, arose by two gene-duplication events in the ancestor of all vertebrates (**Figure CS5.1a**). However, all three *Tas1R* genes appear to have been deleted from the western clawed frog genome, and the sweet-receptor gene *Tas1R2* has apparently been lost from the chicken genome (although many birds, such as hummingbirds, have the ability to detect sweetness, electrochemical studies of chicken tongues suggest they cannot taste sugar or saccharine, and chickens show no preference for sweet foods). *Tas1R2* is in the process of becoming a pseudogene in the cat lineage: a deletion of part of exon 3 and stop codons in exons 4 and 6 show that *Tas1R2* from cats, tigers, and cheetahs cannot produce a functioning T1R2 receptor protein[2]. It seems likely that loss of sweet taste reception does not decrease the ability of a meat-eater to feed itself successfully, so is selectively neutral in the cat lineage[2]. The *Tas1R* genes have apparently undergone selection to diversify function, with an excess of non-synonymous changes affecting the N-terminal region of the protein, which is probably associated with the ability to bind sweet substances.

Conclusions

Given that taste is made up of discrete receptors, the duplication and loss of taste receptor genes can add or remove recognition of particular tastes, such as sweetness, without influencing the ability to taste other chemicals. The number of *Tas1R* genes has remained constant across vertebrate evolution, except for losses in a number of species that appear to be 'sweet-blind' (frogs, chickens, cats; **Figure CS5.1b**). The *Tas2R* genes have diversified greatly, and vary dramatically in number between species, possibly to allow detection of a variety of bitter tastes. Members of this gene family show evidence of positive selection on both T1R and T2R proteins, suggesting functional diversification[3].

Limitations

The authors note that their phylogenies suggest some odd relationships between species, such as humans appearing to be more closely related to dogs than to mice. While the focus

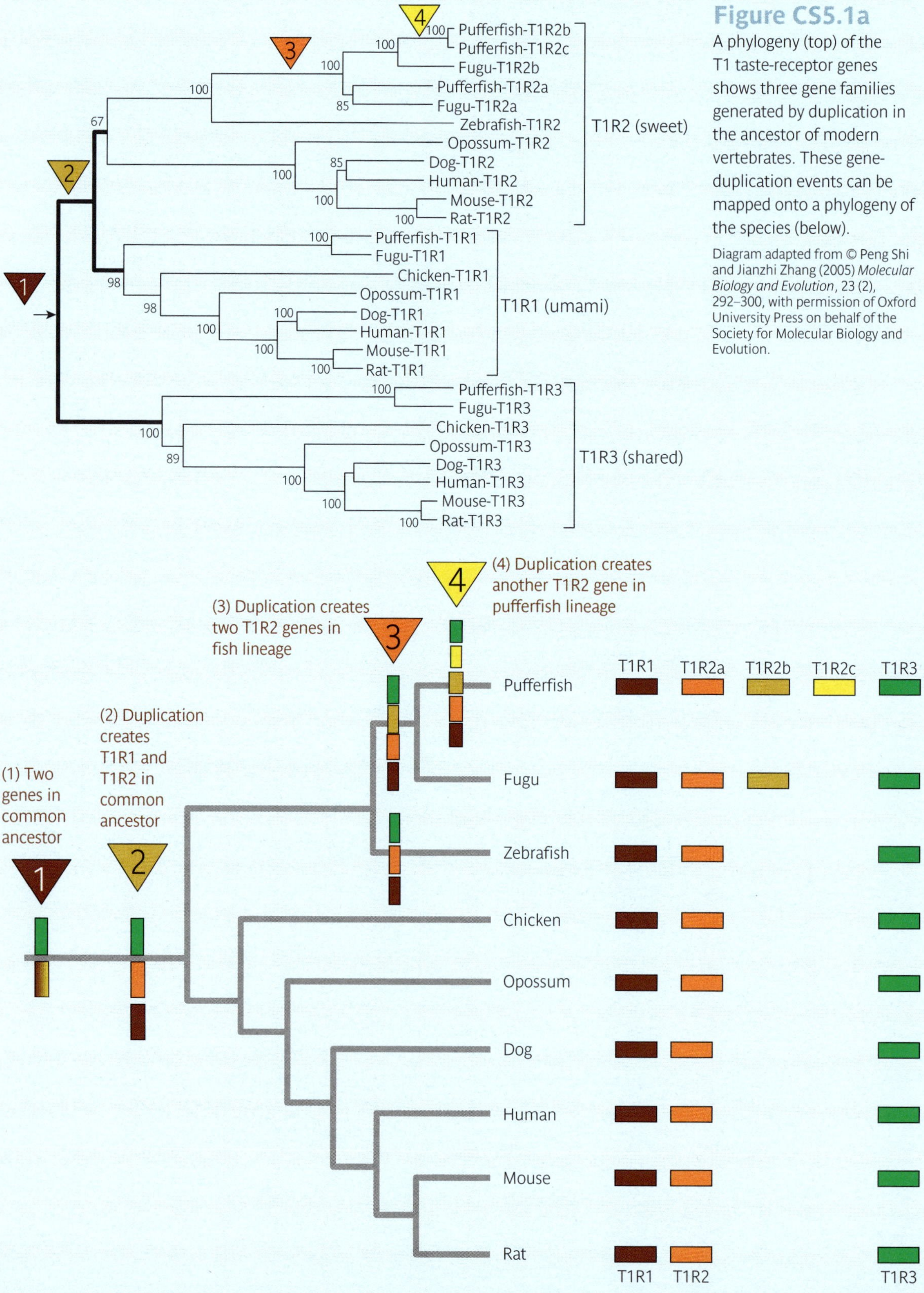

Figure CS5.1a

A phylogeny (top) of the T1 taste-receptor genes shows three gene families generated by duplication in the ancestor of modern vertebrates. These gene-duplication events can be mapped onto a phylogeny of the species (below).

Diagram adapted from © Peng Shi and Jianzhi Zhang (2005) *Molecular Biology and Evolution*, 23 (2), 292–300, with permission of Oxford University Press on behalf of the Society for Molecular Biology and Evolution.

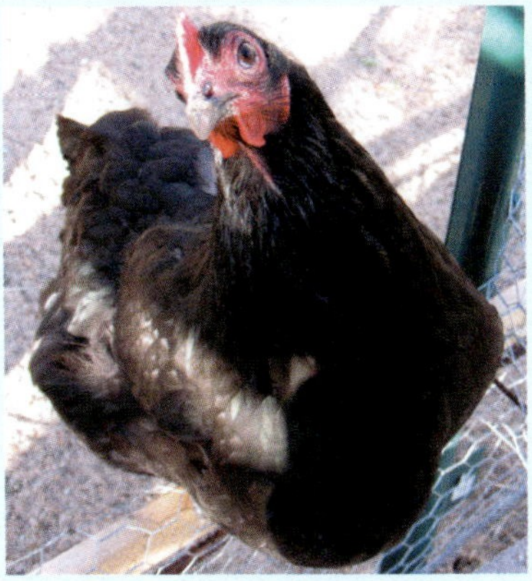

Figure CS5.1b What do these animals have in common? They can't taste sugar. Clawed frogs (*Xenopus*) seem to be missing the *Tas1R* genes that provide sweet taste receptors, in chickens the critical gene seems to have been deleted, and in cats the gene is in the process of decaying into a pseudogene.

African Clawed Frog: Photograph by Peter Halasz.

of this study was the evolutionary history of genes, not species, these results suggest that the story told from the data may need some refinement.

Future work

Surveying genes may provide a fast way of cataloguing taste abilities in species. For example, we could predict the ability of any species to be able to taste sugar by searching for the appropriate genes in its genome. Experimental modification of these genes may provide a useful case study for studies of stimulus–response pathways. For example, researchers created a mouse with a modified T1R2 molecule that binds a tasteless synthetic substance instead of sugar. These mice responded to water containing the tasteless molecule as if it was sugar water[4], showing that stimulating the taste pathway initiated hard-wired feeding behaviour.

References

1. Scott, K. (2004) The sweet and the bitter of mammalian taste. *Current Opinions in Neurobiology*, Volume14, pages 423–427.
2. Li, X. *et al*. (2005) Pseudogenization of a sweet-receptor gene accounts for cats' indifference toward sugar. *PLoS Genetics*, Volume 1, pages 27–35.
3. Shi, P., Zhang, J., Yang, H. and Zhang, Y.-P. (2003) Adaptive diversification of bitter taste receptor genes in mammalian evolution. *Molecular Biology and Evolution*, Volume 20, pages 805–814.
4. Zhao, G.Q. *et al*. (2003) The receptors for mammalian sweet and umami taste. *Cell*, Volume 115, pages 255–266.

Figure 5.6 Males of the side-blotched lizard *Uta stansburiana* come in three kinds, each characterized by a mating strategy and a throat colour. The selective advantage of each morph depends on the relative frequency of the others: in a population with lots of orange-throats it's better to be a yellow than a blue, but when there are a lot of blues around, it is better to be an orange than a yellow. Frequencies of the three morphs fluctuate on a 4 to 5-year cycle. When Barry Sinervo and Curtis Lively discovered this, they 'looked at each other and . . . said "Dude! It's the rock–scissors–paper game!" ' (see Kohn 2004).

Photograph courtesy of Barry Sinervo.

reproductive age, the mutation will be subject to negative selection. But in the 1960s, the Guthrie (heel-prick) test for PKU was introduced, and since then each new born baby in the industrialized world is tested for PKU (see **TechBox 3.2**). Those that test positive are put on a special diet with no phenylalanine, thus avoiding brain damage. In places where the Guthrie test is used to detect PKU and modify the environment accordingly, there is no reason someone carrying the PKU mutation cannot reproduce, so there is no longer negative selection against it.

One aspect of an allele's environment that can vary considerably from one generation to the next is the other individuals in the population. In some situations, the selective advantage of a mutation can change with the frequency of other alleles, for example a particular variant may be advantageous when it is rare, but disadvantageous when common. Frequency-dependent selection can maintain polymorphism: if one allele is only advantageous when it is relatively rare, it will increase by selection only until it becomes too common, then its selective advantage will cease. John Maynard Smith once hypothesized that a population could maintain a polymorphism if the variants were like the 'rock, scissors, paper' game (rock wins against scissors, scissors win against paper, papers wins against rock). Several natural examples of this have since been described (**Figure 5.6**). For example, some strains of the gut bacterium *Escherichia coli* produce colicin antibiotics, which kill competing *E. coli* strains. So colicin-producing strains (C) will clearly have a higher reproductive output in a population of colicin-sensitive strains (S), because fewer Ss will survive to reproduce (**Figure 5.7**, p. 151). Colicin-resistant strains (R) can survive in the presence of colicin, and they avoid the cost of carrying the colicin-producing genes. So, in a population of Cs, Rs have a reproductive advantage. But the modified receptor proteins that confer resistance to colicin are less efficient at other metabolic tasks, so resistant strains have a slower growth rate than sensitive strains. Therefore in a population of Ss, Rs will be at a selective disadvantage.

So now we see that C beats S, S beats R, and R beats C. This non-transitive loop means that if any one strain increases it frequency it will decrease the frequency of the next strain in the chain, which will cause an increase in frequency of the next in the chain, which will cause a decrease in the first strain. This probably explains why all three variants can coexist in natural populations without any variant going to fixation.

Genetic background

The environment of a mutation must also include the genetic context within the individual, because the influence of a mutation on an individual's chance of reproduction can depend on what other mutations are present in the same genome. The genetic environment

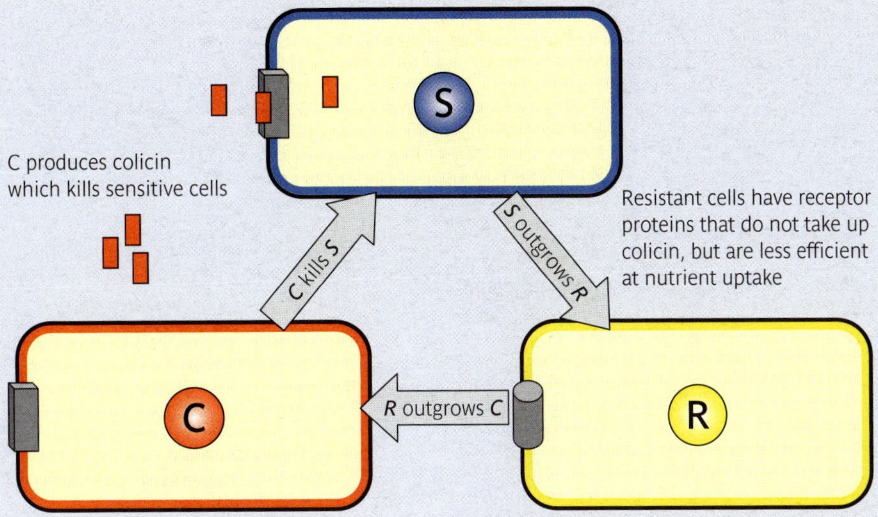

Figure 5.7 Bacteria play rock-paper-scissors: three antagonistic types of *E. coli* can co-exist because their interactions form a non-transitive loop: S beats R, R beats C, and C beats S (see Kerr *et al.* 2002).

Diagram modified with permission from Ben Kerr.

within an individual (more properly referred to as the genetic background) includes mutations at the same locus and at other places in the genome.

We can illustrate the influence of other mutations in the genome on the response to selection using examples from the human haemoglobins (proteins that carry oxygen in the blood). There are multiple haemoglobin genes in the human genome, and the effect of a mutation in one gene can be influenced by mutations carried in other globin genes. Each haemoglobin molecular is made of four amino acid chains, two alpha and two beta (**Figure 5.8**). There is a family of related genes that produce these alpha and beta chains. Mutations of these genes can cause a range of illnesses, collectively termed haemoglobinopathies, due to inadequate transport of oxygen around the body. When considered on a global scale, haemoglobinopathies are the most common single-gene diseases of humans. It is estimated that more than 5% of the world population carries a mutation for a form of haemoglobinopathy. The clinical outcome of mutations in haemoglobin genes depends both on the type of mutation and the genetic background in which the mutation occurs.

Humans normally carry four copies of the alpha globin gene, two on each copy of chromosome 16 (**Figure 5.8**). But because alpha globin genes are tandem (side-by-side) copies, they are prone to deletion by unequal crossing over (see **Figure 3.6**). Individuals with four or three copies of the gene produce enough alpha chains to have sufficient haemoglobin function. But individuals with only two working copies of the alpha globin gene do not produce enough alpha chains, so cannot manufacture sufficient haemoglobin. Without enough haemoglobin to transport oxygen in the blood, individuals with only two functioning alpha globin genes can suffer from anaemia. Someone with only one functioning alpha globin gene will have severe anaemia, and individuals with no working copies of alpha globin will usually die before birth. So the effect of a mutation that ruins or removes one copy of the alpha globin gene will depend on whether there are other working copies of the alpha globin gene in that genome.

Now consider the effect of a new mutation that knocks out an alpha globin gene in a particular individual. Can you tell me what the effect of this mutation will be on the individual: harmless, severe, or lethal? In this particular case, you can't tell me the effect of the mutation until you know the genetic background in which it occurred. If this alpha globin knock-out mutation occurs in an individual with a full complement of working alpha globin genes, it will have no effect on phenotype, and has the same chance of being passed onto the

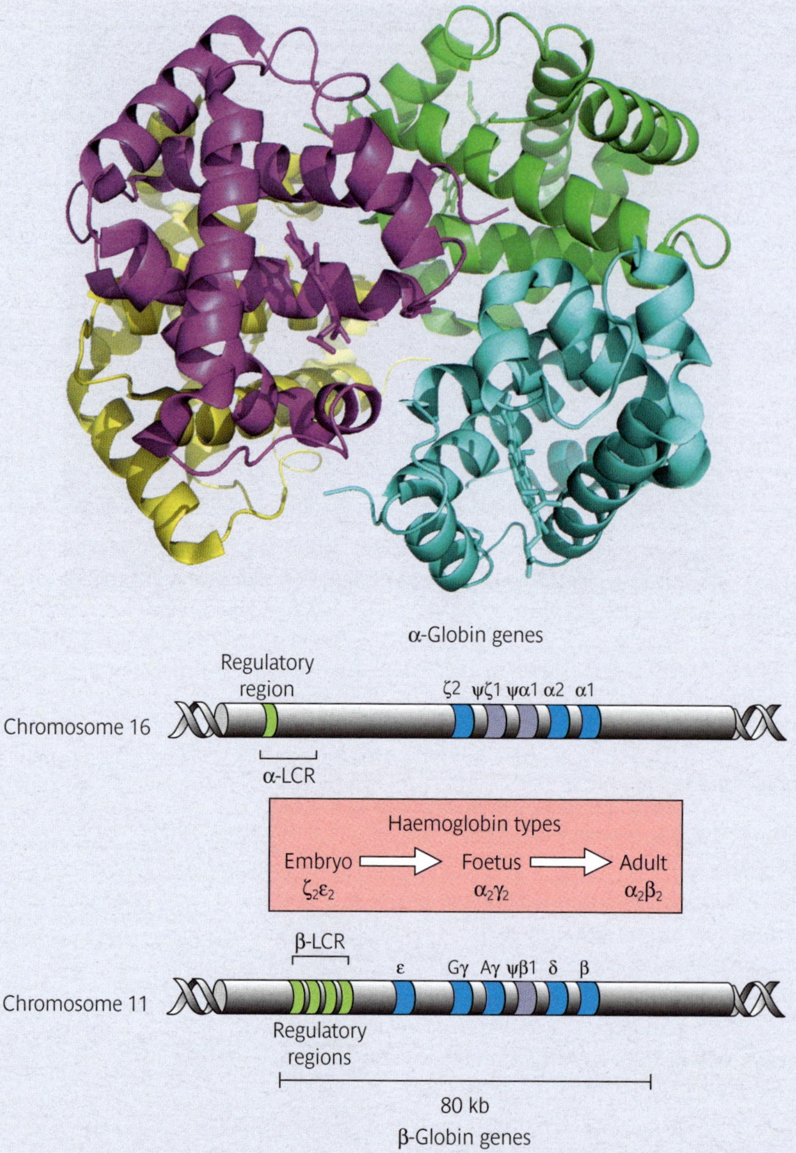

Figure 5.8 Haemoglobin consists of four chains of two types, but the composition of haemoglobin changes throughout development. The genes of the two globin clusters in humans are activated in sequence. In the alpha (α) globin cluster, the zeta (ζ) gene is expressed in early embryogenesis, then alpha 1 and alpha 2 take over. In the beta (β) globin gene cluster, epsilon (ε) is expressed in embryogenesis, gamma (γ) during foetal development, then beta from infancy onward (small amounts of delta (δ) are produced in both children and adults). ψα1, ψζ1 and ψβ1 are pseudogenes.

next generation as any other neutral allele. If the very same mutation occurs in someone with three other defective alpha globin genes, it will kill its carrier, in which case that mutation will not be copied to the next generation.

For an explanation of knock-out mutations see ***TechBox 3.1***

The genetic context of an allele may also include mutations at other loci (sometimes referred to as modifier loci). Because each haemoglobin molecule is made of two alpha chains and two beta chains, any mutation that causes a decrease in production of one type of chain will cause an imbalance in haemoglobin assembly. If a mutation in a beta globin gene results in fewer functional beta chains, there will be fewer complete haemoglobin molecules made, and the excess alpha chains will precipitate out, causing damage to red blood cells. This produces a condition known as beta thalassaemia, which until recently almost invariably led to death before adulthood. But if a person carrying a beta thalassaemia mutation also carries a mutation that decreases the production of alpha chains, there will be less imbalance in production, resulting in less severe thalassaemia: an unusual case of two wrongs making things a bit closer to right.

Linkage

The environment of a mutation also includes its physical location in the genome. Mutations that sit next to each other in the genome tend to be inherited together. We take advantage of the fact that closely occurring alleles are inherited together when we use genetic markers to identify genes associated with particular diseases (**TechBoxes 3.2** and **3.3**). We saw in Chapter 3 that some SNPs are informative markers for disease not because they have any causal role in the disease, but because they occur close enough to the relevant disease allele to always be inherited with it. The corollary of this is that the fate of a particular SNP can depend on neighbouring alleles in the genome.

Say we have a population of slugs, and we want to track allele frequencies at two sites in the slug genome, which are close enough together that alleles at these two loci are almost always inherited together. One of these sites is important for being a well-adjusted slug (say, the active site of an important enzyme), and the other is completely unimportant (perhaps an intergenic sequence that doesn't code for anything). Not surprisingly, the important locus is under strong negative selection so changes are rare, but the unimportant site is free to change so it's polymorphic (**Figure 5.9**).

Most of the mutations in the important sequence will be removed by negative selection, but one lucky slug is the fortunate recipient of a beneficial mutation: a change from A to T happens to increase the efficiency of the enzyme (**Figure 5.10**).

This happy mutation could have occurred in any slug but, by chance, it just happened to occur in a slug that

Figure 5.9

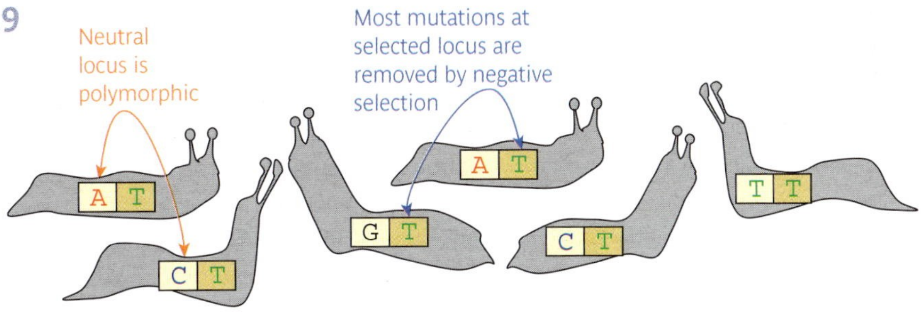

Figure 5.10

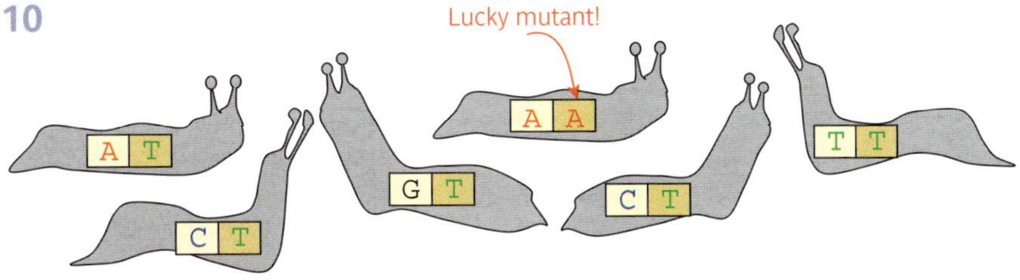

Figure 5.11

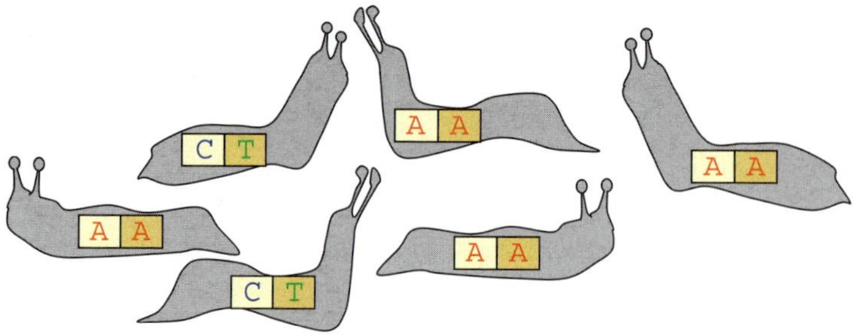

Figure 5.12

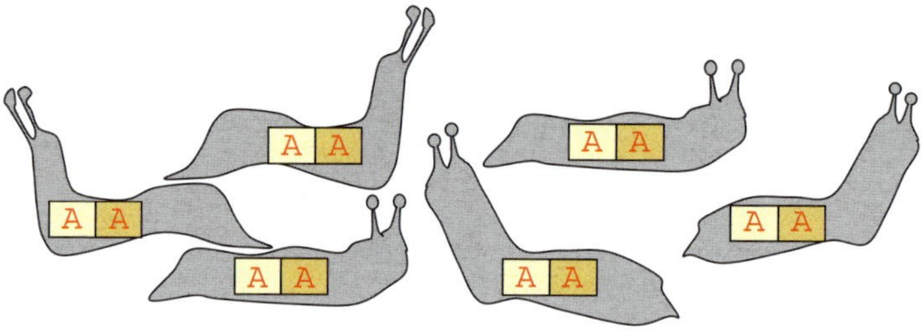

had an A at the neutral locus. The neutral A still doesn't make any difference to the slug, but it has the good fortune to be attached to the selected A that gives the slug an advantage over its comrades. So next generation, this slug leaves more descendants than those with T at the selected locus (**Figure 5.11**).

Eventually, the beneficial mutation goes to fixation, and because it takes the linked neutral allele along with it, the polymorphism at both sites is lost (**Figure 5.12**).

This example illustrates how a mutation occurring next to a beneficial allele can be swept to fixation along with it, hitchhiking on the positive selection of its neighbour (see **TechBox 5.3**, p. 168). Conversely, any mutation occurring next to a negatively selected allele may hitchhike its way to oblivion, being removed from the population when carriers of the negatively selected allele fail to reproduce.

We can see the effects of hitchhiking in patterns of variation across the human genome (**TechBox 5.1**, p. 155). For example, there are a number of alleles of the human beta globin gene that confer some resistance to malaria. One of these, known as HbC, contains a single nucleotide mutation at site 16 in the gene. This mutation, from G to A, changes one amino acid in the beta globin chain from glutamic acid to lysine. HbC appears to be a fairly recent mutation, possibly less than 5000 years old, that is increasing in frequency by positive selection in West Africa, where malaria is prevalent. It has been reported that someone who carries two copies of HbC (i.e. homozygous for HbC at the beta globin locus) is 93% less likely to get sick with malaria than someone with alternative versions of the beta globin gene, and someone with one copy of the HbC allele (i.e. a heterozygote) is 29% less likely to be affected by malaria. In addition to the advantageous change at site 16, HbC differs from other variants of the beta globin gene at a number of other sites. The malaria-resistant mutation seems to have originally occurred on a version of the beta globin gene that happened to have a C at site –2906 (HbA, the 'normal' version of this gene, has a T at this position). Since the C at site –2906 is closely connected to the positively selected A at site 16, it is inherited along with it. This means that as positive selection on the A at site 16 causes the HbC allele to rise in frequency in the population, it takes the C at site –2906 with it. Regardless of the effect of the mutation at –2906, it is hitchhiking to fixation on the malaria-resistant mutation at a neighbouring site.

TECHBOX 5.1 Fitness

KEYWORDS

selection

selective coefficient (s)

population genetics

resistance

selective sweep

FURTHER INFORMATION

Richard Dawkins outlines different definitions of fitness, and the consequences of confusing them, in his book *The Extended Phenotype* (1982), Oxford University Press: see Chapter 5, An agony in five fits.

For an introduction to the algebraic treatment of fitness, see John Maynard Smith's classic text, recently updated: *Evolutionary Genetics* (1998), Oxford University Press.

RELATED TECHBOXES

TB 5.3: Population size

TB 5.4: Detecting selection

RELATED CASE STUDIES

CS 5.1: Sweet and sour (selection on gene copies)

CS 5.2: Time flies (natural selection)

There are many ways of defining fitness. For example, we may use the word fitness to describe the way that organisms are suited to particular aspects of their environment. Alternatively, we may use the term fitness to report the net reproductive output of individuals, or the change in frequency of particular traits in a population. Confusion between the different definitions has led to misunderstandings in evolutionary biology. Here we will consider one particular view of fitness. In this chapter we are concerned with substitution: the process whereby a particular mutation increases in frequency from generation to generation until it replaces all other alternative alleles and becomes fixed in that population. In order to make statements about this process, it is useful to consider the term fitness as representing the effect that the properties of a mutation have on its chance of becoming a substitution in a given population and a particular environment. But remember that you will come across many alternative definitions of fitness, which may be appropriate to different topics.

Probably the best place to start in taking a molecular genetic perspective on fitness is to abandon the phrase 'survival of the fittest', because stating that the fittest are those that survive is not very helpful to our understanding. Instead, we need to think of fitness as a statement about probabilities. If we have a population of copiers which have heritable variation, we say that a particular variant has a higher fitness if it has properties that make it more likely to make more successful copies of itself than alternative variants. The reason we find it useful to have a term with this definition is that we expect fitter variants to increase in frequency in the population with each passing generation. In population genetics, fitness is most commonly expressed algebraically, but in this book the aim is to provide explanations of molecular evolution in words rather than equations (this is not to say that you should make a habit of ignoring equations, which are there for a reason: they can be the clearest statement that can be made about a principle, and you would do well to make sure you understand them).

The important thing to remember when taking a molecular genetic perspective of fitness (that is, one aimed at understanding the process of substitution) is that fitness of a mutation is relative to the other mutations in the population. Remember that the expected relative rate at which a given variant will be copied into the next generation must depend on two things: the effect of the mutation, and the environment in which it finds itself. 'Environment' must be taken here in the broadest sense, to include the external conditions, the composition of the population in which the mutation finds itself (whether there are any other alleles that might do relatively better), and the genetic background (the influence of other mutations in the same genome). The relative selective advantage or disadvantage of each mutation can be considered by comparing the rate at which it is copied to the next generation, with respect to alternative alleles at the same locus.

The relative advantage or disadvantage of a particular mutation is sometimes represented by the selective coefficient, denoted by the letter s. Fitness is (somewhat mystifyingly) often denoted by W. If we say that the fittest mutation in the population has a fitness of 1 ($W = 1$), then all other mutations at the locus must either be of equal fitness ($W = 1$) or must have a lower fitness ($W = 1 - s$, where s is the difference between this type and the fittest type).

Imagine a polymorphic population of flowers, with equal numbers of red, yellow, and blue. We follow this population over some number of generations, noting the survival and reproductive rates of the different morphs. At the end of the observation period we find a population with no yellows, and twice as many blues as reds.

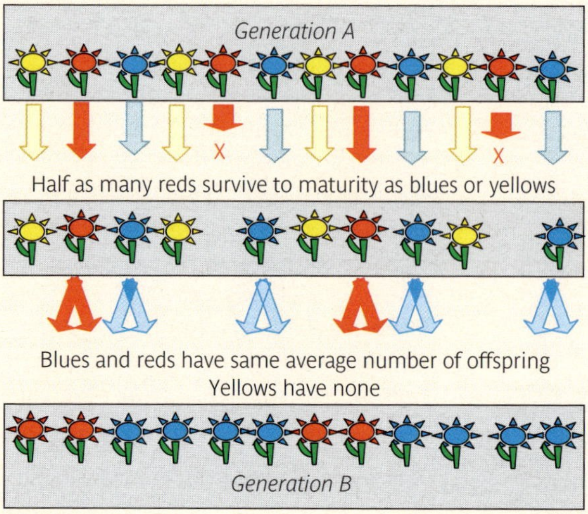

Figure TB5.1

If reds and blues have the same average number of offspring, but half as many reds survive to maturity as blues, then we would say that the difference in the relative reproductive output is 0.5 ($s = 0.5$). So in this case, blue has a W of 1 and red has $W = 1 - 0.5 = 0.5$. If yellows are just as likely to survive as blues but are infertile, then they have no chance of reproducing so they have an s of 1, and W of 0.

Note that just as fitness has various definitions and usages, so does the parameter representing selection. s may be called the selection differential, and it is sometimes reported as the proportional advantage over other alleles in the population (example given below). Furthermore, s is sometimes reported for genotypes, which, for diploid organisms, are a combination of two alleles (the principle is the same but the equations will be slightly different, and we might have to add in another parameter to describe the heterozygotes for traits that are not completely dominant). As with fitness, since there are several different definitions of s, you need to get the intended meaning from the context in which it is written.

The key point is that we need to stop ourselves from thinking of fitness as 'good' or 'bad'. All that matters is relative copy numbers. A mutation that confers antibiotic-resistance may seem like a huge advantage to a bacterium in a hospital. But if it slows the growth of bacterium by reducing the rate of uptake of nutrients, then a resistant strain may still be at a selective disadvantage (**Figure 5.7**, p. 151). If the antibiotic-sensitive strain suffers 80% more mortality than the resistant strain, but each survivor produces 10 times as many descendants as the resistant strain, then it will still reproduce at twice the rate. We could say that the antibiotic-resistant strain had s of 0.5: if there are 100 sensitives for every 10 resistants, all 10 resistants survive but only 20 of the sensitives, then there will still be twice as many sensitives as resistants. So even though we instinctively think of antibiotic resistance as advantageous in the presence of antibiotics, in this particular case it has a lower selective coefficient.

Fitness can be estimated for particular alleles in real populations. For example, an international team including researchers from Laos, Thailand, Vietnam, and Myanmar found that an allele associated with resistance to the antimalarial drug pyrimethamine had increased rapidly in *Plasmodium* (malaria) parasites in South East Asia[1]. They estimated a single origin of the resistant allele, which then spread throughout the region since the introduction of pyrimethamine treatment in the 1970s. They estimated the selection coefficient *s* in two ways – from the decline in efficacy of pyrimethamine treatment, and from the reduction of genetic diversity around the allele (indicating a selective sweep pushing this allele to fixation at the expense of other alleles in the population) – both of which indicated that the resistant allele had around 10% advantage over alternative alleles.

Reference

1. Nair, S. *et al.* (2003) A selective sweep driven by pyrimethamine treatment in Southeast Asian malaria parasites. *Molecular Biology and Evolution*, Volume 20, pages 1526–1536.

Genetic drift

> ***Fitness is a property, not of an individual but of a class of individual. . . . Thus the phrase 'expected number of offspring' means the average number, not the number produced by some one individual. If the first human infant with a gene for levitation were struck by lightening in its pram, this would not prove the new genotype to have low fitness, but only that the particular child was unlucky.***
>
> John Maynard Smith (1998) *Evolutionary Genetics*. Oxford University Press

We have seen how the relative frequency of a particular allele can be increased or decreased by selection. But we also have emphasized that the action of selection is probabilistic. Carriers of an advantageous allele may tend to have more descendants, but this is a statement about average outcomes, not about the fate of each individual with that allele. An individual carrying a mutation that makes its metabolism more efficient at high altitudes may die without issue when it is unluckily buried in an avalanche. Similarly, an individual with a mutation that reduces its oxygen-transport efficiency may just be lucky enough to find a cache of food to keep its offspring alive over winter, despite its metabolic inefficiency. Life is full of surprises, so allele frequencies are always subject to chance events. We use the term 'genetic drift' to describe the way that allele frequencies drift up and down due to random variation in survival and reproductive success.

Genetic drift comes about as the result of the sampling of alleles that occurs every generation when offspring are produced from the parent generation. In general, not all individuals in one generation will produce offspring. So the alleles in one generation are an incomplete sample of the alleles in the previous generation. Chance events may influence which individuals contribute alleles to the next generation and which don't. Understanding the role of chance is so critical to understanding patterns of molecular evolution that it is worth starting from the basics. So let's play with some elementary exercises in probability.

Take a coin out of your pocket. If it is a fair coin and is thrown without bias, then it has an even chance of landing on heads or tails. And, since each coin toss is independent of the last, when you toss it again, you again have a 50/50 chance of getting heads or tails. So, to state the obvious, you expect half of all tosses to result in heads (H), and half to result in tails (T). But any particular sample may deviate from this expectation, simply by chance. Try it now, toss your coin 10 times. I did, and I got 60% tails, 40% heads:

THTTHHHTTT

Now let's turn our Hs and Ts into reproductive entities so that we can start to imagine how populations of template copiers behave. We have a polymorphic population of Ts and Hs, and we allow each one to have two offspring of its own kind, giving 20 offspring:

TTHHTTTTHHHHHHTTTTTT

Not all of them will survive to reproduce, so let's randomly kill 10 offspring (you can use a coin toss to decide who dies). In this instance I killed seven Ts and three Hs:

~~T~~~~T~~H~~H~~~~T~~T~~T~~~~T~~H~~H~~H~~H~~HHT~~T~~T~~T~~TT

Thus in the next generation we are left with 50% H and 50 % tails:

HTHHHHTTTT

So the frequencies have changed in a single generation due to random events. I used this sample to produce a third generation by the same process and got 60% heads and 40% tails (**Figure 5.13**). At no point did Hs have a better survival probability than Ts, they just got lucky.

Figure 5.13

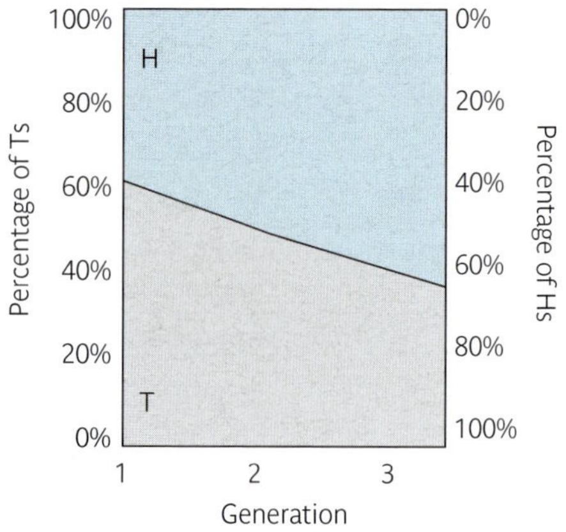

Allele frequencies fluctuate, sometimes up and sometimes down. But how is it possible for random fluctuation to result in substitution? Won't the frequencies just randomly drift up and down forever? Imagine we keep breeding from our Hs and Ts: each generation we randomly choose the survivors who will each have two offspring of their own kind. The frequencies of Hs and Ts in one generation are never very different from the frequencies in the previous generation. But it is possible to occasionally have a run of chance events that all go in the same direction, as we did in the first three generations when the percentage of Hs happened to increase. If it happens that there are more generations in which Hs get over-represented, and Ts under-represented, then the proportion of Ts will undergo a decline, and may even go as far as 0% (**Figure 5.14**).

Substitution occurs because loss of a variant is irreversible (barring a new mutation). Unlike a coin toss, which is endlessly repeatable, in a population of reproducing entities if the frequency of one kind goes to zero, it's Game Over for that variant. There are no more Ts to produce a new generation of Ts. There are only Hs that produce more Hs. Our population is no longer polymorphic. H has gone to fixation at the expense of T. Substitution is the loss of all other variants at a

Figure 5.14

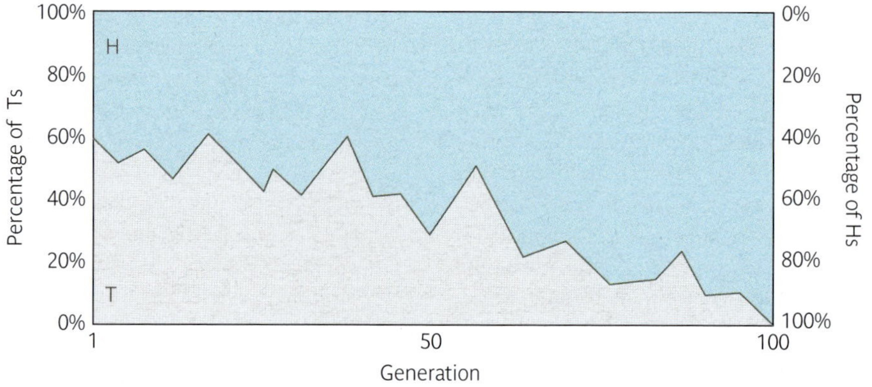

particular genetic locus, leaving only a single type left in the population. (As an aside, extinction of a species represents the irrevocable loss of a large number of unique alleles all at once. Like the loss of single allele, once the genetic inheritance of a species is lost, it's gone forever; see Chapter 6.)

No individual is immune to chance, so all alleles are subject to drift. The effect of drift will be more dramatic when an allele is at low frequency, because it is at great danger of being lost from the population if its few carriers fail to reproduce. So when we consider the fate of mutations we need to consider the effects of both selection and drift. If selection is very strong on a particular allele, then it may overwhelm the effects of drift. But, importantly for understanding molecular evolution, when selection is weak (or absent), drift can be a major determinant of allele frequencies. This is important when we consider the fate of neutral alleles.

Neutrality

Mutations that confer on their carriers a relatively higher rate of successful reproduction will tend to increase at the expense of other variants until all members of the population are descendants of those carrying the advantageous mutation. Mutations that decrease the chance of successful reproduction are likely to dwindle with each generation and eventually disappear. But, as Darwin noted (in the quote at the beginning of this chapter), some heritable changes will make no difference whatsoever to chances of surviving and reproducing. Mutations that have no effect on their own chances of successful reproduction are commonly referred to as being neutral with respect to fitness (see **Heroes 5**).

To illustrate the concept of neutrality, let's consider a hypothetical example. Imagine there is a population of wild geraniums that are polymorphic for flower colour. Some individuals produce red flowers, but some carry a mutation that means they produce a slightly different pigment, so they have dark pink flowers. Both types of flowers are equally attractive to their pollinators, who can't tell the difference between the red and dark pink pigments. So the two types of flowers are equally likely to be pollinated. Both pigments require the same amount of resources to make, so there is no metabolic advantage to producing one pigment rather than the other. In this case, it simply doesn't matter which colour a flower has: there isn't the slightest advantage to having one or the other. The mutation for the pink pigment has no influence on its chances of getting into the next generation. So the evolutionary fate of such a trait cannot be determined by selection.

It is important to note selectively neutral alleles do not imply functionally unimportant sequences. Selective neutrality refers to functional equivalence, such that one variant could be exchanged for another with no effect on fitness. Flower colour is of critical importance to these geraniums. A mutation that ruined the flower pigment gene would result in unpigmented flowers that would not attract pollinators, so could set no seed. Such a mutation would be unlikely to find its way into the next generation: it would be subject to strong negative selection. But pink and red floral pigments are, in this example, functionally equivalent, and will make no difference to the chance of setting seed.

Neutrality is an important concept in molecular evolution, because many heritable changes to the genome have no apparent effect on their carrier's chances of successful of reproduction (**TechBox 5.2**). We have already seen some examples of neutral genetic changes. Microsatellites – runs of short tandem repeats – often occur in parts of the genome that don't contain any genes or regulatory elements. Different numbers of repeats in these sequences may be used to distinguish individuals, but they do not usually make any noticeable difference to an individual's morphology or behaviour. Consequently, an individual with six CA repeats at a particular microsatellite locus may not have any advantage or disadvantage compared to an individual with 10 CA repeats. Even changes to functional sequences can be neutral. Some amino acids in proteins can be exchanged for similar amino acids with no apparent effect. These substitutions generate isozymes – functionally equivalent proteins – that can be detected biochemically, but have no effect on the individual's phenotype. But, as the geranium example above illustrates, a neutral trait does not need to be one that has no effect on morphology or behaviour. It can have a very noticeable effect, such as flower colour, but if it has no influence on relative reproductive success it will be selectively neutral.

See Chapter 3 for an explanation of microsatellites

HEROES OF THE GENETIC REVOLUTION

5

Motoo Kimura

"The late Professor Sewall Wright was my idol when I was young. Soon after graduating from Kyoto University I read Wright's 1931 classic Evolution in Mendelian Populations *and his subsequent papers on random genetic drift and the distribution of gene frequencies. These papers impressed me deeply and, in fact, inspired me to become a theoretical population geneticist."*

Kimura, M. (1991) Recent development of the neutral theory viewed from the Wrightian tradition of theoretical population genetics. *Proceedings of the National Academy of Sciences USA*, Volume 88, pages 5969–5973

NAME

Motoo Kimura (木村資生)

BORN

13th November 1924, Okazaki, Aichi Prefecture, Japan

DIED

13th November 1994, Shizuoka, Shizuoka Prefecture, Japan

KEY PUBLICATIONS

Kimura, M. (1955) Solution of a process of random genetic drift with a continuous model. *Proceedings of the National Academy of Sciences USA*, Volume 41, pages 144–150.

Kimura, M. (1983) *The Neutral Theory of Molecular Evolution*. Cambridge University Press.

FURTHER INFORMATION

Crow, J. (1995) Motoo Kimura (1924–1994). *Genetics*, Volume 140, pages 1–5.

Perspectives on Molecular Evolution web page: *http://authors.library.caltech.edu/5456/01/hrst.mit.edu/hrs/evolution/public/*

Courtesy of the American Philosophical Society Library, USA.

Motoo Kimura was a natural successor to the triumvirate of the neo-Darwinian synthesis: he was turned on to population genetics by Haldane, built his work upon Wright's models of drift and selection, and extended Fisher's work on natural selection. Although his earlier work provided elegant population genetic models for the action of natural selection (he described himself as a committed 'selectionist' during that period), he is best known for formulating and defending the neutral theory of molecular evolution.

In 1944, Kimura entered the Kyoto Imperial University, studying botany as a way of avoiding military service, but focusing on studies of chromosomes (cytogenetics). There he was introduced to mathematical population genetics by reading the papers of J.B.S. Haldane and Sewall Wright. With no way of duplicating the single available copy of Wright's papers, Kimura wrote out his own copies by hand, adding his own notes and derivations as he did so. One of Kimura's first papers (in 1953) added biological depth to Wright's island immigration model, providing a 'stepping stone' model for migration between neighbouring populations instead of assuming random migration from a single large source population.

Kimura went to the United States on a Fullbright Fellowship. On the ship on the way from Japan to the USA, he wrote a paper on fluctuating selection coefficients, which replaced the application of complex differential equations with the more tractable form of equations for heat conduction. This paper was published in the journal *Genetics* in 1954. He later brought about a similar transformation to the description of the process of substitution by applying diffusion equations to the problem. He studied at the University of Wisconsin from 1954 to 1956, under the direction of Jim Crow (who supervised many stars of evolutionary biology, including Joe Felsenstein; see **Heroes 7**). It is remarkable to note that Kimura published his mathematical solution for neutral genetic drift, then extended his models to allow multiple alleles, mutation, migration, and selection, and wrote papers on

Fisher's fundamental theorem and selection on linked loci, all during the 2 years it took him to complete his PhD.

Kimura returned to Japan and spent the remainder of his career at the National Institute of Genetics in Mishima. His contributions to molecular genetics were wide-ranging. His 660 papers (and six books) include mathematical treatments of meiotic drive, epistasis, genetic load, inbreeding, sexual reproduction, and recombination. Kimura was also responsible for many models that have become standard kit for population geneticists, such as the 'infinite alleles' and 'infinite sites' models (developed with Jim Crow). However, Kimura is best known for developing, expounding, and passionately defending the neutral theory of molecular evolution (**TechBox 5.2**).

Kimura developed the neutral theory as an explanation of the patterns emerging from the first protein sequence data (at that time limited to the amino acid sequences for a dozen or so proteins from a small number of species, a far cry from the billions of nucleotides of DNA sequence now available in GenBank). Just as Darwin had explained many puzzling facts of phenotypic variation – such as the geographic and temporal association of similar organisms, adaptation to the environment, and hierarchical structuring of phenotypic similarities in similar species – with a single elegant theory (evolution by natural selection), so Kimura explained many puzzling facts about molecular variation – such as a steady rate of molecular change, the apparent randomness of substitutions, an extremely high rate of molecular change, and surprisingly high frequency of polymorphic loci – with a single elegant theory (that most change at the molecular level is driven by drift not selection). Many of the analyses described in this book are ultimately based on the neutral theory (or its offspring, the nearly neutral theory; **TechBox 5.2**).

The role of chance

If positively selected mutations increase in frequency until they reach fixation (replace all other variants), and if negatively selected mutations disappear from the population, then what happens to mutations that are neither positively nor negatively selected? The relative proportions of functionally equivalent variants can drift up and down, simply due to chance. And, given thousands of generations over which chance events operate, eventually the relative proportions may wander all the way down to zero (so that a variant disappears from the population) or all the way up to 100% (a variant becomes fixed in the population so all members carry the same variant).

To illustrate the change in frequencies of functionally equivalent variants, let's go back to that hypothetical population of geraniums: some were pink, some were red, and it made no difference to their chances of survival. Imagine a quiet hillside on which a patch of geraniums is growing. There are equal numbers of red and pink variants (**Figure 5.15**).

Figure 5.15

Lo and behold, along comes a moose and lies down on the patch of geraniums, creating random carnage: 60% of the geraniums are crushed.

Figure 5.16

Sequence continues on page 164/165.

TECHBOX
5.2

Neutral theory

KEYWORDS
selection
substitution
mutation
nearly neutral theory
Kimura
Ohta
polymorphism
pseudogenes

FURTHER INFORMATION

Kimura, M. (1983) *The Neutral Theory of Molecular Evolution*. Cambridge University Press.

RELATED TECHBOXES

TB 3.3: SNPs
TB 6.3: Population size

RELATED CASE STUDIES

CS 5.1: Sweet and sour (pseudogenes)
CS 5.2: Time flies (detecting selection)

> *"The neutral theory asserts that the great majority of evolutionary changes at the molecular level as revealed by comparative studies of protein and DNA sequences, are caused not by Darwinian selection but by random drift of selectively neutral or nearly neutral mutants. The theory does not deny the role of natural selection in determining the course of adaptive evolution, but it assumes that only a minute fraction of DNA changes in evolution are adaptive in nature, while the great majority of phenotypically silent molecular substitutions exert no significant influence on survival and reproduction and drift randomly through the species."*
>
> Kimura, M. (1983) *The Neutral Theory of Molecular Evolution*. Cambridge University Press

Just as observations of phenotypic variation in natural populations were critical to the formation of Darwin's theory of natural selection, so observations of variation at the molecular level were instrumental in the formation of the neutral theory. The theory of natural selection was developed with phenotypic variation in mind: the physical or behavioural traits that could be observed by a naturalist, such as Darwin patiently dissecting his barnacles. The mathematical theory of natural selection took a population genetic approach, imagining the fate of mutations swimming in the gene pool. But, by and large, imagine was all the early population geneticists could do, for there was very little raw data against which the theory could be tested. Genetic variation in natural populations could only be guessed at, or inferred indirectly by observing the fluctuating proportions of phenotypic variants, such shell banding patterns in snails (**Figure 5.3**). Richard Lewontin, one of the pioneers of the study of genetic variation, described early population genetic theory as 'like a complex and exquisite machine, designed to process a raw material that no-one had succeeded in mining'[1].

The molecular genetic revolution changed that. Sequencing techniques and protein electrophoresis allowed scientists their first glimpse at the genetic variation that underlay population genetics. And what they saw was surprising and, at first, inexplicable. A large proportion of the genetic loci examined showed detectable genetic variation. In some populations, at least a third of loci were polymorphic. If selection was the prime driver of substitution, then we would only expect to see a polymorphism if we just happened to observe the population while one mutation was going (rapidly) to fixation, so on the whole we would expect most sites to be uniform in all members of the population. Having so many polymorphic sites under selection would also reduce the average fitness of the population. If all of that variation was subject to selection, then the chance of any individual being lucky enough to receive the fittest version of every single polymorphic locus would be very low, so all members of the population would suffer a reduction in fitness at some loci (this is often referred to as the genetic load of a population).

A number of researchers, most notably Motoo Kimura (**Heroes 5**, p. 160), explained this excess of variation by postulating that almost all of the observed genetic variation was selectively neutral. If these different alleles were all functionally equivalent, they would (as Darwin foresaw) be left to fluctuate in frequency until one mutation became fixed by chance. Since neutral alleles have no advantage or disadvantage over each other, each one has the same chance of going to fixation. So to estimate the rate of substitution of neutral mutations we need to know how many neutral mutations are produced each generation and what percentage of them is expected to go to fixation. Both of these quantities depend on the population size. The smaller the population, the more likely it is that random sampling of alleles will result in the loss of all alleles but one, so the percentage of neutral mutations going to fixation will go up as population size goes down. But, the smaller the population is, the fewer mutations are produced each generation because there are fewer genomes to undergo mutation. A larger population produces more mutations, but the probability of any one mutation being fixed is lower because random sampling error is less likely to result in fixation. Increasing population size increases the number of mutations but decreases the proportion of mutations that will become substitutions. This leads to the beautiful conclusion that, for purely neutral mutations, the effect of population size rather neatly cancels out, to leave the substitution rate determined solely by the mutation rate[2].

This simple model was used to explain not only the high level of polymorphism in natural populations (if most mutations are neutral they will not be removed by selection), but also the apparently constant rate of molecular evolution (if substitution rate is driven by the mutation rate, then as long as the mutation rate remains constant, substitutions will accumulate at a fixed rate over time). The neutral theory also led to predictions about the patterns of molecular evolution across the genome. If most substitutions were driven by positive selection, then we would expect parts of the genome most critical to survival and reproduction to accumulate the most substitutions. But if molecular evolution was predominantly driven by drift, then the least important sites would change the fastest because they were not subject to selection. The observation that the fastest changing sites in the genome are ones that make no apparent impact on phenotype, such as microsatellites in non-coding DNA, non-functional pseudogenes, and silent sites in protein coding sequences, supports a prominent role for drift in molecular evolution.

The neutral theory was further refined by Tomoko Ohta, a colleague of Kimura's. Where the neutral theory was based on the assumption that the majority of mutations were either

Figure TB5.2 Tomoko Ohta.

Reproduced courtesy of the Dibner Institute for the History of Science and Technology/ Burndy Library.

strongly deleterious or strictly neutral, Ohta considered the effect of having many mutations with only slight effects on fitness[3]. The majority of these 'nearly neutral' mutations would be slightly deleterious, so in a very large population, they would be removed by selection. But in small populations, random sampling effects have a greater influence on allele frequencies, and so drift can overwhelm selection for mutations of small selective coefficients. So the proportion of nearly neutral mutations going to fixation should rise as population size decreases. We can measure this effect in real populations. If we assume that, for a protein-coding gene, synonymous changes (that do not change the amino acid sequence) are likely to be neutral, but non-synonymous changes (that do change the amino acid sequence) are more likely to fall into the 'nearly neutral' category, then we would expect smaller populations to have a higher ratio of non-synonymous to synonymous substitutions. This effect has been observed in mammals with small population sizes[3], species confined to islands[4], and even tiny populations of bacteria and fungi sequestered in the guts of insects[5].

In addition to having a revolutionary impact on the way we consider evolution at the genomic level, the neutral and nearly neutral theories form an important basis for many of the analyses described in this book. However, as with most models, admitting the usefulness of these neutral theories of molecular evolution is not equivalent to proclaiming them 'The Truth'. For many years there was a hearty debate between 'selectionists', who considered that the patterns of molecular evolution were best explained by the action of natural selection, and 'neutralists', who considered the neutral theory to provide a more plausible explanation. The field is less black and white these days, so the terms 'selectionist' and 'neutralist' are now rarely used. Although an early paper described neutral evolution as 'non-Darwinian', there is no contradiction between neutral models of molecular evolution and Darwinian evolution: the neutral theory recognizes that all adaptations are produced by natural selection, but considers that these account for only a small percentage of observed substitutions. However, there is still a vigorous debate about the relative contributions of selection and drift to molecular evolution, and this is the focus of much ongoing research.

References

1. Lewontin, R.C. (1974) *The Genetic Basis of Evolutionary Change*. Columbia University Press.
2. Kimura, M. (1983) *The Neutral Theory of Molecular Evolution*. Cambridge University Press.
3. Ohta, T. (1995) Synonymous and nonsynonymous substitutions in mammalian genes and the nearly neutral theory. *Journal of Molecular Evolution*, Volume 40, pages 56–63.
4. Woolfit, M. and Bromham, L. (2005) Population size and molecular evolution on islands. *Proceedings of the Royal Society: Biological Sciences*, Volume 272, pages 2277–2282.
5. Woolfit, M. and Bromham, L. (2003) Increased rates of sequence evolution in endosymbiotic bacteria and fungi with small effective population sizes. *Molecular Biology and Evolution*, Volume 20, pages 1545–1555.

The moose doesn't care if it lies on red or pink geraniums. It doesn't even notice. So it crushes geraniums with no regard to their flower colour, and the survivors are a random sample of colours from the previous generation (**Figure 5.17**).

Figure 5.17

Once the moose has gone, the uncrushed survivors are pollinated and set seed, repopulating according to their kind. Red geraniums have red offspring, pink geraniums have pink offspring, and, though some have more offspring and some have less, neither colour has consistently more offspring than the other.

Figure 5.18

The following year is a drought, and the ground is so dry that few geraniums survive. The drought selects for plants with thick cuticles, long roots, and good moisture retention. None of these traits have any connection to flower colour, and neither red nor pink is better at surviving drought than the other.

Figure 5.19

When the drought finally breaks, the rain collects in hollows in the ground. The geraniums set seed. The seeds drop on the ground, near the parent plant. Any seed that drops on a damp hollow has a good chance of germinating. Seeds on dry ground do not grow.

Figure 5.20

The next year, the seeds that germinated on the damp hollow grow up and set seed (**Figure 5.21**).

Figure 5.21

Through a series of chance events – being squashed by a moose, subjected to a drought, seeds falling on dry ground – the pink variant has disappeared. In the first generation, the frequency of pinks was 50%. After the moose attack, it was 40%. After the drought it was 33%. Of the seeds that set after the drought, the percentage of pinks was 0%. With respect to flower colour, we have gone from a polymorphism (pink + red) to a homogenous population (all red). There is no more possibility of fluctuation in percentages now. No matter how you sample this population, whether they get eaten by grasshoppers or buried under snow, all the offspring will have red flowers. The pink variant has been lost, so the red variant has gone to fixation.

Let's review the roles of chance and selection in this hypothetical scenario. The indifferent moose was a random disaster. It caused a lot of destruction, but may not have had a selective effect on survival: it is hard to imagine selecting for moose-resistant geraniums (though perhaps the ability to re-sprout when crushed would help). The drought was also a chance event, but surviving the drought was not entirely due to chance. Those plants better able to withstand desiccation were more likely to survive, so the post-drought population had, on average, stronger cuticles and longer roots than the predrought population: a clear case of selection. Chance also played a role in the recovery after drought because some seeds were lucky enough to fall on fertile ground, while others were not (though perhaps certain adaptations might increase the chances of distributing over a wider area and having more chance of falling in a damp hollow). But, throughout all these events, there was no selection on flower colour. At no point did the reds have a better survival chance that the pinks. So why did the reds go to fixation, if they weren't any better? This came about simply because the population was finite, not all individuals reproduced successfully, so not all variants in the population were passed on to the next generation.

This kind of process is going on all the time throughout the genome. Think, if we may get so personal, about the polymorphic sites in your own genome. Let's say you have a SNP somewhere on chromosome 4: one copy of chromosome 4 carries an A at this position, and the other carries a T. Now (getting very personal) when you make gametes, each gamete has an even chance of getting a chromosome 4 with an A or a chromosome 4 with a T, and both are equally viable. Very few of your gametes will go on to form an embryo, but those that do are a random sample, some have A and some have T. Imagine you have five children, two A-carriers and three T-carriers. You have just skewed the gene frequencies in favour of T, even though T was no better than A. Next generation the frequencies might even out again (say, if one of the T-carriers joins a monastery and doesn't have offspring) or it might get more skewed (if one of the T-carriers has 14 children). This random sampling happens every generation for every polymorphism. If the SNP has no effect on reproduction, then it will fluctuate in frequency. Eventually, by pure chance, the fluctuation will hit 0% of A and 100% of T (or vice versa). At that point, the polymorphism is gone, so there is no more fluctuation in frequencies of A and T.

Population size

The effect of drift on allele frequencies will be greatest when the number of parents contributing alleles to the next generation is small. We can see the effect of sample size in our simple coin tossing experiments. Pick up your coin again, if you didn't lose it under the furniture in the last exercise, and go and stand in the middle of the room you are currently in (if you are lucky enough to be reading this book on a beach, then just draw yourself a room-sized square in the sand and stand in the middle of that). Toss your coin and take a step to the left if you get heads, or a step to the right if you get tails. Toss it again, and take another step, left for heads, right for tails. Since you have an even chance of getting heads or tails, chances are you will oscillate around the midpoint of the room. But you just might get a run of heads and move off to the left, or you might get a run of tails and move off to the right. If you keep doing this for long enough there is a chance that this random walk will take you all the way to the edge of the room at which point you stop. How many coin tosses did it take you before you hit the left or right wall (**Figure 5.22**, p. 167)? Now stand in the middle of a football stadium and repeat the exercise. Now how long will it take for your random walk to reach all the way to the edge?

Translating this somewhat tedious exercise into molecular genetics, mutations that have no effect on fitness are not subject to selection, so they will be randomly sampled every generation, and their frequencies will fluctuate up and down. In a small population, a run of random changes is more likely to result in substitution because it takes fewer steps in the same direction to get to 0 or 100%. In a large population it would take a really unlikely run of changes in the same direction to take you all the way to 0 or 100%. Given plenty of generations even unlikely events will happen, so we expect substitution by drift to occur occasionally in large populations, but less often than in small populations. So now we get to a very important conclusion. We have seen how the fate of a mutation depends on its influence on the chance of reproduction, and on the environment in which it finds itself. Now we can see that the fate of a mutation is also dependent on the size of the population in which the mutation occurs.

Because selection is a matter of probabilities, not certainties, even a selectively advantageous trait can be lost by chance events. If a mutation has a slight advantage, but it is in a small population where allele frequencies vary each generation just due to chance, then the positive influence of selection on the frequency of that allele could be easily derailed by random sampling. Imagine running the geranium example again, but this time allowing pink geraniums to have a 1% selective advantage. In a large population, the occasional geranium may get squashed by a rampaging moose, but this will have relatively little impact on overall proportions of reds and pinks, so pinks would steadily rise in frequency each generation until they replaced reds. But in a small population, a slight selective advantage translates into a small number of extra offspring. In a population of 100 individuals, there might be one extra pink. An unlucky accident could easily erase this selective advantage: the moose's foot might just fall on that extra pink geranium. So the fate of a mutation depends not only on its effect on the individual's chance of reproduction, but also on the size of the population in which it finds itself (**TechBox 5.3**).

Genetic drift occurs because not all variants present in the parent population will make it into the next generation. So what we really mean by 'population size' is

Figure 5.22 Most molecular evolution textbooks contain a standard diagram showing the consequences of a random walk in a small population. However, I decided it was more fun to employ a team of small children to generate a similar distribution with coloured chalk, each child tossing a coin and moving to the left when they got a head, to the right when they got a tail (they were bribed by being allowed to keep the dollar coin they used to generate the random walk). You can see that although most lines wander around the 50% mid-line, two lines hit the right-hand side (representing fixation of T and loss of H) and one reaches the left-hand side (representing fixation of H and loss of T). If you are the serious type you may wish to consult a population genetics textbook for a proper diagram. Otherwise, you can try this at home yourself, and at least you will get more fresh air than writing a computer simulation (though, in reality, the 3-year-old couldn't draw straight lines and kept shouting 'I won!' at random junctures, the 4-year-old took the coin and ran away, and only the 6-year-old had the patience to persist to fixation).

the number of parents that successfully get their alleles into the next generation. This will depend not only on the number of individuals in the population, but on the relative mating success of those individuals. There are many situations in which only a small fraction of the individuals in a population have the chance to contribute alleles to the next generation. We saw an environmental example of this in the geranium example, where the seeds of only a few individuals fell upon the damp hollow and germinated. The mating system of a species will also influence the number of successful parents (see **Figure TB5.3**, p. 169). For example, there may be thousands of individuals in a honey bee hive, but only one female and a handful of males will give rise to all of the next generation. The relatively small number of parents may explain why eusocial bees and wasps (those with high reproductive division of labour) have a relatively fast rate of substitution. We can see the genetics consequences of reduction in the number of parents when we consider how inbreeding influences the process of substitution.

We will revisit the effect of population size on rates of molecular evolution in Chapter 8

Inbreeding

When I was a teenager my older sister gave me some advice. She said 'If you ever find yourself in the situation where you are the last woman on Earth, and there is a single remaining man, whatever you do, do not touch him! If you start the human race again from just two people, your descendants will be so horribly inbred that it is better for the human race to die out at that point than to carry on'. This is, of course, just the kind of advice a growing girl needs.

TECHBOX 5.3 Population size

KEYWORDS

effective population size (N_e)

substitution

selection

drift

mutation rate

polymorphism

population

bottleneck

FURTHER INFORMATION

Lynch, M. (2007) *The Origins of Genome Architecture*. Sinauer Associates.

RELATED TECHBOXES

TB 8.3: Detecting rate variation

TB 8.4: Molecular dating

RELATED CASE STUDIES

CS 2.2: More moa (defining populations)

CS 8.2: Molecular detective (estimating molecular dates)

Population size is a constant concern of conservation biologists, and there are various techniques for estimating the total number of individuals in an interbreeding set of organisms from different kinds of survey data. Molecular geneticists are also concerned with population size, but have a rather different perspective. From the standpoint of molecular evolution, we are primarily interested in the movement of alleles from one generation to the next. Since not all individuals in a population will survive and breed, only a subset of the population contributes alleles to the next generation. So rather than estimating the total population count (N) which represents all individuals in the population, geneticists are generally interested in the effective population size (N_e), which represents only those that will contribute alleles to the next generation. Effective population size is an important component of models of molecular evolution, because it is a measure of the degree to which allele frequencies will be influenced by random sampling or, in other words, how effective selection is likely to be at a particular locus. Effective population size may be influenced by many different factors, such as sex ratio bias or variation in offspring numbers between individuals.

Imagine two populations of birds, both of which contain the same number of individuals, and equal numbers of males and females. In one species, males and females form long-lasting pair bonds, mating almost exclusively with each other throughout their lives. In the other species, males display together in leks. Females come to the lek, choose their favourite male and mate with him, then go off and raise the chicks on their own. Fashion being what it is, the females all tend to prefer the same males, so in this species a small number of fashionable males get all the matings, and most males miss out (**Figure TB5.3**, p. 169). Now let's think about the fate of a slightly advantageous mutation occurring in a male in each of these species; say, a modification of the globin gene that makes him slightly less likely to die from avian malaria. In the monogamous species, most males who survive to maturity get to pair up with a female, so a novel advantageous mutation in a male has a good chance of making it to the next generation. But in the lekking species, very few males get the opportunity to mate. As it happens, our male with the antimalarial mutation is, alas, not one of lucky few fashionable males, and he dies without reproducing. So this particular mutation will disappear from the population when its unfortunate male carrier dies. Alternatively, if a mutation occurs in one of the fashionable males who gets all the matings, then that mutation will end up in a large proportion of the next generation even if it is slightly deleterious (say, it slightly reduces shell thickness in his daughters' eggs). So unequal mating opportunities increase the influence of random sampling error on allele frequencies, and thus decrease the effectiveness of selection at promoting advantageous alleles or removing deleterious alleles.

In this example, the slightly advantageous mutation for malaria resistance was lost because it was unlucky enough to occur in a male who lacked the fashionable accoutrements to triumph in the mating arena, and the slightly deleterious mutation for thinner eggshells increased in frequency because it was lucky enough to occur in a fashionable male who had lots of offspring. We can see that having the good fortune to be inherited along with a

Figure TB5.3 Metrosexuals of the bird world: this gorgeous chap belongs to a species known as the Lesser Bird of Paradise (the Greater Bird of Paradise is just as fancy, but bigger). Male Lesser Birds of Paradise form leks, each occupying a specially prepared branch in a mating arena. When a female approaches, the males shimmy and shake, showing their magnificent plumage to best advantage. Despite all their efforts to look good, few males get to mate. In one lek, observed over 18 days in which the males spent several hours each day displaying, there were 99 visits by females resulting in 26 matings. All but one of those matings were with a single male. Most of the males had nothing to show for all that expenditure on appearances (see Beehler 1983[1]).

Photographer: Roderick Eime, 2004.

positively selected allele, or the bad luck to be bundled with a negatively selected allele, can make an important difference to the eventual fate of a new mutation.

We can take this same principle down to the genomic level to examine the influence of genetic linkage on the fate of mutations. If a slightly advantageous new mutation happens to be linked to an allele that reduces its carrier's chance of reproduction, then negative selection on the linked allele may prevent the positive allele from going to fixation (a phenomenon known as background selection). Alternatively, if a slightly deleterious mutation occurs next to a favourable allele, then it may be swept to fixation by positive selection acting on the linked locus (known as hitchhiking). So just as skewed mating frequencies reduce effective population size at the level of individuals, genetic linkage reduces the effective population size at the level of genomic loci.

Estimating population size from genetic data

Population size influences both the rate of substitution and the types of substitutions that occur, so we can use observed patterns of genetic variation to estimate effective population size. Firstly, we expect population size to influence whether the polymorphic loci we observe tend to have synonymous or non-synonymous alleles. Imagine a gene that codes for a protein. We might assume that a non-synonymous mutation (a nucleotide change that alters the amino acid sequence of the protein) is more likely to be subject to selection than a synonymous mutation (one that doesn't change the amino acid sequence). In a large population, selection will efficiently remove the majority of non-synonymous mutations (most of which will be deleterious), so we would expect that most mutations that remain in a population as polymorphisms will be synonymous. In a small population, selection is not efficient enough to remove non-synonymous mutations of small effect, so there will be more of these floating around in the gene pool. So one way to estimate effective population size is to use SNP data to compare the ratio of non-synonymous polymorphisms to synonymous polymorphisms, on the assumption that there will be a higher proportion of non-synonymous mutations persisting in small populations.

Secondly, we expect small populations to have lower levels of genetic variability than large populations. If a population is reduced to a small number of parents during a population bottleneck, then during that period it will experience rapid substitution due to drift with loss of alternative alleles. Small populations increase the chance of individuals mating with relatives that share a similar complement of alleles, and this inbreeding increases the homozygosity of the population (number of individuals carrying two copies of the same mutation), and reduces the heterozygosity (number of individuals carrying two different alleles at the same locus). Low genetic variability is therefore often taken as an indication of a past or current reduction in N_e.

Thirdly, because small populations lose alleles by random sampling, we expect genetically isolated populations to become fixed for different substitutions. This is useful for diagnosing population subdivision, when an apparently large or wide spread population is actually divided into smaller genetically isolated subpopulations. If the whole population were behaving like an idealized, randomly mating population, then each subpopulation should contain a random sample of the available alleles. But if the population is divided into small, isolated groups, then each of these groups may undergo substitution for different alleles, and loss of all alternative alleles, producing non-random representation of alleles in the different subpopulations. The degree of population subdivision can be gauged by comparing the genetic variation within subpopulations to the variation in the population as a whole (using measures referred to as '*F*-statistics').

It is important to remember that estimation of population size from genetic data is usually based on a number of key assumptions. Most importantly, these models typically assume that most mutations are either neutral or deleterious. If a significant proportion of mutations are advantageous, then an estimate of population size based on measures of genetic variability may be incorrect. This is because, unlike slightly deleterious mutations, the rate of fixation of slightly advantageous mutations increases with N_e, because natural selection can override drift in larger populations and drive advantageous mutations to fixation. If we knew what proportion of mutations were slightly positive or negative, we could estimate the overall effect on rates, but we rarely (if ever) have a way of measuring the distribution of fitness effects of spontaneous mutations. Some estimates of population size from genetic data are based on the assumption that all individuals, lineages, and loci have the same mutation rate. For reasons we will explore later (Chapter 8), even closely related species can have different mutation rates, so borrowing the mutation rate of one species to estimate the population size of another can be problematic. Furthermore, the mutation rate probably varies across the genome, so you cannot safely extrapolate results from one locus to another. These assumptions do not invalidate the use of genetic data to estimate population size, but it's a good idea to be aware of the assumptions you are making and keep reminding yourself of them as you consider the answers you get.

Reference

1. Beehler, B. (1983) Lek behavior of the Lesser Bird of Paradise. *Auk*, Volume 100, pages 992–995.

A common cause of small effective population size is the foundation of a new population from only a few individuals. If a population starts from a small number of parents, then each subsequent generation must mate with close relatives if the population is to persist. But why should inbreeding (mating with relatives) result in unhappy offspring? After all, Darwin married his cousin, and Darwin's children were not noticeably less well adapted than other children in the population. The effect of inbreeding is, like most patterns in molecular evolution, a matter of probabilities, not certainties. Mating between relatives increases the chances of offspring having two copies of the same mutation. If your grandfather carried a particular allele, his children and grandchildren may have inherited the allele from him. If you marry your cousin, your children could get one copy of that allele from their mother (who inherited it from her grandfather) and one copy of the same allele from their father (who also inherited it from the very same grandfather). In this way, inbred lines become progressively more homozygous, as with each generation there is more chance of each offspring inheriting two copies of the same allele, which was ultimately copied from a single ancestor.

Breeders take advantage of this progressive loss of variation in inbreeding by crossing related individuals to produce true-bred lines, pushing desirable traits to fixation. But as we saw in Chapter 3, most mutations are deleterious. The increased risk of having offspring that are homozygous for deleterious traits contributes to 'inbreeding depression' (loss of fitness due to shared ancestry of parents). Both of these processes are observable in pure-bred dog breeds, each of which is characterized by a suite of desirable traits and also a set of characteristic illnesses. For example, Newfoundland dogs were kept by fishing communities in Canada, and were often used to pull in fishing nets. Newfoundland dogs have a number of heritable traits that adapt them to this lifestyle, such as good swimming ability, webbed paws and a thick waterproof coat. But the breed is also characterized by a number of less desirable heritable traits, such as being prone to bad hips and bladder stones.

We can extrapolate the effects of inbreeding to all small populations. Each mutation begins at low frequency, so in a large, randomly mating population there is relatively little chance that two individuals with the same mutation will mate and produce offspring homozygous for that mutation. But the chance of getting two copies of a particular mutation will be greatly increased if there is a tendency to breed with related individuals. The likelihood of mating between relatives increases the smaller the population gets, so the chance of offspring being homozygous for any particular allele rises as population size drops. Eventually, for any given locus, all individuals are homozygous for a particular variant, and all other variants are lost. At this point, all individuals in the population carry a copy of the same mutation that originally occurred in a single ancestral individual. And in a small population there are relatively few genomes that might undergo mutation, so fewer new alleles are generated (**TechBox 5.2**, p. 162). So, over many generations, small populations will undergo rapid substitution, and therefore loss of genetic variation.

Some conservation biologists worry about the genetic effects of inbreeding in small populations. For example, there is some evidence that inbreeding can reduce fertility in populations reduced to a very small number of individuals, presumably through the increase in homozygosity of deleterious mutations. Concerns have also been raised about the long-term effects of loss of heterozygosity in small populations, on the grounds that a low pool of standing variation may limit the ability of an endangered population to respond to future environmental chance. However, it is difficult to evaluate the hypothesis that reduced genetic variation increases extinction risk, because small population size itself is a primary risk factor for extinction. Small populations are more likely to go extinct for the same reason that neutral alleles go to fixation in small populations. Random fluctuations in population size are more likely to reach zero for a small population, at which point it's Game Over for the population, as there are no individuals left to produce a new generation. The loss of genetic variation may be an issue for the long-term survival of endangered species, but very often this effect will be overwhelmed by more immediate threats such as habitat loss and competition with introduced species.

Patterns of substitution

We have seen that the process of substitution is influenced by many factors: the effect of mutations on their carriers' chances of reproduction, the influence of the environment, and the size and composition of the population. These factors vary and interact to generate complex patterns of substitution across the genome. We see evidence of these factors acting on substitution rates when we compare DNA sequences. In this section, we will revisit a number of examples from earlier chapters to illustrate how different patterns of substitutions can be produced by the action of selection and drift.

Conservation of sequences is due to negative selection

We saw in Chapter 4 that nearly all organisms have the same amino acid sequence (RFGEME) in the active site of the RNA polymerase beta subunit, an essential enzyme involved in gene expression (see **Figure 4.17**). **Figure 5.23** shows the sequence of this active site in six species.

Figure 5.23

Homo sapiens	GRSRDGGLRFGEME
Rattus norvegicus	GRSRDGGLRFGEME
Drosophila melanogaster	GRARDGGLRFGEME
Neurospora crassa	GRARDGGLRFGEME
Oryza sativa	GRKYGGGIRFGEME
Escherichia coli	GKAQFGGQRFGEME

The DNA sequence that codes for the RFGEME part of the enzyme is just as likely to undergo mutation as any other, so why hasn't it changed? RNA polymerase is so critical to cellular function that any inhibition of its function would be disastrous, so we can guess that the chance acquisition of a mutation that alters the active site of this enzyme is likely to put its carrier at a reproductive disadvantage (to put it mildly). When we see a DNA or protein sequence that has a relatively slow rate of change, we can usually invoke negative selection: the sequence is functionally important to the organism's survival and reproduction, and so any mutation that disturbs that function is likely to disappear from the population. Although mutations continue to occur in this sequence, we tend not to see them, because their carriers did not have descendants. So negative selection is evident in the conservation of sequences.

Even within functionally important sequences, some changes are more likely to be harmful than others. You can see in the above alignment that the human (*Homo sapiens*) and rat (*Rattus norvegicus*) have exactly the same amino acid sequence (GRSRDGGLRFGEME). The fly (*Drosophila melanogaster*) and the mould (*Neurospora crassa*) differ at only one amino acid from human and rat (GRARDGGLRFGEME). But when we look at the DNA sequences we see far more changes. Here is the same section of the RNA polymerase gene, this time showing the underlying nucleotide sequence (**Figure 5.24**).

Now we can see that although the amino acid sequences of human and rat are identical, their DNA sequences vary at four positions (see **Figure 4.15**). Human and fly differ by only one amino acid, but their DNA sequences differ at eight positions (**Figure 4.16**). And rat and mould have only one amino acid difference but 14 nucleotide differences. If the DNA sequence specifies the amino acid sequence, how is it possible to have so much change in the DNA sequence, but so little change in the protein?

The answer lies in the redundancy of the genetic code. In this sense, 'redundancy' does not mean useless or superseded, but refers to the way that several different codons can code for the same amino acid. Recall that the genetic code is read in triplets, with three bases coding for a single amino acid (see **TechBox 2.3**). Each triplet is referred to as a codon. There are 64 different codons but only 20 common amino acids, so most of the amino acids are specified by more than one codon. Different codons that specify the same amino acid are referred to as synonymous codons. Just as synonyms are different words with the same meaning (e.g. 'injurious' and 'deleterious' are synonyms that both mean 'bad for the organism'), synonymous codons are different triplets that code for the same amino acid (e.g. ACC and ACT are synonymous codons that both code for threonine).

Figure 5.24

Homo sapiens	GGT AGA TCT CGT GAT GGT GGC CTG CGT TTT GGA GAA ATG GAA
Rattus norvegicus	GGC AGA TCG CGT GAT GGT GGC CTG CGC TTT GGA GAA ATG GAG
Drosophila melanogaster	GGT CGT GCT CGT GAT GGT GGC TTG CGT TTC GGT GAG ATG GAG
Neurospora crassa	GGT CGT GCC AGA GAC GGT GGT CTC CGT TTC GGT GAA ATG GAA
Oryza sativa	GGA AGG AAA TAC GGT GGA GGG ATT CGG TTC GGT GAG ATG GAG
Escherichia coli	GGT AAG GCA CAG TTC GGT GGT CAG CGT TTC GGG GAG ATG GAA

Changing the DNA sequence from one synonymous codon to another will not change the protein it produces, so, on the whole, we expect synonymous changes to be functionally equivalent and not subject to selection (there may be biochemical reasons for preferring some synonymous codons, but we will gloss over that here for the sake of simplicity). Look at the first triplet in the alignment above. Both GGT and GGC code for glycine (G). In the second triplet, both AGA and CGT code for arginine (R). So the nucleotide sequence differs but the amino acid sequence stays the same. If synonymous changes make no difference to phenotype, then we consider them selectively neutral.

In Chapter 8 we will consider rates of synonymous and non-synonymous changes

Neutral substitution rate is determined by the mutation rate

Given what we have learned about selection and drift, we can now make some predictions about the rate of substitution at different sites in the genome. The rate of substitution in selectively neutral sequences (or neutral sites within a particular sequence) should be primarily determined by the mutation rate: the more mutations occur, the more substitutions occur (**TechBox 5.2**, p. 162). So the substitution rate will be faster in neutral sites than it is in sites under negative selection, since none of the mutations in neutral sites are removed by selection. This prediction is borne out by inspection of non-functional copies of genes, known as pseudogenes. We saw in Chapter 3 that gene duplication is an important factor in genome evolution, because spare copies of genes may be free to evolve new functions (**Case Study 5.1**, p. 146). However, if a duplicated gene is excess to requirements, so that it makes no difference to the chance of reproduction whether it functions or not, then any mutation that destroys the function of the sequence will not be subject to negative selection. Since all mutations in the excess copy are selectively neutral, these non-functional gene copies have a very rapid rate of substitution.

The globin gene families are a good example of gene duplication (**Figure 5.8**, p. 152). All of the globin genes are similar enough that we can tell they are homologous, originating from a single ancestral gene. In the human genome, there are a cluster of alpha globin genes on chromosome 16. There are two copies of the alpha globin gene, alpha1 and alpha2, both producing the alpha chains that combine with beta chains to make haemoglobin. A related gene, zeta globin, makes an alpha-like chain expressed in early embryonic development. There are also at least two pseudogenes, which are non-functional copies of the alpha2 and zeta genes (see **Figure 5.25**). Here is an alignment of the first 35 amino acids of the human version of these genes.

Figure 5.25

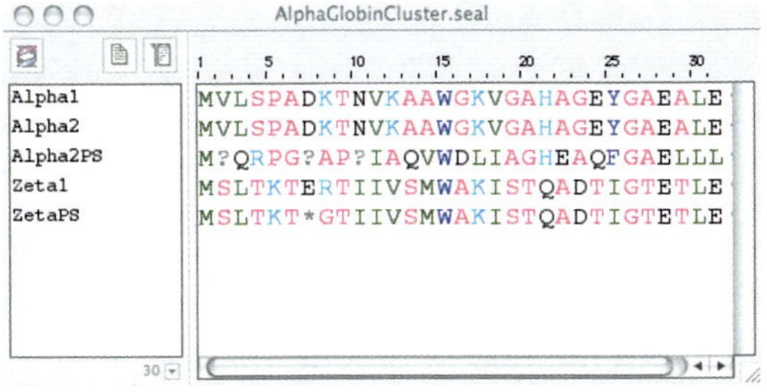

The genes in this alignment are clearly homologous, because they are similar at far more positions than we would expect by chance. The pattern of substitutions in the pseudogenes (Alpha2PS and ZetaPS) is different from the other genes in a number of respects. Critically, we can see that the pseudogenes have collected substitutions that destroy the ability of the sequence to produce a working protein. In the zeta pseudogene, a GAG codon (glutamic acid, E) has acquired a substitution that has turned it into a TAG stop codon (marked with * in the alignment). Construction of an alpha chain from this gene would be terminated at this point, making a useless protein with only the first six amino acids. In the alpha2 pseudogene, insertions and deletions have disrupted the triplet coding structure of the sequence (marked with '?' in the alignment). These insertions or deletions (often referred to as indels) would result in a frameshift mutation that would destroy the amino acid sequence downstream of the mutation (**Figure 5.26**, p. 174).

More generally, we can see that the pseudogenes have a higher rate of substitution than the functional genes. Furthermore, the substitutions in the pseudogenes are not biased towards silent changes that do not change the amino acid sequence specified by the protein. In a working gene, replacement mutations, which cause a change in the amino acid sequence, are likely to change the resulting protein for the worse, so are unlikely to

Figure 5.26

Mutations that ruin protein-coding sequences

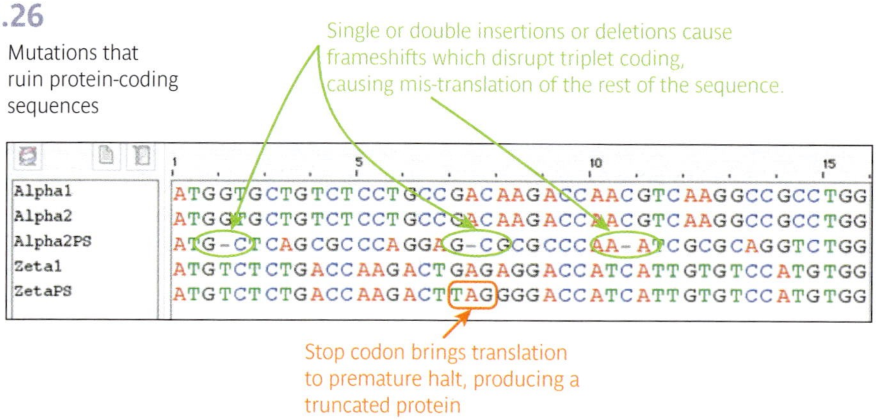

become substitutions. But in a pseudogene, it doesn't matter what kind of mutations occur, because neither silent nor replacement mutations make a difference to their carrier's chances of reproduction.

Because sites evolving as neutral have a substitution rate that reflects the mutation rate, neutral sites are often used to estimate the baseline substitution rate in the absence of selection (**TechBox 5.4**). This is a useful thing to measure, because when you see sites that have slower rate of substitution than the neutral rate, you can assume they are under negative selection. When you see sites with a much faster rate of substitution than the neutral rate, they may well be evolving under positive selection.

Positive selection results in rapid substitution

Negative selection results in slow substitution rates (most mutations are removed). Neutral sites have a higher rate of substitution (mutations are not removed, so by chance some will go to fixation). Sites under positive selection have an even higher rate of substitution, because some mutations are actively promoted (e.g. **Case Study 5.2**, p. 177). Substitution by positive selection will tend to be faster than substitution by drift, because, unlike neutral mutations whose frequencies wander up and down, the frequency of an advantageous mutation is expected to steadily increase.

There have been many studies aimed at identifying sites in the HIV genome that are evolving under positive selection, in the hope that these sites will provide useful targets for antiretroviral therapy or an effective vaccine. In particular, many researchers have targeted CTL epitopes. Cytotoxic T lymphocytes (CTLs), also known as killer T cells, are part of the human immune system. CTL response plays an important role in suppressing HIV replication in the human body. CTLs are built to recognize specific viral protein sequences, called epitopes. A mutation in the HIV genome that occurs in a sequence coding for a CTL epitope may make that virus unrecognizable to the immune system. Any mutation that allows HIV to replicate unrecognized by the immune system will be rapidly amplified, because carriers of that mutation will massively out-reproduce other members of the population. But the selective advantage of a particular mutation can change over time: once the immune system recognizes it, the allele will no longer have a reproductive advantage, and it will be replaced by any new immune-escape mutation that happens to arise at the same locus. So selection for novelty can result in rapid turnover of substitutions at sites in epitopes.

nef is one of the HIV genes that contains CTL epitopes. To look for positively selected sites in the *nef* gene, a team of scientists from Brazil and Britain analysed alignments of *nef* sequences. One of these alignments consisted of *nef* genes from HIV samples from many different infected individuals. Alignments of sequences from different lineages allow us to see how HIV has diversified as it has spread around the globe. In addition, a stupendously high mutation rate, combined with large population sizes and a short generation time,

TECHBOX 5.4

Detecting selection

KEYWORDS

dN/dS

K_a/K_s

MacDonald–Kreitman (MK) test

Synonymous

non-synonymous

polymorphism

selective sweep

FURTHER INFORMATION

Eyre-Walker, A. (2006) The genomic rate of adaptive evolution. *Trends in Ecology and Evolution*, Volume 21, pages 569–575.

RELATED TECHBOXES

TB 5.3: Population size

TB 5.2: Neutral theory

RELATED CASE STUDIES

CS 5.1: Sweet and sour (duplicate genes)

CS 6.2: Time flies (natural selection)

Rather than describe specific methods, which wax and wane in favour, I will here outline three basic approaches to detecting selection in DNA sequences. The first two approaches start by assuming that one class of mutations are always neutral. This neutral class might be all the mutations occurring in particular sequences, such as pseudogenes, introns, or intergenic regions, or it may be the 'silent' mutations in coding sequences, such as synonymous mutations within a protein-coding gene. The rate of substitution in these presumed neutral sites gives a baseline against which to look for the signature of selection. If we observe patterns of substitution that are very different from the expected patterns under neutral evolution, then we might conclude that selection is responsible at least some of the substitutions we observe. In this way, the neutral theory is an important null model in molecular evolution, which we accept or reject based on observations from our data (**TechBox 5.2**, p. 162).

The first approach is to estimate the rates of non-synonymous and synonymous substitutions for a set of aligned sequences (**TechBox 6.3**). Synonymous mutations are phenotypically silent, so they are generally assumed to evolve by drift alone (see **TechBox 5.2**). Non-synonymous changes are more likely to have an effect on survival and reproduction, so they may be under selection. If most non-synonymous mutations are under negative selection they will be removed from the population, so the rate of non-synonymous substitution per non-synonymous site (often termed *dN* or K_a) will be less than the rate of synonymous substitution per synonymous site (*dS* or K_s). In this case the ratio of the two rates (*dN/dS* or K_a/K_s, sometimes given the symbol omega ω) will be less than one. If all changes are neutral, then both kinds of substitution will go to fixation at the same rate (*dN/dS* will be around one). If positive selection is driving many substitutions to fixation faster than they would do so under drift, then *dN/dS* may be greater than one. This kind of test can identify when a particular sequence is under selection in a given lineage (**Case Study 5.1**, p. 146). Or it can be used to identify particular sites that are under selection, by testing whether the ratio of non-synonymous to synonymous substitution rates varies across individual codons. These *dN/dS* type of tests have relatively low power, so a failure to detect selection does not mean that evolution at that locus has been neutral, but a *dN/dS* significantly greater than one is often considered convincing evidence of positive selection.

An alternative approach compares the observed variation within populations to genetic variation between species. When a mutation arises, it initially exists as a polymorphism (one of several alternative versions of that locus in the population), and it only becomes a substitution when all alternative versions have disappeared so all members of the population carry the same mutation (i.e. it has become fixed in that population). We have seen in this chapter that deleterious mutations tend to be removed by selection, so do not become substitutions, and spend relatively little time as transient polymorphisms on the way down. Advantageous mutations can go rapidly to fixation, replacing all other alleles, spending little time as transient polymorphisms on the way up. Neutral mutations are left to fluctuate as polymorphisms, occasionally going to fixation by chance. Because polymorphisms of alleles under selection are expected to be short-lived, our chance of observing them is small. So, on the whole, we expect the majority of observed polymorphisms to be effectively neutral. If positive selection is a significant driver of evolution at a particular locus, then it will be responsible some of the substitutions, but few of the polymorphisms. MK (McDonald–Kreitman) tests take advantage of this imbalance, by comparing the ratio of non-synonymous to synonymous polymorphisms, derived from within-population samples, to the ratio of non-synonymous to synonymous substitutions, derived from comparing sequences to a related species. An excess of non-synonymous substitutions at the species level is taken as evidence of positive selection, because it is assumed they got there by rapid fixation, spending relatively little time as polymorphisms. MK-style tests require data on allele diversity both within a species and between species. Remember that the rate of fixation of mutations of small, selective coefficients (**TechBox 5.1**, p. 155) varies with population size, and this can skew the ratios of non-synonymous to synonymous polymorphisms and substitutions, so MK tests are not reliable for populations that have fluctuated in size (e.g. undergone a recent bottleneck or expansion) (**TechBox 5.3**, p. 168).

Both of these approaches are generally only applied to protein coding sequences, because they rely on defining synonymous (silent) and non-synonymous (potentially selected) changes. Genomic data offer an alternative approach, though its application has been limited thus far. One of the signatures of positive selection is a reduction in nucleotide diversity, not only in the locus under selection but also in any linked sequences. Selective sweeps might be detected by identifying regions of the genome with unexpectedly low nucleotide diversity. A recent survey of haplotype diversity in humans (see **TechBoxes 3.2** and **3.3**) identified a number of regions that potentially bear the signature of positive selection[1]. The putatively selected regions include genes involved in metabolism (e.g. alcohol dehydrogenase), morphology (e.g. skin pigmentation), and fertility (e.g. regulation of the female immune response to sperm).

I won't bother reviewing the available software for these methods, because it changes so rapidly that anything I say here will be out of date by the time you read it. Read the latest papers in the scientific literature to see what people are using, and ask around for the best programs. It must be said that, at the time of writing, some programs for detecting selection from sequence data are famous for reducing even the most robust scientists to tears, so you may need much patience, many soothing cups of hot tea, and a stern constitution to tackle this (but its worth it in the end, honestly it is).

Reference

1. Voight, B.F., Kudaravalli, S., Wen, X. and Pritchard, J.K. (2006) A map of recent positive selection in the human genome. *PLoS Biology*, Volume 4, page e72.

allows HIV to evolve so rapidly that we can witness evolutionary change in the virus within a single infected individual. So this study of the *nef* gene also included alignments of sequences from a time series of samples taken from a single patient during the course of their infection. In all of these alignments, there were particular sites in the *nef* gene that had a rate of substitution far greater than that expected under neutral evolution, so the researchers concluded these sites were probably evolving under positive selection.

For example, in the alignment of sequences of the *nef* gene on p. 180 (**Figure 5.27**; taken from the database at www.hiv.lanl.gov), you can see that one particular site has a much greater rate of substitution than neighbouring sites: the HIV strains shown here tend to differ from

CASE STUDY 5.2

Time flies: using preserved specimens to trace the evolutionary history of insecticide resistance

KEYWORDS

selective sweep
selection
ancient DNA
mutation
variation
polymorphism
adaptation

RELATED TECHBOXES

TB 2.4: DNA extraction

TB 4.2: DNA amplification

RELATED CASE STUDIES

CS 5.1: Sweet and sour (selection on duplicated genes)

CS 2.2: More moa (ancient DNA techniques)

Hartley, C.J., Newcomb, R.D., Russell, R.J., Yong, C.G., Stevens, J.R., Yeates, D.K., LaSalle, J. and Oakeshott, J.G. (2006) Amplification of DNA from preserved specimens shows blowflies were preadapted for the rapid evolution of insecticide resistance. *Proceedings of the National Academy of Sciences USA*, Volume 103, pages 8757–8762

“*The phenomenon of insecticide resistance is proving an informative model for studying microevolutionary processes in eukaryotes because it represents a rapid contemporary acquisition of a major new biochemical phenotype and it generally has a relatively simple genetic basis.*” [1]

Background

The evolution of insecticide resistance has been frequently used to study the dynamics of selection at the molecular level, due to the impact of resistance on human economies and health, and because the rapid timescales and large populations make it a tractable experimental system. One important question, with both theoretical and practical significance, is whether the response to a new selective regime tends to come from standing variation (promoting alleles already in the population), or from the generation of new mutations that happen to arise after selection is applied. In the case of insecticide resistance, it is usually assumed that alleles conferring resistance are costly to individuals, so they will not persist in untreated populations, and therefore must arise after treatment starts.

Aim

This study focused on two species of sheep blowfly, *Lucilia cuprina* and *Lucilia sericata*, that are agricultural pests around the world. Organophosphates (OPs) have been used to control these pests since the 1950s, but resistance to OPs arose within a decade of their introduction. Now virtually all *L. cuprina* in Australasia are OP resistant, and resistance is common in *L. sericata*. Resistance to different forms of OPs (marketed under different names) occurs through changes to the *αE7* gene that codes for the esterase 3 enzyme (E3). A single amino acid change, from glycine to aspartate at residue 137 of the E3 protein, gives a high degree of resistance to the OP insecticide diazinon. This mutation is referred to as Asp-137. A different single amino acid change, from tryptophan to leucine at residue 251 of E3 (Leu-251), provides a similar degree of resistance to the OP malathion. The aim of this study was to determine the distribution of Asp-137 and Leu-251 mutations in populations of *L. cuprinia* and *L. sericata* with different histories of OP exposure.

Methods

Blowfly specimens were taken from collections made in Australasia, Africa, Europe, Asia, North America, and Hawaii, including stored specimens of *L. cuprina* that were sampled

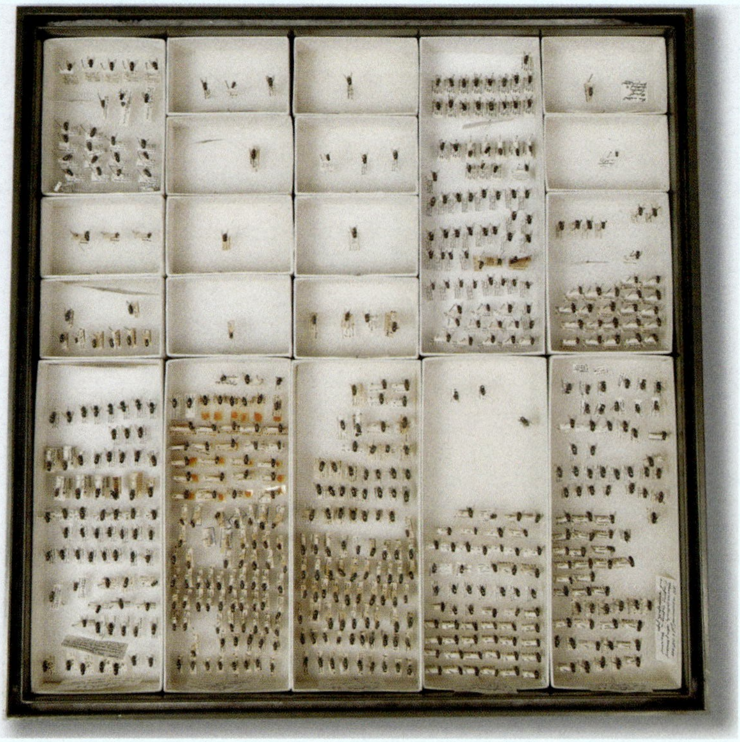

Figure CS5.2 Pole dancing for science: pinned specimens of the sheep blowfly *Lucilia cuprina* from the Commonwealth Scientific and Industrial Research Organisation (CSIRO) in Australia.

in Australasia before the introduction of OP insecticides in 1950 (**Figure CS5.2**). For the fresh samples, DNA was extracted from the specimens and PCR was used to amplify a 1.2-kb region encoding part of the *αE7* esterase gene. For the old samples, sterile 'ancient DNA' techniques were used to extract DNA (e.g. in a lab that was not used for other E3 amplifications to prevent cross-contamination) and two short regions, each around 150 bp spanning the site of a key resistance mutation (Asp-137 and Leu-251), were amplified.

Results

There was no evidence of the allele that confers diazinon resistance (Asp-137) in the pre-1950 specimens, which suggests either that this allele arose by mutation after the introduction of OP insecticides, or that it was at much lower frequency in the pre-OP populations and therefore was undetected in these samples. The Asp-137 mutation allows the E3 enzyme to break down OPs, but they reduce the enzyme's effectiveness at other tasks, so the diazinon resistance mutation confers high benefits in the presence of OP insecticides, but it carries a high cost. It is therefore less likely to be maintained in a non-treated population, but could rapidly go to fixation if it arises in a population exposed to OPs. The authors observed lower nucleotide diversity both at the E3 locus and in surrounding loci in their samples of post-OP *L. cuprina* from Australasia, which they interpret as the signature of a selective sweep due to the strong selection for the diazinon resistant mutation. The malathion-resistance (Leu-251) mutation tells a different story. Two out of 21 pre-OP flies sampled carried this resistant allele, indicating that it was present in the population long before the flies were treated with OP insecticides (another resistant allele, Thr-251, was also detected in the pre-OP samples). The Leu-251 mutation retains more of the E3 enzyme's other activities, so the malathion-resistance mutation is presumably less costly, which may explain why it can persist in a population that is not treated with insecticides.

Conclusions

This study provides examples of both selection on standing variation (malathion resistance) and selection on apparently novel mutations (diazinon resistance). Furthermore, insecticide treatment resulted in selection for the same mutation (Asp-137) independently in two different species: *L. cuprina* and *L. sericata*. The existence of the malathion-resistant allele at noticeable frequencies in the pretreatment fly populations indicates that, contra to a common assumption, resistance alleles do not necessarily carry a high fitness cost (or at least have some benefit even in the absence of insecticide)[2]. This means that we should expect a rapid evolution of resistance in populations carrying these alleles, and resistance will not necessarily disappear when treatment ceases.

Limitations

This study was limited to relatively few specimens (particularly due to the low success rate of extracting useable DNA from old, pinned specimens). Only 5 to 27 individual flies were sampled from each category, which means the estimates of allele frequencies are not very precise. On a semantic note, the term 'preadaptation' is prone to misinterpretation, because it implies that allele frequencies somehow anticipated a future selective regime. While that is not the meaning the authors intended, it is important to remember that selection has no foresight, it can only act on the here and now. Selection coefficients can change over time and, when they do, a mutation already existing in the population (previously either neutral, or only slightly deleterious or beneficial) may suddenly give its carriers a selective advantage: its selective coefficient has taken a sudden leap due to a change in the environment.

Future work

These findings are consistent with results from other studies of insecticide resistance, demonstrating that resistance is commonly based on the same restricted set of mutations selected in different species under insecticide treatment[2]. For example, one of the mutations identified in this study, Ser-251, has also been found to confer malathion resistance in Turkish houseflies (*Musca domestica*)[3]. Since mutations conferring resistance tend to occur at predictable sites[4], it may be possible to use genomic surveys for resistance alleles to judge the likelihood of a population developing resistance to particular insecticides.

References

1. Hartley, C.J., Newcomb, R.D., Russell, R.J., Yong, C.G., Stevens, J.R., Yeates, D.K., LaSalle, J. and Oakeshott, J.G. (2006) Amplification of DNA from preserved specimens shows blowflies were preadapted for the rapid evolution of insecticide resistance. *Proceedings of the National Academy of Sciences USA*, Volume 103, pages 8757–8762.
2. ffrench-Constant, R.H. (2006) Which came first: insecticides or resistance? *Trends in Genetics*, Volume 23, pages 1–4.
3. Taşkin, V. and Kence, M. (2004) The genetic basis of malathion resistance in housefly (*Musca domestica* L.) strains from Turkey. *Genetika*, Volume 40, pages 1475–1482.
4. Claudianos, C., Russell, R.J. and Oakeshott, J.G. (1999) The same amino acid substitution in orthologous esterases confers organophosphate resistance on the house fly and a blowfly. *Insect Biochemistry and Molecular Biology*, Volume 29, pages 675–686.

each other at this site, whereas at the other sites they tend to have the same amino acid. This site, which occurs within a CTL epitope, was identified as being under positive selection in an analysis of HIV samples from Europe and North America (Figure 5.27).

When we look at the DNA sequences for this site we can see that, unlike sequences evolving under negative selection, there are more changes in the first codon position than in the third (Figure 5.28).

This bias of changes to the first codon position also suggests that changes are being promoted by selection, because it seems that mutations that change the amino acid have more chance of going to fixation than those that have no effect on the protein sequence of the epitope.

The effect of codon position on substitution rates will be covered in Chapter 8

Figure 5.27

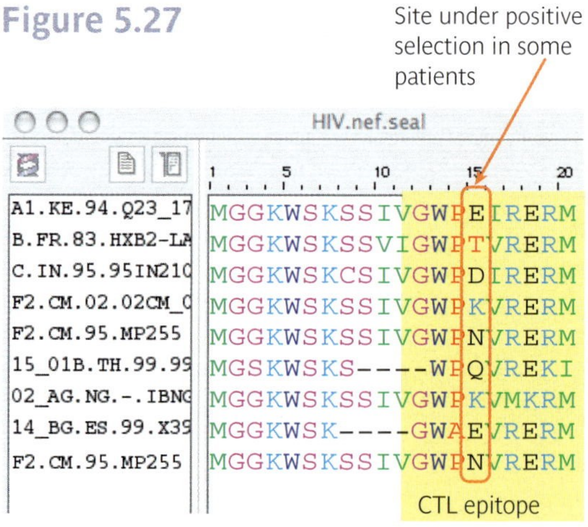

Figure 5.28

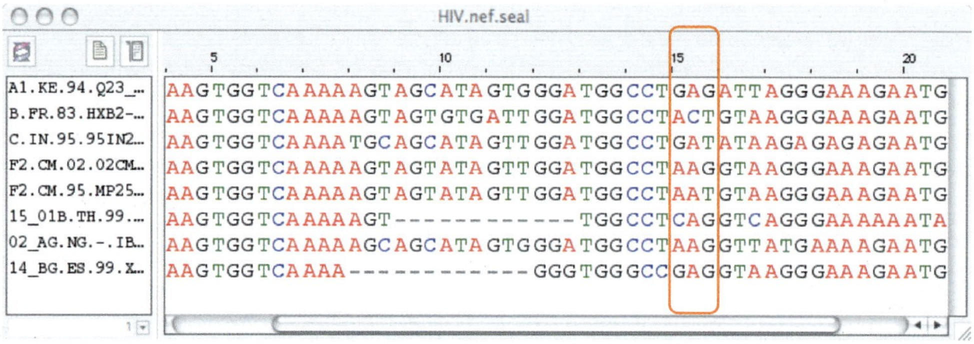

Are humans still evolving?

". . . *Endless forms most beautiful and most wonderful have been, and are being, evolved.***"**

Darwin, C. (1872) *The Origin of Species by Means of Natural Selection, or the Preservation of Favoured Races in the Struggle for Life*, 6th edn. John Murray

I once attempted to learn Latin, but, hampered by a striking ineptitude for languages, I seem to have retained only two Latin phrases in my memory. My favourite is 'auribus teneo lupum': I hold the wolf by the ears. This phrase describes being in a tricky situation: while you are holding the wolf by the ears, it can't bite you, but in order to run away, you need to let go of the ears. . . . and then the wolf can bite you. Admittedly, that's not much to show for a year of Latin classes (the other phrase I remember – 'Sum bos triturans' (I am the threshing ox) – is of no use whatsoever). But 'auribus teneo lupum' nicely describes the trepidation I feel when faced with discussions of human evolution, because they are prone to misinterpretation, liable to offend, and hampered by deeply held beliefs. However, I think that the examples we have used in this chapter, which have centred on human diseases, lead us naturally to the broader topic of molecular evolution in our own species. Humans make a useful case for reviewing what we have learned about molecular evolution, because we will have to consciously step back from the phenotype and ask ourselves what is happening at the level of the genotype. The point of the following discussion is to summarize and reinforce the conclusions of this chapter: molecular evolution is critically dependent on the environment (including external conditions, population size and composition, and genetic background), so the fitness of a particular variant can only be meaningfully judged with respect to the particular situation in which it is found.

There is a common notion that human beings stopped evolving in the Pleistocene, because when they developed tool use, symbolic communication, and abstract thought, they rose above natural selection and thenceforth all evolution was cultural and not genetic (sadly, this argument is often used to claim an innate propensity for uncivilized behaviour). But if we use Darwin's definition of evolution – descent with modification – then we could only say that humans had stopped evolving if we knew that allele frequencies were no longer going to fluctuate. As far as I can see, the only way to achieve such evolutionary stasis at a molecular genetic level is to have such a severe and sustained degree of inbreeding that all variation is removed from the population, producing a true-breeding line of genetically identical clones (then somehow prevent the occurrence or establishment of any new mutations). In addition to producing a very dull species, I suspect the human race would expire from the effects of homozygous recessive diseases long before such homogeneity was achieved.

We began this chapter by considering the ubiquity of variation in natural populations. Genetic variation is fundamental to evolution. It is the continuous generation of variation by mutation, and the substitution of some fraction of these mutations by selection and drift, that drives molecular evolution. SNP databases show us there is plenty of genetic variation in human populations (**TechBoxes 3.2** and **3.3**). In fact, we can confidently predict that, with few exceptions, each of us has a unique genome. What happens to this variation? Just as in any other organisms, the fate of alleles in human populations is determined by selection and drift. As with any other population of template copiers, any heritable difference that reduces the chance of reproduction (such as a deleterious mutation in the alpha globin gene) will tend to decrease in frequency, and any heritable difference increasing the chance of making copies of itself will tend to increase in frequency (such as a beta globin allele that provides protection against malaria). And, as in any other organism, chance plays an important role in allele frequencies in human populations.

The role of chance in allele frequencies is particularly evident for inbred populations, such as those founded by a small number of individuals. For example, mutations that cause the recessive genetic disease Ellis–van Creveld syndrome (EVC, otherwise known as six-fingered dwarfism) are extremely rare in most populations, so it is unlikely that two people with EVC mutations will come together and have an affected child. But EVC is very common in a particular Amish population, who tend to only marry members of their own community. All 50 people with EVC in this population are descendants of a single couple who joined the community over 250 years ago. Twelve per cent of this population now carry this mutation as unaffected heterozygotes.

Founder effects in allele frequencies are also evident on a larger scale. PKU alleles are found at higher frequency in people of Irish descent, wherever they are found in the world. In some populations with a high frequency of a particular disease-causing mutation, genetic testing may be used to inform partner choice. For example, members of Orthodox Jewish communities can use a screening service that advises on the incompatibility of couples if they both carry an allele for the same disease. This contemporary case of active intervention in human breeding in order to reduce the incidence of heritable disease brings us to a topic rarely discussed in polite company, yet one which may help us bring important concepts in molecular evolution into sharp focus. Now that we have the wolf of human evolution by the ears, why don't we gather our courage and leap straight into its jaws. Let us consider the far more dangerous topic of eugenics.

Can we manipulate human evolution?

R.A. Fisher, one of the founders of population genetics, was very deeply concerned about human population genetics, and this concern motivated much of his work (**Figure 5.29**). He was a strong proponent of eugenics,

Figure 5.29 R.A. Fisher, one of eugenics most enthusiastic promoters, wore very thick glasses, yet still considered himself fit enough stock to have nine children. Still, like John Maynard Smith, in Fisher's case poor eyesight may have been a selective advantage as it preventing him from going to war and getting shot.

Reproduced courtesy of Fisher Memorial Trust.

which is the active intervention into human breeding to improve heritable traits. Having worked in agricultural genetics, Fisher stated that 'if the methods of the stock-yard were applicable to mankind the human race could be improved in any desired direction, within a short historical period, to an extent exceeding existing differences between widely different races'. Fisher gave a clear voice to the decades-old concern that, in post-industrial society, natural selection was being turned on its head. Since Victorian times, many people felt that the people with the least desirable traits, particularly 'feebleness of mind', were often those that had the most offspring, so with every passing generation those of poor stock would out-reproduce their betters. Would feeble-mindedness thus go to fixation at the expense of higher intelligence, leaving the human race full of blithering idiots? (Some cynics might suggest this has already happened. Several times.)

More recently, some evolutionary geneticists have expressed fears that modern medicine will weaken the human race by allowing a build-up of deleterious mutations, through providing the means for individuals who would once have died young to survive and reproduce. For example, the great evolutionary biologist Bill Hamilton worried that the availability of Caesarean sections means that individuals who would otherwise have died in childbirth can now survive, and potentially pass on a propensity for obstructed labour to their own offspring. If so, the frequency of traits associated with Caesareans (say, small pelvises or large-headed babies) could increase in the population with each passing generation. Should we worry that the human race could become unable to bear children without surgery? H.J. Muller, a pioneer in the study of mutations, bleakly predicted the accumulation of so many deleterious mutations that the human race would slide into decay: 'instead of people's time and energy being mainly spent in the struggle with external enemies of a primitive kind such as famine, climatic difficulties and wild beasts, they would be devoted chiefly to the effort to live carefully, to spare and to prop up their own feeblenesses, to soothe their inner disharmonies and, in general, to doctor themselves as effectively as possible. For everyone would be an invalid, with his own special familial twists'.

Many, many books have been written on the perceived rights or wrongs of eugenic thinking, and it is not my intention here to examine these issues in detail, nor to promote any particular conclusion. I just thought it was an interesting case in point to think about some of the concepts that we have covered in taking a molecular genetic view of natural selection. Firstly, when it comes to humans we have an instinctive feeling for what we mean by 'fitness': a fit individual is someone gorgeous, clever, talented, healthy and, above all, desirable. But this is not the molecular genetic definition of fitness. At the beginning of this chapter we saw that natural selection is a numbers game, simply the consequence of differential reproduction of heritable variants. How do we identify the fittest variant? It is the one that increases in frequency with respect to other variants in the population. So, in evolutionary terms, we would identify any trait associated with a higher chance of reproduction as having a selective advantage. If it was true that the feeble-minded out-reproduced their brainy betters, and if feeble-mindedness was heritable, then from a genetic point of view we would have to view 'feeble-mindedness' as a positively selected trait, because it resulted in its carriers having more offspring of their own kind.

We also saw that fitness can only be measured with respect to a particular environment. Mutations that cause phenylketonuria (PKU) are strongly deleterious in humans that ingest phenylalanine, but selectively neutral when phenylalanine is absent from the diet. The only measure of fitness is whether a carrier of a particular mutation in a particular environment tends to have more or fewer offspring than other members of the population. In an environment where surgical intervention is unavailable, carriers of a mutation that causes larger-than-average baby heads may reproduce less due to death in childbirth, but the mutation may be effectively neutral in an environment where surgery is available. It may be that the selective value of a trait will change over time: if civilization collapses and hospitals disappear, then carriers of big-baby-head alleles might reproduce less. But evolution has no foresight, and mutation frequencies cannot prepare for future change. The only meaningful molecular genetic measure of fitness is the influence of a given mutation on its own frequency in subsequent generations. Selection operates relative to current conditions.

In some cases the influence of the environment may outweigh any underlying genetic variation in human phenotypes. We can see the effect of environment by considering one of the favourite concerns of eugenicists, variation in human intelligence. Intelligence has some degree of heritability, in that children are more likely to resemble their parents than they are to resemble random members of the population. And there are genetic factors that can influence intelligence (for example, genetic diseases such as PKU can cause mental retardation). However, these observations do not necessarily imply that the heritability of intelligence is due to genetic factors. Smart people have a tendency to raise smart children, even if those children are genetically unrelated to their parents. We know that environmental factors can reduce intelligence (for example, alcohol-related brain damage) or increase intelligence (for example, education). Nonetheless, there has been at least one attempt to set up a sperm bank exclusively based on donations from those perceived to have high intelligence, such as Nobel Prize winners, in order to create a genetic resource for selectively breeding from the smartest individuals. Could we increase the average intelligence of the population by such selective breeding? Even if we could, we would be wasting our time. Whether or not there is genetic variation in levels of intelligence, the observation of significant increases in measures of intelligence in some populations over the course of a single generation demonstrates the strong influence of environment on intelligence, for it cannot be due to change in allele frequencies. This leads to the very practical conclusion that, if your goal is to improve human intelligence, you would guarantee higher returns on your investment of time and money by improving the educational environment of as many people as possible than you would from breeding from Nobel Prize winners.

Conclusions

The study of heritable variation has provided the building blocks of evolutionary theory. Darwin demonstrated that variation was ubiquitous, and showed that cumulative selection of small variants could lead to the formation of phenotypically distinct populations. He emphasized the continuous accumulation of tiny modifications, such that populations were in a constant but insensible state of flux. The study of variation at the genetic level revealed very high levels of genetic variability in most populations and the development of mathematical models of population genetics demonstrated that even mutations of very small effect on fitness could go to fixation in large populations.

But not all changes at the molecular level are driven by selection. Random sampling of alleles can drive some alleles to fixation by chance, particularly in small populations. Because of this, small populations tend to undergo rapid substitution and loss of genetic variation. Genetic drift (random fluctuation in allele frequencies) is the prime determinant of allele frequencies for mutations that have little or no impact on chances of survival and reproduction. We can use our expectation of the rate of substitution of neutral mutations to interpret patterns of substitution in the genome. Sites under negative selection, for which most mutations are harmful, will tend to evolve more slowly than neutral sites, but sites under positive selection, where some mutations are actively promoted, will often evolve faster than neutral sites.

One of the tenets of the modern evolutionary synthesis, which is the refinement of Darwin's theory in light of molecular genetics, is that there is no qualitative difference between the evolutionary changes that happen with populations (microevolution) and the differences that distinguish different species (macroevolution): differences between species are simply the outcome of a long period of population divergence. The accumulation of genetic differences in populations through the process of substitution occurs continuously, and the longer two populations are separated the more different substitutions they acquire, so the more they diverge from each other. In the next chapter, we will learn how to compare DNA sequences from different populations in order to identify consistent genetic differences between them.

Further information

If you want a general book on natural selection and evolution, then why not start with:

Darwin, C. (1859) *The Origin of Species by Means of Natural Selection, or the Preservation of Favoured Races in the Struggle for Life*. John Murray.

In his many books on evolution, Richard Dawkins provides many crystal-clear descriptions of the actions and consequences of natural selection:

Dawkins, R. (1989) *The Selfish Gene*, 2nd edn. Oxford University Press.

If you prefer your population genetics with equations, then this classic text is a good place to start:

Maynard Smith, J. (1998) *Evolutionary Genetics*. Oxford University Press.

Marek Kohn gives a highly readable account of the life and science of some of the key figures mentioned in this chapter:

Kohn, M. (2004) *A Reason for Everything: natural selection and the English imagination*. Faber and Faber.

Origin of species

6

Or: how do I align DNA sequences?

"In considering the Origin of Species, it is quite conceivable that a naturalist, reflecting on the mutual affinities of organic beings, on their embryological relations, their geographical distribution, geological succession, and other such facts, might come to the conclusion that species had not been independently created, but had descended, like varieties, from other species."

Charles Darwin (1872) *The Origin of Species by Means of Natural Selection or the Preservation of Favoured Races in the Struggle for Life*, 6th edn. John Murray

What this chapter is about

Species are formed when genetically isolated populations accumulate changes that make them distinct from all other such populations. DNA sequence data can be used to identify species, and to place these species within a broader systematic framework that reflects patterns of descent with modification. To uncover evolutionary relationships, we must distinguish characteristics that are similar by descent (homologies) from those that have evolved independently in different lineages (analogies). Because sequence alignment is the process whereby homologies are distinguished from analogies, it is the most important part of any evolutionary analysis of DNA sequences.

Key concepts

Evolutionary biology: taxonomy and systematics

Molecular evolution: homology

Techniques: sequence alignment

Naming names

Biologists are very particular about names. The procedure for naming a new species is arcane, and has changed relatively little in the three centuries since the system of binomial nomenclature was established (**TechBox 6.1**, p. 187). Each species that has been formally described in the scientific literature is given a unique, double-barrelled name, in the form *Genus species* (in italics or underlined, with a capital for the genus name and a lower case for the species name). A genus is a group of related species, so binomial nomenclature is similar to the practice of referring to people by Firstname Familyname. I share the family name Bromham with my parents and sisters, but I am the only member of my family with the first name Lindell, so the name Lindell Bromham is a unique double-barrelled identifier for me. Similarly, there are many members of the genus *Canis* (dogs and wolves), but *Canis simensis* refers only to the Ethiopian wolf, and *Canis rufus* refers specifically to the red wolf (**Figure 6.1**).

Figure 6.1 Two members of the genus *Canis* (dogs and wolves): the red wolf (*Canis rufus*) and the Ethiopian wolf (*Canis simensis*). Both of these species are critically endangered. For information on research and conservation of the Ethiopian wolf, see *www.wildcru.org*.

TECHBOX 6.1 Taxonomy

With at least 1.5 million named species (and as many as 30 million still to be described), a sensible and accessible system of classification is an essential foundation for biology. The taxonomic system used today is the direct descendant of the brilliantly simple strategy set out by the Swedish botanist Carolus Linnaeus in his *Systema Naturae*, first published in 1735. Linneaus is most strongly associated with the binomial naming system (though he was not its originator), whereby each species is known by its genus and species name (e.g. *Homo sapiens*). More importantly, Linneaus established the principle of hierarchical classification. Species are grouped together into genera on the basis of shared key characteristics, and related genera are grouped together into families, which are grouped into orders, then classes, then phyla, and finally into the (currently) five kingdoms into which the living world is divided. So as you travel down the taxonomic hierarchy, from kingdom to phylum, to class, order, family, and genus, you find collections of ever more closely related species.

As an example of the modern application of Linneaus' hierarchical taxonomy, here is the classification of the mountain ash, the tallest flowering plant species in the world (**Figure TB6.1a**), according to the International Code of Botanical Nomenclature.

Taxonomic level	Formal name	Common name
Kingdom	Viridiplantae	green plants
Phylum (Division)	Embryophyta	land plants
Subphylum	Tracheophytina	vascular plants
Class	Angiospermopsida	flowering plants
Order	Myrtales	myrtles and allies
Family	Myrtaceae	myrtles
Genus	*Eucalyptus*	gum trees
Species	*regnans*	mountain ash

However, although the basic principle of hierarchical classification is a foundation of modern biology, there is no one agreed set of taxonomic divisions. Authorities may differ in the taxonomic divisions used, or the names given to levels in the taxonomic hierarchy. There is often disagreement are about what name a species should be known by, or in which higher group a given species should be placed. For example, the classification of mountain ash looks quite different on Wikispecies (Plantae>Magnoliophyta>Magnoliopsida>Myrtales>Myrtaceae>*Eucalyptus regnans*). Taxonomy is in a constant state of revision, and the increasing use of DNA sequence data to define taxonomic units has resulted in a great deal of rearrangement of taxonomic hierarchies.

KEYWORDS

species
classification
genus
tribe
family
systematics
Linnaeus
binomial nomenclature

FURTHER INFORMATION

There are many online databases that allow you to enter a species name, or higher taxonomic group, to get information and photographs, and you can click up or down through the hierarchy. Note that all of these databases use slightly different taxonomic systems.

Wikipedia (*http://en.wikipedia.org/wiki/*): an open-access editable database. The Tree of Life (*http://tolweb.org/*): an academic project, currently limited in scope but growing.

Entrez Taxonomy (*www.ncbi.nlm.nih.gov/entrez*): all species for which there are sequences on GenBank (which is an awful lot, but not everything: see **TechBox 1.1**).

RELATED TECHBOXES

TB 1.1: GenBank
TB 6.2: What is a species?

RELATED CASE STUDIES

CS 6.1: Barcoding nematodes (DNA taxonomy)
CS 2.2: More moa (defining taxa with DNA)

Figure TB6.1a A stand of *Eucalyptus regnans* (mountain ash, left), just under a century old, growing above a treefern understorey. *E. regnans* is considered by many to be the tallest species in the world, because trees felled by loggers in the late 1800s were found to be (or at least had been) taller than any other tree known, including the Californian redwoods. However, thanks to the dauntless spirit of those who felled the mighty mountain ashes with hand-saws, the tallest currently surviving trees are just under 100 metres. One of the largest living mountain ashes, unimaginatively named 'The Big Tree' (right) is shown here with a human for scale, although admittedly not a very big human.

Photograph: Lindell Bromham.

There are two important benefits of a hierarchical taxonomy based on relatedness. Firstly, it is a logical and manageable way of organizing information about the ever-increasing number of known biological species. Secondly, and more importantly, taxonomy based on the natural hierarchy of the evolved world can be used to predict the degree of similarity of any two species. For example, without any special knowledge of the organisms, I can make an educated guess that *Secale* (rye) and *Triticum* (wheat), which are both members of the tribe Triticeae, are likely to be more similar to each other than either is to *Poa* (bluegrass), which is a member of the tribe Poeae (**Figure TB6.1b**, p. 189). But *Triticum*, *Secale*, and *Poa* are all members of the family Poaceae, so they are likely to be more similar to each other than they are to papyrus (*Cyperus papyrus*), which is in the family Cyperaceae. These species are all contained within the order Poales, one of a dozen families of monocots (Class Lilliopsida). All monocots, including these species and the banana (*Musa*), are more similar to each other than they are to all dicots, such as a eucalypt tree.

If our hierarchical classification accurately represents patterns of descent then, in general, we expect most heritable characteristics to broadly follow the hierarchy of similarity (although there will always be exceptions). This applies to all levels of biological

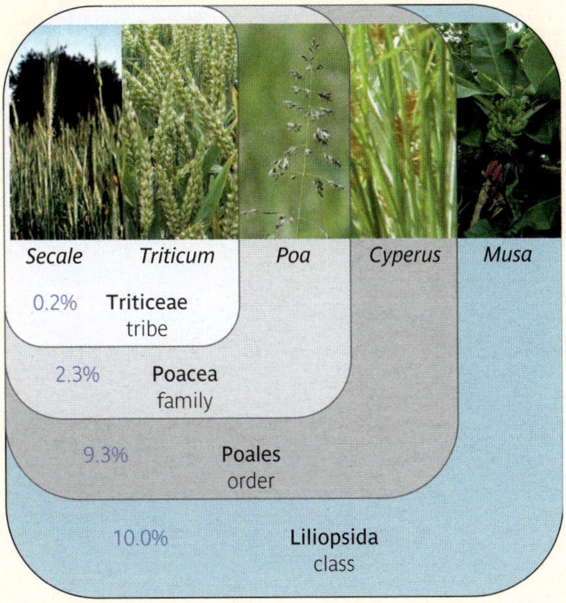

Figure TB6.1b Classification forms a nested hierarchy. The tribe Triticeae is contained within the family Poacea which is contained within the order Poales, which is in the class Liliopsida. The blue numbers record the average percentage difference between the *rBCL* sequences of these species.

Poa photograph: James Lindsey.

organization, including the genome. So, for example, the taxonomic relationships between these five plant species are reflected in the relative percentage difference between sequences of the *rBCL* gene (**Figure TB6.1b**). The *rBCL* gene codes for the protein ribulose 1,5-bisphosphate carboxylase/oxygenase, a photosynthesis enzyme better known as RuBisCO. RuBisCO is possibly one of the most abundant proteins on Earth. As an important enzyme in the carbon-fixing metabolic cycle in green plants and algae, RuBisCO plays a key role in global carbon cycles.

Scientific names are designed to standardize the labels given to species. Technically, a species can have only one scientific name, although it may be known by any number of 'common names' (the names used in everyday discussions). So, for example, the scientific name of the thylacine is *Thylacinus cynocephalus* (which translates to something like pouched dog-headed animal). But the thylacine has been known by many different common names including Tasmanian wolf and Tasmanian tiger, reflecting its recent distribution on the Australian island of Tasmania, the tiger-like stripes on its hindquarters, and its wolf-like appearance and hunting behaviour (**Figure 6.2**). But thylacines are not wolves or tigers. They are marsupials, belonging in the same taxonomic group as kangaroos and koalas. If the dog-headed thylacine looks like a wolf and acts like a wolf, then why is it grouped with the very un-wolflike marsupials?

Throughout this chapter, we are going to use the thylacine (**Figure 6.2**) to illustrate some important points about biological classification. First we are going to review the process of species formation, from a molecular perspective. We have seen that mutations arise in individuals (Chapter 3 and 4), but that, through the process of descent, a mutation can spread throughout an entire population and become a substitution (Chapter 5). At this point, all members of the population carry a copy of a mutation that arose in a single ancestral individual. In other words, alleles spread through populations by interbreeding. If there is some barrier to

Figure 6.2 The thylacine, or Tasmanian tiger, *Thylacinus cynocephalus*. You can find more photographs and movies of captive thylacines at *www.arkive.org*.
© London Zoological Society.

reproduction between one group of individuals and another, then they cannot exchange alleles, so any mutation that arises in one group will not become fixed in the other group. So isolated populations accumulate substitutions that are shared by all members of that population, and make them distinct from all other such populations.

Then we are going to look at how biologists go about classifying organisms into named, and readily identifiable, species. Although most biologists instinctively feel that biological diversity can be divided into distinct units, consisting of similar individuals differing in some consistent and important way from members of all other species, it is surprisingly tricky to pin down exactly what a species is. Nevertheless, there are many important reasons why we need to be able to assign individuals to particular species, so we will begin by considering how we define and recognize species. This is the role of taxonomy, which is the process of describing and naming species and other taxonomic groups. In particular, we are going to look at how DNA sequences are increasingly being used to identify and delineate species.

The primary goal of taxonomy is to provide reliable and consistent means of identifying different species. But modern taxonomy generally aims to do more than this. In addition to outlining characteristics that distinguish members of one species from those of all other species, a species definition will generally also include characteristics that the species shares with other related species, allowing species to be placed into broader taxonomic groups. Shared characteristics allow species to be assigned to a genus of closely related species, a family of related genera, an order of related families, and so on (**TechBox 6.1**, p. 187). Grouping species along lines of relatedness provides a useful classification system, because it is predictive: close relatives are likely to be more similar than more distant relatives.

But not all shared characteristics are evidence of descent. Unrelated species can appear very similar, like the thylacine and the wolf. If we group together species on the basis of traits that they have independently acquired, rather than inherited by common descent, then we will fail to uncover the true evolutionary relationships between species. So we are going to need to distinguish characters that are similar because they were inherited from a common ancestor (homologies) from characters that are coincidentally similar due to convergent evolution (analogies). The importance of identifying characters that are shared by descent is just as critical for molecular data as it is for morphological or behavioural traits. In DNA sequence analysis, we establish homology by sequence alignment. All other analyses described in this book rest upon the sequence alignment being a reliable indicator of evolutionary descent.

Genetic divergence

New species are created when existing populations are separated, and become progressively more distinct the longer they are separated from each other. From a molecular evolution perspective, we can view the process of speciation as one of genetic divergence. In very broad terms, there are three key steps in the process of genetic divergence. Firstly, when a population splits to produce two or more separate populations, each inherits a sample of the genetic variation of the parent populations. So they must start off with similar genomes. Secondly, genetic change is continuous, so the longer these populations are separated from each other, the more substitutions they will accumulate. Thirdly, different populations will tend to acquire different substitutions. So related populations start the same, but get progressively more different over time. At some point, these populations become so different that we decide to call them different species.

The mechanisms and consequences of speciation are among the most hotly debated topics in evolutionary biology, and there is a rich and voluminous literature on the topic. Clearly, we cannot review it here. The main point that we need to consider here is that once two populations have become separate to the point where they cannot interbreed, a mutation in one population cannot be passed into the next population, and so they will continue to accumulate independent sets of substitutions. Let's review some of the things that we have learned in the last few chapters to remind us why populations will get progressively more different from each other once they have become isolated.

We saw in Chapters 3 and 4 that mutations arise constantly, such that many individuals carry unique changes to their genomes. Mutations are chance events, so a mutation that occurs in one population might not arise in a neighbouring population. Of course, it is possible for the same mutation to arise in different populations (**Case Study 5.2**) but we expect each population will have a different set of mutations arising each generation. Chance also plays a role in determining which of these mutations persist and which are lost. We saw in Chapter 5 that genetic drift, arising from random variation in allele frequencies each generation, can cause some mutations to be lost from the population, and drive other mutations to fixation. This is particularly true for mutations with little or no effect on fitness, each of which has only a small chance of going to fixation. If two populations started with exactly the same set of neutral mutations, a small proportion of these will go to fixation by chance, so it is unlikely that exactly the same mutations will be fixed in each of the populations (if you and your friend both pull three lotto balls out of a bag, there is only a small chance that you will both randomly select exactly the same numbers).

But it is not only the random effect of mutation and drift that drives genetic divergence between populations. Selection can drive different substitutions in different populations, for three different reasons. Firstly, two neighbouring populations may experience different selective pressures. Populations of deer mice living at high altitude have a higher frequencies of a haemoglobin allele that aids oxygen transport at altitude. If gene flow between two populations of deer mice was cut by an avalanche or a highway, then we might expect selection to promote the fixation of different haemoglobin alleles in lower and higher altitude populations. Secondly, if two populations are under exactly the same selection pressure, they may 'solve' the same problem with different mutations. For example, there may be more than one mutation that confers resistance against insecticides, so different populations may have different substitutions that confer resistance (**Case Study 5.2**). Thirdly, even if the external environment of neighbouring populations is identical, one of the most important aspects of an alleles 'environment' is likely to differ: the presence and frequency of other alleles in the population.

Variation in mate choice provides an example of a genetic divergence mechanism driven by the genetic composition of a population. Imagine a population of frogs in which males vary slightly in the length of mating calls and females vary in their mate-call preferences. Now place a barrier to mating down the middle of your imaginary frog population. It might just happen that in one isolated half of the population there was a slightly higher proportion of females carrying the 'I like long calls' allele than the other. In this half of the population, males with long calls have a slightly higher chance of mating than short-call males. So we should expect, next generation, there will be an increased frequency of long-call males and long-call-preferring females in this isolated subpopulation. Which means that next generation there will be even more opportunities for long-call males to mate, and so the allele frequencies will continue to rise, potentially leading to the fixation of both the long-call and long-call-preference alleles.

The upshot of these three influences – the randomness of mutation, the random component of substitution, and different responses to selection – is that if you took a population, divided it in half, and let each half evolve independently of the other, then after some time you would see that they had both acquired different substitutions. Now what will happen if we bring the two halves of the population back together again? One possibility is that they will mate with each other and produce heterozygote offspring carrying alleles from both subpopulations. If these heterozygotes are just as good at reproducing as their parents, then eventually the two populations will completely mix together, once again sharing a common pool of alleles. But if these hybrid offspring are in some way disadvantaged, then there will be fewer successful offspring carrying alleles from both subpopulations (see p. 66 for an example). If individuals that mate with members of the other population are less likely to have well-adjusted offspring, then any mutation that gives its carrier a tendency to mate only with members of its own population will have a selective advantage, by preventing its carrier from wasting its reproductive effort producing hopeless hybrids. Thus selection may drive the substitution of mutations that contribute to reproductive isolation mechanisms, favouring traits that increase the chance individuals mate with their own kind. This process is known as reinforcement, because it favours the evolution of traits that reinforce the reproductive separation of the populations. Examples of reproductive isolation mechanisms include different mating calls or separation of flowering times.

When two populations become reproductively isolated, whether for physical or behavioural reasons, they cannot share any mutations, so their genomes will continue to acquire different substitutions, and they will get progressively more different. Taking a Darwinian view of molecular evolution (Chapter 5), we expect the process of genetic divergence to operate continuously, so that the longer two populations are separated the more different they will become. So we ought to be able to use the unique substitutions that accrue in separate populations to identify members of a particular population. Furthermore, because each population carries the genetic inheritance of its ancestors, we ought to be able to use patterns of shared substitutions to reconstruct the evolutionary history of these populations. But although we may recognize the general processes that cause species to form and differentiate, it is not a straightforward matter to say exactly what is a species, and what isn't. So now we need to consider the ways that species are defined and identified by biologists.

Defining species

> *"There are almost as many concepts of species as there are biologists prepared to discuss them."*
>
> Isaac, N.J.B., Mallet, J. and Mace, G.M. (2004) Taxonomic inflation: its influence on macroecology and conservation. *Trends in Ecology and Evolution*, Volume 19, pages 464–468.

Taxonomy is the bedrock of biology. Many of the case studies in this book rest critically on the ability to assign an individual organism to a particular species. For example, recognition and treatment of infections relies on the ability to correctly identify the infective agent (e.g. **Case Study 6.1**, p. 193). Species identification also plays an important role in many legal proceedings, such as the trade in endangered species (e.g. **Case Study 1.2**). Conservation efforts rest critically on the recognition of species as unique biological forms that should be protected and preserved (see **Case Study 6.2**, p. 217). More broadly, if we want to study or communicate ideas about biological diversity, we need units of measurement with which we can quantify species richness and monitor changes in biodiversity (e.g. **Case Study 4.1**).

If our primary goal in naming species is to provide a reliable and useful way of delineating real biological units, then it would be helpful to have a clear and practical definition of what a species actually is. Unfortunately, this is far from a simple issue. The definition of a species is the subject of a long-running debate in biology, and there have been a great variety of species concepts devised, all aiming to provide a neat way of dividing biodiversity into meaningful groups (**TechBox 6.2**, p. 196).

At the broadest level, many biologists feel that a species is a natural unit of biodiversity, which can be separated

CASE STUDY 6.1

Barcoding nematodes: what you don't know can hurt you

KEYWORDS

taxonomy
alignment
parasite
identification
voucher
databases
cryptic species

FURTHER INFORMATION

This nematode taxonomy database can be found at *http://nematol.unh.edu*.

RELATED TECHBOXES

TB 4.2: DNA amplification

TB 6.2: What is a species?

RELATED CASE STUDIES

CS 1.2: DNA surveillance (species identification)

CS 4.1: Glorious mud (measuring biodiversity with DNA)

De Ley, P., De Ley, I. T., Morris, K., Abebe, E., Mundo-Ocampo, M., Yoder, M., *et al.* (2005) An integrated approach to fast and informative morphological vouchering of nematodes for applications in molecular barcoding. *Philosophical Transactions of the Royal Society London B*, Volume 360, pages 1945–1958

> *In terms of the amount of morbidity and mortality inflicted on the human population, it can easily be claimed that parasites and their vectors are among the most dangerous groups of animals on earth. . . . Barcodes offer distinct opportunities for the study of parasite taxa that are largely under-represented in the record of animal species.* [1]

Background

The Nematoda (roundworms) is a diverse animal phylum containing with over 20,000 described species and possibly five times as many species still undescribed. There are both free-living and parasitic nematodes, including many commercially important parasites of both animals and plants. Correct identification of nematode species is essential for treating infections and for screening agricultural imports for infectious agents. But identifying nematode species can be extremely difficult, due to the relative paucity of distinctive morphological characteristics, and their often complex life cycles involving different forms living in number of host species (**Figure CS6.1**). Many diagnostic keys for nematode identification are based on adult characters, sometimes requiring observation of both sexes, and this may not be practical for identifying material from infections[2]. DNA sequence analysis has allowed the diagnosis of previously unidentified nematode infections. For example, an unknown parasitic infection found in the brain of an AIDS patient was identified more than 7 years after his death as the larva of a cestode normally only found in the stomach[3].

Aim

The aim of this project was to establish an accessible database linking DNA sequence data to voucher specimens, which preserve or record the organism from which the sequence was taken for future reference. The voucher-based approach explicitly connects DNA barcoding to 'traditional' taxonomy, by including enough information on morphology, ecology, and location to allow species assignments to be reviewed in future (unlike sequence-only databases). Videos make the morphology of the voucher specimen more widely accessible than physical, museum-based vouchers.

Methods

Live nematodes were collected from soil, litter, and marine environments, then mounted on microscope slides. A video of each specimen was made across multiple planes of focus, in order to record important diagnostic characteristics. After filming, each worm was cut into several pieces and digested to release DNA. To increase the amount of DNA in the

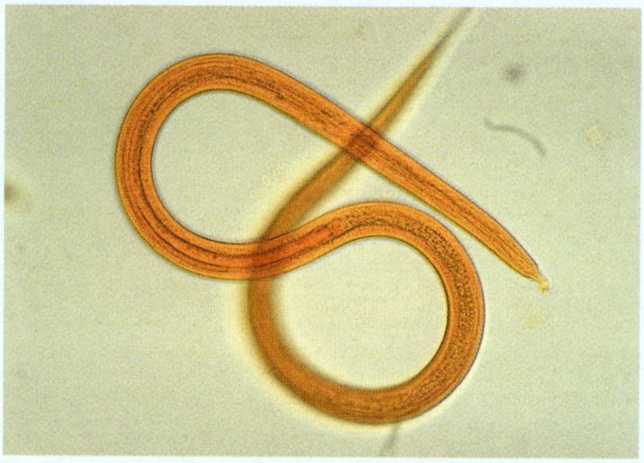

Figure CS6.1 Free-living larva of the nematode *Strongyloides*, which can enter the human circulatory system through the skin, move to the lungs, then to the intestine where they produce eggs. The eggs can reinfect the host or pass out of the body to re-enter the free-living cycle. Nematode larvae often lack key diagnostic features of adult flatworms, and can be difficult for a non-expert to identify. DNA barcoding databases may take the guesswork out of diagnosing nematode infections.

Photograph: CDC/ Dr Mae Melvin.

samples, the researchers used genome amplification kits with bacteriophage Phi29 DNA polymerase to copy all DNA in the sample using strand displacement (**TechBox 4.2**). They used PCR to amplify a section at the 5′ end of the large ribosomal RNA gene (the D2D3 expansion segment of 28S rRNA). This sequence was chosen because it contains both highly conserved regions, which provide robust primer sites (see **TechBox 4.3**), and highly variable sections, which vary even between cryptic species. Each sequence was checked for close matches in GenBank using BLAST (**TechBox 3.4**). Sequences were initially aligned using an online automatic alignment program (MAFFT), then regions of ambiguous homology were trimmed from the alignment (see **TechBox 6.3**, p. 211).

Results

It seems unlikely that a single gene will be able to identify all nematode species. The D2D3 regions of the 28S gene distinguished most of the voucher specimens, but sequence differences did not always correspond to accepted species boundaries. An alternative sequence commonly promoted for DNA barcoding nematodes, the internal transcribed spacer (ITS) that occurs just upstream of 28S, has a sufficiently high level of variation to distinguish most species. However, ITS can also vary within species, and possibly even within individuals[2]. Mitochondrial protein coding genes, although a common target of barcoding studies, may be less reliable for nematodes, perhaps due to unusual behaviour of their mitochondria (e.g. nematode mitochondria may undergo recombination). The researchers also found that alignments with ambiguously aligned sites gave demonstrably erroneous results, such as placing invertebrates from other phyla within the nematodes.

Conclusions

DNA barcoding cannot replace the traditional methods of describing nematode species. But, when combined with morphological and ecological information, DNA sequence databases can provide an invaluable aid to identifying nematode species. Barcoding using a combination of different marker genes would appear to be the most robust approach. Correct alignment is essential for reliable identification: computer-generated alignments must be inspected for regions of poor homology.

Limitations

DNA barcoding is only as good as the reference database: it cannot be used to identify species not already catalogued (however, phylogenetic analysis of DNA sequence data may reveal the affinities of an unknown species[3]; see Chapter 7). Barcoding will also be most reliable for species whose genetic diversity has been well surveyed.

Future work

DNA barcoding is being developed for many taxonomic groups, though there is currently a debate over formats and protocols. In particular, DNA barcoding may prove useful in monitoring international trade in biological products, such as detecting trafficking of endangered species (such as tiger parts for use in traditional medicines), or checking the ingredients of food products (e.g. identifying mixed species in meat products)[4].

References

1. Besansky, N.J., Severson, D.W. and Ferdig, M.T. (2003) DNA barcoding of parasites and invertebrate disease vectors: what you don't know can hurt you. *Trends in Parasitology*, Volume 19, pages 545–546.
2. Powers, T. (2004) Nematode molecular diagnostics: from bands to barcodes. *Annual Review of Phytopathology*, Volume 24, pages 367–383.
3. Olson, P.D., Yoder, K., Fajardo, L.G., Marty, A.M., van de Pas, S., Olivier, C. and Relman, D.A. (2003) Lethal invasive cestodiasis in immunosuppressed patients. *Journal of Infectious Diseases*, Volume 187, pages 1962–1966.
4. Teletchea, F., Maudet, C. and Hanni, C. (2005) Food and forensic molecular identification: update and challenges. *Trends in Biotechnology*, Volume 23, pages 359–366.

from all other such units on the basis of reproductive isolation (members of one species cannot interbreed with members of a different species) and/or unique adaptations (members of a species share the same set of features that make them suited to their particular location and way of life). However, these key aspects of population biology and ecology of a species are not always easy to measure in practice. A species may contain individuals that vary markedly in appearance or behaviour, yet are still members of a single interbreeding gene pool (**TechBox 6.2**). Conversely, members of genetically isolated populations could be identical in morphology, behaviour, and ecology. It could take a substantial amount of genetic analysis to prove that a recognized species never interbred with any other, particularly as genetic isolation is not an all-or-nothing phenomenon. In fact, many recognized species undergo low levels of gene flow with other species (**TechBox 6.2**). For example, there are over 700 species of gum tree (*Eucalyptus*; **TechBox 6.1**, p. 187), adapted to a wide range of niches from alpine areas to the desert. *Eucalyptus* species have recognizably distinct morphological and ecological traits, but many eucalypts can form the occasional hybrid with other species.

A formal taxonomic description of a species is an account of the unique characteristics that allow the members of a species to be distinguished from all other species. Rather than using direct observation of breeding and ecology to delimit species, most taxonomy is based on more easily measured characteristics, such as leaf shape, bone structure, or biochemical components of cell walls. Ironically, the defining characteristics of a species are usually not features we would easily notice, such as the iridescent blue colour of a beetle or acorn-collecting habits of a squirrel. Instead, the formal classification may be based on less exciting characters such as number of bristles on the forelimbs, or the shape and size of holes in the skull.

It may seem odd that species could be classified using such apparently unimportant characters, but it follows from the principle that, if all species have descended from shared ancestors, then related species will have a great many shared characteristics. So we should expect to be able to look at nearly any characteristic and find it is more similar between close relatives. However, this principle will be least robust for characteristics that are adaptations to particular niches or locations, such as particular colour patterns or foraging behaviours, because these are just the sort of characteristics we expect to be shaped by selection acting on particular species (which is a shame, because these are usually the most interesting and obvious characteristics of a species). The use of DNA sequences to identify species is essentially an extension of this principle. Members of

TECHBOX 6.2 What is a species?

KEYWORDS

taxonomy

DNA barcoding

biological species concept (BSC)

phylogenetic species concept (PSC)

interbreeding

cryptic species

reproductive isolation

Endangered Species Act

FURTHER INFORMATION

The international conservation union IUCN publishes the definitive list of species threatened with extinction: see *www.iucn.org*.

RELATED TECHBOXES

TB 6.1: Taxonomy

TB 6.4: Cloning and conservation

RELATED CASE STUDIES

CS 4.1: Glorious mud (DNA hybridization)

CS 6.2: Keeping the pieces (DNA and conservation)

> *"No one definition has as yet satisfied all naturalists; yet every naturalist knows vaguely what he means when he speaks of a species."*
>
> Darwin, C.R. (1859) *On the Origin of Species by Means of Natural Selection, or the Preservation of Favoured Races in the Struggle for Life*. John Murray.

The debate over what, if anything, is a species has been rumbling along for more than a century. As the number of possible 'species concepts' continues to increase, it could be argued that the main outcome of the debate has been to conclusively demonstrate that there is unlikely to ever be a single, universally acceptable definition that encompasses all of the entities we instinctively refer to as species. The most intuitive species concept is to group together recognizably similar individuals that differ consistently from members of all other such groups. For example, the crimson rosella (*Platycercus elegans*) and yellow rosella (*Platycercus flaveolus*) are clearly morphologically distinct, have different calls, and occupy different habitats (**Figure TB4.2a**).

But physical dissimilarity may not always define natural groups. Many interbreeding populations contain morphologically distinct forms. For example, females of the African

Figure TB6.2a The crimson rosella and yellow rosella.

mocker swallowtail butterfly (*Papilio dardanus*) come in 14 distinct forms (**Figure TB6.2b**). A single gene controls the striking different forms of female, many of which mimic unrelated species that taste nasty[1]. Populations of this species, found throughout sub-Saharan Africa, can contain up to six morphologically distinct types.

Conversely, many long-separated species may be physically indistinguishable. *Chiloglottis* orchids attract male wasp pollinators by mimicking a female wasp (**Figure TB6.2c**). The orchids use specific chemicals to attract different species of wasp pollinators. So coexisting individuals can belong to two reproductively isolated populations, because the pollen from the type that attracts one wasp species will not be carried to the other type, which attracts a different wasp species. Genetic studies have revealed that, in some areas, there may be several different orchid species present, indistinguishable to the human eye but smelling very different to their insect pollinators[2].

Figure TB6.2b Female forms of the mocker swallowtail butterfly: beside the name of each form is the name of the allele that causes that pattern (indicated by an H with a superscript).

Figure TB6.2c Sexually deceptive orchids, such as this *Chiloglottis* flower, attract wasp pollinators (*Neozeleboria cryptoides*) through a combination of visual and chemical signals that make the flowers sexually irresistible. A male wasp, blinded by love, attempts to copulate with the flower, and in the process collects a bundle of pollen on his back. Then the disappointed wasp goes off to try his luck elsewhere, taking the pollen to other sexually deceptive orchids.

Instead of relying of morphological similarity, we could consider that a species is a group with a common genetic heritage, sharing a distinct set of alleles through interbreeding. This is the basis of the Biological Species Concept (BSC), one of the most well-known and widely accepted species definitions. It makes evolutionary sense: if two populations are prevented from interbreeding with each other, then they will have different evolutionary fates. A new mutation can become fixed in one interbreeding population, but it cannot be shared with the other population, so reproductively isolated populations will gradually diverge from each other. Eventually, this accumulation of genetic changes will lead to the physical or ecological distinctness that we will recognize as species.

Yet the BSC will not always delineate groups that we would like to recognize as species. For example, the crimson rosella and yellow rosella (**Figure TB6.2a**, p. 196) would not be considered 'good species' under the BSC because, where their distributions overlap in nature, they can interbreed to form viable hybrids. In fact, these rosellas form part of a ring species, a series of morphologically distinct types with a wide-spread distribution, in which each population can interbreed with its near neighbours[3]. Furthermore, the BSC cannot be easily applied to the majority of organisms, because it leaves no way of classifying asexual organisms, or populations that exchange genetic material through lateral gene transfer rather than sexual reproduction, or populations that maintain distinct characteristics despite genetic exchange with other populations.

The practical difficulties of applying the BSC have led to the increasing use of a Phylogenetic Species Concept (PSC), which considers a species to be a natural group whose members share a unique evolutionary history. The PSC was originally formulated on the basis of shared physical traits that indicate common descent. But, increasingly, analysis of molecular data is used to demonstrate that a given group forms a distinct evolutionary lineage (**Case Study 6.2**, p. 217). Conversely, the PSC has been used to reject species status for populations that do not form a single unique lineage. For example, the Cape Verde kite

(*Milvus milvus fasciicauda*) is considered to be one of the rarest birds of prey, yet analysis of DNA sequences from the current population and from historical museum specimens, shows that individuals from this population do not form a distinct lineage, and instead should be considered to be members of the more widely distributed red kite (*Milvus milvus*)[4].

The evolutionary bases of the BSC and PSC highlight a general problem with species definitions. Even if reproductive isolation occurs suddenly – for example, when continuous forest is subdivided by logging – we expect the genetic changes associated with speciation to accumulate gradually. If we accept that species change over time, then where do we draw the line between a species and its modified descendents? Is *Homo sapiens* a different species to *Homo floresiensis*, the diminutive people who inhabited the Indonesian island of Flores until 12,000 years ago? Can we be sure that *Homo erectus* could not have interbred with a (fairly open-minded) member of our own species[5]?

At this point the reader may, as many thousands have done before, give up in despair and state that there is no universal species concept, so we should stop fussing about it. That would be an attractive proposition, were it not for the great practical importance of defining species, both in terms of scientific research and management of natural resources. Biologists who study patterns of biodiversity rely on species lists being accurate, for example to identify areas with high numbers of endemic species that should receive special environmental protection, or to study the driving forces behind high species diversity in the tropics[6]. Species definitions also have important legal implications. For example, the US Endangered Species Act is influenced by the BSC, so a species can be removed from the protected list if it is found to interbreed with any other population. Genetic analysis has demonstrated that there is ongoing hybridization between the endangered North American red wolf (*Canis rufus*; **Figure 6.1**, p. 186) and coyotes (*Canis latrans*). This had led to calls to remove the red wolf from the endangered species list, and to cease spending such large amounts of money on the red wolf recovery programme.

The most pragmatic approach to the mire of species definitions is to consider that, just as there are many processes that lead to species formation, so there are many different kinds of species. A species definition that works for one group may not create clear divisions in another group. We should use utility as our guide: what do we want to define species for? If we are concerned about conservation, we should, in an ideal world, make sure we give protection to any unique groups that we would be sorry to lose, whatever we decide to call them.

References

1. Nijhout, H.F. (2003) Polymorphic mimicry in *Papilio dardanus*: mosaic dominance, big effects, and origins. *Evolution and Development*, Volume 5, pages 579–592.
2. Mant, J., Bower, C.C., Weston, P.H. and Peakall, R. (2005) Phylogeography of pollinator-specific sexually deceptive *Chiloglottis taxa* (Orchidaceae): evidence for sympatric divergence? *Molecular Ecology*, Volume 14, pages 3067–3076.
3. Irwin, D.E., Irwin, J.H. and Price, T.D. (2001) Ring species as bridges between microevolution and speciation. *Genetica*, Volume 112–113, pages 223–243.
4. Johnson, J.A., Watson, R.T. and Mindell, D.P. (2005) Prioritizing species conservation: does the Cape Verde kite exist? *Proceedings of the Royal Society of London B*, Volume 272, pages 1365–1371.
5. Sterelny, K. and Griffiths, P.E. (1999) *Sex and Death: an introduction to the philosophy of biology*. University of Chicago Press.
6. Isaac, N.J.B., Mallet, J. and Mace, G.M. (2004) Taxonomic inflation: its influence on macroecology and conservation. *Trends in Ecology and Evolution*, Volume 19, pages 464–469.

a species will tend to share particular DNA sequences, even in genes that are not in any way associated with the morphology or behaviour unique to that species.

DNA taxonomy

Molecular data are, in many ways, an ideal form of information for identifying species. In Chapter 5, we considered how populations acquire genetic differences. Mutation makes individuals unique, by creating a set of alleles that characterize a particular genome. By sequencing loci that vary between members of a population, we can identify individuals from molecular data (Chapter 3). Similarly, substitution makes populations unique. We expect different mutations to go to fixation in different populations, and if those populations cannot interbreed then they cannot exchange alleles. The longer populations remain genetically isolated, the more unique substitutions they will acquire. Therefore we ought to be able to use consistent genetic differences between populations to identify genetically diverged populations that represent distinct species.

The process of substitution is discussed in Chapter 5, and the process of genetic divergence will be covered in more detail in Chapter 7

There are many advantages of using DNA sequences to identify species. Most importantly, DNA is universal to all living things, so molecular taxonomy can be applied to any kind of organism. Molecular data can be used to distinguish species that have relatively few distinctive morphological characteristics, such as viruses (**Case Studies 7.1** and **8.2**) and bacteria (**Case Study 4.1**). If you consider each nucleotide in the genome to be a potential diagnostic characteristic, then even the smallest viral genome has thousands of directly comparable characters, far more than you could derive from physical features of the organism. And, since the genome is contained in almost every cell, DNA taxonomy can be applied to partial specimens, from living or dead organisms (**Case Studies 1.1** and **2.2**). Even tiny samples of biological material can yield enough DNA to make an unambiguous identification. This is invaluable for species that cannot be directly observed, so has been useful for monitoring populations of elusive organisms (e.g. **Case Studies 1.2** and **2.1**), and solving many case studies in cryptozoology (the study of mysterious or mythical creatures; see **Case Study 1.1**). DNA taxonomy can also help to identify species whose different life-cycle stages are morphologically dissimilar, for example allowing plant species to be identified from seeds or animal species from larvae (**Case Study 6.1**).

One of the practical advantages of molecular taxonomy, though it might ultimately be regarded as a disadvantage for biology as a whole, is that it requires a great deal less training to be able to identify species using DNA data. It can be incredibly frustrating to peer down a microscope and try to determine whether a pickled beetle has a diagnostic character such as a 'pronotum with transverse furrow and with or without lateral depressions', or to follow a botanical key only to find that you are missing one of the main diagnostic characters, because you only have a leaf and cannot count petal number. But someone who has been trained to perform DNA extraction, amplification, and sequencing would technically be able to apply those skills to any biological sample (though in practice, lab protocols differ somewhat between sample types). Furthermore, DNA sequencing is becoming increasingly efficient and cost effective.

All of these advantages of DNA-based identification have led to attempts to develop universal 'DNA barcodes' – loci in the genome that can be sequenced to reliably identify any biological sample to species level. DNA barcoding is not a new technique, as it uses the same procedures described throughout this book: sequencing a specific locus then looking for matches in a DNA database. The term DNA barcoding really describes an attitude: that it is both possible and desirable to identify and catalogue species using DNA sequence databases. One of the goals of DNA barcoding is to provide unambiguous diagnostic characters that identify species, yet can be read automatically, without requiring years of specialist training. Although the holy grail of barcoding is to find a short gene segment that can be used to unambiguously identify any species on Earth, it may be impossible to find a universal barcode that will work for all taxonomic groups (see **Case Study 6.1**, p. 193).

Defining species with DNA

It is important to make the distinction between using DNA to identify species and using DNA to define species. If we know that all members of a certain species share particular bases in a specific DNA sequence, then we could use that sequence to prove an organism belonged to that particular species. In **Figure 4.19**, we saw that members of the species *Bacillus* *anthracis* all appear to have the same sequence for a particular region of the RNA polymerase II beta subunit gene, so this sequence is likely to provide a useful diagnostic test for the presence of anthrax.

More controversially, DNA analysis is increasingly being used to define species. We saw that taxonomy often relies on using easily measured characters as a proxy for reproductive isolation or ecological distinctness. DNA sequences are a very convenient way to measure similarity. For example, in **Case Study 4.1**, we saw how the practical difficulties of counting bacterial species that cannot be cultured in the lab can be bypassed by measuring the diversity of genomes in soil samples. Not only do DNA sequences provide a convenient taxonomic character, but they can also reveal distinct populations that would not have been detected using morphological data alone.

Since substitutions accumulate continuously, the DNA sequences of isolated populations will become increasingly different, even if the external morphology remains the same. Genetic analysis has led to the recognition of many 'cryptic species', which are clearly reproductively separate populations despite a lack of morphological divergence. Many zooplankton (minute, aquatic organisms) have been considered to belong to cosmopolitan species, found all over the globe, but DNA analysis has revealed many ancient cryptic species, whose lack of morphological differentiation belies substantial genetic diversification (see also **Case Study 8.1**). For example, DNA sequences taken from the rotifer *Brachionus plicatilis* collected all over the world show that this species actually consists of at least seven distinct lineages, sometimes coexisting in the same habitats, that may have been separated for tens of millions of years (**Figure 6.3**).

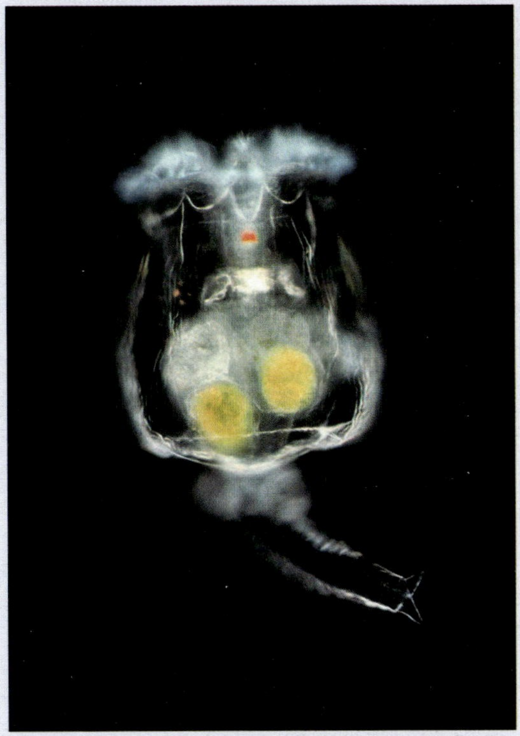

Figure 6.3 The rotifer *Brachionus plicatilis* is a globally distributed species. Rotifers are microscopic aquatic animals. They were formerly known as wheel animalcules on account of the circular band of cilia that powers their locomotion, and draws food into their mouths. Rotifers are famous for their powers of abstinence: in their dormant state, they can go without food or water for extended periods, and some lineages of rotifers have apparently gone without sex for tens of millions of years. You can find a pictorial guide to rotifers (and other microscopic organisms) at *www.micrographia.com*.

The relationship between genetic and morphological divergence is discussed in Chapter 8

The relative ease of DNA analysis, compared to traditional taxonomy, has spurred a huge increase in the use of DNA sequences to define species' limits. However, the molecular definition of species is problematic in practice because there is no clear cut-off for what is a 'real' species and what is a genetically diverse, interbreeding population. If we want species to represent distinct populations characterized by unique traits, then identifying differences in gene sequences may be of little value if we do not know how those differences relate to other important features such as ecology and behaviour. So DNA is not an instant solution to taxonomy. And, as we will see later in the chapter, molecular data are not free from the problems of determining which similarities are the result of shared ancestry and which are merely superficial cases of convergent evolution. But molecular data do provide a very useful and accessible tool for estimating relatedness, which is becoming more and more valuable in systematics. Systematics is the discipline of placing taxonomy within an evolutionary framework, to group related species in a hierarchical pattern.

Systematics

A hierarchy is a useful way of organizing the very large, and increasing, number of recognized species. This hierarchy of shared characters can be used to design a taxonomic key – a nested set of questions that will progressively narrow down the number of possible species to which a given organism might belong. But the taxonomic hierarchy is more than just a fancy filing system. The systematic hierarchy should reflect the actual pattern of evolutionary descent that produced the species. This provides a logical and natural system of classification. More importantly, an evolutionary classification has great predictive power, because we expect patterns of relatedness to generate patterns of similarity at all levels of biological organization. Since we expect the hierarchy of differences in the DNA sequences to be broadly indicative of how closely related two individuals are, whether these are individuals of the same species or different species, DNA sequence comparison is a very useful tool for estimating relatedness. Measures of relatedness, in turn, allow us to make predictions about similarities and differences, because we expect closely related individuals or species to have more in common than those which are more distantly related.

The hierarchy of similarities arising from the copying process at the heart of evolution is discussed in Chapter 4

Despite initial appearances, thylacines are actually a very good example of the hierarchical nature of biological characteristics. Although aspects of their external appearance may be similar to wolves, almost everything else about thylacines is more similar to other marsupials. Marsupials are one of the three major divisions of mammals. These three groups can be distinguished by a number of defining characteristics, the most obvious of which is reproductive mode. Monotremes (platypuses and echidnas) lay eggs. The embryos of placental mammals (such as horses and bats) are carried internally until an advanced stage of development. But the embryos of marsupials (such as possums and kangaroos) are born at a very early stage and undergo most of their development in a pouch on the outside of the mother's body.

Although they have the external appearance of placental wolves, thylacines have all of the classic, defining characteristics of marsupials, including pouch-based development. In fact, the name *Thylacinus* is derived from a Greek word for pouch, though in this case it may equally refer to the pouch around the testes on male thylacines, an unusual aspect of their biology. But thylacines also have other, less obvious, key marsupial characteristics. Female thylacines, like all other marsupials, have two vaginas (and most male marsupials have a forked penis to match). As with other marsupials, the penis on male thylacines is behind the scrotum, not in front of it as in placental mammals. There are also skeletal characteristics that separate marsupials like thylacines from placentals like wolves, such as the epipubic bones (attached to the pelvis), and small curved triangles that stick out of the jawbones (formally referred to as medially inflected angular processes of the lower jaw).

Just as most physical characteristics of thylacines are more similar to other marsupials than they are to wolves, so too is their DNA. **Figure 6.4** (p. 203) shows part of an alignment the 12S ribosomal RNA gene, which codes for an RNA molecule that makes an essential part of the genomic system.

Four of the species included in this alignment – thylacine, quoll, dunnart, and marsupial mole – are marsupials. The other four are placentals: dog, cat, shrew, and European mole. The sequences from the marsupials are all more similar to each other than any of them is to any of the placental sequences (**Figure 6.5**, p. 204). The same cannot be said for the external morphology of these species, which varies greatly between the four marsupials. Not only that, but the morphological features of each of the marsupials is far more similar to one of placental species than it is to the other marsupials. Why are the patterns of similarity in the DNA sequences not reflected in observable features of the organisms?

The eight species in **Figure 6.5** illustrate the concept of convergent adaptation, where separate biological lineages independently evolve similar adaptations to similar niches (i.e. ways of living). For example, both the marsupial mole and the European mole are adapted to a life of burrowing through soil hunting for worms and grubs: both have the typical mole-like characteristics of elongated body, flat, paddle-like feet for digging, no external ears, and only rudimentary eyes. But marsupial and placental moles don't share their mole-like

Figure 6.4

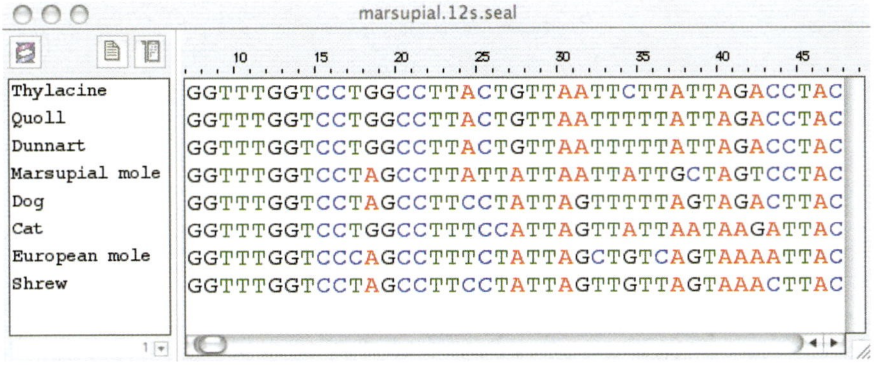

characters because they both inherited these characteristics from a mole-like ancestor. Both have independently evolved very similar adaptations to the same mole lifestyle. The same can be said for the marsupial dunnart and placental shrew, which are independently adapted to eating insects and other small animals, and the nocturnal predators, the marsupial quoll and the placental cat (Figure 6.5).

Clearly, it would be possible for convergent adaptations such as these to confuse taxonomy, causing species to be grouped on the basis of superficial similarity rather than true relatedness. However, if we look at enough different characteristics, particularly concentrating on those characters not associated with adaptation to a particular niche, then we ought to be able to see the pattern of hierarchical differences resulting from the copying process at the heart of evolution. This hierarchy of similarities should apply to all inherited traits, including DNA sequences. For example, we can see that the DNA sequence from the thylacine is more similar to the sequences from the other marsupial species than it is to the morphologically similar dog (Figures 6.4 and 6.5). This illustrates the principle that DNA data are expected to unite those species that share common ancestry, rather than uniting species that have a superficial resemblance.

Because the entire genome is copied every generation, we expect the sequence of any gene to be more similar between close relatives than it is between more distantly related individuals or species. The gene we choose to analyse does not have to be connected with characteristics that might contribute directly to the biological separation of species, such as coloration or foraging behaviour. In fact, most biologists choose to sequence 'housekeeping' genes that are associated with fundamental metabolic pathways or with the genomic information system. These genes are expected to steadily accumulate changes as they are copied down the generations. But, because they do essentially the same, fundamental job in all species, they are less likely to accrue bursts of change in response to selective events, such as coping with environmental change or adapting to a new way of life.

Chapter 8 covers some considerations in selecting sequences

All of the genes we have looked at so far are housekeeping genes. Both the RNA polymerase II beta subunit gene and the 12S rRNA gene make essential components of the cellular system that translates RNA messages into proteins. These genes are essential in all organisms (bacteria have a different version of the 12S gene, but it performs essentially the same role). Other housekeeping genes may be involved in metabolism (e.g. cytochrome *b*, which forms part of the electron transport chain that generates energy) or cell structure (e.g. histones, which form the 'beads' around which DNA is wrapped). Not all evolutionary analyses rely on housekeeping genes, but it is a common strategy when the aim is to uncover evolutionary relationships, because it is a handy way of finding homologous sequences common to all species, and avoiding the problems of analogous changes due to adaptation to a specific niche.

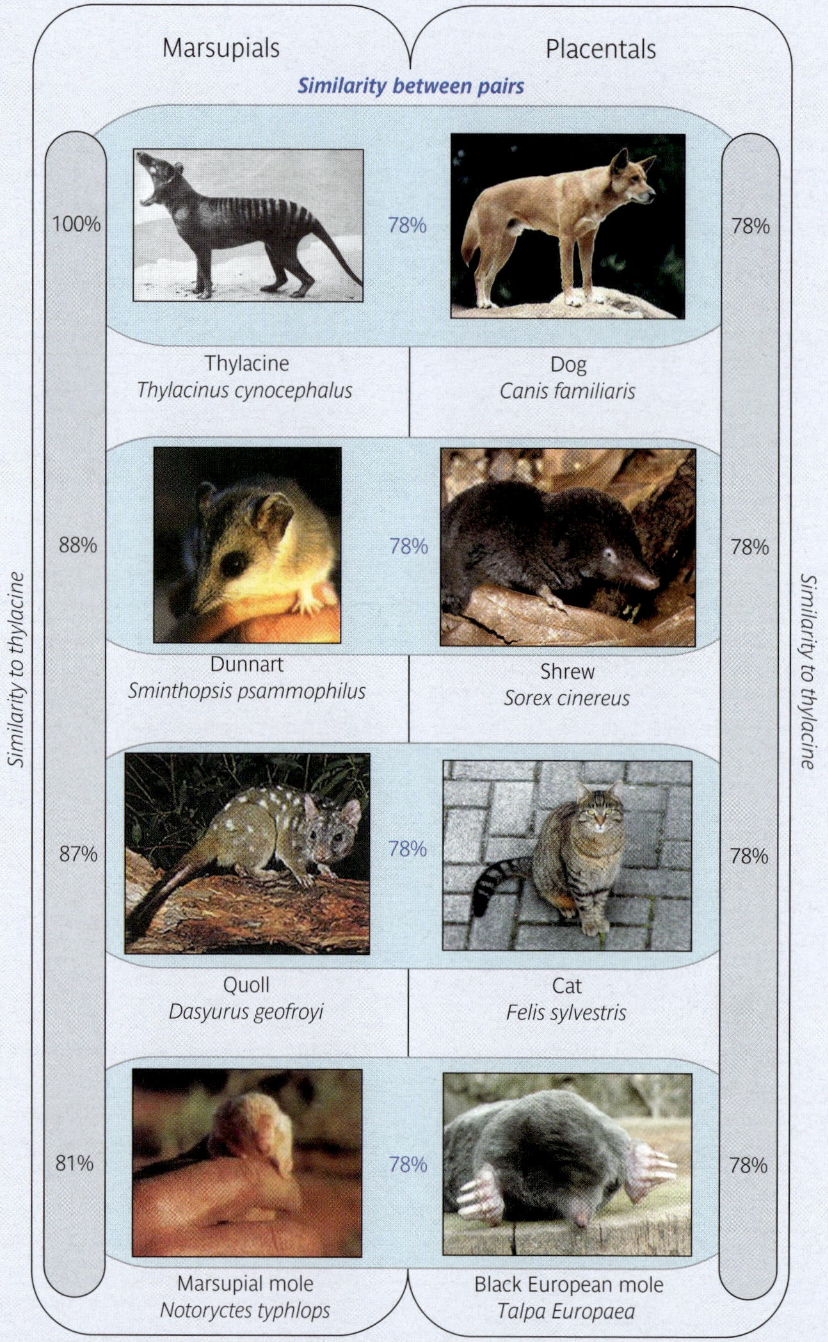

Figure 6.5 The percentage similarity between DNA sequences for the 12S rRNA genes for eight mammals species. Marsupial species are on the left and placentals are on the right. The grey bars show the percentage similarity of each species to the thylacine 12S sequence. The blue bar shows the percentage similarity between morphologically similar species pairs, such as thylacine and dog, or quoll and cat. The placental mammals are all equally distant from the marsupial species. For an explanation of why the genetic distance does not increase linearly with time since divergence, see Chapter 8.

Homology: shared by descent

"*Numerous cases could be given of striking resemblances in quite distinct beings between single parts or organs, which have been adapted for the same functions. A good instance is afforded by the close resemblance of the jaws of the dog and the Tasmanian wolf or* Thylacinus, *animals which are widely sundered in the natural system.*"

Charles Darwin (1872) *The Origin of Species by Means of Natural Selection*, 6th edn. John Murray.

The case of the thylacine illustrates an important point in evolutionary biology: to uncover evolutionary relationships it is necessary to distinguish two types of similarities. Similarity by descent is referred to as homology. The pouches of the thylacine, marsupial mole, dunnart, and quoll are homologous: these animals all have pouches because they are all descended from the same pouched ancestor. This ancestral marsupial, which lived over 80 million years ago, must also have had the other traits shared by all marsupials, such as a double vagina and epipubic bones. By definition, a homologous trait is one that evolved only once, though it may have been inherited by very many species. Therefore shared homologous traits are evidence of shared ancestry. But although they can only be gained once, homologous traits may be lost during evolution. For example, the northern quoll (*Dasyurus hallucatus*) does not have a pouch: although a fold of skin develops around the teats, the young must hang on for dear life when their mother goes hunting.

Not all similar traits are evidence of descent. The last common ancestor of the thylacine and the wolf was not the slightest bit wolf-like. Instead, it was probably a small shrew-like insectivore. The wolf-like jaws, teeth, and hunting behaviour evolved separately in both the canid and thylacine lineages (and, incidentally, also in the long-extinct borhyaenid lineage of American marsupials). Similar traits that have evolved independently in separate lineages are often described as analogies, to distinguish them from homologies. Analogous traits have evolved more than once. While analogies are very interesting they are, by definition, traits that are counter to the hierarchy of similarities. So while analogous traits are critical for understanding the processes of adaptation to a particular niche, they are no help at all in uncovering descent.

Classification is based on homology

Loss of homologous traits, or gain of analogous traits, can confuse classification. One of the advantages of molecular taxonomy is that it can reveal when species are erroneously grouped together on the basis of analogous similarities. For example, honeybees and stingless bees were traditionally grouped together on the basis of shared morphological characters, and because their way of life was so similar: both make hives where large numbers of non-reproductive workers store pollen and honey and raise the queen's offspring (**Figure 6.6**). But DNA sequence analysis has been used to suggest that honeybees are more closely related to the solitary (non-social) orchid bees than they are to the eusocial stingless bees. It is therefore possible (though not entirely certain) that eusociality has evolved multiple times in bees. Similarly, molecular data also imply that eusociality has arisen multiple times in shrimps, in rodents, and in wasps (in fact, sociality has arisen independently in more than 25 different biological lineages, including beetles, termites, and ants). Furthermore, molecular data show that eusociality has been lost in multiple lineages in ants and bees. So molecular data reveal a more complex pattern of the evolution of eusociality, with multiple gains and losses. This pattern would have been more difficult to detect from morphology alone.

DNA analysis can also reveal surprising affinities between apparently unrelated species. This is particularly helpful for organisms that have lost the usual defining homologous traits of a group. For example, parasites often evolve a very simple morphology, jettisoning characteristics needed for independent life, such as feeding apparatus (or even their stomach). Some intracellular parasites take this simplification to extremes, even losing their mitochondria (energy-generating organelles). Microspordia are intracellular parasites that infect a wide range of species: the resulting diseases have a large economic impact on industries from apiculture to aquiculture. Because of their apparently simple morphology, microsporidia were, in the 1980s, grouped with other simple single-celled parasitic taxa into a new kingdom called the Archezoa. But DNA sequence analysis has shown that the similarity

Figure 6.6 Molecular data suggest that the very similar hive-based societies of these two types of bees – honeybees (*Apis mellifera*; a) and stingless bees (*Melipona*; b) – are analogous (evolved independently) not homologous (similar by descent). While most of the global honey industry is based on honeybees, Mayan Indians have traditionally kept stingless bees for honey. However, the indigenous *Melipona* hives are now threatened with replacement by imported *Apis mellifera*, which, combined with habitat loss, has raised fears for the long-term survival of many stingless bee species.

Courtesy of Bassdrum Books Ltd, New Zealand.

between these parasites is due to superficial resemblance, not shared ancestry. Microsporidia are actually dramatically reduced fungi, so they belong in the fungal kingdom with their mushroom relatives. DNA analysis demonstrated that many other groups of unicellular parasites that were originally thought to be primitive lineages have actually evolved from more complex forms by a process of simplification. It seems the kingdom Archezoa was based on analogy not homology, so this grouping has now been abandoned.

Alignment establishes homology

The distinction between homologous (shared by descent) and analogous (independently evolved) traits is just as critical when using molecular data to uncover relationships. In fact, it is even easier to be fooled by analogy in DNA sequences, because there are only four possible character states. An adenine will look just the same whether it is a new mutation or inherited from an ancestor: an A is an A.

Look at the alignment of the 12S rRNA gene in **Figure 6.4**, p. 203. We have already seen that, for this gene, the marsupial species have more nucleotides in common with each other than with any of the placental species. Now we can look more closely at the patterns of similarities. For example, at position 31, we can see that the marsupials all have an A, while the placentals all have a G (**Figure 6.7**, p. 207).

There are several possible explanations for the presence of A at this position in all four marsupial species. The marsupial mole, quoll, thylacine, and dunnart may all have inherited it from a common ancestor that had an A at position 31, in the same way that they all inherited the pouch and the epipubic bone. In this case, we could say the A was homologous: shared by descent. Alternatively, the common ancestor of the marsupials could have had, for example, a G at that position, but subsequently there was a change from a G to an A in each of the four lineages. In this case, the As would be similar by analogy: they had evolved independently many different times. You may think 'who cares if it's homologous or analogous? An A is still an A!'. But the reason we wish to know whether shared nucleotides are similar due to homology or analogy is that only homologies reveal relationships.

Figure 6.7

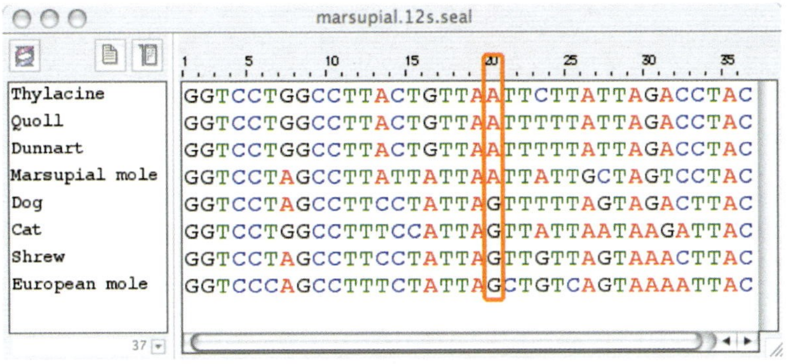

The only way we can tell whether a shared A is a homology or an analogy is to consider it in the context of the rest of the sequence. This is similar to recognizing that the wolf-like characters in thylacines are analogous to those in placental wolves because they are found in an otherwise entirely marsupial context. For DNA sequences, this context is revealed by alignment, which is the process of arranging DNA sequences so that all of the homologous nucleotides are lined up in columns.

Sequence alignment

Alignment is, without a doubt, the most important stage in any evolutionary analysis of DNA sequences. This is because alignment is the step that establishes homology, which is the basis for recognizing and understanding changes in DNA sequences over time. Incorrect alignment will lead to false conclusions. But alignment is also very difficult, due to the limited number of possible character states in DNA (or protein) sequences.

For physical characteristics, it will often be possible to distinguish homology from analogy by careful inspection. At first sight, the thylacine's teeth seem remarkably similar to a wolf's, with their prominent canines, and sharp, slicing molars. But closer inspection reveals important differences: the molar teeth of the thylacine, for example, differ in size, number and structure from those of the wolf. But this approach won't work for nucleotides: no matter how closely we look at an A, it will look the same as all other As. Instead, we rely on making inference of homology from probability: it is more likely that all four marsupials inherited the same A at the same position than it is that each one of them just happened to evolve the A independently. Weighing up the probabilities of different possible evolutionary explanations for the nucleotide sequences is the basis of sequence alignment.

In Figure 6.7, we can be quite confident that we are looking at homologous sequences, because they are clearly far too similar to be due to chance. Over half the bases in the sequence are the same for all of the sequences, far greater than we would expect if the sequences were not related. But not all parts of the 12S gene can be so easily aligned. Let's compare a different segment of the gene, and to simplify the example we will include just three species: two marsupials (thylacine and quoll) and one placental (dog; Figure 6.8).

Here we can see that the sequences are still roughly similar, and yet there is enough variation in the nucleotide sequence that we could align the bases in a number of alternative ways. When we compare the nucleotides that are aligned against each other in Figure 6.8, we see that in all but two cases, the thylacine sequence has the same base as the quoll. When we compare aligned nucleotides between the thylacine and dog, we note that there are six differences. So we could conclude from this alignment that the thylacine is more closely related to the quoll than it is to the dog.

Figure 6.8

```
Quoll       ACC-TAATTAGAATACGCTAAAAA----GAGGAG
Thylacine   ACC-TAATACGAATACG-TAAAAAA---GAGGAG
Dog         ACCATA-TTAACTTAA-CTAAAACACAAGAGGAG
```

Figure 6.9

```
Quoll       ACC-TAATTAGAATACGCTAAAAA---------GAGGAG
Thylacine   ACC-TAAT--AC-GAA--TA---CG-TAAAAAAGAGGAG
Dog         ACCATA-TTAACTTAA-CTAAAACACAA-----GAGGAG
```

However, this is not the only way of aligning these sequences. We can see that the alignment contains gaps (shown as a grey dash), where a given sequence is missing a base. By inserting more gaps, we could make more bases match. So we could have chosen to align these sequences as shown in Figure 6.9.

And now a curious thing has happened. For exactly the same sequence data, but with a different alignment, the thylacine sequence is now more similar to the dog (three differences) than it is to the quoll (five differences). The two species that share analogous physical traits are now united by analogous nucleotides. In this case, most biologists would be happy to consider the first alignment (Figure 6.8) to be a more believable assignment of the homology of those sequences than the second alignment (Figure 6.9). But this example illustrates that different alignments can change the conclusions reached from examining DNA sequence data, so we need a way of choosing between alternative alignments.

Homologous sites

We need to do two things in order to distinguish a meaningful (homologous) match from a random (analogous) match. Firstly, we need to establish homologous sites. A site is a position in a sequence. Each site could hold any one of the four nucleotides (or a gap). When we align DNA sequences we are attempting to find sites that were all originally copied from the same sites in the common ancestor of all the sequences in the alignment. Then, once this is achieved, we can identify changes that have occurred along particular lineages.

The 12S genes of the marsupial mole, thylacine, dog, and cat were all ultimately copied from a single ancestral 12S gene in their shared mammalian ancestor. So the 12S gene, taken as a whole, can be considered homologous (similar by descent). The 12S gene has just under 1000 sites. When we look at the alignment in Figure 6.4 (or Figure 6.7), the sites are labelled by number at the top. We can see that the sites are homologous, because they all have very similar sequence of nucleotides: all of the species share at least 50% of the nucleotides in this alignment. Just as we argued the RFGEME motif in RNA polymerase II beta would be unlikely to have arisen by chance in all species (Chapter 2), so too we consider that the match between mammal species for this stretch of the 12S gene is far too similar to be simply due to chance. Instead, we assume that these sequences are similar because they were all copied from the same ancestral copy of the gene.

Homologous nucleotides

Once we are sure that we are looking at homologous sites, we can begin to consider which specific nucleotide states are homologous; that is, which shared nucleotides reflect shared ancestry. Since that last common ancestor of marsupials and placentals, the 12S gene has been copied over a hundred million times. Not surprisingly, it has accumulated various changes in different lineages. So, although the sites are clearly homologous, the sequences are no longer identical. Let's focus on a shorter segment of the 12S alignment to illustrate the types of differences we can observe between sequences.

Firstly, we can see in Figure 6.10 (p. 209) that some species have unique changes, which are not shared with any of the other sequences in the alignment. Only the European mole has a C at site 32, all other species

Figure 6.10

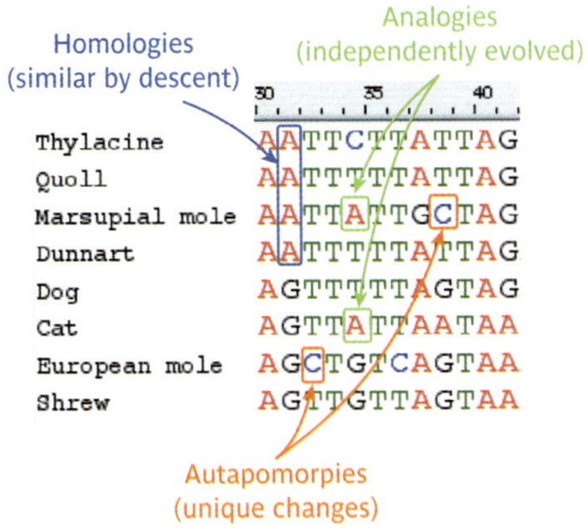

have a T, so it is fair to conclude that, at some time in the history of the European mole lineage, the T at site 32 was changed to a C. Similarly, only the marsupial mole has a C at site 38. A nucleotide substitution in only one species in the alignment is sometimes referred to as an autapomorphy. Autapomorphies are useful for two things. Firstly, just as unique physical characters are a key part of species taxonomy, unique sequence changes are exactly what we need if we want a diagnostic 'DNA barcode' to identify a species (see **Case Study 6.1**, p. 193). Secondly, autapomorphies can tell us about relative rates of change in particular biological lineages. For example, the marsupial mole has a relatively high number of autapomorphies, indicating either that it has a long, independent evolutionary history, or that it has a particularly rapid rate of change (or both).

We will look at rates of sequence change in Chapter 8

But autapomorphies are no help in establishing relationships because, by definition, they are not shared by any of the other species in the alignment, so we can't use them to group related species together. To classify species, we need to group them on the basis of shared characters. More specifically, we need to distinguish homologous and analogous shared characters. All marsupials in this alignment share an A at site 31, which can reasonably be interpreted as a homology, inherited from the marsupial ancestor. But not all shared nucleotides are evidence of shared ancestry. This short section of the alignment contains at least one example of an analogy: two species having independently evolved the same nucleotide (**Figure 6.10**). The cat and the marsupial mole share an A at position 34, but we can guess that their shared common ancestor did not have an A at this position, because none of the other descendants of that ancestor have an A. It seems most likely that at some point in the marsupial mole's history this site changed to an A, and, entirely independently, this site also changed to an A in the cat lineage. We can recognize this A as an analogy from the context of the rest of the alignment: at most sites, cats are similar to dogs not to marsupial moles, so this shared A between the cat and marsupial mole is anomalous.

Changing, adding, and taking away

There are two kinds of sequence change to consider when aligning DNA sequences: substitutions and indels. A nucleotide substitution occurs when, at a given site in the sequence, a particular base is exchanged for another base. For example, in the alignment opposite (**Figure 6.11**), you can see that at position 836, a G has been substituted for an A in the quoll lineage.

An indel, on the other hand, occurs when an extra base is inserted, or a base is deleted from a sequence. Indels may be any size, ranging from insertions or deletions of a single nucleotide to the loss or gain of large sections of DNA. In this alignment we can see that the cat lineage appears to have gained an extra T at position 831.

Insertions, duplications, and deletions are introduced in Chapter 3

So to align sequences that are similar, but not identical, we have to consider both nucleotide substitutions (one

Figure 6.11

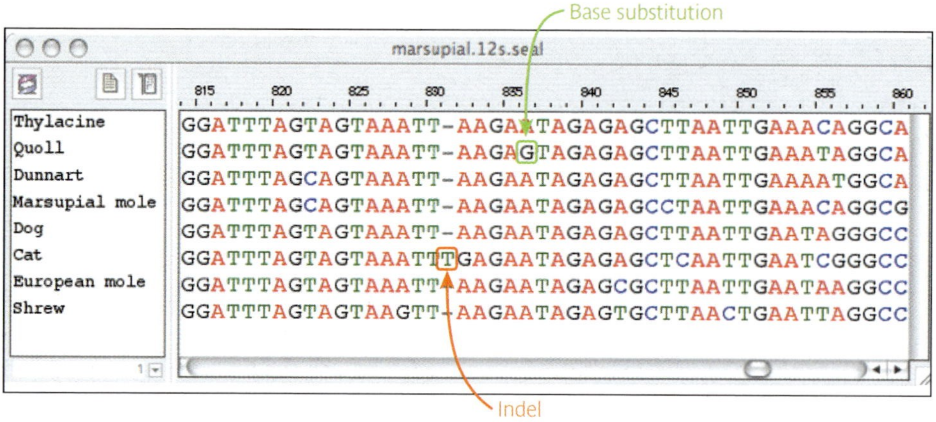

base is exchanged for another) and indels (bases are inserted or deleted). There will be some cases when it will be obvious which sites are homologous, and which sites have undergone nucleotide substitutions, insertions, or deletions. But there will be other cases when the sequences seem quite dissimilar, and could have arisen by a number of different scenarios. In this case we will be forced to decide which alignment is more likely. And this is where things get tricky.

Manual versus automated alignment

How do we decide upon the correct alignment for a set of sequences? There are two approaches. One is to align sequences manually, often with the help of a computer program that makes it easy to slide the nucleotides back and forth (just as a word processing package makes it easy to rearrange words on a page). A biologist can use a manual sequence editor to produce different possible sequence alignments, then use their understanding of molecular evolution to judge which alignment is most likely to represent homologous characters. The alternative alignments in **Figures 6.8** and **6.9** were created manually, and most biologists would judge the first alignment (**Figure 6.8**, p. 208) to be more reasonable than the second alignment (**Figure 6.9**, p. 208).

Manual sequence alignment has a number of advantages (**TechBox 6.3**, p. 211). One advantage that is often overlooked is that biologists of all kinds benefit from close observation of nature, to develop 'a feeling for the organism' (a phrase coined by the pioneering geneticist Barbara McClintock; **Heroes 3**). Close observation of sequence data is essential for developing a feeling for DNA and an appreciation of molecular evolution. The more you look at alignments, the more you become familiar with the kinds of changes that are most likely to occur. This feeling for DNA sequence evolution is the key to recognizing a reasonable alignment. But manual sequence alignment is time-consuming, so many biologists prefer to use the second option: automated sequence alignment. This is where an algorithm (an ordered series of instructions) is used to create and evaluate alternative alignments according to some objective criteria. The major advantage of using an algorithm to align sequences is that you can then get a computer to do all the hard work for you.

Morphological characters are so complex that it would be a challenge to develop a computer program that could compare them directly. But DNA sequences are very simple, and just the kind of information that computers are built to handle: strings of repeating characters. It is a relatively simple matter to write a computer program that can read two DNA sequences and tell you how many sites have the same nucleotide. This score could then be used to choose between alternative alignments. We could score an alignment, change it slightly, then score it again, repeating this process until we have found the highest-scoring version of the alignment.

Scoring alignments

How do we make sure our score will be highest for the most biologically realistic alignment? We could use a very simple score that gives the alignment one point

TECHBOX 6.3

Sequence alignment

KEYWORDS

automatic alignment
manual alignment
gap penalty
affine
penalty function
substitution
homology
indel

FURTHER INFORMATION

There are many freely available alignment programs, here are just three examples. The alignments in this book were all produced using the manual editor Se-Al which is available from *http://tree.bio.ed.ac.uk/software/seal/*. There are many publicly accessible server-based automatic alignment programs, such as CLUSTALW (*www.ebi.ac.uk/clustalw*), and T-Coffee (*www.tcoffee.org*).

RELATED TECHBOXES

TB 2.3: Genetic code
TB 3.4: BLAST

RELATED CASE STUDIES

CS 1.2: DNA surveillance (alignment against a database)
CS 6.1: Barcoding nematodes (importance of correct alignment)

The aim of multiple sequence alignment is to find the arrangement of a set of DNA sequences that maximizes the chance of comparing homologous sites. There are many alignment programs freely available, either as downloadable programs or web interfaces for remote servers, and new improved methods are published regularly. Rather than discussing any particular program currently available, it is best to simply consider what multiple alignment programs are capable of, and where their limitations currently lie. You do not have to choose between manual and automated alignment: in many cases, the best strategy will be a combination of both, generating an automated alignment which is then edited manually.

Automated alignment has two basic components: the score (a numerical function designed to rate the biological reasonableness of an alignment) and a search strategy (a way of changing the alignment to create alternative arrangements whose scores can be compared).

1. **Score:** The score reflects the goodness of match between a set of aligned sequences, according to some defined model that expresses the probability of different types of sequence change. The score for the whole alignment is essentially a weighted sum of scores for each site in the alignment plus a penalty for each gap. For a multiple alignment, the overall score is calculated either as a sum of scores between all pairs of sequences in the alignment, or from each sequence to a consensus sequence[1]. The key features of the penalty function are:

- **Substitution score:** Most nucleotide alignment programs use a basic score of +1 for a match and −1 for a mismatch. But, particularly for protein-coding sequences, some mismatches may be considered more likely to occur than others. For example, exchanging one hydrophobic amino acid for another may be less disruptive to a protein than changing it to a hydrophilic amino acid (**TechBox 2.3**). So the penalty function may give positive scores for exact matches, intermediate scores for conservative differences between similar amino acids, but negative scores for non-conservative differences or gaps. Amino acid alignment programs often use a matrix of substitution probabilities, such as PAM or BLOSUM, derived from the frequency of substitutions in known sequence alignments.

- **Gap penalty:** Gaps – sites where one or more sequences have no base, representing an insertion or deletion event – will generally accrue a negative score, to discourage matching nucleotides by introducing too many gaps. A linear gap penalty scores all gaps equally, so three one-base gaps would have the same penalty score as one three-base gaps. An affine gap penalty includes a score for opening a gap plus a length-dependent score for extending the gap. In this way, affine gap penalties favour extension of existing gaps over the introduction of new gaps, for example preferring one three-base gap over three one-base gaps.

The penalty function should reflect what we know about the evolution of these sequences. When you use a multiple alignment program, it is usually possible to input different values for parameters that reflect the model of sequence evolution. Altering these parameters can produce different alignments.

2. **Search:** It is not possible to score and compare all possible alignments for a set of sequences. For example, for two sequences of 300 bases, there are over a googol of possible alignments[2] (that is, more than the number of particles in the universe). Therefore, any program needs an efficient heuristic method for searching alternative alignments. Most multiple alignment methods are progressive[3]: they begin by making pairwise alignments which are used to cluster similar sequences into a 'guide tree', which is then used to select the most closely related pair of sequences to be aligned against each other. These aligned pairs are then held constant in a profile alignment that is then aligned against the next most closely related sequence, and so on, until all sequences are aligned. Progressive alignment dramatically reduces the search space (and therefore the computation time), but does not guarantee to find the best alignment: an incorrect guide tree may lead to a poor end result. The reliability of progressive alignment can be improved by using more information to make profile alignments (such as structural information or libraries of short local alignments[4]) or iterative refinement (randomly sampling sequences to create and score alternative profile alignments[5]).

Manual sequence alignment allows you to employ your understanding of molecular evolution to choose the best alignment. For example, you would be reluctant to introduce a one- or two-base gap in a protein-coding sequence, because that would cause a frameshift, destroying the translation sense of the rest of the protein. This should be obvious in a manual alignment editor which allows you to toggle between nucleotide and amino acid sequences. But most automatic alignment programs do not penalize frameshifts, and this can lead to unrealistic alignments of protein-coding sequences. Manual alignment editors also make it easy to delete sections of the alignment where there has been too much sequence change to allow confident assignment of homology. The limitations of manual alignment are that the process can be long and tedious, and it is best suited to relatively small datasets (say, fewer than 100 sequences which are less than 10,000 bases long).

Always check your alignments: Remember that an alignment program is just following orders. It will find the best possible alignment for any set of sequences, given a set of rules and a scoring function. What the program cannot do is make an assessment of the biological meaningfulness of the resulting alignment. If you put in randomly generated sequences, or sequences from unrelated genes, the program will still give you an alignment, even though it will in no way reflect homology between those sequences. Therefore, while you can ask a machine to do the hard work of alignment for you, only you can assess the results. This is why manual inspection of computational alignments is essential for any analysis of DNA sequences. Whether manually or automatically generated, you must look very carefully over all parts of an alignment to make sure there are no areas where one or more sequences are out of alignment. More specifically, you must always check alignments for:

1. **Frameshifts and stop codons:** indels in protein-coding sequences can destroy the translation sense of the protein. If there are stop codons within protein-coding sequences then your alignment is unlikely to be correct (see **TechBox 2.3**, p. 47). Note that frameshifts and stop codons do not apply to sequences that do not code for a protein, such as RNA genes, pseudogenes, introns, or intergenic DNA.

2. **Excessive introduction of gaps:** automatic alignments can sometimes insert gaps in apparently nonsensical places, for example isolating single bases in the middle of long gaps. Try modifying the gap penalty, editing by hand, or removing these regions altogether.

3. **Regions of poor alignment:** sequences that have had a high rate of nucleotide substitutions or indels may be effectively randomized, so that the historical signal in the sequence is lost. Such regions, which do not contain the story you wish to read, should be excluded from your analysis. There is no golden rule for deciding which regions to exclude.

You must ask yourself if you can be sure that each position in your alignment represents true homology, because if it doesn't any inference from that position is spurious. You can repeat your analysis with and without the questionable regions to test whether it has any effect on your results (see **Case Study 6.1**, p. 193).

References

1. Batzoglou, S. (2005) The many faces of sequence alignment. *Brief Bioinformics*, Volume 6, page 6–22.
2. Eddy, S.R. (2004) What is dynamic programming? *Nature Biotechnology*, Volume 22, page 909–910.
3. Higgins, D. (2003) Multiple alignment. In *The Phylogenetic Handbook: a practical approach to DNA and protein phylogeny*, Salemi, M. and Vandamme, A. (eds), pages 45–71. Cambridge University Press.
4. Notredame, C., Higgins, D.G. and Heringa, J. (2000) T-Coffee: A novel method for fast and accurate multiple sequence alignment. *Journal of Molecular Biology*, Volume 302, pages 205–217.
5. Katoh, K., Kuma, K., Toh, H. and Miyata, T. (2005) MAFFT version 5: improvement in accuracy of multiple sequence alignment. *Nucleic Acids Research*, Volume 33, pages 511–518.

for every matching nucleotide and subtracts one point for every mismatch. Or we could have a more sophisticated scoring system where different kinds of mismatch were awarded different scores. For example, programs to align protein sequences commonly use a matrix that assigns higher scores to mismatches between chemically similar amino acids, on the grounds that changing to a functionally equivalent amino acid will be a less serious structural change to a protein (**TechBox 6.3**, p. 211).

But making matches at all costs is not reasonable, otherwise we could just slide letters across until we found a matching one, creating two perfectly matching sequences with an absurd number of gaps. For example, compare the following sequences from the thylacine and dog 12S sequences (**Figure 6.12**):

Figure 6.12

Thylacine	CAAGTTTCCGCGC
Dog	CAAGCCTCCACGC

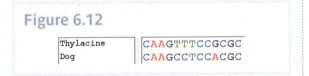

There are three mismatches between these sequences, suggesting three substitution events. But if we could introduce as many gaps as we liked, then we could make all the bases match (**Figure 6.13**).

Figure 6.13

Thylacine	CAAGTT---TCCG-CGC
Dog	CAAG---CCTCC-ACGC

The second alignment has a higher match score, but we might consider it a less likely explanation of the evolution of these sequences because we would have to infer at least four insertion or deletion events. This is why most alignment programs penalize gaps, to discourage indiscriminate inclusion of indels.

We should also expect the probability of a gap occurring in a sequence to be related to the biochemical function of the sequence. For example, the product of the 12S gene is a RNA molecule that assumes a tertiary structure by twisting around itself: in some places two parts of the strand pair with each other, in other places the single strand loops out (**Figure 6.14**). Indels that disrupt the matching of bases along a paired region are likely to be more damaging than indels that extend or reduce the length of a loop. For protein-coding sequences, three-base indels, which add or remove a whole codon (**TechBox 2.3**), are more likely to persist that one- or two-base gaps which introduce a frameshift, destroying the translation sense of the rest of the protein.

How to make the best alignment

There are many advantages of automated alignment programs: they are fast, can spot shared sequences that a human might miss, and they can handle very large amounts of data. Best of all, they can be asked to complete what is often a tedious task without (on the whole) complaining. However, you may be surprised, disappointed, or pleased to hear this, but computers do not

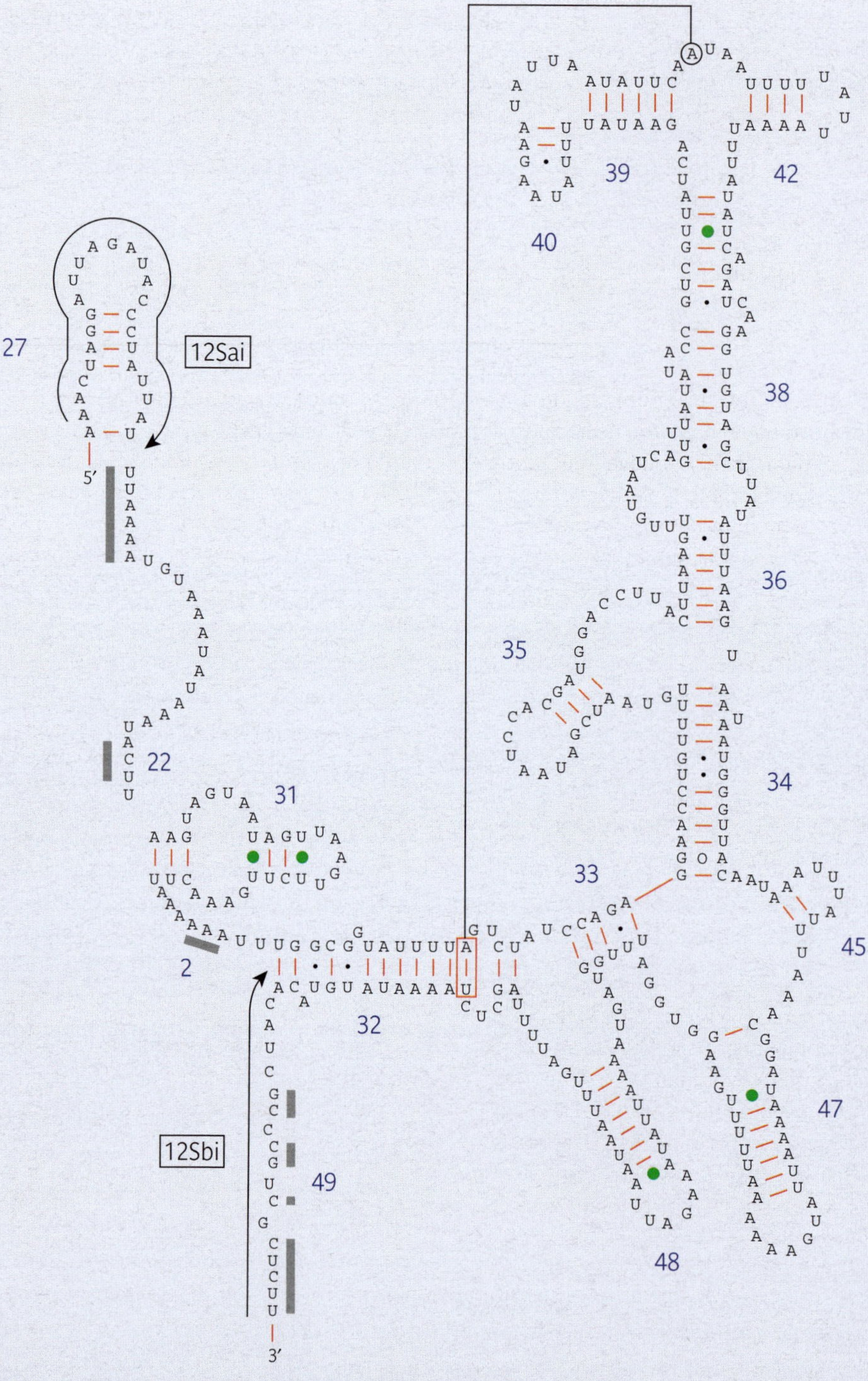

Figure 6.14 RNA molecules, such as the ribosomal RNA produced by the 12S gene, can form functional secondary structures by base pairing between different parts of the RNA polynucleotide strand. This diagram of part of the 12S RNA molecule from a fruit fly (*Drosophila*) shows double-stranded stems made of paired bases and loops made of single strands of bases.

From: Page, R.D.M. (2000) Comparative analysis of secondary structure of insect mitochondrial small subunit ribosomal RNA using maximum weighted matching. *Nucleic Acids Research*, Volume 28, No. 20, 3839–3845. © 2000 Oxford University Press.

make better biologists than humans do. Biologists are generally better able to judge what is a reasonable hypothesis for the evolution of DNA sequences than a computer is. Therefore, computer-generated alignments should never be used in an evolutionary analysis without careful inspection of the resulting alignment. This is because a computer can produce the alignment that best satisfies the criteria it has been given, but only a human can tell whether those criteria have produced the most biological meaningful arrangement of the DNA sequences.

Biologists who use very large amounts of sequence data complain that manual inspection of alignments is impractical. But if you have made the decision to analyse DNA sequence data, then you have to accept that part of the process is making sure your data are correct, and this may involve many hours staring at sequence alignments. After all, you would have little time for a biologist who, bored with lab work, said 'it's simply too tedious to check for contamination in my samples, so I'm not going to bother'. Biologists who use DNA sequence data but are reluctant to spend time on the alignment phase of the analysis need only remind themselves of the basic maxim: 'rubbish in, rubbish out'.

Manual inspection allows you to correct the alignment in places where the algorithm has produced an erroneous alignment, or to exclude from your analysis regions where so many substitutions or indels have occurred that it is now impossible to establish homology. Unlike the objective algorithmic approach, the decision to correct or exclude portions of an alignment is necessarily a subjective one: two people could produce different alignments for the same set of sequences. The principle of objectivity is held in high regard in science, and rightly so: we would have little confidence in the results of a study if the outcome depended on which scientist performed the experiment. But an objective algorithm will not always produce the best sequence alignment, and subjectivity is not necessarily a bad thing if it arises from experience and knowledge of molecular evolution and analyses. It may be that computer programs will one day be as good at alignment as humans are, and I for one will celebrate when they are. But until that day, it would be highly inadvisable to trust a computer-generated alignment without checking it very thoroughly, using your very own superior wetware (the human brain).

The exclusion from analysis of highly variable regions of the alignment is discussed in Chapter 8

The fate of the thylacine

There are many reasons why all biologists should learn about thylacines, not just because they are an excellent case study in convergent evolution. More importantly, the thylacine was one of the most beautiful and fascinating mammal species on Earth. It is heartbreaking to think that no-one reading this book will ever see a thylacine alive. For, although some optimistic souls still hope that they may be rediscovered in the wilds of Tasmania one day, the thylacine officially went extinct on 7th September 1936. That is the day on which the last known thylacine died in a Tasmanian zoo.

Many different factors may have contributed to the thylacine's demise. That they went extinct due to the influence of European settlers seems an inescapable conclusion. From 1803 to the present day, these settlers cleared the forests of Tasmania for agriculture or forestry, reducing the thylacine's habitat. European settlers may also have inadvertently introduced a distemper-like disease that decimated native animal populations. But, most tragic of all, thylacines were actively hunted: killed by dogs, strangled in snares, poisoned, and shot. The vendetta against the thylacine was justified as a stock-protection measure, though there is little evidence that thylacines ever caused major losses of poultry, sheep, or cattle (in fact, it seems likely that most predatory attacks on sheep were due to dogs, another destructive influence brought by the settlers). Nonetheless, private grazing companies, landholder associations, and the Tasmanian government itself all offered bounties for thylacine kills. These bounties may have been the nail in the coffin for a species already in severe decline.

HEROES OF THE GENETIC REVOLUTION

6

Craig Moritz

"Whatever approach we adopt to recognise species or intraspecific units, we should recognize the dangers inherent in the sometimes necessary imposing of thresholds or boundaries on a genealogical continuum of evolutionary divergence."

Moritz, C. (2001) Units in conservation genetics: what is real, what is useful? Abstract for *Conservation Genetics in the Age of Genomics*, American Museum of Natural History. Available at http://symposia.cbc.amnh.org/archives/conservation-genetics/

NAME

Craig Moritz

BORN

18th September 1958, Sydney, Australia

CURRENT POSITION

Director of the Museum of Vertebrate Zoology at the University of California, Berkeley, USA.

KEY PUBLICATIONS

Moritz, C., Dowling, T.E. and Brown, W.M. (1987) Evolution of animal mitochondrial DNA – relevance for population biology and systematics. *Annual Review of Ecology and Systematics*, Volume 18, pages 269–292.

Moritz, C. (1994) Defining evolutionarily significant units for conservation *Trends in Ecology and Evolution*, Volume 9, pages 373–375.

FURTHER INFORMATION

http://ib.berkeley.edu/labs/moritz/

Figure Hero 6
Craig Moritz.

Photograph courtesy of Conrad Hoskin.

Craig Moritz has pioneered the use of mitochondrial sequence data to study both the generation and the conservation of biodiversity. Moritz's work typically centres on 'herps' (herpetofauna; amphibians and reptiles). I have seen him dive hands-first into the undergrowth to catch a lizard that I had not even seen. But although lizards are his first love, his research has broad significance for the use of molecular data in evolution and ecology.

Moritz's early research used molecular analysis to explore the evolution of parthenogenetic species of lizards, whose populations consist entirely of females. Bynoe's geckoes (*Heteronotia binoei*) are members a complex of cryptic species, spread across the arid zone of Australia, which includes both sexual and asexual populations. Moritz used mitochondrial DNA sequences to demonstrate that the parthenogenetic populations of *Heteronotia* probably arose from a single population then rapidly expanded, challenging the hypothesis that sexual species should have a selective advantage over asexuals in harsh and unpredictable environments.

Moritz used his insights into the evolution of the mitochondrial genome to contribute to the development of DNA sequence analysis in evolution and ecology. His books and reviews set the standard for the field. In particular, Moritz has promoted an evolutionary perspective on contemporary conservation issues, looking not only at the current patterns of genetic diversity, but using the genetic data to infer patterns and mechanisms of species formation. His work has frequently set the terms of debate concerning the use of molecular data in conservation, particularly for prioritizing conservation effort on groups or areas containing the maximum amount of unique evolutionary history. The term 'evolutionarily significant unit' (ESU) is commonly associated with Moritz because, although he did not invent the concept, he did much to explore and promote its application in setting conservation priorities.

Moritz was one of the founding members of the Cooperative Research Centre for Tropical Rainforest Ecology and Management (CRC-TREM) in Australia, which brought together experts in both science and management. With collaborators from biology, geology, and mathematical modelling, Moritz has contributed to the development of a multidisciplinary picture of the history of the Australian Wet Tropics. His molecular zoology lab at the University of Queensland compared the phylogeographic patterns for a range of species, including frogs, lizards, birds, snails, and beetles, linking genetic patterns to the history of habitat expansion and contraction inferred from the fossil record and climatic models (see **Case Study 6.2**, p. 217). He used these studies to promote a dynamic approach to conservation, aiming not simply to protect existing species but to preserve the potential for future evolutionary diversification.

CASE STUDY 6.2

Keeping the pieces: a dynamic approach to conservation in the tropics

KEYWORDS

phylogeography

Evolutionarily Significant Unit (ESU)

phylogenetic diversity (PD)

speciation

biodiversity

refugia

endemism

RELATED TECHBOXES

TB 7.1: Distance methods

TB 6.2: What is a species?

RELATED CASE STUDIES

CS 8.1: Same but different (cryptic species)

CS 4.1: Glorious mud (DNA hybridization)

Moritz, C. and McDonald, K.R. (2005) Evolutionary approaches to the conservation of tropical rainforest vertebrates. In *Tropical Rainforests: past, present and future*, Bermingham, E., Dick, C.W. and Moritz, C. (eds), pages 532–556. University of Chicago Press

> *"If there is one message from the various studies on rainforest history and evolution . . . , it is that we cannot take a static approach to conserving biological diversity in the Wet Tropics, as conditions suitable for particular species move around the landscape. Managing this dynamic system will be a major challenge . . ."* [1]

Background

The Wet Tropics of northeastern Queensland, an 'archipelago' of upland rainforest patches, are the last remnants of a formerly widespread habitat that covered much of Australia until the mid-Miocene (around 10 million years ago). The Wet Tropics are a hotspot of species endemism, containing over a thousand species found nowhere else in the world. Nearly four hundred of these endemic species are currently considered rare, threatened, or endangered. Although the Wet Tropics cover less than 0.1% of the Australian continent, they contain nearly a quarter of Australia's endemic reptile species and a third of Australia's endemic frogs. Research in high-diversity rainforests such as these are important both for theoretical and practical reasons, to understand the mechanisms of biodiversity generation and to devise programmes for its conservation.

These two issues – generation and conservation of biodiversity – can be combined by taking an evolutionary perspective which considers species as ever-changing entities: products of past influences and subject to future change.

Aim

Genetic data, when combined with information on landscape change derived from fossil record and climatic modelling, can be used to understand both the patterns of current species richness and the processes underlying species formation. The use of genetic data allows identification of Evolutionarily Significant Units (ESU): populations that may not differ physically but represent distinct genetically isolated evolutionary lineages[2]. The aim of this study was to combine genetic data from many different species to gain an overview of speciation processes in the Wet Tropics, and to develop management goals that maximize the current species richness and preserve the potential for future speciation.

Methods

The authors combined estimates of mitochondrial sequence diversity from previous studies to gain an overview of the distribution of vertebrate diversity across the Wet Tropics. Firstly, they identified regions with putative ESUs: areas containing populations that differed from all other related populations by virtue of unique sequence differences. They inferred that the genetic uniqueness of the populations in these areas was due to lack of genetic exchange with other areas. Then, to gain a measure of the degree of isolation, they estimated the number of substitutions unique to each ESU, combined across all sequences examined. The isolation of the area was then calculated as the average number of substitutions separating the ESUs in that area from populations in any other area. They also calculated an index of phylogenetic Diversity (PD), a tree-based measure that sums path lengths along a phylogeny (see Chapter 7) in order to identify the set of taxa preserving the maximum amount of the genetic history of the group under investigation[3].

Results

Deep divergences between mitochondrial sequences (generally greater than 10%) suggest that most vertebrate species in the Wet Tropics are considerably older than the Pleistocene, the last great period of climate change (approximately 2 Myr ago). Therefore, on the whole, the Wet Tropics are 'museums' that preserve vertebrate species rather than 'cradles' that create them. Furthermore, sister species (closest relatives) tend to occur in adjacent rainforest patches, implying that the contraction of the rainforest was marked by extinction of local species rather than migration of more wide-spread species into the Wet Tropic refugia. The genetic data suggest that rainforest vertebrates were restricted to isolated refugia when the rainforest contracted during late Pleistocene, then expanded again as the climate warmed in the Holocene (the last 10,000 years). Thus the present genetic diversity reflects past habitat distribution: for example, all of the species studied showed a major genetic disjunction across the Black Mountain corridor, which separated rainforest refugia in the past but is today connected by continuous habitat. Areas of the Wet Tropics with the highest isolation values were not those with the highest current species richness.

Conclusions

Because the genetic data suggest that the endemic vertebrates of the Wet Tropics are survivors of a former age, the authors conclude that any conservation strategy should aim to 'keep all the pieces': even where ESUs show little physical differences they should be preserved as unique gene pools that cannot be recovered once lost. Rather than basing conservation strategies wholly on current patterns of species richness, the authors argue that we should aim to protect future biodiversity by identifying regions with the greatest representation of unique lineages. They also argue that we should attempt to identify regions of ongoing speciation, particularly areas of range expansion, secondary contact between previously isolated populations, or environmental gradients.

Limitations

As with all molecular descriptions of biodiversity, the significance of genetic divergence as an indicator of meaningful differences between populations is open to debate[4], particularly the usefulness of neutral genetic markers as a prediction of future biodiversity generation. Furthermore, the use of genetic data here is primarily descriptive: no statistical tests are used to identify significant differences in mitochondrial DNA branchlength. However, this may be ameliorated by the use of multiple taxa to identify patterns consistent across many taxa.

Future work

The impact of future climate change may be explored using climatic modelling, potentially identifying areas of future conservation significance[5]. Finer-scale genetic data, combined with morphological and behavioural data, can be used to investigate the process of species formation in contact zones between previously isolated Wet Tropics refugia[6].

Figure CS6.2 Where the rainforest meets the reef: much of the Wet Tropics of northeastern Australia is protected as a World Heritage Area, but ongoing management decisions are critical to conserving the unique flora and fauna.

Reproduced courtesy of Marcel Cardillo.

References

1. Moritz, C. (2005) Rainforest history and dynamics in the Australian wet tropics. In *Tropical Rainforests: past, present and future*, Bermingham, E., Dick, C.W. and Moritz, C. (eds), pages 313–321. University of Chicago Press.
2. Moritz, C. (1994) Defining "Evolutionarily Significant Units" for conservation. *Trends in Ecology and Evolution*, Volume 9, pages 373–375.
3. Crozier, R.H. (1997) Preserving the information content of species: genetic diversity, phylogeny, and conservation worth. *Annual Review of Ecology and Systematics*, Volume 28, pages 243–268.
4. Chaitra, M.S., Vasudevan, K. and Shanker, K. (2004) The biodiversity bandwagon: the splitters have it. *Current Science*, Volume 86, pages 897–899.
5. Williams, S.E., Bolitho, E.E. and Fox, S. (2003) Climate change in Australian tropical rainforests: an impending environmental catastrophe. *Proceedings of the Royal Society of London B*, Volume 270, pages 1887–1892.
6. Hoskin, C.J., Higgie, M., McDonald, K.R. and Moritz, C. (2005) Reinforcement drives rapid allopatric speciation. *Nature*, Volume 437, pages 1353–1356.

The thylacine was mourned as soon as it was lost. It was officially declared a protected species in July 1936, 59 days before the last, lonely thylacine died in captivity. Whereas previously there was a modest bounty for scalps, there is now a very substantial reward for anyone who can produce a live thylacine. This stupid and irreversible loss of such an extraordinary species should remind us to think carefully about the destructive consequences of human activity. It will not, alas, be the last preventable extinction of an irreplaceable species.

Movies of the last thylacine show a beautiful animal, pacing its cage, opening its jaws to display its distinctive gape (probably a threat warning; indeed, this thylacine bit the photographer on the buttock shortly afterwards). These movies never fail to fill me with a great sense of sorrow and loss. I would dearly love to see a live thylacine, so it is tempting to hope that there is some chance of success for the thylacine cloning project launched in 1999, based on extracting DNA from a thylacine pup preserved in a jar of alcohol in 1866 (**TechBox 6.4**). However, the Australian Museum officially abandoned the project in 2005, stating that the DNA was too degraded. Although other research groups intend to continue attempts to clone the thylacine, they face apparently insurmountable technical obstacles. However, in the rapidly moving field of molecular genetics, it is unwise to ever make sweeping claims about what it will be possible to achieve in the future. So we can only hope that unexpected advances in ancient DNA sequencing and cloning technology will one day cause us to look back and laugh at our earlier cynicism, as we stroke our resurrected thylacine puppies.

Case Studies 2.2 and 3.2 give examples of sequencing ancient DNA

TECHBOX 6.4 Cloning and conservation

KEYWORDS

- nuclear transfer
- extinct species
- biobanking
- Genetic Resource Banks
- endangered species
- gaur
- mammoth
- thylacine
- genome assembly

RELATED TECHBOXES

TB 2.4: DNA extraction

TB 1.2: DNA sequencing

RELATED CASE STUDIES

CS 2.2: More moa (Ancient DNA from extinct species)

CS 5.2: Time flies (ancient DNA from museum specimens)

In the broadest sense, cloning refers to reproduction where the genetic material of one individual is copied to make a new genetically identical individual (note that the word 'cloning' is also used to describe the process of making multiple copies of a gene). Many organisms produce natural clones: for example, tuberculosis bacteria divide by fission, new strawberry plants can grow from a runner (extended stem) from one parent, and identical twins are genetic clones that have developed from a single fertilized egg (**Figure TB6.4a**). The technology of reproductive cloning generates an embryo from the genome of one parent, rather than by fusing gametes from two different parents. Typically, this is achieved by nuclear transfer. For example, Noah, the first cloned gaur (*Bos frontalis*), was created in 2001. The gaur is a species of a wild woodland ox from south Asia which is currently classified as vulnerable to extinction. Noah was created by taking the nucleus from a gaur skin cell, then inserting it into an egg cell from a domestic cow. Sadly, Noah died when only 2 days old (the cause of death – dysentery – may have been unconnected to the cloning procedure).

Cloning is now being explored as a last-ditch form of captive breeding for endangered species. In populations with very few remaining individuals, cloning could increase the number of individuals in the population (but, by definition, it will not increase the genetic diversity of the population). Conveniently, cells for cloning can be moved long distances, and even frozen for later use. There are now several Genetic Resource Banks that store cells from endangered species for future reproductive cloning[1]. In this sense, cloning technology can be seen as an extension of current assisted breeding strategies, such as sperm banking and artificial insemination. However, at this point in time, the utility of cloning for rescuing endangered species is low. It currently requires around 100 nuclear transfers to achieve one or two successful pregnancies[2]. As cloning technology improves, successes are likely to become more frequent, but it will be difficult to fine-tune the procedures for rare animals

Figure TB6.4a Human clones are already with us. Identical twins arise when a single embryo splits to form two genetically identical embryos. Unlike clones produced by recent technology based on nuclear transfer, natural clones share both their nuclear and mitochondrial genomes.

Reproduced courtesy of Tom Cardillo.

that cannot be easily experimented on. For most endangered species recovery programmes, cloning seems unlikely to offer significant advantages over commonly applied methods of assisted reproduction, such as artificial insemination.

But assisted reproduction is not an option for extinct species. Could cloning be used to resurrect extinct species using genetic material from preserved specimens? Cloning programmes have been reported for at least two extinct species – thylacine and woolly mammoth – though many scientists are highly sceptical of the chances of success. Current reproductive cloning requires intact nuclei from live donor cells, which is unlikely to be an option for extinct species (attempts to extract whole gametes from frozen mammoths have so far met with no success). Instead, current programmes to clone extinct species are resting their hopes on extracting and sequencing DNA, assembling the entire genome, then somehow persuading an unrelated living cell to use the genome to develop an embryo. There are many barriers to the success of such attempts:

- **DNA:** Recent advances in pyrosequencing technology (**TechBox 1.2**) have resulted in a large amount the mammoth genome being sequenced[3]. But only a handful of genes have been sequenced from the thylacine as the DNA is too degraded to allow useful amounts of the genome to be amplified.

- **Genome:** even if all of the DNA from the thylacine could be sequenced, assembling it into a working genome would be a major challenge. Even the human genome is not 100% complete, as it has not been possible to reconstruct highly repetitive regions, such as those around the centromeres. Even if the genome of an extinct animal could be fully assembled, the task of creating artificial chromosomes that would function within a living cell would be an enormous challenge.

- **Development:** A reconstituted genome would need to be placed in a cell that could interpret the genetic information and use it to build a functioning embryo. A cloning programme for an extinct species must necessarily use an egg from a related species. Even if development proceeds, this effectively makes the cloned individuals a hybrid, with the nuclear DNA of one species, and the mitochondrial DNA of a different species.

- **Gestation:** By definition, an extinct species has no living mothers to gestate a cloned embryo. The mammoth project plans to use a female elephant, even though elephants are separated from mammoths by millions of years of evolution. The closest relative of the thylacine is the Tasmanian devil, half the size of the thylacine and strikingly dissimilar in

Figure TB6.4b The impressive gape of the thylacine is often described as a 'threat yawn'. Attempts to clone the thylacine (*Thylacinus cynocephalus*) from a specimen preserved in alcohol have thus far met no success. Photograph: National Institutes of Health, US.

morphology and ecology (although development of young in a pouch may make marsupials more amenable to cloning than placentals).

- **Environment:** concerns have been raised over the wisdom of recreating a species that may have gone extinct through loss of habitat, though this need not be a practical problem if the resurrected clones are destined for life in captivity (as demonstrated by the presence of polar bears in tropical zoos).

From the current state of technology, the prospect of using preserved DNA to resurrect extinct species seems remote, though of course it is impossible to predict future technological advances. Cloning for conservation seems more likely to succeed for still-living but endangered animals, particularly those with external development such as amphibians and fish.

References

1. Holt, W.V., Pickard, A.R. and Prather, R.S. (2000) Wildlife conservation and reproductive cloning. *Reproduction*, Volume 127, pages 317–324.
2. Roslin Institute (2002). *Somatic Cell Nuclear Transfer (Cloning) Efficiency*. Available at www.roslin.ac.uk/downloads/webtablesGR.pdf.
3. Poinar, H.N. *et al*. (2006) Metagenomics to paleogenomics: large-scale sequencing of mammoth DNA. *Science*, Volume 311, pages 392–394.

Conclusions

In this chapter we have considered the importance of homology in establishing evolutionary relationships. We can trace the evolutionary history of species by considering traits that are shared through common descent. In order to read the history of species, we must consider traits that are truly homologous (inherited from a common ancestor) not analogous (independently evolved). Distinguishing homology from analogy is just as critical for DNA sequences as it is for morphology. Alignment, which is the process of establishing homologous sites, is therefore the most important step in any analysis of DNA sequences.

Here, we have considered how alignment is essential for using DNA sequences to identify and classify species. Alignment reveals unique sequences that can act as 'DNA barcodes' for rapid and reliable identification of species. Correct alignment is also essential for classification (grouping species in to a hierarchy of similarity that reflects evolutionary history), which requires the identification of homologous characters, inherited from a common ancestor. In the next chapter we will see how alignments of DNA sequences allow us to reconstruct the history of life.

Alignment rests critically on the concept of homology, or similarity by descent. By definition a homologous trait has evolved only once. The corollary of this is that once genetic information is lost, it is gone forever. When a species goes extinct, it takes with it the unique genetic information that informed its development and recorded its history. The dream of resurrecting extinct species from the DNA they have left behind is a distant hope: best not to lose them in the first place.

Further information

Many books have been written on the tragic loss of the thylacine, for example:

Paddle, R. (2001) *The Last Tasmanian Tiger: the history and extinction of the thylacine*. Cambridge University Press.

You can find an introduction to speciation and species concepts in most evolutionary biology textbooks, for example:

Ridley, M. (2003) *Evolution*, 3rd edn. Blackwell Publishing.

There is a chapter on sequence alignment in:

Higgins, D. (2003) Multiple alignment. In *The Phylogenetic Handbook: a practical approach to DNA and protein phylogeny*, Salemi, M. and Vandamme, A. (eds), pages 45–71. Cambridge University Press.

7

Tree of life

Or: how do I construct a phylogeny?

"The time will come, I believe, though I shall not live to see it, when we shall have fairly true genealogical trees of each great kingdom of Nature"

Charles Darwin (1857) Letter to Thomas Henry Huxley

What this chapter is about

In this chapter we first illustrate the process of population divergence, noting how isolated populations become gradually more different from each other at the genomic level. We then show how DNA sequences sampled at the end point of this process carry the historical signal of the pattern of population divergence. We can use alignments of DNA sequences to draw evolutionary trees that display the similarities between related lineages and the paths of descent of species. These phylogenies can be used to test hypotheses about the evolutionary past and processes.

Key concepts

Evolutionary biology: diversification

Molecular evolution: genetic divergence

Techniques: phylogenetics

Phylogeny reflects history

On the Origin of Species was a best seller. The first edition sold out on the day of publication, and the second edition sold out 6 weeks later. *The Origin* went through six editions in Darwin's lifetime, and has remained in print ever since. The phenomenal success of *The Origin* is certainly not due to lavish illustration, because there is only a single black and white picture in the whole book. Not a drawing of a fancy pigeon, not a sketch of a fossilized giant ground sloth, not even a portrait of the author. The sole illustration in *The Origin* is a simple line drawing (**Figure 7.1**), showing the diversification of an imagined set of species.

In order to explain the diversity of life, Darwin had to demonstrate not just how one species could be transformed into another by natural selection, but how this process could give rise to a great diversity of organisms. To do this, he needed to show how natural selection could produce two separate species from a single stock. The one and only diagram in *The Origin* depicts the following evolutionary story to illustrate the process of diversification: 11 related species (A to L) each occupy a particular habitat (**Figure 7.1**). Each species is a group of individuals that vary slightly from each other. Through the process of natural selection (Chapter 5), advantageous variations tend to be preserved and propagated. Darwin suggested variations at the extremes of a distribution – those that are most different from the rest of the population – will often be those that flourish: 'the more diversified the descendants from any one species become in structure, constitution and habits, by so much will they be better enabled to seize on many and widely diversified places in the polity of nature, and so be enabled to increase in numbers'. Eventually sufficient variation accumulates in these divergent lineages to warrant them being called varieties or species. After a long interval of time has elapsed (at point XIV in the diagram), some lineages have gone extinct (such as D), some have persisted without change (such as F), and some have given rise to many descendent lineages (A gave rise to eight new lineages). The net result is that there are more lineages at time XIV, and they occupy a greater range of variation. This process could have taken place over any time scale – each horizontal line

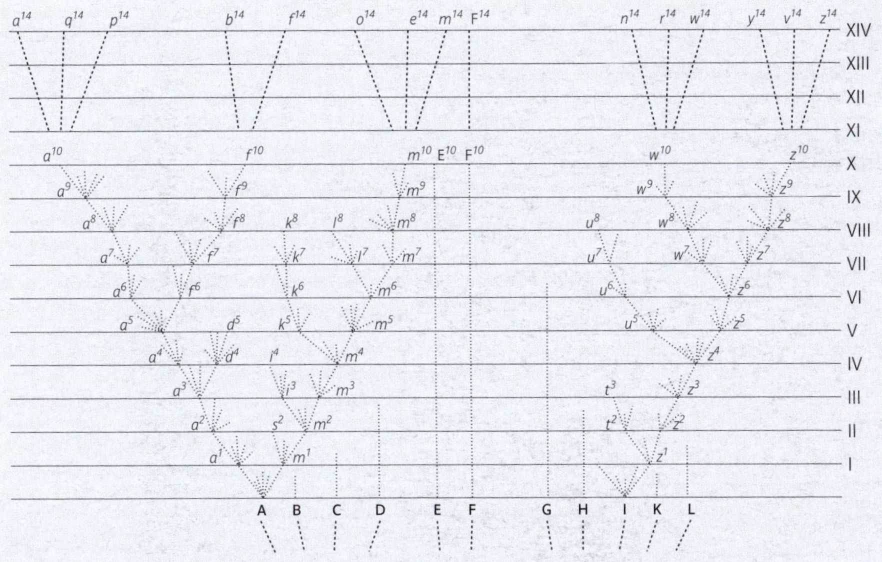

Figure 7.1 Darwin's tree, the only illustration in *The Origin of Species*.

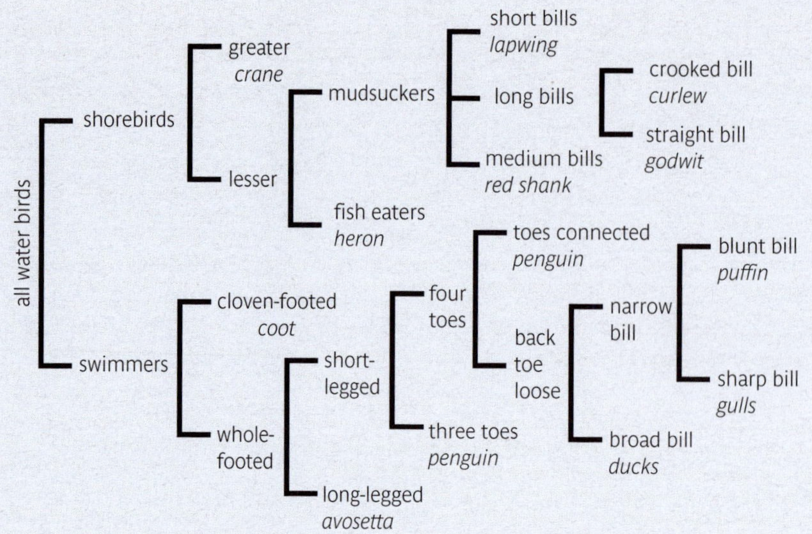

Figure 7.2 John Ray used nested sets of shared characters to divide species into groups containing ever-more similar species. Part of his classification of birds, published with his student Francis Willughby in 1642, is shown here as a branching diagram, but Ray wrote out his classifications in the same manner as identification keys today, as a table of nested categories, rather than as a branching diagram.

Based on information given in Bedall (1957).

might represent one thousand generations, or one million generations, or successive geological strata. And the lineages could represent any level of biological organization, varieties, species, genera, families, and so on. Darwin proposed that the process of diversification is essentially similar for all.

Diversification is a fundamental process of evolution, and it is usually represented by a branching diagram, referred to as an evolutionary tree, or phylogeny. But although Darwin's one and only diagram should perhaps be recognized as the first modern evolutionary tree, people had been drawing branching diagrams to represent biological diversity for centuries. This is because scientists had long recognized that biological diversity has a hierarchical nature, and they used this observation to create logical and useful classifications (see **TechBox 6.1**). For example, long before the discovery of evolution, the diversity of bird species had been variously divided into nested categories (forgive the pun). Characteristics used to divide bird species into natural groups included morphology (e.g. Pliny used the shape of the foot), behaviour (e.g. Aristotle divided birds into those that took dust baths or water baths or both or neither), and ecology (e.g. ground birds, river birds, etc.).

The pioneering systematist John Ray produced an ornithological classification, in 1676, which resembled modern taxonomy in that species were arranged in a hierarchical list. This list is structured by using some key characteristic to divide birds into groups, then another character chosen to divide each of those groups into smaller groups, and so on (**Figure 7.2**). For example, Ray divided all birds into land and water birds; water birds were divided into those that swim and those that feed by water; those that swim are divided into those with cloven (separate-toed) feet or whole (webbed) feet; those with whole feet are divided into those with short or long legs; the short legged ones are divided into those with three toes or four toes, and so on. So a pelican is grouped with other birds with four connected toes, which are then grouped with a larger set of four-toed bird, and these are grouped with all of the short-legged aquatic birds. All the short-legged birds are then grouped with all the whole-footed aquatic birds, which are then grouped with all aquatic birds, and so on upwards until we have a group of all birds (**Figure 7.2**).

The hierarchical system of taxonomy used today is outlined in TechBox 6.1

Branching diagrams are a logical way to represent a hierarchical structure, and you will see them in many non-biological situations: for example, 'org charts' illustrating levels of governance in organizations such as universities. But the discovery of evolution provided an explanation of why biodiversity, in particular, has a hierarchical organization. And this is what makes Darwin's dull-looking diagram special. It does not just describe the hierarchical nature of biological diversity, it provides an explanation for it. Descent with modification gives rise to a nested hierarchy of individuals, populations, species, and lineages, in which close relatives tend to be more similar to each other than more distant relatives. Darwin's tree represents the process of diversification which leaves a nested pattern of similarities and differences.

Life as a tree

Phylogenies are a way of organizing ideas about evolution. We can read the information in phylogenies in several different ways. We can use phylogenies to display biological relationships: lineages most closely related to each other will be more closely connected on a phylogeny, because they share a more recent common ancestor. Phylogenies also represent a record of the evolutionary past: we can use the connected paths

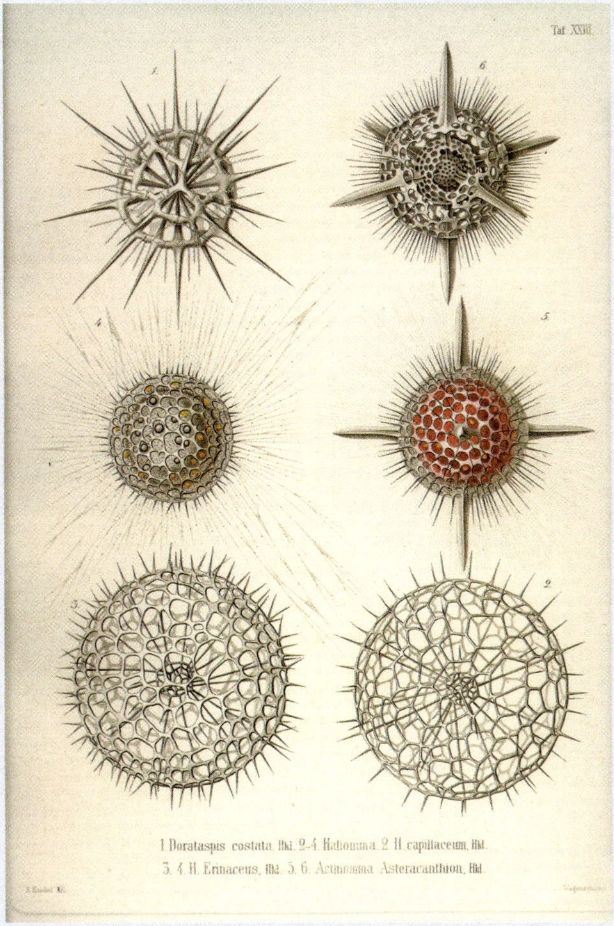

Figure 7.3 Ernst Haeckel used the principle of evolution to develop theories of biodiversity, based largely on patterns of development. He is responsible for generating many terms we use today (such as phylum, phylogeny, and ecology), but perhaps is best known for his beautiful taxonomic illustrations. These drawings were not simply aesthetic, they recorded important information about species that, at that time, could not be photographed. This is an illustration from Haeckel's *Die Radiolarien* published in 1862.

Ernst Haeckel, 1862.

of the phylogeny to reconstruct the evolutionary history of a lineage. More generally, phylogenies allow us to generate and test hypotheses about the patterns, processes, and causes of diversification.

The word 'phylogeny', meaning evolutionary tree, was coined by Ernst Haeckel, a naturalist famous for his exquisite drawings of microscopic creatures (**Figure 7.3**, p. 227). Haeckel had been a doctor until he read Darwin's *Origin of Species*, at which point he devoted himself to the study of evolution. He saw that the detailed drawings he had made reflected patterns of similarity that could be explained by Darwin's theory of diversification from a common origin. More than anyone else, Haeckel is responsible for embedding the iconography of the evolutionary tree into modern biology. For not only did Haeckel describe the relationships between living groups using branching diagrams, he drew these branching diagrams as actual trees (**Figure 7.4**, p. 229).

If you watch a tree grow from seed, you will see it start as a single tiny stem, which then branches into two thin twigs, and as the twigs grow thicker they branch to form more twigs and so on. So the particular pattern of branches and twigs on a tree represents the historical process that gave rise to the tree. Instead of just looking at the tree as a static object, we can consider that the tree we see now represents a time series of branching events, from the original stem (now represented by a thick trunk from which all other branches arise) to the most recently branched twigs. We could follow the series of branching events in two ways: from the bottom up (from the past to the present) or from the top down (from the present to the past).

Reading the tree of life

As an illustration, let's follow the series of branching events from the root of Haeckel's tree to the Echinodermata (starfish and their kin). Before we start, it is important to note that Haeckel's classification is now superseded, so this charming tree-like phylogeny does not agree with contemporary classifications of organisms. Many of the biological groups he includes now bear different names, and many of the relationships he depicts have now been revised. Why choose an out-of-date phylogeny as our example? The classification of the living kingdoms changes on a regular basis. New studies are published and old ideas change. In fact, molecular data have played a major role in rewriting the phylogeny of living species. On the whole, we may feel confident that new phylogenies bring us closer to the true evolutionary history of species. However, I would be willing to bet that any phylogeny of these species I choose today will be superseded sooner or later. Since the point of the following exercise is to understand the way we read phylogenies, we might as well use Haeckel's classic tree as an exemplar (**Figure 7.5**, p. 230).

As we travel from the root of the tree to one tip, we will number the branching points we go through. In phylogenetic parlance, the point at which a branch divides to give rise to two or more new branches is called a node. When we start at the root and move upward, we begin with the ancestor of three kingdoms (multicellular plants and animals, and unicellular protists), which then split to form the branch leading to the plants (node i), and another branch that divided again to give rise to the protist and animal lineages (which split from each other at node ii). The animal lineage branches to give rise to the Coelenterata (jellyfish and their kin; node iii), splits again to give rise to the Vertebrata (bony animals; node iv), and again to give rise to the Mollusca (snails, octopuses, and suchlike; node v), then splits again to form the stem lineages of the Articulata (insects, crabs, and other hard-shelled creatures) and the Echinodermata (node vi). This lineage splits at node vii to give rise to the Holothuroidea (sea cucumbers), Echinida (sea urchins), Asterida (starfish), and Crinoida (sea lilies).

Alternatively, we can begin at one of the tips of the tree and follow a branch back toward the trunk, joining other branches as we go. In his book *The Ancestor's Tale*, Richard Dawkins used the analogy of following a path back in evolutionary time and meeting the ancestors at various crossroads (nodes in the tree), thus being joined by a growing band of evermore distant relatives who travel down their own lineages to meet you at the common ancestral nodes. Dawkins began his evolutionary journey with *Homo sapiens*, but let's be a bit more exotic. We will begin our journey from the tip of Haeckel's tree to the root with the single-celled amoeboid *Vampyrella* (**Figure 7.6**, p. 230).

In Haeckel's tree, *Vampyrella* groups with a strange collection of single-celled beasties that he terms the Monera. This taxonomic group is now essentially defunct: Haeckel's Monera includes mostly amoeba, but also some organisms now thought to be only

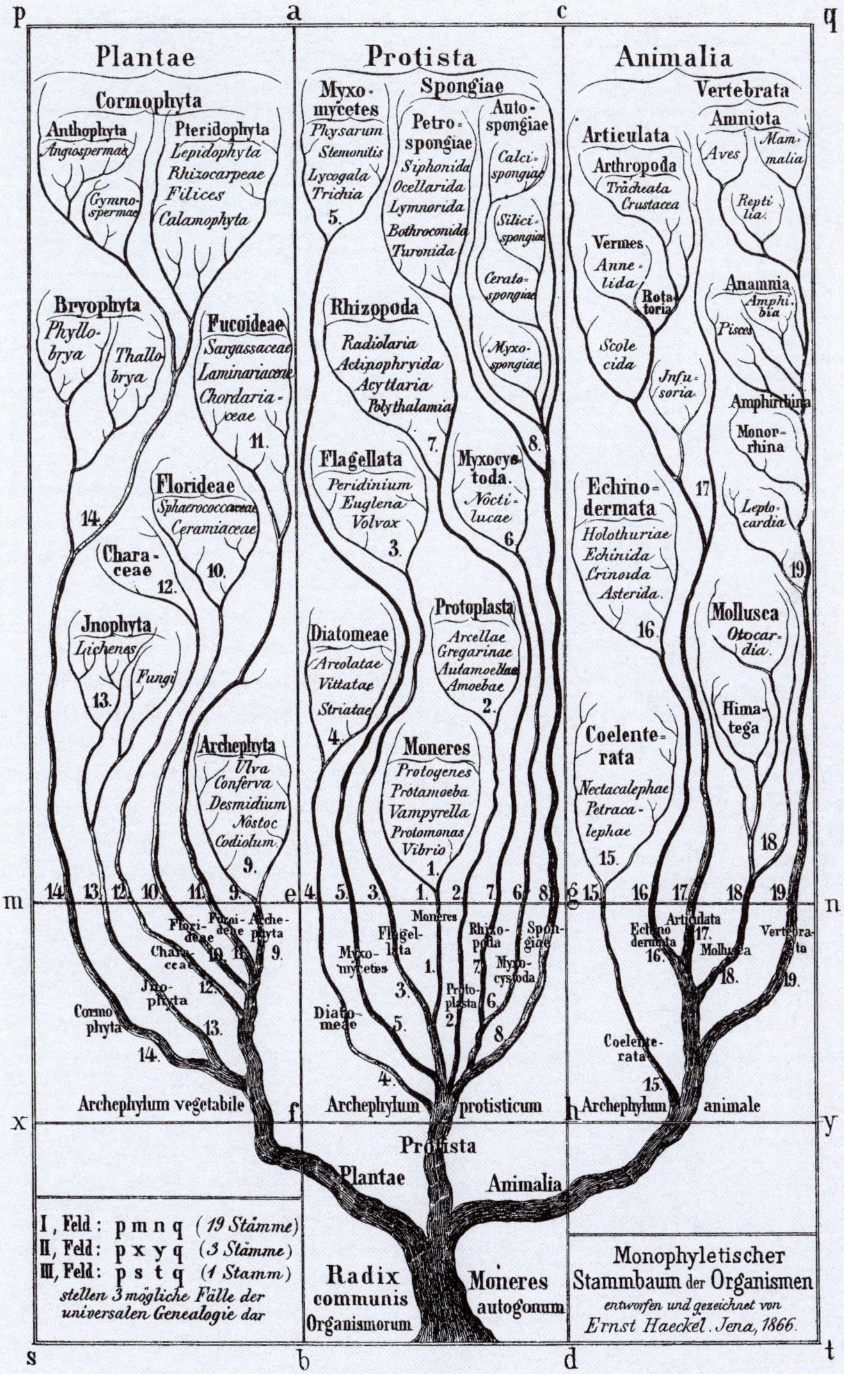

Figure 7.4 One of Ernst Haeckel's evolutionary trees. Most of the relationships depicted here have since been revised, and continue to be revised. There is an ongoing and vigorous debate about the relationships at the deeper levels of the tree of life. Molecular data make a valuable contribution to this debate, but must be considered in the light of evidence from cell biology, palaeontology, and other biological fields.

Ernst Haeckel, 1866.

Figure 7.5

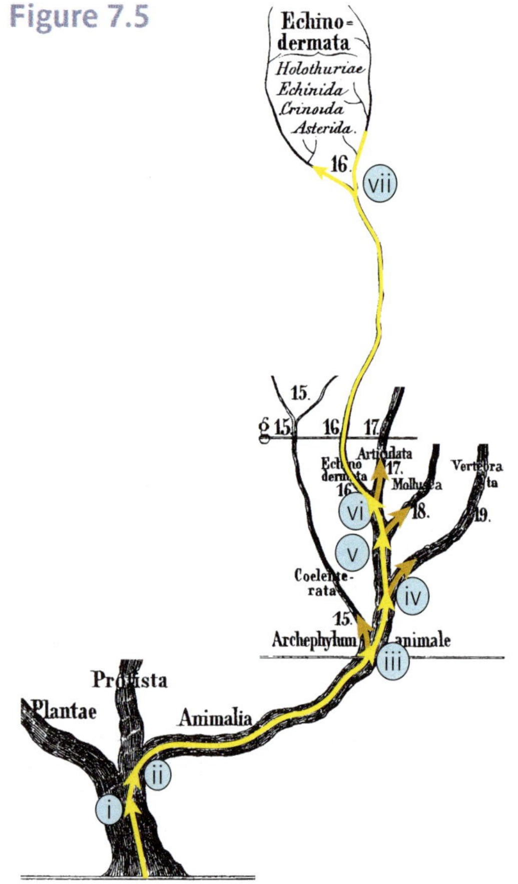

Based on a diagram by Ernst Haeckel, 1866.

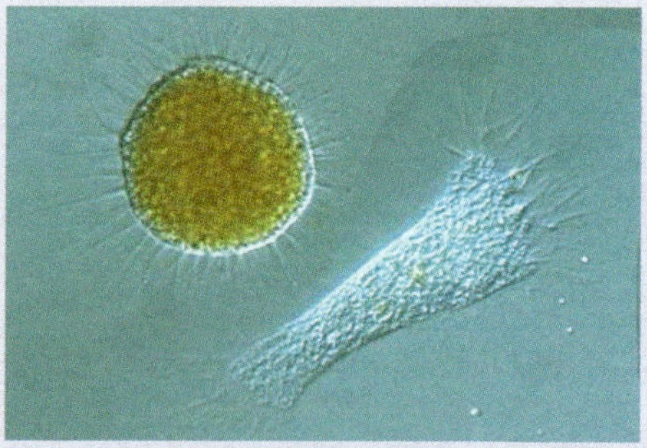

Figure 7.6 *Vampyrella* is a protist (single-celled organism) that lives in bog pools. Its name is presumably derived from its habit of stabbing hapless algal victims with its spines then sucking out their chloroplasts. The taxonomy of *Vampyrella* is uncertain. There are currently no DNA sequences for *Vampyrella* in GenBank (See **TechBox 1.1**).

Reproduced with permission from Yuuji Tsukii, Protist Information Server, http://protist.i.hosei.ac.jp/.

distantly related to amoebae, such as *Vibrio* (bacteria of such diverse habits as causing human cholera and making jelly-fish glow) and *Protomonas* (nowadays referred to as 'pink-pigmented facultative methylotrophs', these bacteria can metabolize some unusual carbon sources such as formaldehyde and methanol). These diverse organisms would not be grouped together today, but again, we are using Haeckel's tree as an illustration of how we derive information on branching points, rather than the latest word on systematics (and in the case of *Vampyrella*, its classification is still uncertain; see **Figure 7.6**).

Travelling down the tree from *Vampyrella* (**Figure 7.7**, p. 231), the first node we meet is the branching point between the moneres and the flagellates (single cells that swim with whip-like tails; node b), then the combined moneres + flagellates lineage is joined by the slime-mould lineage (myxomycetes; node c), then by all of the other protist lineages (node d, including the sponges, which today are considered to be in the animal kingdom), and then the diatoms (node e). As we travel down the tree, each node connects to another related lineage. We keep collecting progressively more distantly related lineages until we arrive at the root, the ancestor shared by all the lineages in the tree (node g).

Molecular phylogenetics

Phylogenies can be based on any information that reveals similarity or descent. Haeckel's phylogeny is based on his observations of the morphology and development of living species. Some phylogenies use palaeontological information to reconstruct the relationships between extinct species (**Figure 7.8**, p. 232). But of course the subject of this textbook is the way we use DNA sequences to reconstruct evolutionary history, so naturally that is what we will focus on here.

Molecular data have a number of important advantages for estimating phylogeny. Firstly, the genome

Figure 7.7

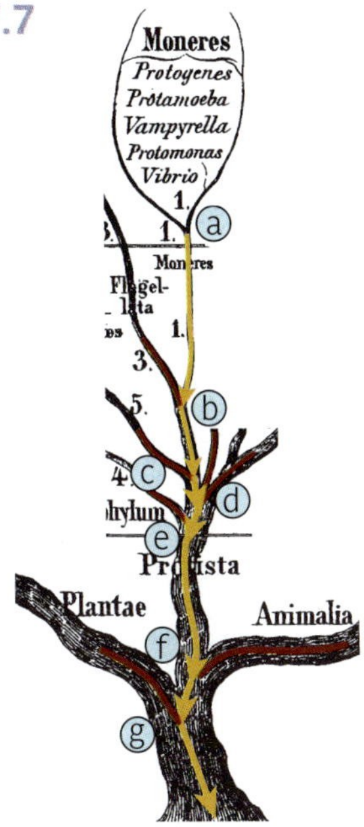

Based on a diagram by Ernst Haeckel, 1866.

provides a unifying framework for estimating phylogeny, because DNA sequences can be compared between all species, at all depths of evolutionary divergence. As we saw in Chapter 2, all living species share the same genomic system. All genomes are written in the directly comparable language of DNA (or RNA) bases, so DNA sequences can be compared between all organisms, from individuals (**Case Study 1.2**), to populations (**Case Study 2.1**), to species (**Case Study 2.2**), to genera (**Case Study 6.2**), to families (**Case Study 7.2**), all the way up to kingdoms (Chapter 6). For closely related organisms, we compare rapidly changing sites of the genome (e.g. using microsatellites to track paternity; **Figure 3.18**). For distantly related organisms we will use genes for highly conserved enzymes (e.g. comparing RNA polymerase II beta across different kingdoms; **Figure 2.2**). But there is something to be compared between all living organisms when we look at the level of the genome. This is not true of morphological characters: the phenotypic characters that are informative for one set of organisms may be meaningless for another. The phylogeny of dinosaurs in **Figure 7.8** was generated by analysis of the similarities and differences in fossil bones: for example, skull shape, leg bone dimensions, and so on. We could not use the same set of characters to classify starfish (that have no bones) or lungfish (that have different bones).

Secondly, DNA provides a record of evolutionary history that is essentially independent of many other sources of historical information, such as the fossil record, comparative morphology, or biogeography. DNA sequences are therefore very useful for uncovering the evolutionary history of organisms that have no fossil record, such as viruses (**Case Studies 5.1**, **7.1**, and **8.2**). Furthermore, the process of substitution continues whether species undergo a great amount of physical and ecological change, or whether they barely change at all. So DNA evidence can be very useful in deciphering the origins of species whose morphology gives little clues to their origins, such as highly reduced parasitic taxa (**Case Study 7.2**, p. 236), or species that have undergone little physical change (**Case Study 8.1**).

We will look more closely at the decoupling of morphological and molecular evolution in Chapter 8

Thirdly, DNA sequence data are ideally suited to statistical analysis. Even the smallest genome contains a very large number of essentially independently evolving characters. We can describe a model of evolution that states the probability of one base changing to another at any given site in the sequence, and we can use this model to weigh up the likelihood of different alternative phylogenies. It is very difficult, if not impossible, to construct a statistical model that adequately describes the complexity of morphological evolution.

I want to emphasize that by listing the advantages of molecular data, I am not rejecting the value of other forms of phylogenetic information. DNA is not, and should not be, the sole source of phylogenetic information, for a number of reasons. Firstly, although all life on Earth is based on DNA, DNA data are not available for all species. This may be because DNA samples have not been collected (e.g. *Vampyrella*; **Figure 7.6**, p. 230), are difficult to collect (e.g. some soil bacteria; **Case Study 4.1**) or, in the case of most extinct species, impossible to

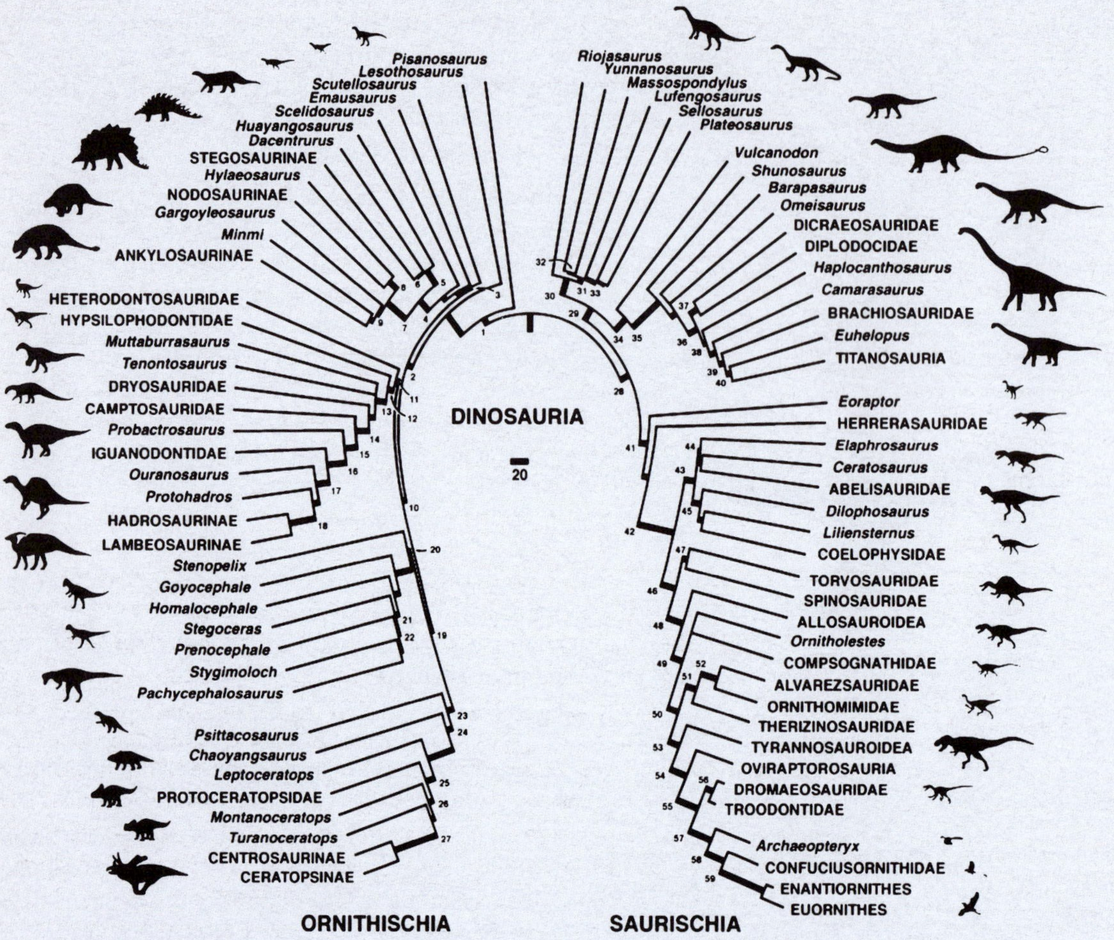

Figure 7.8 This phylogeny of dinosaurs was arrived at by parsimony analysis of hundreds of skeletal characteristics, each coded using single digits to represent character states. For example, one of the characters used to analyse the Ornithischia (a group including the ceratopsids and stegosaurs) was 'premaxillary tooth number', which in this dataset has five possible states (3, 4, 5, 6, or 7 teeth). From Sereno (1999).

collect (e.g. dinosaurs; **Figure 7.8**). Secondly, phylogenies based on DNA may not always be correct. It is quite common for two different research groups to publish contradictory molecular phylogenies for the same group of species. How can you tell which one is correct? Where possible, evolutionary hypotheses generated using molecular data should be compared to those derived from other sources of information, such as fossils, biogeography, and morphology. Thirdly, we can derive a great deal of information from molecular phylogenies, but they cannot tell us everything. All the DNA in the world will not tell you there were once pterodactyls with 10-metre wing spans, nor can DNA phylogenies tell us what the first mammals looked like. This book aims to illustrate how molecular data can be used to uncover evolutionary history and processes, but I do not claim that molecular data are the only, nor even the best, means of doing so.

CASE STUDY 7.1

Up the river: polio vaccine and the origin of HIV

KEYWORDS

bootstrap
maximum likelihood
Bayesian inference
cross-species infection
identification
virus
emerging disease

RELATED TECHBOXES

TB 7.2: Maximum likelihood
TB 7.3: Bayesian phylogenetics

RELATED CASE STUDIES

CS 2.1: On the origin of faeces (DNA surveys)
CS 8.2: Molecular detective (phylogenetic epidemiology)

Worobey, M., Santiago, M.L., Keele, B.F., Ndjango, J.B., Joy, J.B., Labama, B.L., *et al*. (2004) Origin of AIDS: Contaminated polio vaccine theory refuted. *Nature*, Volume 428, page 820

> *"The new facts in the case still tend to be widely separated and none by itself amounts to a proof; however, taken together, the steady trend and accumulation has become very impressive. At the very least, the OPV hypothesis of the origin of AIDS now merits our acute attention."* [1]

Background

HIV (human immunodeficiency virus) is an emerging disease. Until the early 1980s it was unknown; now it infects around 0.6% of the global population (over 40 million people) and has been responsible for an estimated 25 million deaths. Where did it come from and what caused its rapid spread throughout the world? The most widely held hypothesis is that HIV originated from primate viruses that infected people in Africa who hunted monkeys and apes, and then it spread from Africa through personal contact and needle use, accelerated by the explosive growth of global travel. But there is an alternative hypothesis, vociferously championed by a handful of journalists and biologists (including the great evolutionary biologist Bill Hamilton who died as a result of an expedition to the Congo to collect some of the samples analysed in the study described here). The OPV (oral polio vaccine) hypothesis is that HIV spread via live polio vaccines produced using chimp kidney tissue infected with simian immunodeficiency virus (SIV). The OPV hypothesis stirred heated controversy, with detractors frustrated that it was given any credence whatsoever, and supporters claiming a conspiracy by the mainstream scientific community to deny them a voice[2]. The OPV hypothesis makes a number of predictions that can be tested against observation, for example that early CHAT polio vaccine samples should contain both HIV and chimp DNA (so far all tested vaccine samples have been found to be HIV- and SIV-free, and to contain monkey DNA but no chimp DNA[3,4]).

Aim

This study set out to test whether the source of the HIV pandemic was the chimpanzee kidneys supposedly used to manufacture the CHAT polio vaccine. In particular, they tested one prediction that the OPV hypothesis makes about the topology (branching order) of a phylogeny containing SIV and HIV sequences: that SIV sequences from the source pool of chimpanzees used to generate the vaccine at the Kisangani lab should be at the base of the HIV phylogeny (that is, closest to the root).

Methods

Samples of chimpanzee faeces and urine were collected in the Democratic Republic of the Congo, from forests near Kisangani, site of the laboratory that administered the CHAT

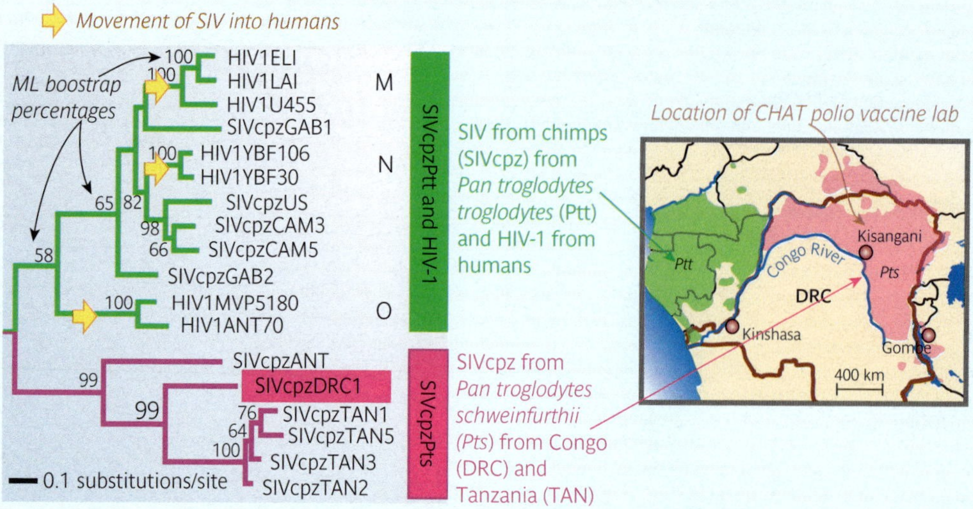

Figure CS7.1a Maximum likelihood phylogeny of sequences of HIV/SIV genes (*gag* and *nef*) from a range of human and chimpanzee populations. HIV-1 sequences are nested with the clade of chimpanzee SIV (SIVcpz) sequences, suggesting that HIV-1 is derived from SIVcpz. Furthermore, this phylogeny supports previous studies showing that SIVcpz has entered human populations on multiple occasions. However, the SIVcpz sequence from Kisangani is apparently distantly related to HIV-1 so this population of SIV is unlikely to be the ultimate source of HIV-1. M, N, and O are subtypes of HIV-1.

Adapted by permission from Macmillan Publishers Ltd: Worobey, M., Santiago, M.L., Keele, B.F., Ndjango, J.B., Joy, J.B., Labama, B.L., Dhed'A, B.D., Rambaut, A., Sharp, P.M. and Shaw, G.M. (2004) Origin of AIDS: Contaminated polio vaccine theory refuted. *Nature*, Volume 428, page 820.

vaccines (proponents of OPV hypothesis claim that this laboratory also manufactured the CHAT vaccine, but this claim is disputed). PCR amplification demonstrated that SIV was present in these wild chimps. These SIV sequences were manually aligned with other SIV and HIV sequences from a range of locations. Phylogeny of these sequences was estimated using maximum likelihood with bootstrap (**TechBoxes 7.2** and **7.4**), and with Bayesian inference (**TechBox 7.3**).

Results

Consistent with previous studies, this phylogenetic tree suggests that HIV-1 has moved from chimps into humans on multiple occasions (**Figure CS7.1a**). But the phylogenies group the chimp SIV from near Kisangani with sequences from infected chimps from Gombe National Park in Tanzania, not with the HIV-1 sequences. This suggests that the Kisangani chimps were not the source of HIV in humans.

Conclusions

Transfer of an animal virus through an infected vaccine is a serious threat that requires rigorous testing and strict safeguards. Testing hypotheses concerning vaccine safety, and communicating these results to the concerned public, is critical to community acceptance of vaccination programmes. This result strongly suggests that HIV-1 was not derived from chimpanzees local to the laboratory that administered the CHAT polio vaccine. This result on its own may not be enough to disprove the OPV hypothesis, but it adds to a body of evidence that weighs heavily against it. Phylogenetic analysis has played an important role in this debate, for example showing that CHAT samples tested to date were not derived from chimps[3,4] and that the origin of the HIV-1 pandemic probably predates the CHAT vaccination campaign[5].

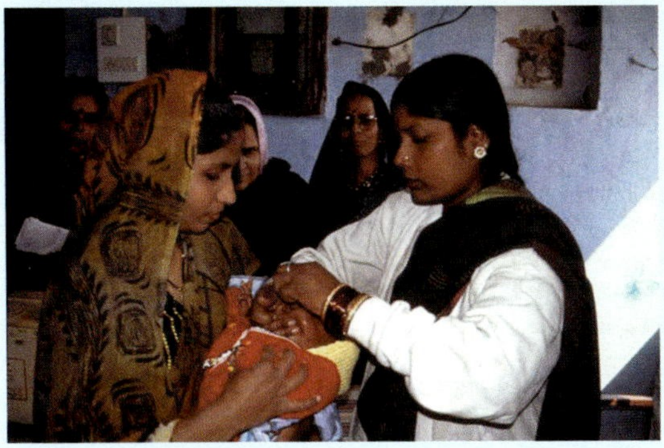

Figure CS7.1b The development of the orally administered polio vaccine, here being administered to a baby in India, was a breakthrough that offered the World Health Organization the chance to make polio the second communicable disease to be eradicated by vaccination (after smallpox). But vaccination programmes rely on public confidence. For example, in 2003 a regional government of Nigeria stopped the polio vaccination programme after rumours circulated that the vaccines were infected with HIV and contained covert anti-fertility drugs. Although vaccination resumed in 2004, the halt in the programme caused a polio outbreak that spread through Nigeria and to at least five other countries[7]. The eradication of polio has thus been delayed, but will hopefully be achieved in the near future. You can follow the progress of the campaign on *www.polioeradication.org*.

Photograph: Chris Zahniser, B.S.N., R.N., M.P.H.

Limitations

While this study provides convincing evidence that HIV-1 is not derived from chimps living in the area of the Kisangani laboratory, it cannot rule out the possibility that the polio vaccine was developed on SIV-infected chimp or monkey tissue taken from animals from another area. However, there is currently no solid evidence to suggest this was the case. Due to the brevity of this one-page paper, there are no details of the analyses undertaken to test the robustness of the result, apart from bootstrap percentages.

Future work

One of the wider issues to emerge from the debate is the role that scientists and the media play in public debate concerning vaccines. On the one hand, scientists argue that the media should not give unwarranted attention to hypotheses with little empirical support, as this creates unnecessary suspicion of vaccination programmes. On the other hand, failure to openly discuss ideas and clearly communicate the evidence for and against each hypothesis risks claims of a cover-up, which undermines public confidence. Loss of public confidence can have serious effects. For example, a press conference by a medical researcher who claimed a link between the measles-mumps-rubella (MMR) vaccine and autism caused a drop in vaccination rates in the UK, which preceded an increase in measles outbreaks[6] (see also **Figure CS7.1b**).

References

1. Hamilton, W.D. (1999) Foreword. In *The River: a journey back to the source of HIV and AIDS*. Hooper, E, ed. Harmondsworth.
2. Cohen, J. (2000) Forensic epidemiology: vaccine theory of AIDS origins disputed at Royal Society. *Science*, Volume 289, pages 1850–1851.
3. Blancou, P. *et al*. (2001) Polio vaccine samples not linked to AIDS. *Nature*, Volume 410, pages 1045–1046.
4. Berry, N. *et al*. (2001) Vaccine safety. Analysis of oral polio vaccine CHAT stocks. *Nature*, Volume 410, pages 1046–1047.
5. Hahn, B.H., Shaw, G.M., De Cock, K.M. and Sharp, P.M. (2000) AIDS as a zoonosis: scientific and public health implications. *Science*, Volume 287, pages 607–614.
6. Jansen, V.A.A. *et al*. (2003) Measles outbreaks in a population with declining vaccine uptake. *Science*, Volume 301, page 804.
7. Jegede, A.S. (2007) What led to the Nigerian boycott of the polio vaccination campaign? *PLoS Medicine*, Volume 4, page e73.

CASE STUDY 7.2

Non-tree-like flowers: horizontal gene transfer between parasitic plants and their hosts

KEYWORDS

maximum likelihood

long-branch attraction

parsimony

Bayesian inference

horizontal gene transfer

organelle

FURTHER INFORMATION

You can find the alignments and phylogenies from this study on TreeBase (*www.treebase.org*) Study Accession S1177.

RELATED TECHBOXES

TB 7.2: Maximum likelihood

TB 7.4: Boostrapping

RELATED CASE STUDIES

CS 5.2: Viruses within (recombinant genomes)

CS 6.1: Barcoding nematodes (DNA taxonomy)

Nickrent, D., Blarer, A., Qiu, Y.-L., Vidal-Russell, R. and Anderson, F. (2004) Phylogenetic inference in Rafflesiales: the influence of rate heterogeneity and horizontal gene transfer. *BMC Evolutionary Biology*, Volume 4, page 40

> “*Determining the photosynthetic relatives of Rafflesiales has long presented a challenge owing to the extreme reduction and/or modification of morphological structures that have accompanied the evolution of this lineage. Molecular phylogenetic approaches, although providing great promise in resolving such questions, also come with their own set of challenges that includes losses of some genes, substitution rate increases in other genes, and horizontal gene transfer.*”[1]

Background

Rafflesia produces the largest known flowers, up to a metre in diameter and 10 kilograms in weight. By comparison, the rest of the plant is insubstantial: no leaves, stems, or roots. This is because *Rafflesia* is entirely parasitic on a vine (*Tetrastigma*) so its body consists of a fungus-like haustourium that grows within its host's tissues. Parasitic lineages are often difficult to place in phylogenies, because the parasitic lifestyle commonly results in loss of characteristics associated with free-living species. In many cases molecular data can help place these taxa, but loss of the characters may extend to the genome itself. Most molecular phylogenies of plants have been based on chloroplast genes that make proteins essential to photosynthesis. Parasitic plants that have lost the ability to photosynthesize (holoparasites), such as *Rafflesia*, may undergo reduction of the chloroplast genome. *Rafflesia* appears to have lost the gene most commonly used in molecular phylogenetics of flowering plants, *rBCL* (**TechBox 6.1**). Furthermore, rates of molecular evolution in the remaining genes tends to be higher in holoparasites, complicating phylogenetic inference. Long branches with many substitutions can be falsely grouped together, particular in parsimony analyses, a phenomenon known as long-branch attraction.

Aim

Previous attempts to place members of the family Rafflesiales in the phylogeny of flowering plants have produced variable results, so this research team aimed to use multiple genes from three separate genomes (nuclear, mitochondrial, and chloroplast) to clarify the relationships between four distinct groups of plants within the Rafflesiales group: the 'small-flowered clade' (Apodanthaceae), 'large-flowered clade' (Rafflesiaceae), 'inflorescence clade' (Cytinaceae), and 'hypogynous clade' (Mitrastemonaceae).

Methods

They collated available DNA sequences from 106 different species across a wide range of plants, making manual alignments of the mitochondrial genes *matR* and *atp1*, nuclear gene small subunit ribosomal RNA (SSU rRNA), and two chloroplast genes, *rBCL* and *atpB* (not available for the parasitic taxa). They analysed these sequences using three different

Figure CS7.2a *Rafflesia* flowers take up to 12 months to grow then flower for less than a week. Because they are unisexual, a flower will only be pollinated if a flower of the opposite sex flowers within fly-flight distance in that 5 to 7-day window. Given that *Rafflesia* are rare (and getting rarer), this is fairly unlikely. I was once staying in a village in Sumatra when word got round that there was a *Rafflesia* in flower. We hiked up the rim of the crater lake with a group of locals who were as keen to see it as we were. It was a magnificent beast of a flower and, despite a reputation for stinking like a rotting buffalo carcass in order to attract flies to act as pollinators, the *Rafflesia* we saw smelt just fine.
Photograph: Lindell Bromham.

methods: maximum parsimony (MP), maximum likelihood (ML; **TechBox 7.2**), and Bayesian inference (BI; **TechBox 7.3**), assessing the strength of support for clades using bootstrap for MP and ML trees (**TechBox 7.4**). For the ML analysis, the researchers used likelihood ratio tests (**TechBox 7.2**) to help them choose the simplest suitable substitution model, in order to reduce computation time. Because it was such a large dataset, these researchers adopted a number of other strategies to reduce the computation time. For example, they estimated parameter values on an MP tree, used these values to search for a set of trees using ML, then re-estimated parameter values on these ML trees and used these values to perform another ML search. For the SSU dataset, they reduced computation time by constraining parts of the tree to known relationships (that is, they evaluated only trees which showed certain clades identified in previous analyses). They then solved the unknown parts of the tree, including the placement of the Rafflesiales taxa. To test the possible effect of long-branch attraction, they simulated the evolution of sequences along a tree modelled on the SSU tree, then used MP and ML to see if they recovered the model tree correctly.

Results

The phylogenetic position of the four groups of Rafflesiales varied depending on the gene analysed and the method used. The ML and BI trees places the members of the Rafflesiales group in different places in the tree, suggesting that they are not a natural group but have independently come to resemble each other. Parsimony analysis grouped all four Rafflesiales groups together; however, there was low bootstrap support for this clade. But when MP analysis was conducted with only one parasite group at a time, each group came out in same position as in the ML tree. This suggests that MP is falsely grouping these lineages together, a conclusion supported by the simulation analyses. MP on the simulated data did not correctly recover the model tree, the majority brought together unrelated clades with long branches. ML fared better, but still only recovered the correct relationships in just over half of the simulated datasets. However, the position of some of the groups depended on the gene analysed. For example, the *matR* tree (**Figure CS7.2b**) groups the 'small-flowered' species of Rafflesiales, *Pilostyles* and *Apodanthes*, with the Curcurbitales (melons and their kin). But the phylogeny based on a different gene, *atp1*, groups *Pilostyles* with its host plant, the legume *Pisum*.

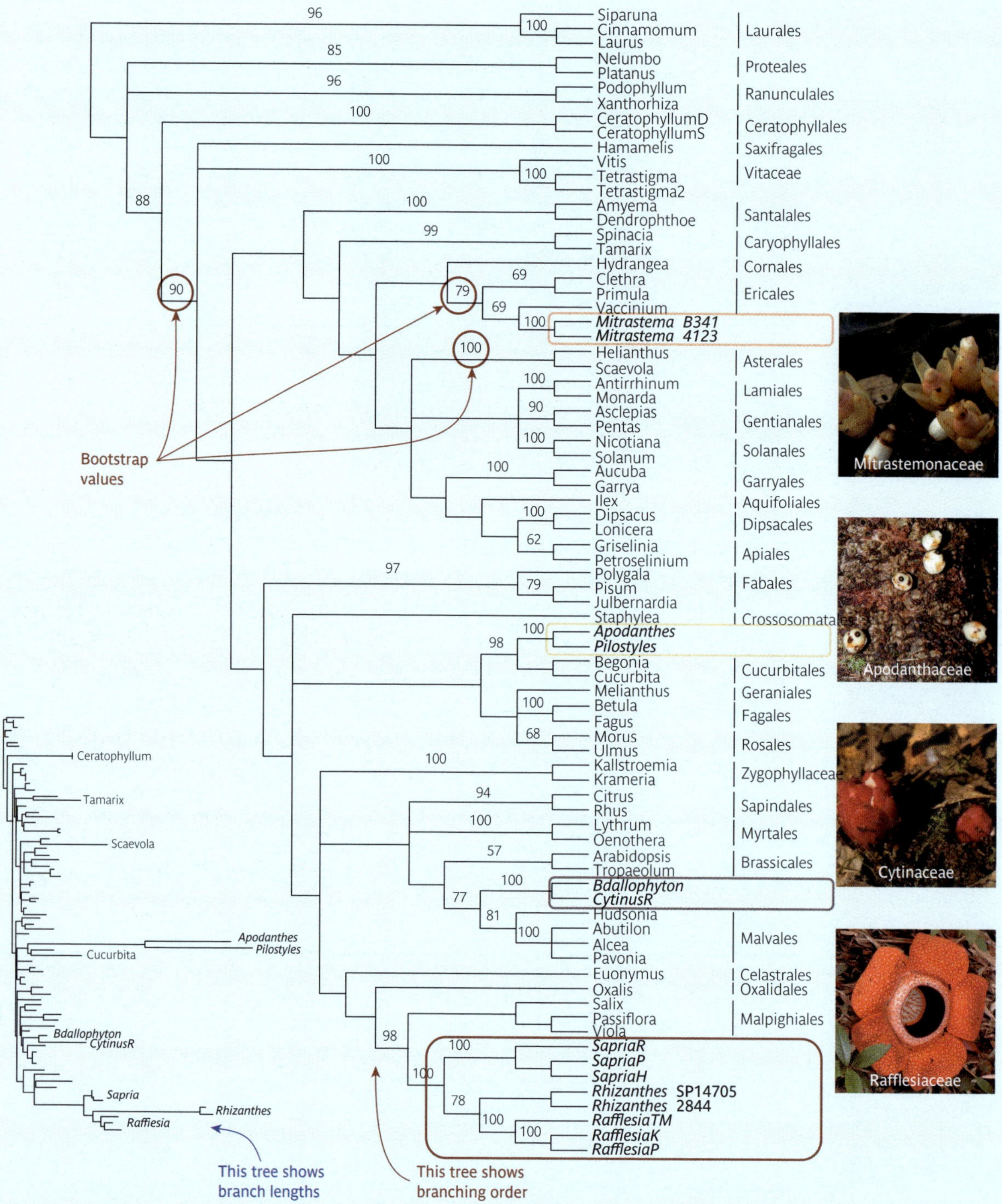

Figure CS7.2b A phylogeny of the four groups of Rafflesiales based on a maximum likelihood analysis of the *matR* mitochondrial gene. The bootstrap values give an indication of the relative strength of support for different clades. The tree in the left-hand corner demonstrates the uneven rates of change: the lineages leading to three of the Rafflesiales groups have a much higher substitution rate than most lineages in the tree. *Mirastema*: courtesy of Hu Research; *Apondanthes*: © Darwin Initiative; *Cyntinus*, photograph: M. Koltzenburg; *Rafflesia*, photograph: Steve Cornish.

Conclusions

These results suggest that the four main groups of the Rafflesiales are not closely related. They have included in the same family because they have similar lifestyles, but these lifestyles evolved independently in four separate lineages. Several features of the molecular evolution of these groups make their placement in molecular phylogenies problematic. Firstly, the fast substitution rate in the parasitic lineages seem to caused them to be falsely grouped together. Secondly, the grouping of a parasite (*Pilostyles*) with its host (*Pisum*) in the *atp1* tree suggests horizontal gene transfer between species, presumably facilitated by the intimate relationship between parasite and host tissues. This phenomenon has also been reported for *Rafflesia*, which groups with its host *Tetrastigma* in phylogenies based on a different mitochondrial gene (*nad1B-C*)[2].

Limitations

The trees produced from different genes are not entirely congruent, either with each other or with generally accepted hypotheses of plant relationships. While one aspect of the disagreement may be due to unusual patterns of genomic evolution (see above), other indications suggest that this is simply phylogenetic uncertainty. Most of the trees also had low support, as indicated by low bootstrap values and low Bayesian posterior probabilities. This means that it is problematic to interpret the differences between them, which could be due simply to noisy data.

Future work

This phylogeny is very broad scale, covering all flowering plants. Further sequencing will help give a finer-grained picture of the evolution of the Rafflesiales. A recent study has used additional sequencing within the Malphigiales to clarify the closest relative of *Rafflesia*, suggesting that they fall within the tiny-flowered Euphorbiaceae[3].

References

1. Nickrent, D., Blarer, A., Qiu, Y.-L., Vidal-Russell, R. and Anderson, F. (2004) Phylogenetic inference in Rafflesiales: the influence of rate heterogeneity and horizontal gene transfer. *BMC Evolutionary Biology*, Volume 4, page 40.
2. Davis, C.C. and Wurdack, K.J. (2004) Host-to-parasite gene transfer in flowering plants: phylogenetic evidence from Malpighiales. *Science*, Volume 305, pages 676–678.
3. Davis, C.C., Latvis, M., Nickrent, D.L., Wurdack, K.J. and Baum, D.A. (2007) Floral gigantism in Rafflesiaceae. *Science*, Volume 315, page 1812.

Substitutions reveal relationships

I have recently fulfilled a long-held ambition to have a pet onychophoran (**Figure 7.9**, p. 240). Onychophora is an enigmatic group of caterpillar-like animals commonly known as velvet worms (or peripatus). They are beautiful little creatures but, admittedly, they do not make very responsive pets. My velvet worm, Perry, lives in a lunchbox in the fridge and sulkily refuses to eat the termites I lovingly provide for her. The closest she has come to interacting with me in any meaningful way was once, when I surprised her, she spat glue at me

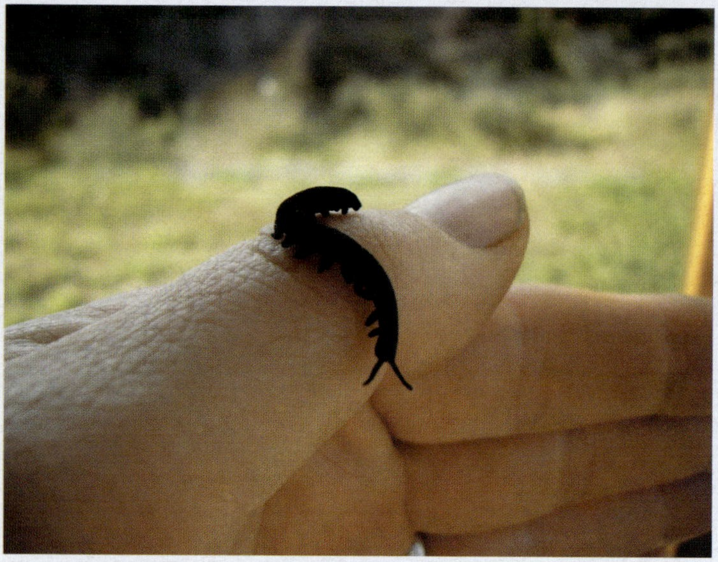

Figure 7.9 My pet velvet worm, Perry.
Photograph: Lindell Bromham.

(velvet worms are predators and they hunt by shooting a jet of superglue at their prey to immobilize them). If labradors are at one end of the spectrum of loyal and affectionate pets, then I guess my onycophoran Perry must be at the other end. But, although she is wholly indifferent to me, I love her because she is blue and velvety and moves with an entrancing rippling gait on her soft little legs. Perry is a member of the species *Euperipatoides rowelli*. This species is named after Dave Rowell, a biologist who has been studying velvet worms in Tallaganda State Forest in Australia as a case study in genetic divergence.

We are going to use a mythical population of velvet worms to allow us to follow the process of genetic divergence as populations divide and diversify over time (Figure 7.10). We will track just one tiny part of that genome, a six-base sequence. Like most of the genome, this sequence is the same for most members of the population. But it just so happens that one of our onychophorans is a mutant, because its fifth base is a G instead of a C. By chance, this mutation is neutral. It has no effect on its carriers' chances of reproduction.

Neutral alleles, that make no difference to fitness, are discussed in Chapter 5

If all goes well, the G-mutant will meet a member of the opposite sex, at which point the male may slap a parcel of sperm on the female's skin, then the female's tissues will dissolve to allow the sperm to pass down into her body cavity, and from there find their own way to her ovaries. In this way the G-mutation may make it into some little baby velvet worms, that emerge fully formed from their mother's body. And if all goes particularly well, some of these baby velvet worms will eventually have babies of their own, and the G-mutation will be perpetuated down the generations.

Onycophorans do not travel far. Most of the G-mutant velvet worms will spend their whole life in a single rotting log. Their soft bodies lose moisture quickly so they risk desiccation if they attempt to cross the open spaces between logs. But when it is wet, some brave individuals venture forth across the forest floor in search of new logs to colonize. If one of our mutants makes it to a new log, it will bring its G-allele with it, and if it finds a mate and successfully reproduces, then that G-allele can spread throughout a new rotting log. In this way, generation by generation, log by log, a new mutation can spread throughout the forest. Thus an interconnected population shares a common pool of alleles.

However, one summer a fire burns through the forest, torching the forests along the ridge but leaving the wet forest in the gullies untouched (Figure 7.11). Velvet worms survive in the unburnt gullies but are unable to cross the dry ridges to get to the neighbouring gully. Now, a velvet worm from Scribbly Gum will never get

Figure 7.10

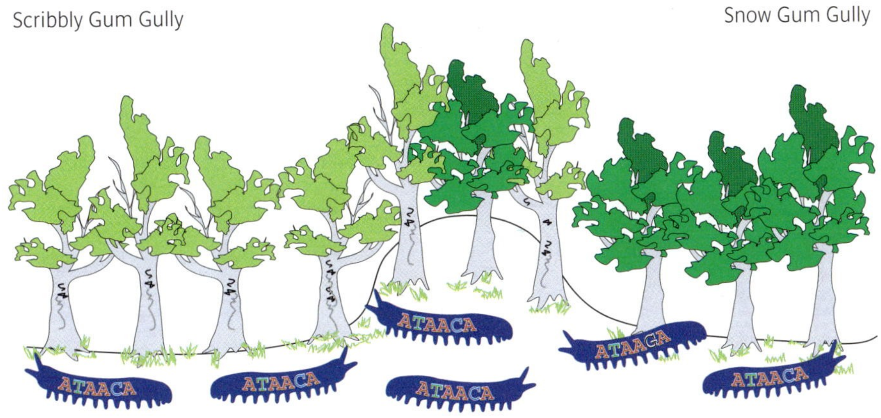

Figure 7.11

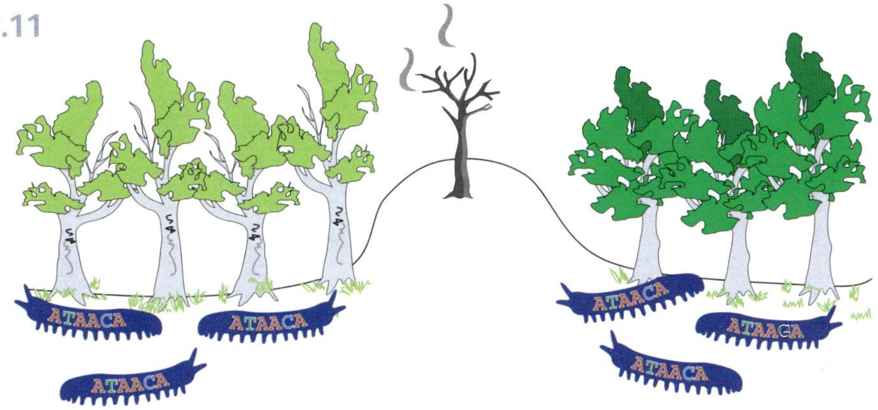

close enough to a velvet worm from Snow Gum to exchange gametes. And so the G-allele that is found in Snow Gum Gully is never going to end up in the offspring of a Scribbly Gum Gully velvet worm. The populations have become genetically isolated from each other.

Over time, it happens that, by chance, the G goes to fixation in the Snow Gum population (**Figure 7.12**, p. 242). A new mutation, from A to T, arises in the Scribbly Gum population. This new mutation influences the production of a pheromone. Pheromones are chemicals that influence the behaviour of other individuals (examples include the chemical trails laid by ants, scent marking of territories by carnivores, or mating pheromones released by moths). Some onycophorans release pheromones that attract other members of their species. The change from an A to a T in this pheromone gene changes the chemical signal, so that the A-allele velvet worms are not attracted to T-allele velvet worms. Therefore the T-allele is unlikely to end up in the offspring of an A-allele velvet worm.

But the T-allele pheromone is more powerful than the A-allele version, so T-allele carriers are more effective at finding each other, so on the whole they mate more often and have more offspring. Eventually the T-allele goes to fixation in the Scribbly Gum population by positive selection due to this reproductive advantage. Now when the forest on the ridge grows back and velvet worms can once more move through the valley (**Figure 7.13**, p. 242), the A-allele velvet worms from the

Figure 7.12

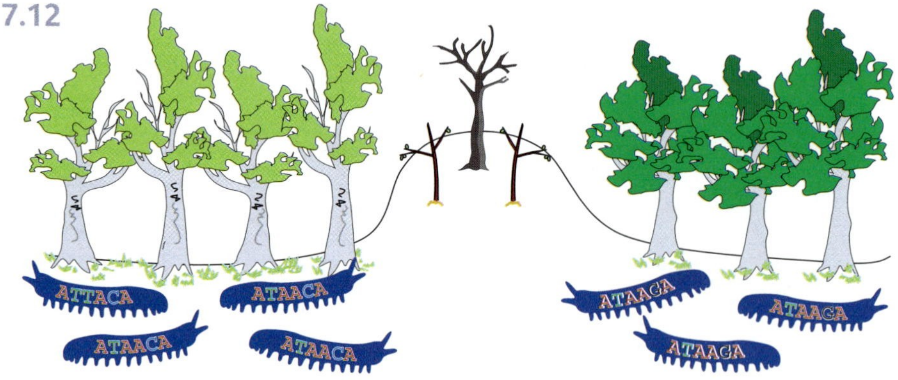

Figure 7.13

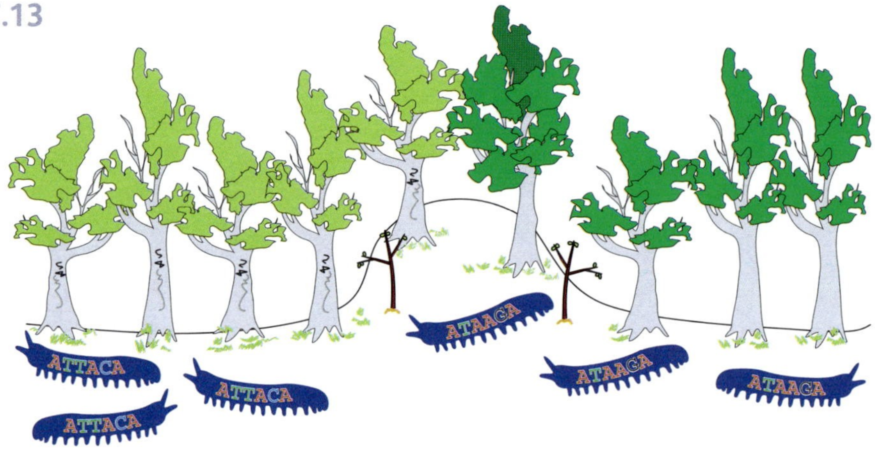

Snow Gum Gully are not attracted to the T-allele velvet worms from the Scribbly Gum Gully. So although the individual velvet worms can now move from one gully to the next, the alleles from the Snow population can no longer enter the Scribbly population.

A storm sweeps through the forest, knocking down old trees in the Snow Gum Gully. The damage is extensive. A tiny population of velvet worms survives in a lucky stand of Snow Gums, but they cannot cross the open ground to the ridge, nor reach the Scribbly Gum Gully (**Figure 7.14**). So now we have three separate populations, each genetically isolated from the other two. The two Scribbly Gum populations are separated by a biological mechanism (different pheromones) and the Snow Gum population is isolated by a physical factor (can't cross open ground). Now any mutation in one population cannot move to a different population. Each population will continue to accumulate different substitutions, by drift or selection. For example, the T-allele Scribbly velvet worm population undergoes further selection-driven molecular change affecting habitat choice. The remnant Snow Gum population accumulates another substitution by drift (being such a small population, drift is an important driver of substitutions in this population).

Now let's put all of these events together in one diagram. For simplicity, we will take away the velvet worms and just look at their DNA sequences (**Figure 7.15**). Every split in the population is marked by a division in the lines of descent.

Figure 7.14

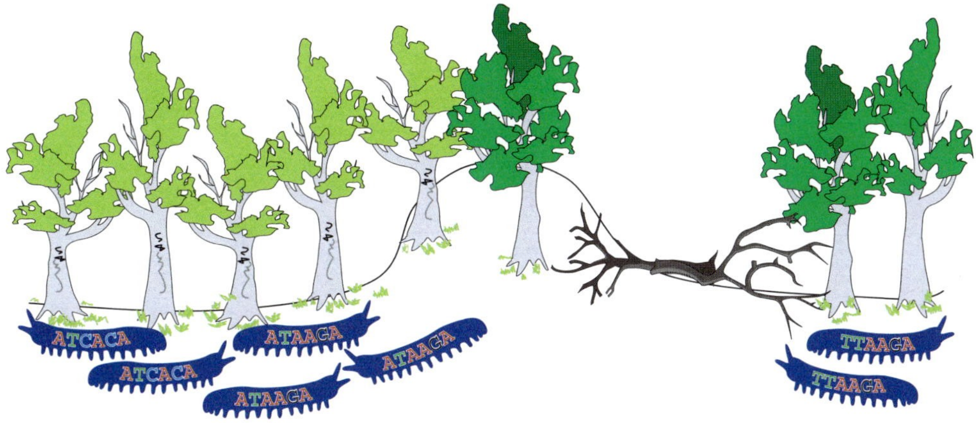

Figure 7.15

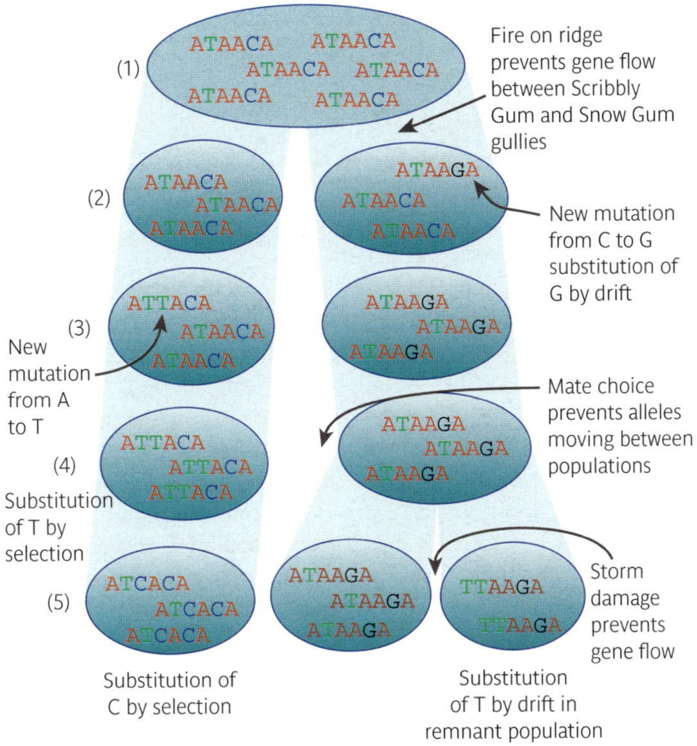

We can simplify this diagram even further (**Figure 7.16**, p. 244), simply marking the substitutions on lines of descent.

We have produced a branching diagram that represents the evolutionary history of these populations. Such a diagram is usually referred to as an evolutionary tree, or, more formally, a phylogeny.

Figure 7.16

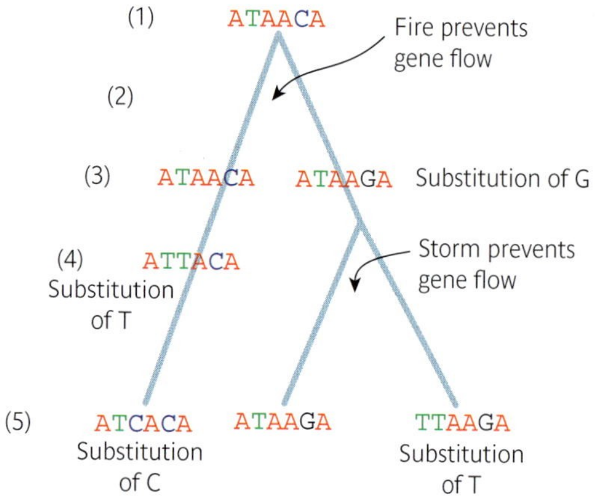

If we let these three populations evolve even longer, so that they accumulate more substitutions and undergo further population subdivisions, we might see the tree grow like this (Figure 7.17):

Figure 7.17

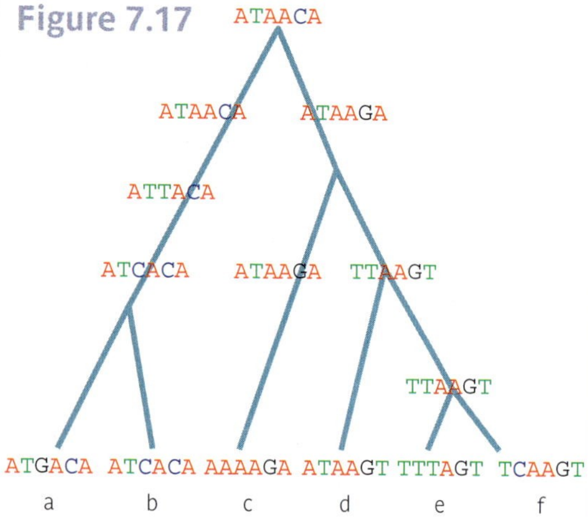

Before we go on, it is helpful to have a standard set of terms for describing trees. We are going to call the end-states of the phylogenetic tree 'tips'. For DNA-based phylogenies, the tips are sequences from our alignment, but we expect these tips to represent the taxa from which the sequences were sampled (taxon is an all-purpose term for 'biological group' that might be a population, a species, a phylum, etc.). Nodes are the branch-points, where a single line of descent splits to give rise to two or more lineages. We can interpret nodes as being the common ancestors of lineages (technically tips are nodes too). Branches, also known as edges, are the lines that connect nodes together (or tips to internal nodes). So a branch (or an edge) is any of the lines in a phylogeny. A clade is all of the descendants of a particular node. Because phylogenies represent a hierarchy of relationships, clades are nested within phylogenetic trees, just as taxonomy is nested (Figure 7.2, p. 226). In this case, the tip f is nested within the e–f clade, which is nested in the d–e–f clade, which is nested in the c–d–e–f clade.

Figure 7.18

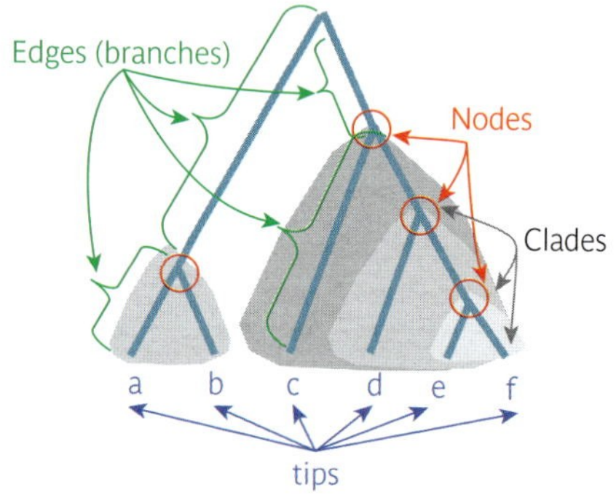

Phylogeny reconstruction

Let's quickly recap what we have learned so far. Mutations occur in individuals (Chapters 3 and 4). Some mutations become substitutions carried by all members of a population (Chapter 5). Closely related populations have more similar DNA sequences than more distantly related populations, so when we align DNA sequences we can usually identify hierarchies of shared substitutions (Chapter 6). Now we are going to explore how we can use these hierarchical patterns of shared substitutions to reconstruct evolutionary history.

In the example above, we were able to follow the substitutions through evolutionary time, as the populations divide and diverge. This is equivalent to watching Haeckel's tree start from a single shoot, then grow more and more branches. But we can rarely watch evolution happen. In most situations, all we can directly observe is the end points of the process: the DNA sequences at the tips of the tree. So the aim of phylogenetics is usually to take information from the twigs to reconstruct the series of branching events leading back to the root.

If we collected some onychophorans from our hypothetical populations (Figure 7.17), then took them back to the lab and sequenced their DNA, could we accurately reconstruct the series of evolutionary events that produced the present day genetic data? We would have six sequences, sampled from populations a to f. Our first task would be to align those sequences (Figure 7.19). In this case we can align six homologous sites.

Figure 7.19

```
  123456
a ATGACA
b ATCACA
c AAAAGA
d ATAAGT
e TTTAGT
f TCAAGT
```

In Chapter 6, we saw that when we compare the patterns of nucleotides at homologous sites in an alignment, we can recognize different patterns of similarity and difference (Figure 6.10). We can categorize these sites by the kind of information they give us about evolutionary relationships.

Chapter 6 explains the importance of correct alignment to recover evolutionary history from DNA sequences

Some sites contain the same base in all sequences. For example, all sequences in this alignment have an A in the fourth site (Figure 7.20). The most likely explanation for this pattern is that all of these lineages inherited this A from a common ancestor (see Chapter 6), so we refer to the similarity between them as ancestral. Sites that are the same in all taxa do not reveal relationships within the group. If you are fond of polysyllabic jargon, then you might like to know that shared ancestral character states are known as symplesiomorphies.

Figure 7.20

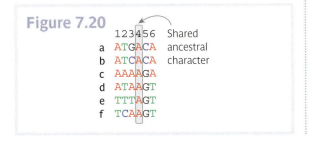

```
  123456   Shared
a ATGACA   ancestral
b ATCACA   character
c AAAAGA
d ATAAGT
e TTTAGT
f TCAAGT
```

Some sites in the alignment have a particular nucleotide state that is found in one sequence but not in any of the others. For example, at position 2, most populations have T but in population c this was substituted with an A and in population f the T was substituted with a C (Figure 7.21). Because they have arisen anew in a particular population, we refer to these substitutions as being derived. Unique derived changes, found in one sequence but not shared with any others, are also known as autapomorphies (Figure 6.10). Unique derived characters tell us about evolution in specific lineages, but, again, they do not help us split the sequences into related groups.

Figure 7.21

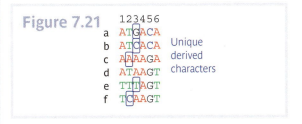

```
  123456
a ATGACA
b ATCACA   Unique
c AAAAGA   derived
d ATAAGT   characters
e TTTAGT
f TCAAGT
```

So shared ancestral characters do not help divide our sequences into related groups, and neither do unique derived characters. What we need is sites that give us information about which populations are more closely related to each other. In other words, we need characters that split the sequences into groups, just as Ray used physical and behavioural characteristics of birds to split them into related groups (Figure 7.2).

When a population divides to give rise to two or more new populations, each new daughter population inherits the genetic variation of the parent population. A unique derived substitution in the parent population is now shared between two related populations. And if one of these populations splits again, then that shared substitution will be passed on again. We can see this process of inheritance of substitutions in the example above. At stage 3 the Snow Gum population underwent a substitution of a G at site 5 (Figure 7.12). When the storm divided the Snow Gum population into two populations at Stage 5, both populations inherited this G (Figure 7.14). As the lineages continued to diverge, that G was inherited by their descendants (lineages c, d, e, and f: Figure 7.17). The process of descent leaves a trail of shared derived characters, each of which arose in a particular lineage (derived), then was inherited by its descendants (shared). So sites where a particular nucleotide state is shared by some but not all of the sequences in our alignment can provide information on descent. Shared derived characters are also known as synapomorphies.

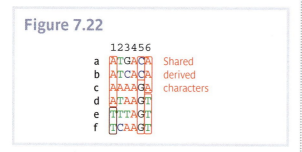

Figure 7.22

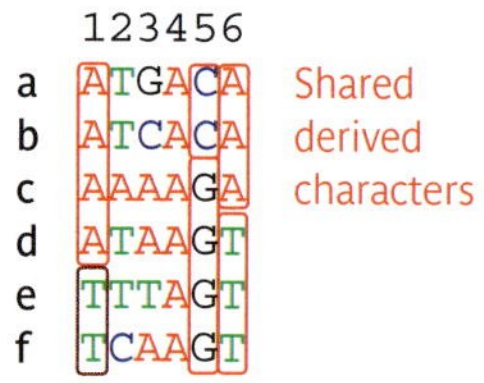

Splits

Shared derived characters carry information about evolutionary history because they split the sequences into related groups. Look at site 1. We can see that populations a, b, c, and d all have A and populations e and f have T. So this site splits our population into two groups: the A group (a, b, c, d) and the T group (e, f).

Figure 7.23

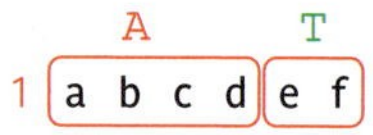

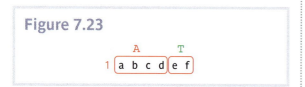

We could interpret this split as a branching event – the lineage that led to populations e and f diverged from the rest of the lineages and on the way an A was changed for a T. In other words, e and f share a more recent common ancestor than either does with a, b, c, or d, and we mark this on our tree with a node connecting e and f, and an edge along which the substitution occurred.

Figure 7.24

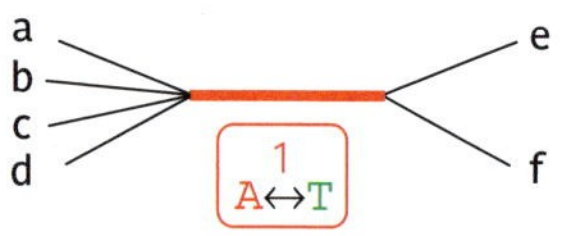

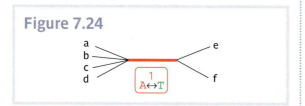

We can interpret site 5 in a similar way. Populations a and b are characterized by having a C at this site, and populations c, d, e, and f are characterized by having a G, so site 5 splits populations a and b from all the others.

Figure 7.25

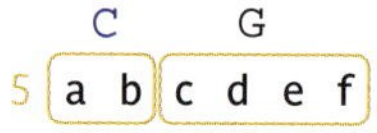

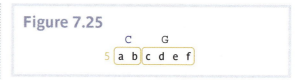

Just as before we can interpret this split as an edge in our tree.

Figure 7.26

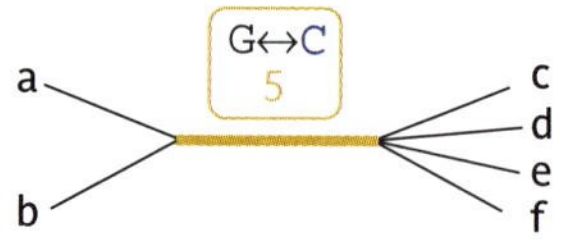

But now we can add this information to what we already know from site 1. Let's take the diagram we drew from site 1 (e and f split from the rest) and add to it the split inferred from site 5 (a and b split from the rest).

Figure 7.27

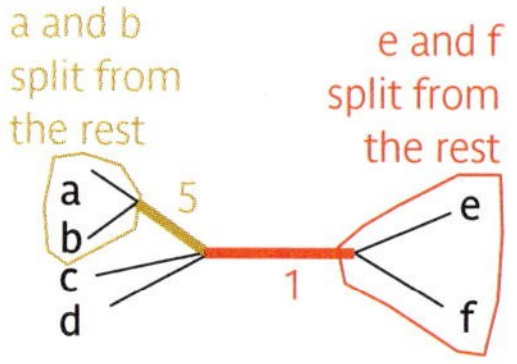

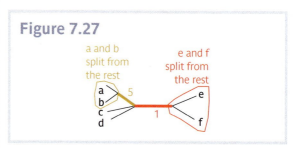

We have combined the splits from two different nucleotide sites. Now let's add site 6, which splits populations a, b, and c from populations d, e, and f.

Figure 7.28

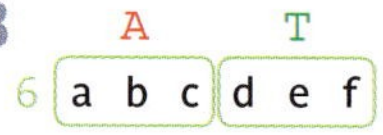

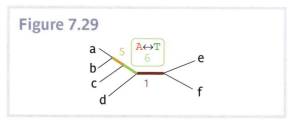

We already know that site 1 splits e and f from the rest, site 5 splits a and b from the rest, and now we can put an edge between the a–b–c group and the d–e–f group.

Figure 7.29

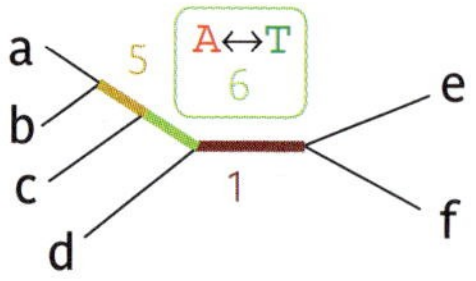

On their own, each of the splits (1, 5, and 6) simply divided the populations into two groups, but taken together they give us a series of nested branching events. So now we have uncovered the relationships between these six populations simply by considering the information in the DNA alignment.

Reading trees

Why doesn't this tree (**Figure 7.29**) look like the one in **Figure 7.17**? Trees contain information about descent. But that information can be displayed in a variety of ways. In this section we will spend some time rearranging our phylogeny so we can see that trees that look very different from each other may contain the same information. Firstly, we need to distinguish between rooted and unrooted trees. In Haeckel's and Darwin's trees (**Figures 7.1** and **7.4**), we could begin at the root (ancestral lineage) and, following the branching patterns forward in time, reach the tips (descendant lineages). With an unrooted tree, we have all the branching events that divide the taxa into groups but we don't know which edge represents the ancestral lineage, from which all the others arose. In other words, unrooted trees don't have a starting point.

Figure 7.29 is an unrooted tree: it shows the relationship between lineages, but it doesn't reveal to us which of these series of splits happened first. To make **Figure 7.29** look like the tree in **Figure 7.17**, we need to give it a starting point (root), by identifying the edge along which the first split happened. In this case, we happen to know that the very first split divided the lineage leading to a and b from the lineage leading to c, d, e, and f. On the unrooted tree, this is the edge marked 5. Imagine that our unrooted tree diagram is actually a mobile made of string. Pick up the unrooted tree on the edge marked 5 (**Figure 7.30**).

Figure 7.30

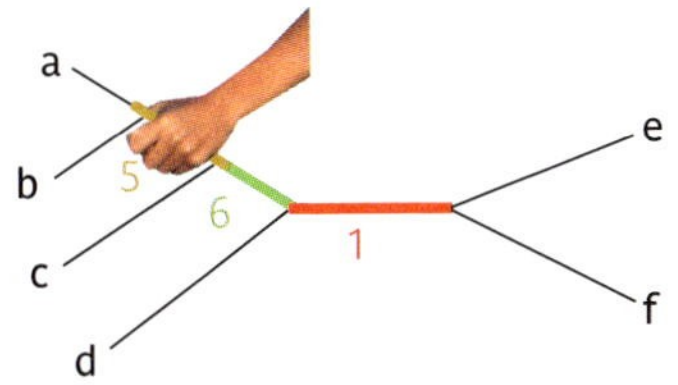

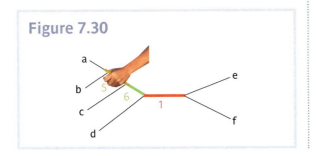

Let the rest of the branches hang down, as if the letters on the tips were heavy (**Figure 7.31**).

Figure 7.31

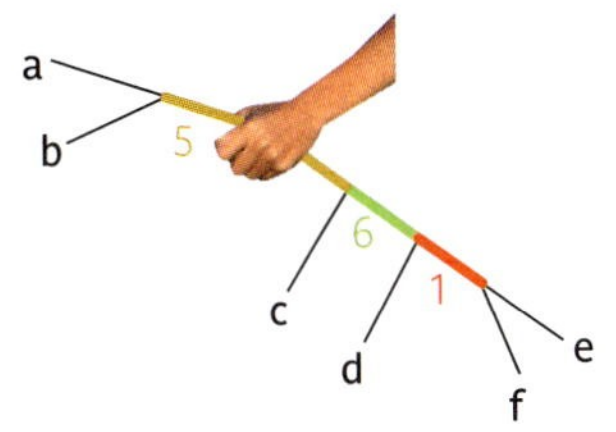

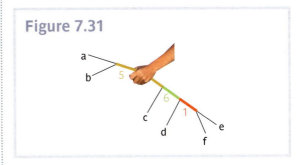

And now we have a rooted tree that looks like **Figure 7.17**. Our tree now begins at the root with a branch point that splits a and b from the lineage leading to c, d, e, and f (**Figure 7.32**).

Figure 7.32

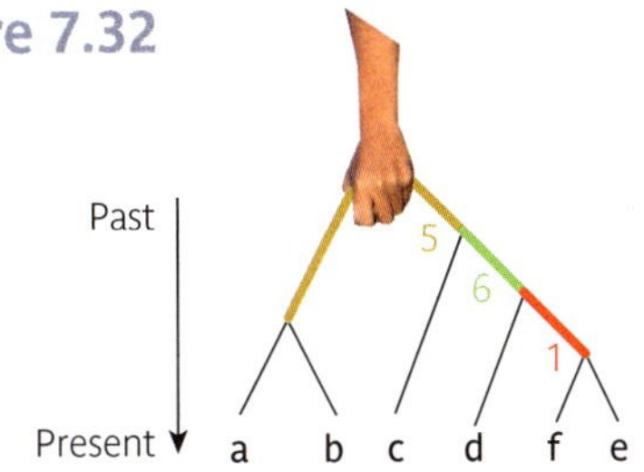

We have added the direction of time to our phylogeny. The information about which splits occurred is exactly the same, but we have now added information about the order in which the splits happened.

If we didn't know when the root was, we could orient the tree in a different way, yet still preserve all the actual splits (**Figure 7.33**, p. 248). Try picking the tree up at a different point – say, at the edge marked 6 – and letting the tips hang down again. The order of nodes has changed but the splits are all still the same.

We can also change the order of the populations along the tips of the tree and still preserve the branching order (**Figure 7.34**, p. 248). Imagine that we let each node in our phylogeny mobile swing around. We will change the order of the labels along the bottom, but the branching order will be the same.

The information provided by the splits is the same in all these trees. You can check this by starting at a population at the tip of the tree and asking which nodes you

Figure 7.33

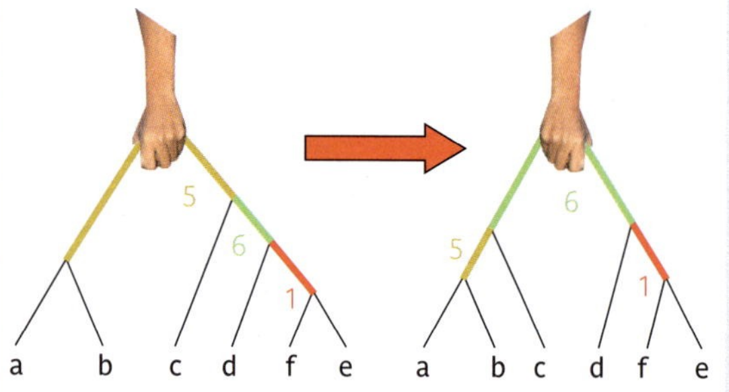

travel through to get to the root. Try starting at f. The first node you come to is the split between e and f, then the next node splits the e–f lineage from d, then the next node splits e, f and d from a, b, and c, and then you reach the root.

Here (Figure 7.35) are three more alternative ways of displaying the information in Figure 7.17, p. 244.

Satisfy yourself that each one of these trees carries exactly the same information about branching order by starting at any tip and following the lineage through to the root, checking that you get the same series of splits in each of the three trees. Remember that you can think of picking the trees up by the root and letting the tips hang down, so that the nodes can spin around. If you let each of these trees spin from their roots then you will get the same order of nodes, even when the order of the tips changes.

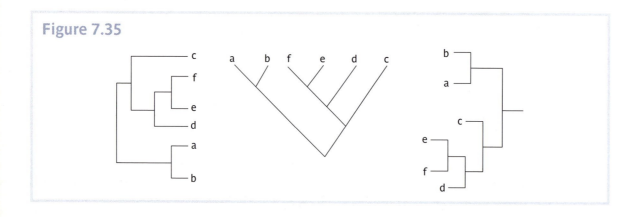

Figure 7.34

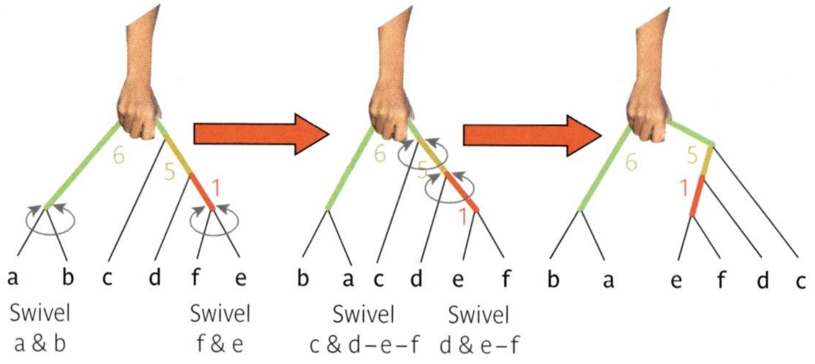

Figure 7.35

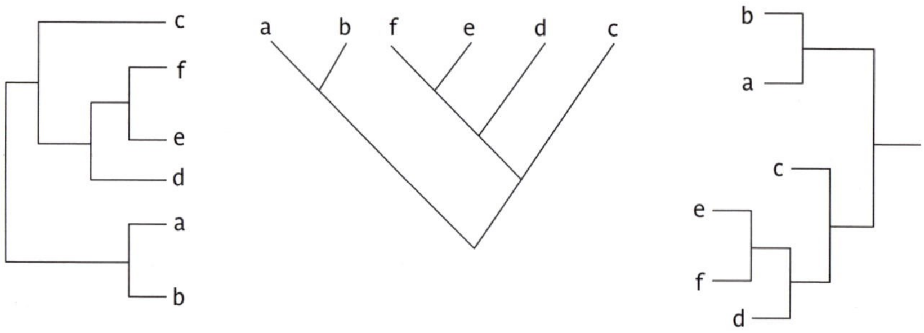

Trees that reflect similarities

We have shown how the pattern of substitutions in an alignment of homologous sequences can be used to reconstruct the historical process that created the sequences. This was easy to do with such a small alignment: only six sequences with six aligned sites. But most molecular phylogenies are based on sequences with hundreds or thousands of aligned sites. There are two key elements that make it possible to infer phylogenies from much larger datasets: automation and statistics. We will start by looking at automation. The rise of molecular phylogenetics coincided with the rise of computers (see **Heroes 7**, p. 252), and it is difficult to see how molecular phylogenetic inference could have developed without computing.

Computers are very good at solving problems involving tedious, repetitive tasks: they have good memories and they don't get bored. But computers cannot think for themselves. For a computer to solve a problem, it needs to be given an unambiguous set of instructions that break the task down into a series of relatively simple steps. If you can describe the process of constructing phylogenies as a series of computational steps (an algorithm), then you can get a computer to do it. An algorithm is a set of instructions that can be followed to solve a problem. We use a range of types of algorithms in our lives, for example when you follow the set of instructions that make up a recipe. Rather than considering any particular method, we are going to consider a few fundamentally different approaches to solving phylogenies using algorithms.

Distance methods

The simplest approach to estimating phylogenies from DNA sequence data is to start with the proposition that the process of descent leaves a hierarchy of similarities, so similarities tell us about patterns of descent. If we compare DNA sequences sampled from different populations, then we expect sequences sampled from more closely related populations to be more similar to each other than they are to sequences from more distantly related populations (Chapter 6). We can use these measures of differences between sequences to draw a phylogenetic tree by clustering sequences together in the order of least different (therefore probably most closely related) to most different (therefore probably most distantly related). Phylogenetic methods that use measures of difference between sequences to draw evolutionary trees are generally known as distance methods (**TechBox 7.1**, p. 253).

Let's go back to our hypothetical onychophorans, but this time we will take a slightly longer alignment (you can see that the first six sites are the same ones as the alignment in Figure 7.19, p. 245).

Figure 7.36

```
a  ATGACAATATGACAGACA
b  ATCACAATATGACAGACA
c  AAAAGAACAAAAGAATGA
d  ATAAGTACATAAGTAAGT
e  TTTAGTACATAAGTAAGT
f  TCAAGTACATAAGTAAGT
```

Count how many sites differ between a and b. There is just one difference between a and b out of 18 sites in the alignment: a has a G at the third position whereas b has a C. Now compare a to c: 10 differences out of 18. So on this information alone we would guess that a is more closely related to b than it is to c. The assumption we are making is that a is more similar to b than to c because a and b share a more recent common ancestor.

To automate the process of using measures of sequence similarity to draw a tree, we need an algorithm describing a simple, ordered set of instructions. Our first set of instructions must be to estimate the genetic distance between each of the sequences in our alignment, and record these distances in a useful format. Start by comparing a with every other sequence in the alignment, and recording the proportion of sites at which they differ. First compare a to a (obviously a is identical to a so it differs at 0 sites), then a to b (different at 1 out of 18 sites, which expressed as a proportion is 0.06), then a to c (10 out of 18, or 0.55), and so on.

Figure 7.37

	a	b	c	d	e	f
a	0	0.06	0.55	0.56	0.61	0.66

Repeat the process for sequence b. We have already compared a and b (distance = 0.06), so we don't need to do it again. So now we record the proportional difference between sequence b and the other sequences.

Figure 7.38

	a	b	c	d	e	f
b	–	0	0.55	0.61	0.61	0.67

We can do the same for every sequence in the alignment, building up a matrix where each line represents the difference between one sequence and all the other sequences. This matrix is triangular because we don't need to record any comparisons twice. For example, by the time we get to the last row, we have already compared f to all other sequences and don't need to do it again.

Figure 7.39

	a	b	c	d	e	f
a	0	0.06	0.55	0.56	0.61	0.66
b		0	0.55	0.61	0.61	0.67
c			0	0.33	0.39	0.39
d				0	0.28	0.22
e					0	0.17
f						0

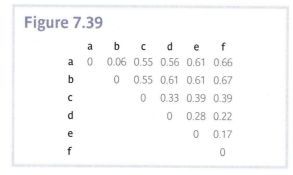

Now that we have measured and recorded genetic distances between our sequences, we need a set of instructions for turning this information into a tree. Most distance methods build a tree step by step, first finding the most similar sequences and clustering them together, then repeating the process until all the sequences are joined to each other. Look at the matrix (**Figure 7.39**). We can see that the most similar sequences are a and b because they differ at only 6% of sites. So they get joined together first.

Figure 7.40

The next most similar sequences are e and f, so we join those two together next.

Figure 7.41

The next most similar entries in the matrix are the difference between d and e (0.28) and d and f (0.22). So we can infer from that that the lineage that connects e and f is most closely related to lineage d.

Figure 7.42

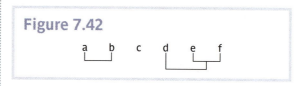

We continue to look for the next most similar lineage and group them together: c differs from d, e, and f by around 30%, so we assume it is the sister lineage to the d–e–f clade.

Figure 7.43

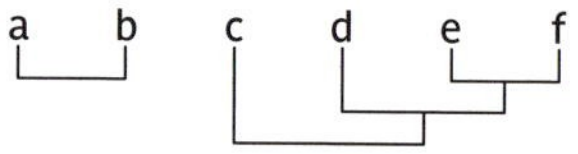

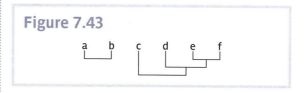

Then c, d, e, and f are all roughly the same distance from a and b, so we are placing the split between these two groups at the root of the tree.

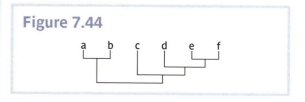

Figure 7.44

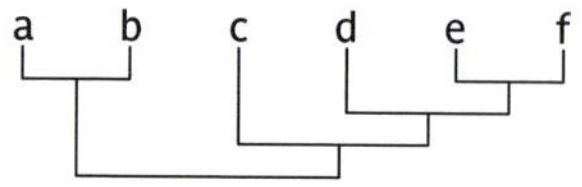

And now we have a tree! Compare this tree to **Figure 7.17** (p. 244) and you will see that we have correctly uncovered the evolutionary history of these populations using nothing more than the sequences we sampled at the end point of the process.

I'm sure you will appreciate that we have used a very simple example. Not only did we use short sequences for a small number of populations, but we even took a very simple approach to estimating phylogeny. In practice, most distance methods use more sophisticated means of estimating genetic distances, and more complicated ways of using those distances to build a tree (see **TechBox 7.1**, p. 252). But the basic approach we have taken illustrates the essential elements of most phylogenetics programs based on genetic distance: begin with a set of aligned sequences, generate a distance matrix that represents the genetic difference between the sequences, then use these distances to progressively cluster the sequences into a phylogeny.

Distance methods are usually relatively simple and fast and they will, in many cases, correctly infer the evolutionary history of a set of sequences (**TechBox 7.1**). However, there are two main problems with distance methods. The first problem is a practical one: they tend to return an incorrect phylogeny under several common scenarios (for example when rates of molecular evolution vary between lineages; see **TechBox 7.1**). The second problem is more philosophical: distance methods are essentially non-evolutionary. A distance tree is just a way of displaying information about similarities and differences. It may reflect evolutionary relationships, because descent with modification tends to leave a hierarchical pattern of differences. But just because we can draw a tree from a distance matrix does not mean we have necessarily uncovered the true evolutionary history of those sequences. In fact, you can take any set of distances and display them as a tree. You could stop half a dozen shoppers as they left the supermarket and compare the contents of their shopping trolleys, use this information to create a measure of difference (say, one minus the ratio of the number of shared items to the total number of items compared), then draw a tree that shows which shoppers bought the most similar groceries. This would in no way imply that the reason two shoppers had similar trolley contents was because they were somehow more closely related to each other.

Distance methods describe patterns of similarity in the molecular data. In this sense, distance methods are similar to the early classification systems, such as those of Ray (**Figure 7.2**, p. 226) or Linnaeus (**TechBox 6.1**). These pre-evolutionary schemes described the hierarchical pattern of similarities between species without giving an explanation of why biological diversity was organized that way. These classification systems work well in cases where similarity is a good indicator or evolutionary history. But, just as similarity between morphological characteristics is not always a sign of shared ancestry (such as the thylacine and the wolf in Chapter 6), similarity between DNA sequences is not always a reliable indicator of evolutionary history. We will now discuss several alternative approaches to estimating phylogenies from molecular data. These approaches, including parsimony, maximum likelihood, and Bayesian inference, aim not just to describe but also to explain the patterns in the data. Just as Darwin's tree depicted the process of diversification, these methods use the observable molecular data to reconstruct the path of evolution.

HEROES OF THE GENETIC REVOLUTION

7

Joe Felsenstein

"He has indeed promoted the conceptual unification of biological science through his efforts to develop and use theoretical tools in evolutionary biology."

M. Slatkin (1995). Sewall Wright award: Joseph Felsenstein. *American Naturalist*

NAME

Joseph Felsenstein

BORN

9th May 1942, Philadelphia, Pennsylvania, USA

CURRENT POSITION

Professor, Departments of Genome Science/Biology, University of Washington, Seattle, USA

KEY PUBLICATIONS

Felsenstein, J. (1981). Evolutionary trees from DNA sequences: a maximum likelihood approach. *Journal of Molecular Evolution*, Volume 17, pages 368–376.

Felsenstein, J. (2004). *Inferring Phylogenies*. Sunderland, Massachusetts, Sinauer Associates.

FURTHER INFORMATION

Interview with Joe Felsenstein on *www.blindscientist.genedrift.org/2007/04/19/sciview-scientific-interviews-part-1/* (SciView: scientific interviews, part 1 April 19th, 2007).

Figure Hero 7

Reproduced courtesy of Joe Felsenstein.

Joe Felsenstein has been a key player in setting problem-solving in molecular phylogenetics in a robust statistical framework. He began developing these tools when the sequence databases were beginning to get larger and computers were starting to get smaller (early versions of PHYLIP were developed on pioneering microcomputers developed by his younger brother Lee Felsenstein). As the number of DNA sequences exploded, and molecular phylogenies have invaded every aspect of biology, Felsenstein's statistical tools provided a robust statistical framework for producing and interpreting phylogenies.

Felsenstein completed his PhD in 1968 with the great evolutionary biologist Richard Lewontin at the University of Chicago. After a short postdoctoral position in Edinburgh, he joined the University of Washington in Seattle, where he has been ever since. Although he is best known for his work on the statistical inference of phylogeny, Joe Felsenstein has published in very wide range of topics. His early papers were in the field of theoretical population genetics, covering topics as diverse as mutation accumulation, speciation and the rate of loss of genetic variation from populations, but later he also examined models of ecological diversity and macroevolutionary change.

Felsenstein's contributions to molecular phylogenetics have defined the field. He introduced statistical techniques that have now become standard, such as bootstrapping sequence data (**TechBox 7.4**). In particular, he developed and promoted the use of maximum likelihood (ML) (**TechBox 7.2**), developing algorithms that made ML tractable for phylogeny estimation. Felsenstein has also played a key role in testing the performance and reliability of phylogenetic methods, demonstrating the conditions under which particular methods will return the wrong tree. For example, he showed that parsimony will be consistently misleading if evolutionary rates vary widely between lineages, a set of conditions now known as the 'Felsenstein Zone' (which he has commented is like having a black hole named after him). Importantly, he did not just demonstrate the logic of applying these statistical tools to phylogenetics, he also developed ways of making them easily applicable to biological data. He created one of the first really useful phylogenetics packages, PHYLIP, which remains one of the most widely distributed phylogeny programs, with over 20,000 registered users. Felsenstein has championed accessibility of both the theory and the tools of molecular phylogenetics. PHYLIP is freely available as long as it is not used for commercial gain (though, during apartheid, Felsenstein would not permit PHYLIP to be distributed to South Africa).

As you will no doubt discover, much of the molecular phylogenetics literature is rather dull. And one of the things that makes it dull is the lack of a personal voice: most papers are written in an anonymous style that completely obscures the person behind the paper.

One of the things I like about Joe Felsenstein's writing is that it breaks this mould: in each paper, or book, or website, it feels very much like Joe is speaking directly to the reader. His commonsense approach has been summarized in 'Felsenstein's Law': anything that is true is obvious. This approach is also evident in his long-standing war against cladistic systematists (Felsenstein claims to have 'founded the fourth great school of classification, the It-Doesn't-Matter-Very-Much school.') His magnum opus, *Inferring Phylogenies*, has been described as 'truly majestic', 'an outstanding achievement', and 'a classic by the time it was published'. Felsenstein was pleased to see a reviewer draw a parallel to the classic text on theoretical population genetics by Crow and Kimura (see **Heroes 6**): 'I modelled my level of presentation and writing style in great measure on theirs, and hoped to have an effect similar to theirs. It is gratifying to see that connection made.'

TECHBOX 7.1 Distance methods

KEYWORDS

immunological distance
DNA hybridization
neighbour joining
UPGMA
cluster
minimum evolution
multiple hits
ultrametric

FURTHER INFORMATION

Page, R.D.M. and Holmes, E.C. (1998) *Molecular Evolution: a phylogenetic approach*. Blackwell.

RELATED TECHBOXES

TB 8.1: Estimating branch length
TB 8.3: Detecting rate variation

RELATED CASE STUDIES

CS 1.2: Whale meat (DNA surveillance)
CS 4.1: Glorious mud (DNA hybridization)

Distance methods begin with a matrix of the genetic distances between a set of sequences, then find the tree that best describes the genetic distance data[1]. There are two separate steps in this process: first, estimating distances and, second, describing these distances as a branching diagram. It's important to remember that these two steps are essentially separate from each other. There are many ways to generate a matrix of genetic distances, but once you have that matrix, you can apply any means of deriving a tree from that distance data.

Distance data

Some types of molecular data begin as a matrix of distances. For example, in **Case Study 4.1**, genetic distance was estimated by measuring the binding strength between hybridized DNA of different strains of soil bacteria. The following phylogeny of horses is based on immunological distances (**Figure TB7.1a**; **Table TB7.1**). The authors created antibodies that would bind specifically to proteins from nine different horse species (bizarrely as it may seem in this postgenomic age, these antibodies were created by injecting horse proteins into rabbits)[2]. Antibodies are specific to particular antigens, so they will bind most strongly to a protein that is identical to the proteins they were created against. Strength of binding is therefore a measure of protein similarity. Here, they generated a distance matrix by testing each of the horse proteins against each of the antihorse rabbit antibodies. They also placed the extinct quagga (**TechBox 3.4**) in this phylogeny by testing ground up quagga skin against the nine different antihorse antibodies. This may seem to be a very involved way to produce a molecular phylogeny (and not much fun for the rabbits), but it is worth noting that the results are similar to those produced by more recent phylogenies using mitochondrial DNA sequences from several quagga specimens[3].

For both DNA hybridization and immunological distance, the primary data are distances. But these techniques are rarely used today. Nowadays, the most common approach is to estimate genetic distance from a sequence alignment. The simplest way to estimate genetic distance between two sequences is to count the number of positions in the

Table TB7.1 Correlation coefficients (*r*) of immune reactions of quagga and extant zebras

	Plains skin	Quagga skin	Quagga distance (*D*) D = 100 (1–*r*)
Plains zebra skin		0.75	25
Plains zebra serum	0.95	0.72	28
Mountain zebra serum	–0.92	–0.40	140
Grevy's zebra serum	–0.96	–0.68	168

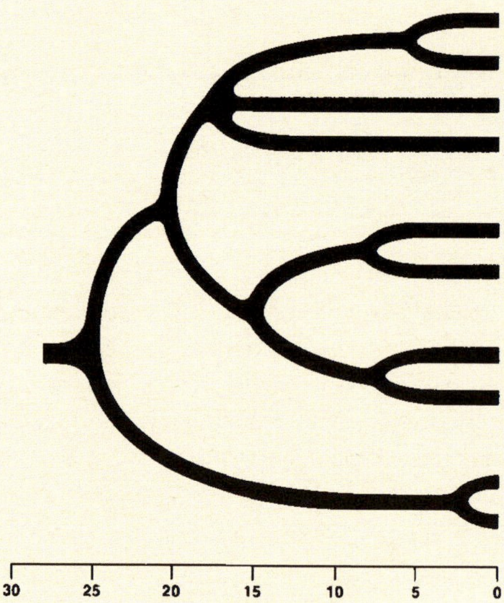

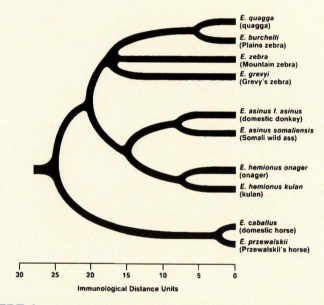

Figure TB7.1a Phylogeny of genus *Equus* based on immunological distances (see **Table TB7.1**).

alignment where they differ. But when we build phylogenetic trees that represent evolutionary history, we are interested in distances as a representation of the amount of evolutionary change that has occurred on these lineages since they split from each other. For sequence data, we may miss some of those changes when we do a count of differences due to the problem of 'multiple hits'. If a single site changes multiple times, we can only observe a single difference. There are a number of corrections that aim to estimate the number of substitutions that have occurred from the observed differences (**TechBox 8.1**). The key point for this box is that, however the genetic distances are estimated, distance-based phylogenetic methods replace the original data (whether biochemical reactions or alignments of sequences) with a distance matrix describing those data.

Clustering algorithms

Here is a very general outline of a common, distance-based approach to phylogenetics. Note that this algorithm does not describe every distance method, but is intended only to give an overview of a basic approach.

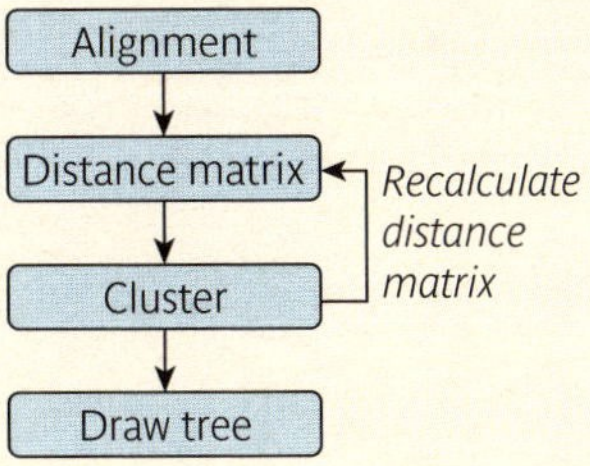

1. **Alignment:** construct an alignment of your sequences so that you know you are comparing homologous sites (**TechBox 6.3**).

2. **Distance matrix:** count the number of differences between each sequence and every other sequence, and convert these count data to a distance measure – this may be as simple as calculating the proportion of sites at which two sequences differ, or it may involve a more sophisticated correction for multiple hits (**TechBox 8.1**).

3. **Cluster:** use distances to progressively join together the sequences with the smallest distance between them. Here are two common clustering methods:
   - UPGMA: the most similar sequences are clustered, then this cluster is treated as a single lineage in each subsequent clustering step. The distance matrix is recalculated after each step to give the average distance between each sequence and this new, conjoined pair of sequences, then the smallest distance in this new matrix is used to cluster again, and so on. This algorithm assumes that the rate of change to be equal in all lineages (under this assumption, distances are said to be additive, and the tree is ultrametric).
   - Neighbour joining (NJ): like UPGMA, similar sequences are clustered then replaced by a single node in the distance matrix. However, NJ does not assume absolute rate constancy. The distance between each sequence and all other sequences is used to come up with a 'corrected' (average) distance.

4. **Draw tree:** the branching diagram in **Figure 7.44** (p. 250) indicates which lineages are more similar to which others. We could also have drawn the branches so that they were proportional to the amount of difference between the sequences. For example, we could draw an ultrametric tree displaying average distances, so that each path from tip (sequence) to root (basal split in the data) was the same (in this ultrametric tree, each tip is 0.6 from the root).

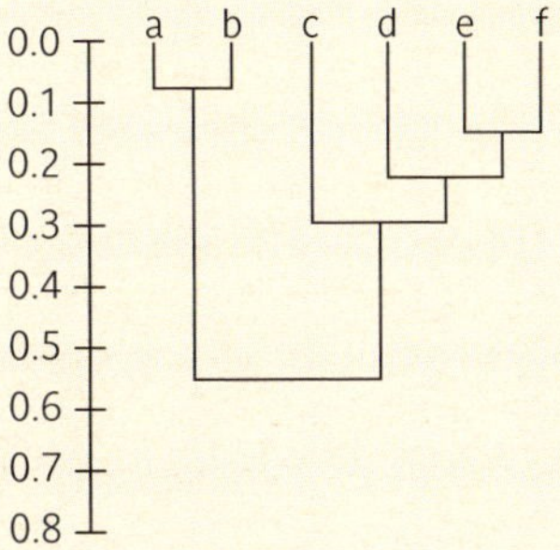

Or we could have shown the relative amounts of change on each lineage. For example you can see from the distance matrix in **Figure 7.39** that the distance between d and e is greater than between d and f, implying that there has been a faster rate of change in the lineage leading to e (**TechBox 8.3**). So some sequences are separated from the root by more changes than others. We draw this as uneven heights of the tips of the tree. This should not be confused with a statement about time: in this case, all tips are sampled in the present, so f is not older than e just because it is lower on the tree (see **TechBox 8.1**).

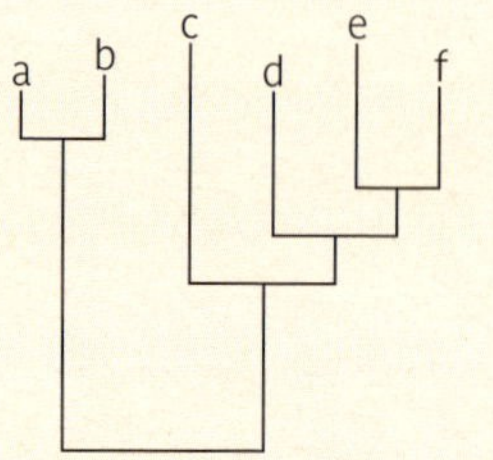

Minimum evolution

There is an alternative way of deriving phylogenetic trees from distance data that does not rely on stepwise clustering and readjustment of the distance matrix. Like clustering methods, minimum evolution (ME) starts with a distance matrix. However, instead of grouping tips two at a time, ME essentially sums the distances along the branches of each possible phylogeny, then selects the phylogeny with the lowest total distance (otherwise known as the shortest tree). Because it involves calculating total branch-lengths for every possible tree, ME is generally slower than clustering methods.

Advantages and disadvantages

Distance methods are often described as being 'quick and dirty'. Most phylogenetics programs will return a distance tree almost immediately, whereas you may have to wait hours, days, weeks, or months for other methods. It is common practice to use a distance-based analysis to give you a good starting point for your investigation. For example, you might estimate a NJ tree to see whether the patterns of similarity in the data are as you expected, or whether there are certain lineages for which the results seem counter-intuitive and therefore might require more investigation. Indeed, many other phylogenetic methods use a neighbour joining tree as a starting point for searching tree space. But, increasingly, distance trees are viewed as being poor cousins to the optimality-based phylogenetic methods, such as maximum likelihood (**TechBox 7.2**) and Bayesian methods (**TechBox 7.3**), because distance methods are prone to return an incorrect tree for some data sets, most importantly when rates of change are not the same in all lineages.

References

1. Felsenstein, J. (1984) Distance methods for inferring phylogenies: a justification. *Evolution*, Volume 38, pages 16–24.
2. Lowenstein, J.M. and Ryder, O.A. (1985) Immunological systematics of the extinct quagga (Equidae). *Cellular and Molecular Life Sciences*, Volume 41, pages 1192–1193.
3. Leonard, J.A., Rohland, N., Glaberman, S., Fleischer, R.C., Caccone, A. and Hofreiter, M. (2005) A rapid loss of stripes: the evolutionary history of the extinct quagga. *Biology Letters*, Volume 1, pages 291–295.
4. Sibley, C.G. and Alquist, J.E. (1990) *Phylogeny and Classification of Birds: a study in molecular evolution*. Yale University Press.

Figure TB7.1b Birds from the Girona *Tapestry of Creation*. One of the first large molecular phylogenies ever constructed was a DNA hybridization tree of all bird families[4]. This tree was known in phylogenetic circles as 'the tapestry' because, when it was first displayed at a scientific conference, it took up a surprisingly large amount of wall area.

Trees as evolutionary paths

One way to describe the process of evolutionary change that gave rise to the observed sequences is to map substitution events onto a phylogenetic tree. This is essentially what we did in **Figures 7.24** to **7.29** (p. 246): we interpreted each split in the alignment as a substitution event that had occurred in one particular lineage, then was inherited by its descendant lineages. We could map these substitution events onto particular lineages (edges) in the phylogeny.

Figure 7.45

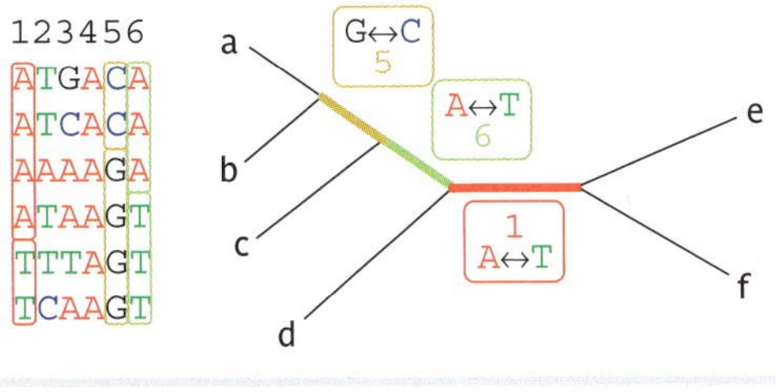

This is a very different approach to the distance-based methods. Firstly, we are not simply describing the end points of the process (which sequences are most similar?), we are aiming to reconstruct the underlying process itself (what series of substitutions created these sequences?). Because of this fundamental difference – process instead of pattern – the approach to solving phylogenies is very different. Firstly, for the distance methods there is (usually) a single tree that describes the pattern of similarities. In contrast, there are usually many alternative ways to create a tree showing a hypothetical series of changes that took place to create the sequences. Secondly, because a distance tree essentially describes the patterns we see, we don't need to know how those changes occurred. But creating a phylogeny that maps evolutionary changes requires some kind of statement about the way DNA sequences evolve. So, to describe our DNA sequences using a phylogeny that maps evolutionary change, we are going to need a way of creating and evaluating different trees, and to do that we will need to have a model of how DNA sequences evolve.

Remember that we don't want to have to do all the work ourselves, we want a computer to do it. The only way a computer can solve this kind of problem is if we give it a set of instructions. We tell the computer to generate one possible phylogeny, then use some kind of rule to score how plausible that phylogeny is as an explanation of our data, then generate a different phylogeny, score that one for plausibility, and keep going. Then, once all possible trees have been compared, the computer can report the tree that got the highest plausibility score. This sounds pretty straightforward, but we need to tell the computer how to do both parts of the process: generate a phylogeny and score it for plausibility.

If this is sounding familiar, you may recall that we described a similar process for telling a computer how to find the best alignment for a set of sequences (Chapter 6): generate an alignment, score it, change it slightly, score it again, and keep going until you have found the alignment with the best score. There are over 10^{100} possible alignments of two 300-bp sequences (**TechBox 6.3**), so any useable computer program needs to have an efficient strategy for searching for the alignment that best describes the data. Similarly, there are a very large number of possible phylogenies. For the six species we have, there are 945 possible rooted trees, or 105 possible unrooted tree. For small numbers of taxa, a computer can do an exhaustive search, scoring every possible tree. But for most datasets, there are so many possible trees that even the fastest computer is not going to be able to evaluate all of them. There are over eight thousand trillion possible rooted trees for 20 sequences, and there are more than an octodecillion possible rooted trees for an alignment of 40 sequences (an octodecillion is a 1 followed by 57 zeros). So, in general, we ask the computer to do a heuristic search, adopting a particular way of generating and testing trees that is likely to find the best tree without having to try every single possible tree.

One such search strategy can be summarized like this: start with a distance-based tree, cut the tree at a randomly chosen edge, then reattach that edge somewhere else in the tree; if the rearranged tree scores better than the starting tree, keep it and use it as the basis of more rearrangements, if it scores lower than the original tree, abandon it and go back to the first tree; keep repeating until you find that no further rearrangements can improve the score of your tree, which you then accept as the best tree for this data. Of course, if your search method does not test every single possible tree, you can't guarantee that one of the untested trees wouldn't have had a higher score. But in practice these algorithms work very well. However, they are slow. This searching process – modifying a tree, comparing it, modifying it again – is the most time-consuming part of the computational approach to phylogenies. The more taxa you have in your tree, the more possible trees there are and the longer it will take to find the best one.

Evaluating alternative trees

> "*A wise man, therefore, proportions his belief to the evidence. . . . All probability, then, supposes an opposition of experiments and observations, where the one side is found to overbalance the other and to produce a degree of evidence, proportioned to the superiority.*"
>
> David Hume (1748) Of Miracles. In *An Enquiry Concerning Human Understanding*

How are we going to score alternative trees in order to decide which is the best? The tree we created in **Figure 7.29** (p. 246) provided a reasonable explanation of the pattern of substitutions seen in the alignment (**Figure 7.19**, p. 245). But it is not the only possible explanation of that pattern of substitutions. In other words, there are other possible trees that could also give rise to the same set of sequences. To illustrate this, I generated an alternative tree by randomly pulling off some branches and reconnecting them elsewhere. I disconnected f from its position next to e and I reconnected it next to b, then I disconnected c and stuck it next to a, and so on.

This is an alternative evolutionary history for our sequences (**Figure 7.46**, p. 259). In other words, the first

Figure 7.46

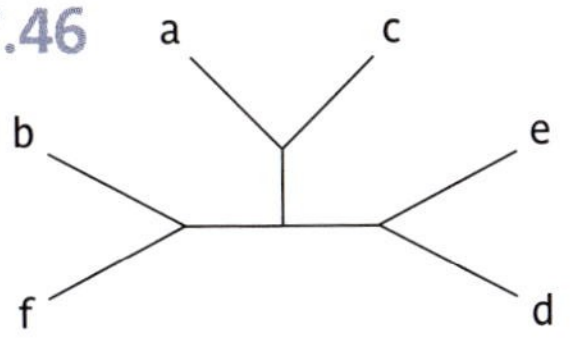

tree was one way that our present day sequences could have been generated, and this is another way. To see how this path would lead to the sequences we see today, we need to map substitution events onto the tree. So, to explain the pattern of nucleotides we see in site 1, we need to show how a, b, c and d all ended up with an A, but e and f ended up with a T. On this tree, e and f aren't grouped together, so we can't say that the reason that they both have a T is that they both inherited it from a common ancestor (unless it was then subsequently lost three separate times, in the b, d and a–c lineages). Instead, we might say that e and f both have a T in position 1 because they both independently acquired a substitution from A to T.

Figure 7.47

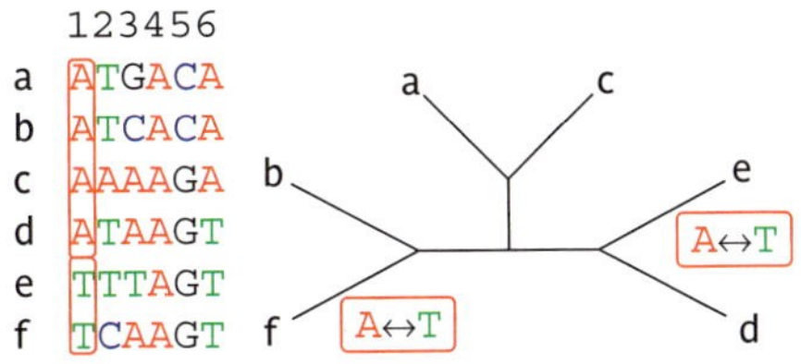

This tree requires two A to T substitutions to explain site 1, whereas the previous tree required only one. We can map the other substitutions on in the same way. For example, a, b, c, and d are not grouped together, so we can't simply say that the ancestor to all four populations underwent a substitution from T to A, so we have to infer multiple substitutions from T to A (Figure 7.48).

Figure 7.48

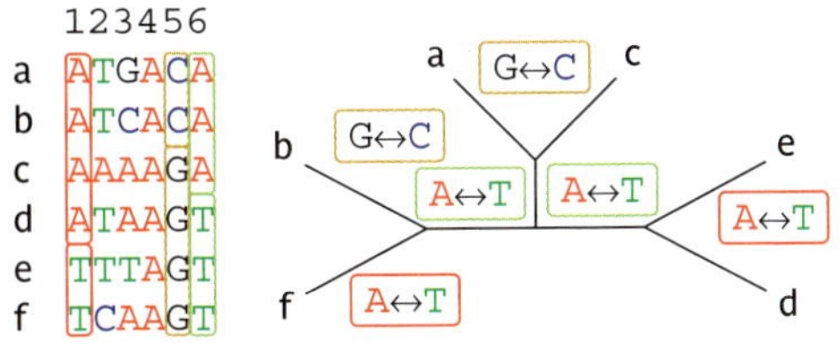

We now have a way of scoring and comparing our alternative trees. We can map substitutions onto our alternative trees, then ask 'how many substitutions does each tree imply happened during the history of these sequences'? We don't need to bother mapping on the shared ancestral characters or the unique derived characters because these will be the same in all possible trees, so we only map on the shared derived characters. Here is a comparison of the number of substitutions we need to map on to our two alternative evolutionary hypotheses, Tree 1 (Figure 7.45, p. 257) and Tree 2 (Figure 7.46), in order to explain the sequence data we observe (Figure 7.19, p. 245; Figure 7.49).

Figure 7.49

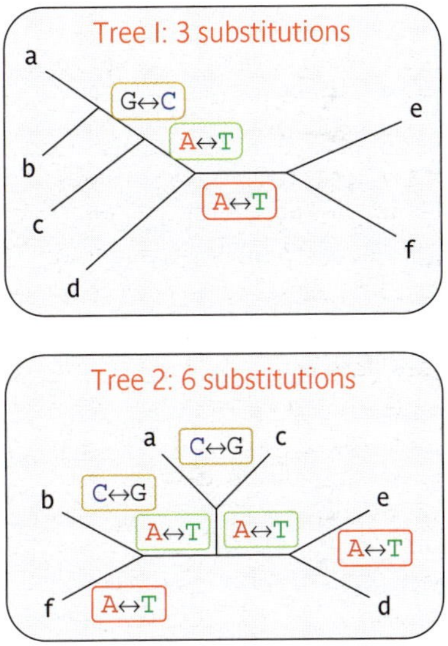

Tree 2 requires that we assume more substitutions events than Tree 1. Which one of these trees do you feel gives the most plausible explanation of the history of these sequences? Given no other information, you might decide that Tree 1 is the most plausible, because it requires us to assume fewer independent evolutionary changes. This is how most phylogenetic methods work: in order to explain our observations (DNA sequences) we generate different possible hypotheses (trees) and we compare their plausibility.

Parsimony

We have just employed the principle of parsimony to decide which tree is more believable). We chose the

TECHBOX 7.2

Maximum likelihood

KEYWORDS

branch lengths

model

parameters

optimization

likelihood ratio test

FURTHER INFORMATION

There is an excellent introduction to maximum likelihood estimation (as a general procedure, rather than specifically for phylogenetics), that requires no previous knowledge of statistics, written by Shaun Purcell and available on *http://statgen.iop.kcl.ac.uk/bgim/mle/sslike_1.html*.

RELATED TECHBOXES

TB 7.4: Bootstrapping

TB 8.1: Estimating branch lengths

RELATED CASE STUDIES

CS 7.1: Up the river (viral phylogenies)

CS 7.2: Non-tree-like flowers (horizontal gene transfer)

In Chapter 5, we saw how R.A. Fisher was a key figure in the development of mathematical approaches to evolutionary genetics (**Figure 5.29**). In fact, Fisher's contributions to statistics were as great as his contributions to biology, and statisticians are sometimes surprised to hear Fisher referred to as a geneticist. Amongst his many achievements, Fisher was responsible for the development of the maximum likelihood (ML) approach, which was later adapted for estimating phylogenies from DNA sequences by Joe Felsenstein (**Heroes 7**).

When we estimate phylogeny from DNA sequences, we start with a set of sequences observed in the present day and ask what is the most probable series of substitutions that happened in the past to produce these sequences. To do this, we need a model of molecular evolution that states the probability of different kinds of substitution events occurring. The parameters that will be included in the model are decided before the analysis commences, but the optimum values of these parameters for this particular dataset are found during the procedure. The aim of a maximum likelihood analysis is to find the set of parameter values that maximizes the probability of the data, given the model you have applied. In phylogenetics, the most important parameter that we vary is the tree itself: we try different topologies (branching orders) and edge lengths (branch lengths) until we find the tree that is the most likely explanation for the sequence data we observe. The model may also allow different kinds of substitutions to have different probabilities of occurring. Since the value of these parameters will influence the likelihood score we calculate for any given tree, we can also optimize these parameters by finding the values that give us the tree with the highest likelihood.

The procedure followed during a maximum likelihood analysis is something like this:

1. **Alignment:** This is probably the most critical step, yet is often given relatively little attention (see **TechBox 6.3**). The outcome of the analysis depends wholly on whether the sequences are accurately aligned such that homologous sites can be compared.

2. **Starting tree:** this could be a randomly selected tree, but a common way to generate a starting tree is to construct a distance phylogeny from the alignment (**TechBox 7.1**).

3. **Optimize likelihood:** Once you have a tree that you wish to estimate the likelihood of, the observed data are mapped onto the tips of that tree, then the parameter values of the model are varied until the values that maximize the likelihood of that tree are found. We can break down this procedure to estimate the likelihood of a given tree as follows:

   (i) For each site: For each column (site) in the alignment, the nucleotide states are mapped onto the tips of the trees. These are the observed end states of a process, but there are many possible sets of substitutions along the tree that could have generated that particular pattern of nucleotides at the tips of the tree. Using the substitution probabilities given in the model, the site likelihood is calculated by summing the probabilities of each of the proposed substitution events, starting at the tips of the tree and working down through the shared nodes until you reach the root.

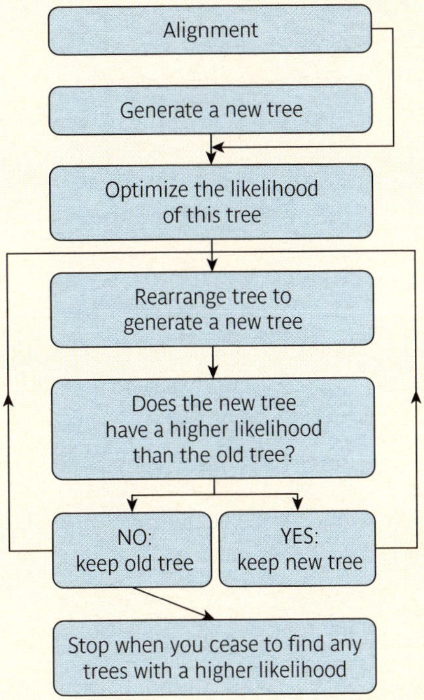

(ii) Over all sites: On the assumption that each site evolves independently, the probability of this tree producing the whole alignment is the product of all of the site likelihoods. This gives one likelihood score for this particular tree given this particular model and a particular set of parameter values. Now it's time to fiddle the parameter values to find their optimum value for this tree.

(iii) Over all parameter values: The value of a free parameter is changed slightly, and the likelihood of the tree recalculated (as in steps i and ii). Remember that the edge lengths of the tree are free parameters, so the change may be to slightly shorten or lengthen one of the branches. Or it may be a change to one of the substitution probability parameters, such as the transition/transversion ratio. Iterate, varying parameter values and calculating the likelihood over and over again. The parameter values giving the highest likelihood are retained.

4. **Rearrange the tree** to generate a new tree. For example, you might cut the tree at one edge and rejoin it somewhere else.

5. **Compare the new tree to the old tree:** Calculate the likelihood of the new tree using the procedure 3 above. If the new tree has a higher likelihood than the old tree, keep it in memory. If the new tree has a lower likelihood than the old tree, forget it, and keep the old tree in memory. So at any given point, the tree you keep in memory is the best one you have found so far.

6. **Rearrange the tree again** and repeat steps 3 to 5. Keep going until you never find a new tree with a higher likelihood than the one you hold in memory. At this point, accept the one in memory as the best tree for this data given the assumptions of the model.

Advantages of maximum likelihood

The reliance on an explicit model of molecular evolution is a strength of maximum likelihood because it allows maximum likelihood to be adapted to different situations and datasets. It also makes it quite clear that the results you get are conditional on the model you use, hopefully prompting you to test the robustness of your conclusion with different models. One way of doing this is to use a Likelihood Ratio Test, which compares the likelihood of the data given two different models, one of which has one (or more) extra parameter. The simpler model is rejected only if the more parameter-rich model is found to be significantly better fit to the data. In fact, likelihood gives a formal statistical basis for testing all aspects of the phylogeny estimation, including comparing specific phylogenetic hypotheses (see **TechBox 7.4**, p. 272).

Disadvantages of maximum likelihood

This iterative procedure can take a long time. The more sequences you have, the more nodes in the tree over which you must calculate site likelihoods. The longer your alignment, the more site likelihoods you must calculate. The more free parameters in your model, the more times you must recalculate the joint likelihood for each tree topology for each parameter value. And since the edge lengths of the tree are also free parameters, the more sequences you have the greater the number of edges in your tree that must be varied in length during your likelihood optimization. So maximum likelihood estimation of phylogenies very easily turns into a computational marathon. The 'hill-climbing' heuristic search method means that you only ever move 'upwards' to a tree with a better likelihood, so it is possible to miss out on the best tree by getting stuck on a local optimum that cannot be improved in a single step. Maximum likelihood is a robust method that has been shown to perform well under many circumstances. However, there are also situations in which maximum likelihood is consistently misleading, so as with all phylogenetic estimation methods, caution should always be exercised when interpreting results.

explanation that required us to infer the fewest number of evolutionary events. A more general form of parsimony is often referred to as Ockham's razor, regarded by many as a key principle of scientific enquiry: when assessing competing explanations for a phenomenon, go with the explanation that requires the fewest number of *ad hoc* assumptions, until such time as it is proven false. *Ad hoc* assumptions are additional assumptions we must make in order to fit a hypothesis to an observation. The reason we might want to minimize the number of *ad hoc* assumptions we make is that they have no other role than to prop up a particular hypothesis, and have no additional supporting evidence. *Ad hoc* assumptions are the leaps of faith we must make in order to believe in a particular hypothesis. It's important to recognize that the principle of parsimony, or Ockham's razor, does not tell us which explanation is more likely to be true. Instead, it is a guide for our behaviour when comparing hypotheses: go with the hypothesis that requires the fewest *ad hoc* assumptions until it is proved false. This prevents us from inventing fancy underlying phenomena when we could just as easily explain what we have observed using mechanisms we already know about.

When we use the principle of parsimony to decide between alternative trees, we are stating that the tree with the fewest inferred substitutions is always the most probable. Parsimony is the leading way of estimating phylogenies from non-molecular data (e.g. **Figure 7.8**, p. 232). The central assumption of parsimony is that evolutionary change is rare enough for our best explanation of shared characters to be common ancestry, not independent acquisition. This may be true for many morphological data sets, particularly for complex traits that seem unlikely to have evolved multiple times. But is this assumption realistic for DNA sequence evolution? Do we expect substitutions to be so rare that a tree with fewer substitutions is always more plausible than one with more substitutions?

Multiple hits

The differences we can observe between aligned sequences may not be all of the substitutions that have occurred in these sequences. In Chapter 6, we discussed the special features of molecular evolution that make it, in some ways, more difficult to interpret than morphological evolution. One problem is that DNA only has four character states. If an A changes to a G then back to an A, we are not going to be able to tell whether we have a different A from the one we started with. An A is an A. And if the A changes to a G then a C, all we will have is the start point (we began with A) and the end point (we now see a C), and we won't know that there was a G in between.

Consider the case of the diverging onychophorans. Here is a simplified version of **Figure 7.16** (p. 244).

Figure 7.50

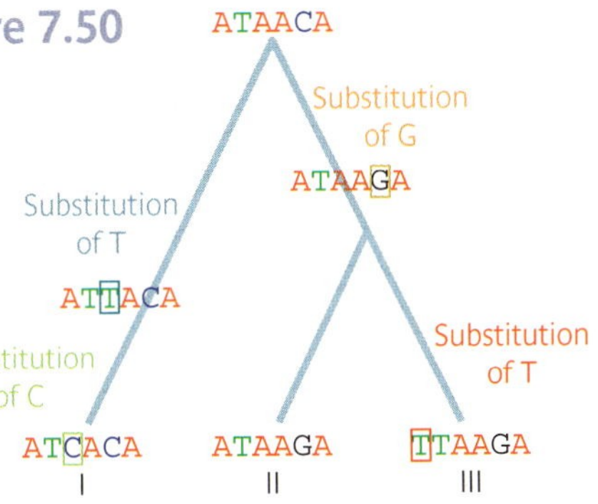

We can draw this phylogeny as an unrooted tree and count the number of substitutions that occurred.

Figure 7.51

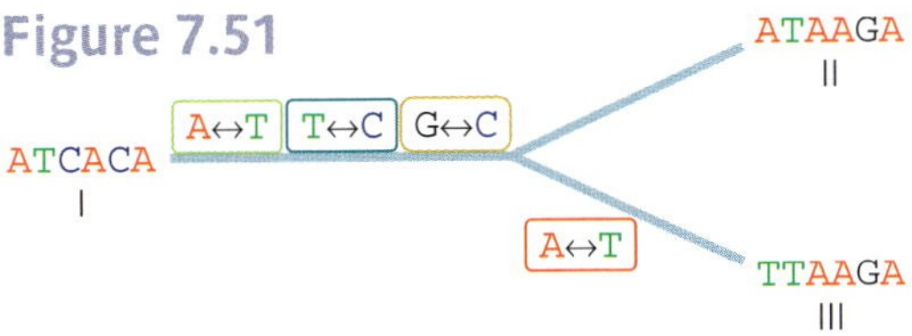

We can see that there have been three substitution events between sequence I and sequence II, and four substitutions between sequence I and sequence III. But if we were trying to reconstruct this phylogeny using only the sequences at the tips, we would miss one of these substitutions. Try constructing a distance matrix from these sequences (recording the absolute number of differences between sequences rather than proportions).

Figure 7.52

		I ATCACA	II ATAAGA	III TTAAGA
I	ATCACA	0	2	3
II	ATAAGA		0	1
III	TTAAGA			0

Our distance matrix suggests there have been only two substitutions between sequence I and II, but we know there have been three. Similarly, our distance matrix records three substitutions between I and II, but there have actually been four. How did we miss one? The same site changed more than once: the third site in the sequences changed from A to T to C. But all we have left to show for this chain of events in the end product, which is a sequence with a C. We would have the same end product whether there had been one change (from A to C) or six changes (from A to T to C to A to G to T to C). Substitutions that occur at the same site, overwriting previous changes, are referred to as 'multiple hits'.

We will return to the subject of multiple hits in Chapter 8

Parsimony specifically ignores multiple hits. If two sequences differ by one nucleotide, then parsimony states that the best explanation is that there has been a single substitution between them. But remember that the most parsimonious solution is not necessarily the true explanation. For molecular data, we recognize there might be extra substitutions that have been covered up by multiple hits. So, how can we account for changes we cannot directly observe? We need some way of saying 'I observe this pattern, but I believe that there may have been more events that I can't directly observe'. Obviously, we could let our imaginations run wild and put in lots of hypothetical changes. For example, we could have mapped the following changes onto our tree (**Figure 7.53**, p. 264).

This tree has exactly the same branching order as Tree 1 (**Figure 7.45**, p. 257), but it involves more substitutions, because many of the substitutions are multiple hits. In Tree 3, e and f both have a T at site 1, not because they inherited it from a common ancestor, but because they both acquired it independently (**Figure 7.54**, p. 264).

Figure 7.53

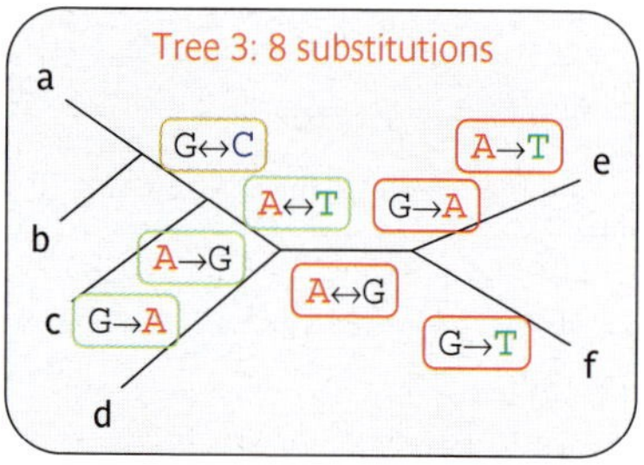

Figure 7.54

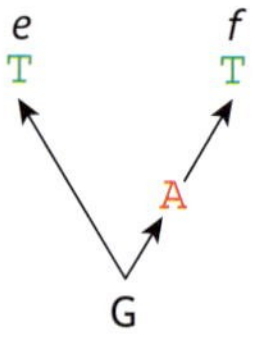

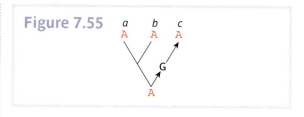

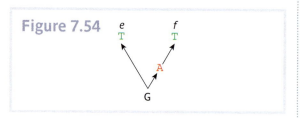

And although a, b, and c all have an A at position 5, this position underwent two substitution events on the lineage leading to c (**Figure 7.55**).

Figure 7.55

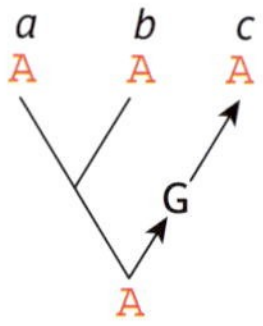

Tree 3 with eight substitutions may seem intuitively less likely than Tree 1 that only requires three substitutions. But Tree 3 is not impossible. Molecular data are such that we can well imagine some nucleotides have undergone a lot of substitution events even if they end up looking exactly the same. If all nucleotide changes are equally likely, then if a particular site changes base more than once, there is a one in four chance that it will return to its original base and therefore be undetectable.

So we have struck two problems in our quest to make an evolutionary tree that is a map of the substitution events. One is that multiple hits erase past substitutions, so we need a way of inferring how many unseen changes are likely to have occurred. And the other problem is that, as we can construct many different possible paths of substitutions that could have given rise to the sequences we observe, we need a way of deciding which possible histories of the data are the most plausible. Both of these problems can be tackled by taking a probabilistic approach to reconstructing phylogeny.

Statistical inference of phylogeny

The following discussion is a very general view of the statistical estimation of phylogeny. Although there are both practical and philosophical differences between the various statistical approaches to solving phylogeny (see **TechBoxes 7.3** and **7.4**, p. 265 and 272), we are going to gloss over those differences here because the essence of the problem is the same for all methods: given that we observe a certain set of sequences in the present day, how can we use what we know about molecular evolution to help us decide which evolutionary tree provides the most plausible version of the evolutionary history of the data?

We employ this kind of probability-based thinking in many ordinary situations. A classic example is the court case, where the judge or members of the jury are asked to decide which is the most likely scenario for past events given only their observations of the end points of this process, plus a set of beliefs about the probability of certain occurrences, and some information specific to this case. For example, they may need to ask themselves: 'given that a woman has been murdered, how likely is it that her husband killed her?'

The dead woman is an observation. It is a known fact that there has been a murder, because we have a corpse with stab wounds in it. What we don't know is the series of events leading to this observation of a murdered corpse. But we have two kinds of evidence we can bring to bear (ignoring, for the moment, other observations

TECHBOX 7.3 Bayesian phylogenetics

KEYWORDS

prior probability

posterior probability

MCMC

credible set

MAP

tree space

uninformative priors

FURTHER INFORMATION

MrBayes is a freely downloadable program for Bayesian Inference of phylogeny, complete with a useful manual and wiki FAQs: *http://mrbayes.scs.fsu.edu/*

RELATED TECHBOXES

TB 7.2: Maximum likelihood

TB 7.4: Bootstrapping

RELATED CASE STUDIES

CS 7.1: Up the river (phylogenies and the origin of HIV)

CS 7.2: Non-tree-like flowers (horizontal gene transfer)

Bayesian methods for inferring phylogenies are conceptually very similar to maximum likelihood (ML) methods (see **TechBox 7.2**, p. 260). In both cases, the analysis explores 'tree space' (the set of all possible trees), moving from one possible phylogeny to another by changing model parameters (e.g. branch lengths, substitution probabilities, tree topology), and calculating the likelihood of each tree. But Bayesian methods differ from maximum likelihood in the use of prior beliefs about the probability of different hypotheses, the method for exploring the set of possible trees, and the way that the plausibility of a given tree is reported.

Bayesian inference asks 'what is the probability that a given hypothesis is true, given our observations?' In the case of phylogenetics, the observation is a sequence alignment, the hypotheses are particular phylogenetic trees, and the estimate of probability is based on a model of the evolutionary process. While ML incorporates prior beliefs about the way the world works in the form of a model of sequence evolution, Bayesian inference goes a step further than this, assigning prior probabilities to alternative hypotheses before we have seen the data. Prior probabilities reflect the chance that any hypothesis is true, regardless of the data. The use of prior beliefs is both the best thing and the worst thing about Bayesian inference. If we do know, before we start, which hypotheses are more likely to be true then we can use this information to dramatically reduce the amount of time needed to search and evaluate different hypotheses. But, given that our prior beliefs will influence the outcome of the analysis, we need to be sure we do not choose inappropriate priors that will lead our analysis astray.

In phylogenetics, we rarely have prior knowledge of which hypotheses are more probable than others before we start the analysis. Therefore, phylogenetic applications of Bayesian inference tend to use uninformative priors for most variables. Uninformative priors are distributions of prior probabilities that do not bias the outcome towards a particular hypothesis, so that the posterior probability simply reflects the likelihood. Typically, parameter values are given 'flat priors' that state that all values (within certain bounds) are equally likely. For example, the prior probability for tree topology is usually set so that all trees are considered equally likely before the data are considered. But not all attempts to set uninformative priors use flat distributions. For example, the prior distribution of edge lengths on each tree may be set using a particular model of lineage generation and loss, such that large departures from the expected tree structure will be penalized and end up with a lower posterior probability. Whatever probability distribution you choose, if your aim is to have uninformative priors, it is critical to test whether the priors you set are influencing the outcome of the analysis. One way to test this is to sample from the prior distribution and compare it to an equivalent sample from the posterior distribution. If the two are the same it suggests the answer you get is being determined by priors and not influenced by the data you are analysing.

There are a number of different Bayesian methods for phylogenetic inference, but the basic approach is this:

1. **Alignment:** as with all methods, if your alignment is not reliable, neither is your phylogeny (**TechBox 6.3**).

2. **Starting tree:** commonly a randomly chosen tree to ensure independence of runs (chains).

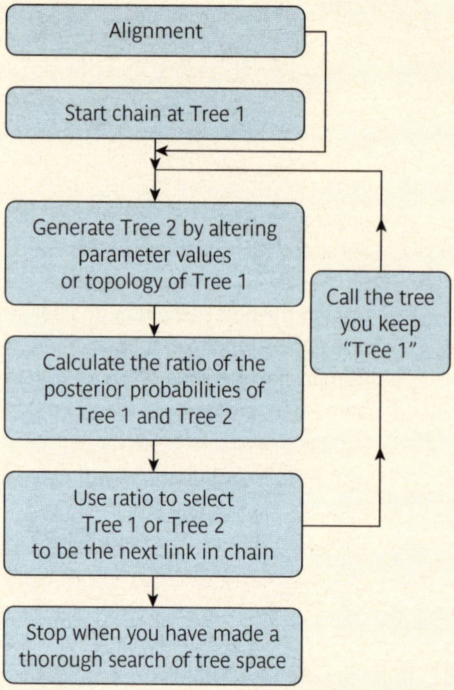

3. **Generate a new tree:** Bayesian methods make use of a procedure called Markov Chain Monte Carlo (MCMC) to explore the set of all possible trees. A Markov Chain describes the movement of a system through a series of states: at each moment the chain may move to a new state or stay in its current state, and the movement is not influenced by past states (the chain has no 'memory' of where it has been). Monte Carlo refers to random sampling of numbers (the name is a reference to a famous casino, highlighting the role of chance). So an MCMC algorithm takes a random walk through tree space. In practice, it does this by randomly altering one or more parameters of the current tree to produce a slightly different tree[1]. The chain might then move to the new tree, or it might stay on the current tree. The chance of it moving depends on the ratio of the posterior probabilities of the two trees.

4. **Calculate the ratio of posterior probabilities of the two trees:** The prior probability is the probability of a hypothesis being true before the data are taken into account. All variables – such as tree topology, branch lengths, and parameters of the substitution model – must have prior probabilities, even if they are uniform (i.e. no value is more probable than any other). The posterior probability is the probability of a hypothesis being true after you have considered it in light of your data. The posterior probability is arrived at by calculating the likelihood of a tree (see **TechBox 7.2**, p. 260), multiplying it by the prior probability of that tree, then dividing by the probability of the data. It would be difficult, if not impossible, to calculate the probability of the data, so estimating the posterior probability of a single tree is generally not possible. But calculating the ratio of posterior probabilities neatly allows the unknowns to be cancelled out, making the calculation tractable.

5. **Movement of chain to new tree is conditioned by the ratio of posterior probabilities:** In ML, you always move to the tree with the higher likelihood (see **TechBox 7.2**). But in Bayesian inference, the chance of moving to the new tree is weighted by the ratio of the posterior probabilities of the current tree and the new tree. So if the new tree has a much higher posterior probability than the old tree, then the chain will probably move to the new tree, but there is a small probability it will stay on the old tree. If both trees have a similar posterior probability, there is a near even chance that the chain will stay with the current

tree or move to the new tree. Whether it moves or stays, steps 3 to 5 are then repeated: generate new tree, calculate the ratio of posterior probabilities, select whether to move or stay based on probabilities.

6. **Stop when MCMC chain has made a thorough search of tree space:** Because the chance that a chain will move to a new tree depends on the difference between the posterior probabilities of the current and new trees, the chain is more likely to stay with better (higher-likelihood) trees. As the chain progresses, it will tend to move on to better and better trees, and the chance that a newly generated tree will have a higher likelihood is progressively reduced. Eventually, the chain will stabilize so that it is almost always staying on only the best trees. How do you know when your chain has stabilized? Currently, researchers tend to employ a variety of ways of establishing when the search is complete, from inspecting plots of posterior probability (often referred to as traces) to analysing properties of samples of trees from the chain. In addition, there are a number of strategies to help ensure that the analysis makes a thorough search of tree space (see below).

Advantages of Bayesian inference: for the sake of brevity, we will ignore the more fundamental (and rather vigorous) debates on whether a Bayesian or ML approach is more logically defensible. Most phylogeneticists who choose Bayesian methods do so for their practical advantages: they offer a relative fast way to return trees with probabilities reported for all nodes[2]. As with maximum likelihood, you can choose the best tree (referred to as the maximum *a posteriori* (MAP) tree), or you can report a 'credible set' of trees by starting at the MAP tree and progressively adding the next-best trees to the set until you have some specified cumulative probability (usually 95%)[3]. Because the MCMC is less likely to move from a tree with a high posterior probability, the chain spends more time on better trees. So you can indicate the support for each node in the tree by randomly sampling trees from the posterior distribution and asking what percentage of trees in that sample contain a particular node. This is conceptually similar to the bootstrap (**TechBox 7.4**, p. 272), but much faster as it does not involve calculating the likelihood of replicate datasets.

Disadvantages of Bayesian inference: the increase in speed is a consequence of sampling fewer trees. After all, a Bayesian method must still calculate the likelihood of each tree considered, just as an ML method does. But unlike ML methods, the Bayesian chain does not attempt to calculate the likelihood a series of nearby trees, nor for all possible branch lengths per tree. Instead, it draws random samples of trees (each with a particular topology and set of branch lengths). So the performance of Bayesian methods compared to ML must depend on how well the tree space is sampled. An MCMC chain can get stuck on local optima just as ML can: strategies can be adopted for jumping off local peaks to explore different parts of tree space, or for running multiple chains from different starting points and seeing if they converge on the same set of trees. Similarly, the percentage values on nodes are reliable only if they are based on a random sample of trees from the posterior. In reality, these samples are often correlated with each other, biasing the posterior probabilities. As with the bootstrap (**TechBox 7.4**), it is important to remember that these percentages do not represent the chance that the tree is 'true', but the strength with which this alignment supports this particular phylogeny when analysed with these particular assumptions about molecular evolution.

References

1. Huelsenbeck, J.P., Ronquist, F., Nielsen, R. and Bollback, J.P. (2001) Bayesian inference of phylogeny and its impact on evolutionary biology. *Science*, Volume 294, pages 2310–2314.
2. Holder, M. and Lewis, P.O. (2003) Phylogeny estimation: traditional and Bayesian approaches. *Nature Reviews Genetics*, Volume 4, pages 275–284.
3. Huelsenbeck, J.P., Larget, B., Miller, R.E. and Ronquist, F. (2002) Potential applications and pitfalls of bayesian inference of phylogeny. *Systematic Biology*, Volume 51, pages 673–688.

such as DNA samples from the crime scene: see Chapter 3). We have our experience of the probability of events, based on our knowledge of how the world works. So, data from previous cases might suggest that a third of all murdered women have been killed by their partners. Furthermore, when we have a murdered woman who was previously abused by her partner, then chance that she was killed by her partner is much higher than one in three (possibly as high as 80%). And we have some relevant information about this particular situation: police recorded a previous assault on this particular woman by her husband.

We can come up with many hypotheses for how we ended up with the observed situation. For example, the defence may present a hypothesis that the woman was killed by an intruder, while the prosecution may present a hypothesis that she was killed by her husband. We combine our knowledge of the world (murdered women have often been killed by their husbands, particularly if they have been abused) with information about this case (this wife was abused by her husband) to judge which of these hypotheses is more likely. But we cannot say which hypothesis is true. Indeed, we cannot rule out that the woman was murdered by aliens conducting a bizarre medical experiment, but we can say that given what we know about the world and the particulars of this situation, it does not seem as plausible an explanation (**Figure 7.56**).

We can apply the same kind of approach to solving phylogeny. We have gone out into the world and sampled some DNA sequences. Barring laboratory errors, these sequences are 'facts'. They are observations about the world. They are, in short, what we know to be true. What we don't know is the exact series of events that gave rise to those DNA sequences. But we can come up with many different hypothetical paths of substitutions that produced the sequences we observe. We want to find the tree that gives us the most plausible explanation for our observations. To do this we need some kind of statement of belief about how sequences evolve: this is our model of molecular evolution.

Models of molecular evolution

> "*Attempts at reconstructing evolutionary trees using computers are leading to a clarification of our basic ideas as to how it should be done. It has become particularly clear that any attempt at producing an evolutionary tree must be based on a specific model, for only then can proper statistical procedures be adopted and only then are the assumptions implicit in the method clear for all to see. . . . As the methods and computer programmes develop from the prototypes of today, it will become possible to handle the vast amount of information in the published data. . . .*"
>
> Edwards, A.F. (1966) *New Scientist*, Volume 19, pages 438–440

The model is a statement of belief about the process of molecular evolution, based on a prior experience. When we weighed up competing explanations in our hypothetical murder case (**Figure 7.56**), we separated our beliefs into the general statements about the world and specific statements about this particular case. We do the same thing for molecular data (**Figure 7.57**, p. 269). The general statements are our model. The specific information that tailors the model to a given situation are the parameter values for the model.

Figure 7.56

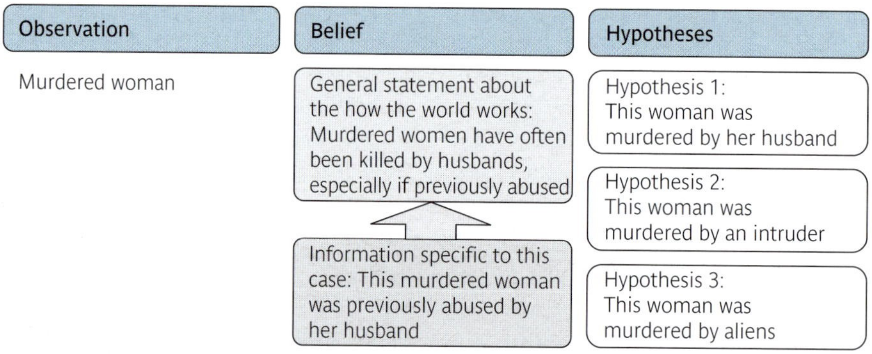
Observation
Belief
Hypotheses
Murdered woman
General statement about the how the world works: Murdered women have often been killed by husbands, especially if previously abused
Information specific to this case: This murdered woman was previously abused by her husband
Hypothesis 1: This woman was murdered by her husband
Hypothesis 2: This woman was murdered by an intruder
Hypothesis 3: This woman was murdered by aliens

Figure 7.57

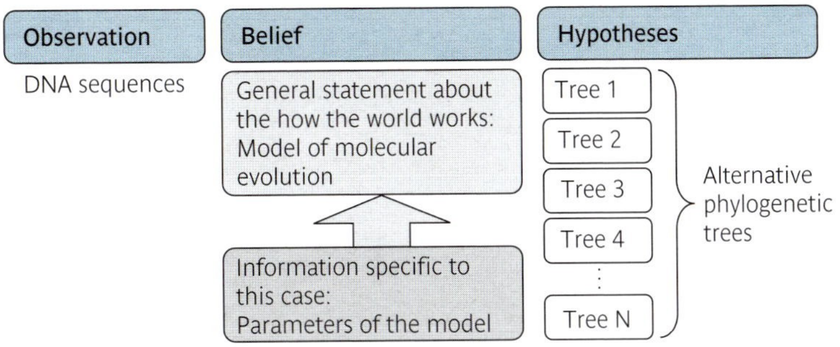

For example, one common aspect of models of molecular evolution is the transition/transversion ratio. Most people who have compared DNA sequences will have noticed that, in general, there are far more transition substitutions (changes from one pyrimidine, C or T, to another or from one purine, A or G, to another) than transversions (changes from a pyrimidine to a purine or *vice versa*: see Chapter 5). So our model could contain the general statement that transitions have a higher probability of occurring than transversions. If we were comparing two possible trees, and one required us to infer six transitions, and the other required us to infer six transversions, then we would generally consider the transition tree to be more probable than the transversion tree.

We can include this knowledge in our model of molecular evolution by having a different substitution rate for transitions and transversions (**TechBox 8.1**). These two rates are now parameters of our model. But what value should these parameters have? Exactly how much more common are transitions than transversions? The degree to which transitions outnumber transversions varies between different sequences. For example, mitochondrial sequences tend to have much higher ratios of transitions to transversions than nuclear sequences. So if we include a parameter for transition/transversion ratio in our model, we are going to have to decide on a value that is appropriate to the particular alignment we are analysing. We could choose a reasonable value based on our prior experience, and plug that into our model (a fixed parameter value). Or we can choose to estimate the value of that parameter from our dataset (see **TechBox 7.2**, p. 260). So we have a general belief about the world (transitions are more common than transversions) and information about this particular case (e.g. there are 20 times as many transitions as transversions for my alignment of a particular mitochondrial gene).

There are many aspects of molecular evolution we may wish to capture in our model. One important observation is that, for most functional sequences, some sites undergo many more substitutions than other sites. If we ignore this fact, then we will underestimate the number of substitutions that have occurred between the sequences we observe. We may also need to take account of differences in base composition. Some genomes have a greater proportion of Gs and Cs than As and Ts. In a GC-rich genome, we expect changes from any nucleotide to G or C to be more common than those to A or T. So a phylogeny with lots of changes from G to C would be more probable than one with the same number of changes from G to A.

Models for estimating amount of molecular change are discussed in more detail in Chapter 8

Molecular evolution is complex. We hardly expect two sequences to evolve in exactly the same manner, because there will always be different factors affecting patterns of substitution: fluctuations in population size, responses to changing environment, interactions between alleles, and so on. Clearly our model of evolution can only ever capture a very, very small part of this complexity. But we should not fall into the trap of thinking that a model is an attempt to fully describe the process of sequence evolution. Not only is that impossible, it wouldn't even be useful. Remember that the sole purpose of our model is to aid us in evaluating competing hypotheses. Our model only needs to be good enough to help us find the right answer. We are aiming for the simplest explanation of our observations that is consistent with all other evidence. Models are approximations of reality, not illustrations of reality.

Likelihood

How do we use our model of molecular evolution to choose between alternative phylogenies? We start with an alignment of sequences. Then we generate one possible phylogeny with these sequences at the tips. Now we calculate all the possible ways that this one tree could give rise to these sequences. For any possible series of substitution events, from the tips to the root, we use our model of molecular evolution to work out the combined probability of that particular series of substitutions. When we estimate the likelihood of a particular tree topology, we have to come up with a single number that summarizes the overall probability that this tree could produce these sequences, given all the possible patterns of substitutions that could have occurred along the tree (see **TechBox 7.2** for a more detailed explanation).

Now we generate another possible tree and repeat the process. In fact, we have to do it all over again for thousands of alternative trees. Then when we have done so, we will have created a set of possible trees, each of which has a likelihood score. At this point we could simply say 'give me the tree with the highest likelihood' and consider that the best possible phylogeny for this data, given this method and these assumptions. In practice, this is what most people do. But the next-best tree might have a likelihood that is only slightly lower than our best tree. If two hypotheses have only a very small difference in some measure of plausibility, is that really enough to reject one and accept the other?

How good is my tree?

You have spent 18 months in the lab, grinding up sweet little velvet worms and extracting their DNA (alas, it is still the case that most DNA-extraction techniques involve the death of the unfortunate DNA donor: for an example of non-destructive sampling see **Case Study 2.1**). You have battled contamination (Chapter 4) and had to develop your own DNA-extraction technique that works on mushed-up onychophorans (**TechBox 2.4**). You have been frustrated by non-specific primers (**TechBox 4.3**), then cried as your PCRs came out blank (**TechBox 4.2**). Finally you produced some useable sequences (**TechBox 1.2**), then spent so long staring at a computer screen aligning your sequences (**TechBox 6.3**) that you see DNA letters every time you close your eyes. One of your colleagues sat beside you at a computer and helped you upload your alignment, type in all manner of perplexing instructions to set the phylogenetics program running (**TechBoxes 7.1**, **7.2**, **7.3**, pp. 253, 260, and 265). Then you waited. And waited. One week. Two weeks. After three and a half weeks you check the computer and it's done. It's found the tree! Yippee, you have the answer!! You jump up and down and shout, print out the tree and wave it at passers-by, and immediately head out to celebrate, although none of your non-biological friends can understand why you are so elated.

At the risk of raining on your parade, at this point I have to remind you that your tree is not 'the answer'. It is 'the working hypothesis'. Remember, the sequences are the facts (if you have faith in your lab skills). The tree is a hypothesis that explains those facts. It may be a very good hypothesis, and it may well be true. But you don't know that. All you know is that given your method, and the assumptions you had to make to apply that method, this is the most likely explanation of your data.

However, you might look carefully at your tree and realize that things are not entirely as you expected. Perhaps one sequence is coming out in a really strange place on the tree (e.g. dinosaurs coming out with the fungi, see Chapter 4). Or you might find that your tree disagrees with a previously published tree in several important respects (e.g. **Case Study 7.2**, p. 236). At this point there are four options: one, defend your tree as the 'true tree' by force of will (irrational as it sounds, this is a surprisingly common option); two, pretend that your tree is fine by ignoring the silly bits and concentrating on the bits you like (another very popular option); three, gather more data and see if it changes your picture (always a good option if available); and four, try to find out just how convincing your tree is for this dataset (in fact, this is not so much an option as highly recommended behaviour).

Testing phylogenetic hypotheses

There is a common misconception that you cannot conduct experiments in evolutionary biology because you cannot directly witness past events or rerun evolutionary history. But we don't need to go back in time to conduct experiments in evolutionary biology. An experiment is a test of competing hypotheses against observations. You start with two or more hypotheses and you use these to make predictions about the observations you expect if the hypotheses were true. For example, researchers who conducted **Case Study 7.1** (p. 233) wished to test the hypothesis that HIV was spread to humans through a contaminated oral polio vaccine grown on chimp kidneys at a laboratory near Kinshasa. Given this hypothesis, they can make some predictions, in particular that human HIV should be more closely related to simian immunodeficiency viruses (SIV) from chimpanzees from the Kinshasa region than to any of the other SIVs. They used phylogenetic analyses of DNA sequences from human and primate viruses to show that this prediction is not met, and therefore they reject the hypothesis of a polio vaccine origin of HIV (**Case Study 7.1**).

But how closely should the data match your predictions before you conclude that your hypothesis is true? An important part of carrying out any experiment is to consider the uncertainty in your observations. There are many reasons why your phylogeny may not reflect the true history of those sequences. What if your tree is wrong? And how can you tell? There are two important ways to test your phylogenetic hypothesis: by asking whether other independent sources of evidence also support this hypothesis, and by examining just how strongly your data support your hypothesis.

Molecular evidence is one source of historical information. The best way to test its reliability is to compare your molecular phylogeny to independent sources of evolutionary information, and see if they all tell the same story. In **Case Study 7.1**, the researchers considered their phylogeny in light of other lines of evidence, such as the failure to find evidence of HIV or chimp DNA in archived polio vaccine stocks. In other cases, independent evidence might come from palaeontology, biogeography, comparative morphology, developmental biology, and so on. If your tree is at odds with all of these lines of evidence, then you might want to take a hard look at your data and methods in order to ask why your answer is different. But often molecular phylogenies are applied to cases where no other lines of evidence are available, so this option may not be possible. Even in the absence of any independent evidence you can ask just how strongly your sequence data support your phylogenetic hypothesis.

There are many informal ways to explore how well your data support your tree. Is your result robust to the assumptions you have made in your method? You could try changing the method you use or the model you employ and see if you get the same result. Did you make decisions during the alignment of your sequences that could influence your result? Maybe you could try cutting out the ambiguously aligned sites and see if you get the same answer. There are also many formal statistical techniques for assessing how well your tree is supported by your data, such as the bootstrap (**TechBox 7.4**, p. 272) or likelihood ratio test (**TechBox 7.2**). More generally, one way to approach this uncertainty about whether your tree is correct is to turn your expectations around. Instead of asking 'is this tree the right tree?', you could ask 'can I safely reject alternative trees?'. This follows a kind of Popperian logic: you can never know a scientific hypothesis is true, but you can make observations that prove it false. You can make use of this approach by setting up experiments designed to test whether a certain hypothesis is supported by the data or not (for example, see the parametric bootstrap in **TechBox 7.4**).

We will consider hypothesis testing in more detail in Chapter 8

However, you will sometimes find that your DNA sequences do not clearly distinguish between alternative hypotheses. This may be because there have been too few substitutions to produce sufficient splits in the alignment to resolve all the groups (low signal; see **TechBox 7.4**). Or it may be that there have been too many substitutions, so that the historical record has been largely overwritten (high noise). But failure to find a single tree to represent your data may not be a problem of either low signal or high noise. It could be that your data carry the signal of the true evolutionary history of your data, but that history is not a simple story of populations splitting again and again to give a clear hierarchy of similarities. When lines of descent do not follow a simple hierarchical branching pattern, then we

TECHBOX 7.4 Bootstrapping

KEYWORDS

parametric bootstrap

alignment

simulated data

likelihood

statistical significance

RELATED TECHBOXES

TB 7.2: Maximum likelihood

TB 7.3: Bayesian phylogenetics

RELATED CASE STUDIES

CS 7.2: Non-tree-like flowers

CS 7.1: Up the river (origin of HIV)

It is a well-known fact that you cannot lift yourself off the ground simply by pulling on your own shoes (feel free to test this hypothesis for yourself right now). The bootstrap statistical technique performs the statistical equivalent of picking yourself up by the bootstraps (**Figure TB7.4**) by testing the signal in the data using only the data itself (thus it shares a common etymology with 'rebooting' a computer. A computer 'boots up' because it has to use one of its own programs to start all of its other programs). The bootstrap was introduced to phylogenetics by Joe Felsenstein (**Heroes 7**) as a way of testing the strength of phylogenetic signal in an alignment[1]. It takes a single alignment, then randomly selects columns to include in a replicate alignment which is similar, but not identical, to the original. Although you end up with the same number of sites in the alignment, any given site from the original alignment may be included once, multiple times or not at all. You can repeat this sampling process again and again to create a large number of replicate datasets, all based on the original dataset but not exactly the same.

Now you estimate a phylogeny from each of your replicate datasets. If you always get the same phylogeny, no matter how many times you resample your data, then you can confidently say that, given your chosen phylogenetic method, these data points unambiguously to particular tree. But what if some of your replicates give rise to different topologies? Say you made 100 replicate datasets, and 99 of them grouped a with b, but one tree out of the hundred grouped a with c. Then you would say that the a–b group had 99% bootstrap support, and you might feel fairly comfortable that this was the best hypothesis for your data. But what if 55 of the replicate alignments grouped a with b, and 45 of them grouped a with c? Now how confident are you that your data support the a–b grouping?

Figure TB7.4 Baron Munchausen is said to have lifted himself out of the sea by pulling on his own bootstraps. Bootstraps are not shoelaces, but the loops at the top of the boot to help you pull them on, as in the iconic Australian footwear, Blundstone Boots (affectionately referred to as Blunnies).

Photograph: Ted Phelps.

A low bootstrap percentage tells you that if you had slightly different data you might get a different tree. Clearly, then, we would have less confidence in a node with low bootstrap value (even though it may be correct).

Just as there are conventions about what makes an experimental result statistically significant (usually a probability of your result occurring by chance of less than 5%), so there is a convention about what level of bootstrap support is acceptable (usually greater than 95%). However, it is up to you to decide what level of support you find plausible. The bootstrap is just a tool for helping you judge how well your data support a hypothesis, it does not tell you whether your hypothesis is likely to be true or not. In fact, it would be circular reasoning to generate a hypothesis from a particular dataset, then test it using the very same dataset. Instead, the bootstrap is testing how strongly your dataset supports that particular phylogeny, given the particular method you are using. If there is a bias in your data or your method, then you might find you have very strong support for the wrong tree. Caution must be exercised when considering bootstrap values generated for very large datasets, such as whole-genome sequences. Because the resampled datasets are so large, the variation between them is relatively small, generating very high bootstrap percentages on virtually all nodes, even if there is a substantial amount of conflict in the data.

Also remember that there are two possible causes of low bootstrap values: low signal or high noise. If there have been very few substitutions in your sequences, then there may be few informative splits. Resampling the data may miss the small number splits that correctly point to the history of the sequences, so some of the replicate datasets could support an alternative tree. Low signal may be ameliorated by collecting more data to increase the sample of informative sites. Alternatively, there could be lots of informative splits in the data, but they do not all support a single tree. High noise occurs when multiple hits overwrite the signal, or parallel substitutions are acquired independently in different lineages, or when evolution is not tree-like (e.g. hybridization). In this case, there is no point adding more of the same sort of data, because it just adds more mess (though getting different data might be helpful: sequencing a different gene, for example; see **Case Study 7.2**).

Parametric bootstrapping

The standard bootstrap is effectively asking 'if my sequences were slightly different, would I still get the same tree?' The parametric bootstrap turns this around to ask 'what is the chance that a different tree could have produced my sequences?'. The problem is that we can't rerun history. But we can achieve the same effect by simulating the evolution of multiple datasets along the same phylogeny. Since we know what the true phylogeny of these sequences is, we can ask how often our phylogenetic reconstruction methods gets the right tree, and how often our data support a different, incorrect, tree.

The most common way to apply a parametric bootstrap is to use an alternative tree (usually the second-best tree, but sometimes a specific alternative hypothesis) to generate replicate datasets, with a program that uses a substitution model to evolve simulated sequences along a tree[2]. These programs usually start with a randomly generated sequence at the root of the tree, then move up the tree, asking at each node the probability that each site will have undergone a substitution. Substitutions are accumulated until the tips are reached, giving the final sequences. This process can be repeated to produce hundreds of simulated datasets, each one of which represents the possible outcome of DNA sequences evolving along this tree, given this particular substitution model. You can then reconstruct the phylogeny of each simulated dataset and see how often you correctly recover the true tree. You can also estimate the likelihood of each dataset on your alternative phylogenies and compare it to the likelihood of the true tree. This gives you a distribution of likelihood differences for these two trees. Now you can test the difference in likelihood between your two phylogenetic hypotheses to ask 'if my alternative hypothesis is really true, what is the chance that I could have got my observed likelihood difference by chance?' (the term 'parametric' refers to using this distribution to test for significance of a likelihood difference).

References

1. Felsenstein, J. (1985) Confidence limits on phylogenies: an approach using the bootstrap. *Evolution*, Volume 39, pages 783–791.
2. Goldman, N., Anderson, J.P. and Rodrigo, A.G. (2000) Likelihood-based tests of topologies in phylogenetics. *Systematic Biology*, Volume 49, pages 652–670.

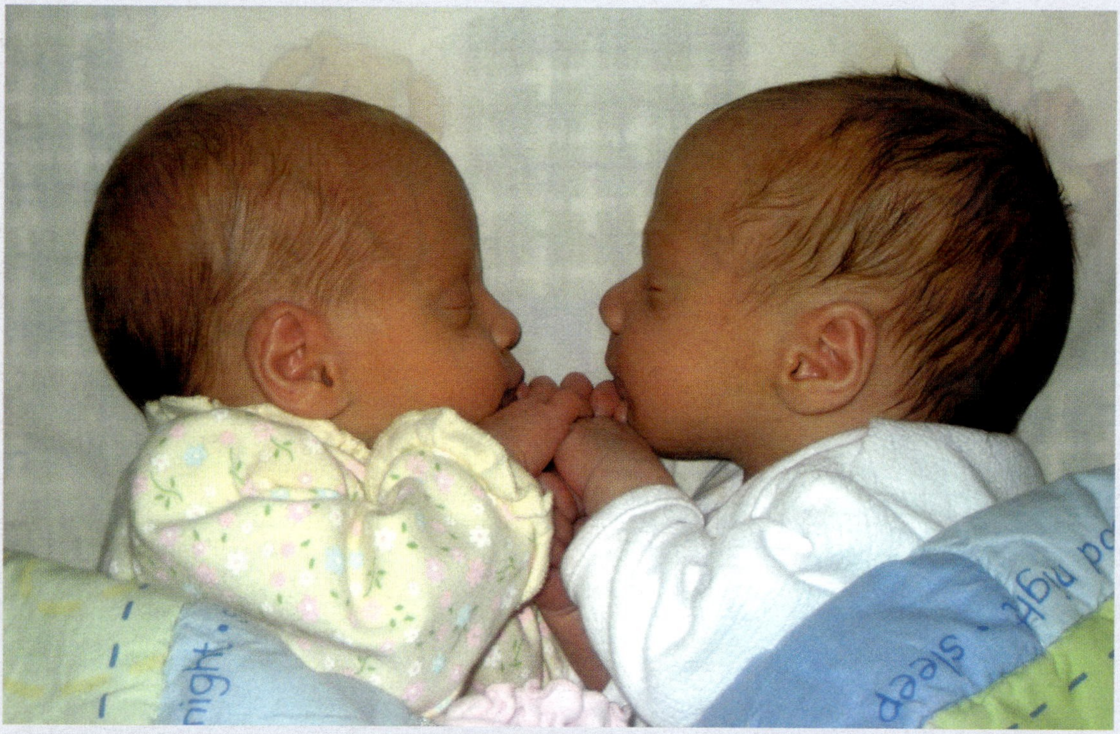

Figure 7.58 Fraternal twins are formed from different gametes so are not genetically identical (in contrast to identical twins; **Figure TB6.4a**), and may have different chromosomal complements. For example, we could be fairly sure that the female twin has inherited an X chromosome from her mother, and the male twin inherited a Y chromosome from his father. But in biology, there are few rules that have no exceptions. It is possible for a mutation in the SRY gene on the Y chromosome to result in the failure to trigger male development. A carrier of a SRY-defective Y chromosome would develop as a female. Alternatively, if the SRY gene moves by recombination to the X chromosome, then an XX embryo carrying the SRY male development trigger would develop as a male (see Andersson *et al*. 1986). Because your chromosomal complement may not match your gender, the International Olympic Committee have now dropped genetic tests as means to deciding cases of disputed gender. Another general rule in biology – that humans inherit their mitochondrial DNA from their mother – also has very rare exceptions: there has been at least one human who inherited mitochondria from his father (Bromham *et al*. 2003).

say that evolution is not wholly tree-like. It is important to keep this in mind if you find that your sequences do not unambiguously point to one phylogenetic tree.

Evolution is not always tree-like

One of the advantages of molecular data as a source of phylogenetic information is that we can, roughly speaking, consider each nucleotide site to be an independent recorder of history. This is an important basis of the statistical inference of phylogeny, since virtually all methods in statistics assume that datapoints are independent of each other. But this independence has an important corollary: if sites have different histories, then their individual records of history will not be the same. Remember that when you make a phylogeny of molecular data you are tracing the history of the DNA sequences themselves. You probably hope that the phylogeny you construct from DNA sequences reflects the evolutionary history of the organisms you got the DNA from. But in most organisms, different parts of the genome have different evolutionary histories, because the genome is not always inherited as a distinct whole. For example, your mitochondrial genes were (almost certainly) inherited from your mother. If you have a Y chromosome, then you (almost certainly) got that from your father (**Figure 7.58**, p. 274). If your mother and father are unrelated, then your mitochondrial sequences have a different history from your Y-chromosome sequences. The separate histories of mitochondrial and Y-chromosome sequences are exploited by researchers wishing to trace human movement using DNA sequence data (see **Case Study 4.2**).

But even sequences on the same chromosome can have different histories, because your genome was inherited from many different sources. You inherited half of your alleles from your mother, and she inherited half of her alleles from her father, who inherited half of his alleles from his mother, and so on. So different loci in your genome will have different ancestries. We have already seen that every allele in a population originated as a mutation in a single individual. If you share an allele with someone else, then both of you have a DNA sequence that was ultimately copied from a single individual. Imagine you compared two loci in your genome with the same loci in your cousin's genome. At one locus, it happens that you and your cousin both have a copy of an allele that originated in your paternal grandfather. These two copies of this allele – your copy and your cousin's copy – have a recent common ancestor, only two generations ago. But at the second locus, you have an allele that you inherited from your mother (who is descended from Tasmanian aborigines) and your cousin has an allele that she inherited from her mother (who is descended from Inuit from Greenland). These two copies of this allele have a much older common ancestor, over fourteen hundred generations ago. Because these alleles have different histories, phylogenies based on each of these loci might look very different.

Many techniques in population genetics, and increasingly in phylogenetics, make use of the fact that alleles can have different histories. For example, a small population will tend to have more related parents, so alleles are likely to have more recent common ancestors. Alternatively, a population with lots of migrants will contain alleles from different populations, so these alleles are likely to have more distant common ancestors. The variation in the trees inferred from different sequences can be useful. But it will create a mess if you try to infer a single phylogeny from sequences with different histories, because there won't be a single tree that describes the path of inheritance of all the sequences. Instead, some splits in the data will support one tree, and some will support a different tree.

The genetic effects of inbreeding are discussed in Chapter 5

This can be most clearly seen in phylogenies of recombinant genomes. For example, the human immunodeficiency virus (HIV) can undergo recombination between strains and this can complicate attempts to reconstruct the evolutionary history of HIV. What's more, it seems that HIV can recombine with the related simian immunodeficiency viruses (SIV). This seems to be how the N-strain of HIV-1 originated. In contrast to the globally distributed M-strain of HIV-1, which infects tens of millions of people worldwide, the N-strain of HIV-1 is very rare, with only a handful of known cases, all from Cameroon. A phylogeny of the genes from the 5′ end of the genome groups the N genome with global M-type HIV-1, but a phylogeny of genes from the 3′ end of the genome groups N-type HIV with SIV from chimpanzees from Cameroon (**Figure 7.59**). This result suggests that N-type HIV is recombinant: the 5′ end appears to be derived from an M-type HIV-1 strain, but the 3′ end is derived from a chimpanzee virus. The

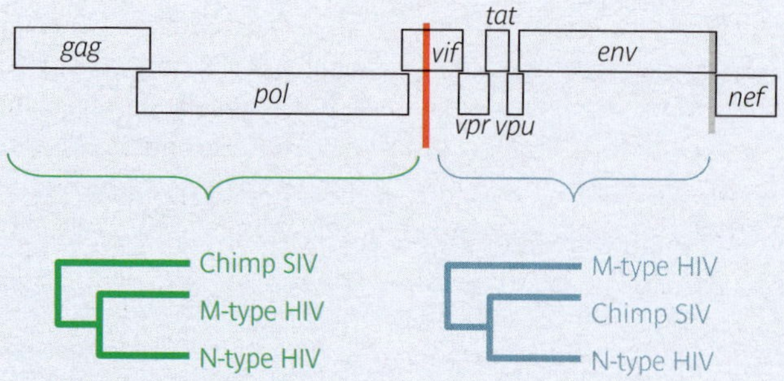

Figure 7.59 The genome of the N-type of HIV-1 includes the *gag* and *pol* genes of HIV-1 but the *env* and *nef* genes of a chimpanzee virus, SIVcpz. Phylogenies of the different genes would therefore revesl different histories. Adapted from Roques *et al*. (2004).

implication is that the N-type HIV-1 genome formed when a single individual – either a human or a chimp – was infected with both HIV-1 and SIV, and that these two different virus genomes recombined to create a new virus (**Figure 7.59**).

Recombination occurs most commonly between closely related genomes, such as members of the same species or similar viral strains. But some lineages swap DNA with more distant relatives. Movement of DNA between the genomes of unrelated individuals is commonly referred to as horizontal gene transfer. This has great practical implications. Many bacteria seem to be able to swap genes, and this can lead to the rapid spread of antibiotic resistance across different types of bacteria. Similarly, there is a concern that hybridization between plant lineages could provide a conduit for herbicide-resistance alleles from domestic plants to spread into wild species. Horizontal gene transfer also creates problems for phylogenetics, because it creates a genome with a mosaic of histories, leading to contradictory phylogenies based on different parts of the genome (**Case Study 7.2**, p. 236). Some biologists suggest that the movement of sequences from one lineage to another in the early stages of the evolution of the biological kingdoms may have been so common that evolutionary history may be better represented with a interconnected network. Consequently, there is currently a vigorous debate about whether the tree of life is really a tree at all.

Which phylogenetic method should I choose?

> *“I have steadily endeavoured to keep my mind free so as to give up any hypothesis, however much beloved (and I cannot resist forming one on every subject), as soon as facts are shown to be opposed to it. Indeed, I have had no choice but to act in this manner, for with the exception of the Coral Reefs, I cannot remember a single first-formed hypothesis which had not after a time to be given up or greatly modified. . . . On the other hand, I am not very sceptical, – a frame of mind which I believe to be injurious to the progress of science. A good deal of scepticism in a scientific man is advisable to avoid much loss of time, but I have met with not a few men, who, I feel sure, have often thus been deterred from experiment or observations, which would have proved directly or indirectly serviceable.”*
>
> Charles Darwin (1887) In *The Life and Letters of Charles Darwin*, including an autobiographical chapter, Volume 1. F. Darwin, ed. John Murray

The field of phylogenetics is surprisingly passionate. Many people working in this area feel very deeply about their favoured method, and alternative viewpoints can make their blood boil. Phylogenetics conferences are sometimes marred by heated exchanges where respected scientists will stand up and shout at each other, interrupting presentations on the results of the latest attempts to uncover the evolutionary tree of some particular group of organisms. The literature gives testament to many long-running feuds, and series of published replies and counter-replies can run for years. These feuds may concern competing solutions to the same phylogenetic problem, but more often than not the most heated conflicts concern not the phylogenies themselves but the methods used to derive them.

One of my first experiences in scientific research was to work for a 'pheneticist' (who used clustering algorithms to produce phylogenies), who spoke in the most vehement terms about the despicable 'cladists' (who used parsimony to produce phylogenies). I then moved universities and was taught by cladists, smugly assured that they had displaced the old-fashioned and patently misguided pheneticists. At my next university, researchers looked down on cladists from a great height, sure that their chosen phylogenetic method (maximum likelihood) was the most superior approach. And now Bayesian methods are the top of the tree. All of this in the decade and a half since I started doing scientific research. The moral of the story is that when someone tells you their phylogenetic method is the ultimate solution, try holding your breath and waiting for the next method, it won't be far away.

The history of phylogenetics could be interpreted as a move from the subjective toward the objective. Subjective hand-drawn phylogenies of systematists have been largely replaced by the algorithmic computer-generated phylogenies. For molecular data, methods that score and rank trees according to a single criterion have been largely replaced by methods that place phylogeny estimation within a statistical framework, making explicit assumptions about molecular evolution and weighing the relative levels of support for different phylogenetic hypotheses. The more sophisticated these methods get, the more computationally intensive they become. But, although the methods may be objective, the scientists that use the methods are not.

The reason I am pointing this out is that there is a tendency in the field to think molecular phylogenetics is a purely objective enterprise: collect the data, put them through the computer, generate the best tree. But, while computers do the hard work, they don't make decisions about what data to collect, nor how to analyse it, and they cannot judge what is a plausible phylogenetic hypothesis, nor interpret the evolutionary history it suggests. All of these things rely on scientists, and, unlike computers who have no vested interest in the answer, there is no living scientist who is truly objective. The human qualities that make people excellent scientists – such as creativity, imagination, enthusiasm, dedication – also tend to make people get rather attached to certain ideas and ways of doing things. And I for one could not imagine science any other way. I can't help thinking that an idealized, truly objective scientist would be a rather dull and unmotivated person.

However, while recognizing it's our humanness that makes science possible and fun, we must be wary of becoming too attached to favourite methods or beloved hypotheses. We should constantly remind ourselves of the principle of Ockham's razor: don't invent fancy explanations when a simple one will do the job, but reject the simple explanation when it is shown to be inadequate. Personally, when I feel my humanness getting the better of my objectivity, I remind myself of Darwin's words: 'I have steadily endeavoured to keep my mind free so as to give up any hypothesis, however much beloved . . . as soon as facts are shown to be opposed to it'.

Conclusions

Will Darwin's dream come true? Molecular phylogenetics offers hope that we might be able to construct a single tree of life. DNA is universal to life on Earth, so potentially offers a unifying framework for uncovering the evolutionary history of all organisms. For example, if anyone ever obtains DNA sequences from *Vampyrella* (**Figure 7.6**, p. 230), there is a good chance we will be able to work out where it fits in the tree of life. The historical record in DNA is essentially independent of changes in the morphology and behaviour of organisms, so the genome continues to record history as species diverge and change in function and form. The fact that DNA has only four states – A, C, T, G – is one of its strengths, because we can model evolution from one state to another as a stochastic process. This gives us all the advantages of a statistical framework in which to generate and test hypotheses. But the simplicity of DNA characters is also a weakness, because we cannot tell when multiple hits have erased the historical signal in the data. Because of this, it is essential to assess molecular data in the light of probabilistic models of substitution, which explicitly evaluate the probability that our data could be explained by a hidden series of substitution events. In the next chapter, we will take a closer look at substitution models, and ask whether we can accurately estimate not just how many substitutions have occurred, but how long it took for them to accumulate.

The historical signal in DNA is not always clear. There may be too few substitutions to place a sequence within a phylogeny, or there may be so many that the historical record has been lost. And, like any other source of data, it is possible to produce conflicting molecular phylogenies for exactly the same species. DNA data will play an important role in trying to achieve Darwin's dream of a complete tree of life, but it is best used in combination with other, independent sources of evidence such as morphology, development, palaeontology, and biogeography. Most systematicsts are optimistic that, although there is often disagreement between conflicting phylogenetic hypotheses, the field is moving towards ever-more accurate phylogenies of most living groups. Even if we never achieve one true tree of life, we will certainly have a lot of fun trying.

Further information

Many of Ernst Haeckel's iconic images are available online:

http://caliban.mpiz-koeln.mpg.de/~stueber/haeckel/kunstformen/natur.html

One of the best books on phylogeny estimation, from the lab to the computer, is now a little out of date:

Hillis, D.M., Moritz, C. and Mable, B.K. (eds) (1996) *Molecular Systematics*, 2nd edn. Sinauer Associates.

This 'instant classic' covers most phylogeny reconstruction methods in detail (and includes a personal account of the conflicts between their proponents), though it is not for the mathematically-timid.

Felsenstein, J. (2004) *Inferring Phylogenies*. Sinauer Assoc.

Tempo and mode

8

Or: how do I estimate molecular dates?

"I am trying to pursue a science that . . . has no name: the science of four-dimensional biology or of time and life."

Simpson, G.G. (1953) *The Major Features of Evolution*

What this chapter is about

The longer two lineages have been separated, the more differences we expect to see when we compare their genomes. Therefore, comparisons between DNA sequences can help us estimate when lineages diverged from each other. To estimate evolutionary time from DNA sequences we need to know the rate of molecular evolution. However, the rate of molecular evolution varies between lineages. This complicates the estimation of time from DNA sequences, but as long as we express the uncertainty in date estimates with honest confidence intervals, we can still use molecular data to test hypotheses in evolutionary biology.

Key concepts

Evolutionary biology: variation in rates of evolution

Molecular evolution: variation in rates of molecular evolution

Techniques: molecular dating

An odd fish

In December 1938, Marjorie Courtenay-Latimer, the curator of the East London Museum in South Africa, spotted an unusual fish in the day's catch of a fishing trawler: 'I picked away the layers of slime to reveal the most beautiful fish I had ever seen. It was five foot long, a pale mauvey blue with faint flecks of whitish spots; it had an iridescent silver-blue-green sheen all over. It was covered in hard scales, and it had four limb-like fins and a strange little puppy-dog tail. It was such a beautiful fish – more like a big china ornament – but I didn't know what it was' (**Figure 8.1**). The trawler captain had likewise never seen one before. Courtenay-Latimer took it back to the museum (much to the chagrin of the taxi driver who did not like the look of the 60-kilogram fish). The strange blue fish was not to be found in any of the books she consulted. When she examined it, it became clear to her that it belonged to a group of fish that were known only from fossils. Her gut feeling was confirmed by a fish taxonomist who identified the strange blue fish as a coelacanth and named it after her: *Latimeria*.

Coelacanths are one of the oldest lineages of bony fish. Their fossil record extends from around 400 million years ago until around 80 million years ago. But coelacanths disappeared from the fossil record towards the end of the Cretaceous, the period from which the last dinosaur fossils are found. Finding a live coelacanth was therefore almost as unexpected as finding a live dinosaur. Not only had the coelacanth emerged from the sea 80 million years after its apparent extinction, it appeared not to have changed much in all that time. *Latimeria* is very similar in morphology to the most recent fossil coelacanths, such as *Macropoma*. The coelacanth represents an extreme form of evolutionary stasis. While the world's climate had gone in and out of ice ages; while the mammals evolved out from under the feet of dinosaurs to dominate many terrestrial, aerial, and aquatic niches; while human populations expanded and wreaked massive changes on global ecosystems, the coelacanth had apparently stayed exactly the same.

But while the coelacanth's fossil record had stopped, and its morphology was in stasis, its genome continued to change. Mutations arise every generation (Chapters 3 and 4), and although most will be removed by selection or lost by chance, a small percentage of these mutations are expected to go to fixation and become substitutions that all members of a population carry (Chapter 5). The process of substitution is continuous, so the longer a population evolves independently, the more unique substitutions it will accumulate (Chapter 7). Many substitutions have no observable influence on morphology (Chapter 5), so they will continue to accumulate even when there is no external sign of evolutionary change. This is what has happened in the coelacanth genome. Lack of change of morphology did not preclude continuous genetic change. You can see evidence of this when you compare DNA sequences from the coelacanth and other fish. In the molecular phylogeny in **Figure 8.2**, the length of the branches represents the amount of genetic change that has occurred along each of the lineages (**TechBox 8.1**, p. 282). The lineage leading to

Figure 8.1 The African coelacanth (*Latimeria chalumnae*) can weigh as much as a person and live as long. They have a jelly-filled rostral organ on the front of the head which is probably used for electro-detection of prey (and may explain their odd habit of standing on their heads: are they scanning the sea floor for tasty morsels?). You can find information about coelacanths, including photographs and videos, on *www.dinofish.com*.

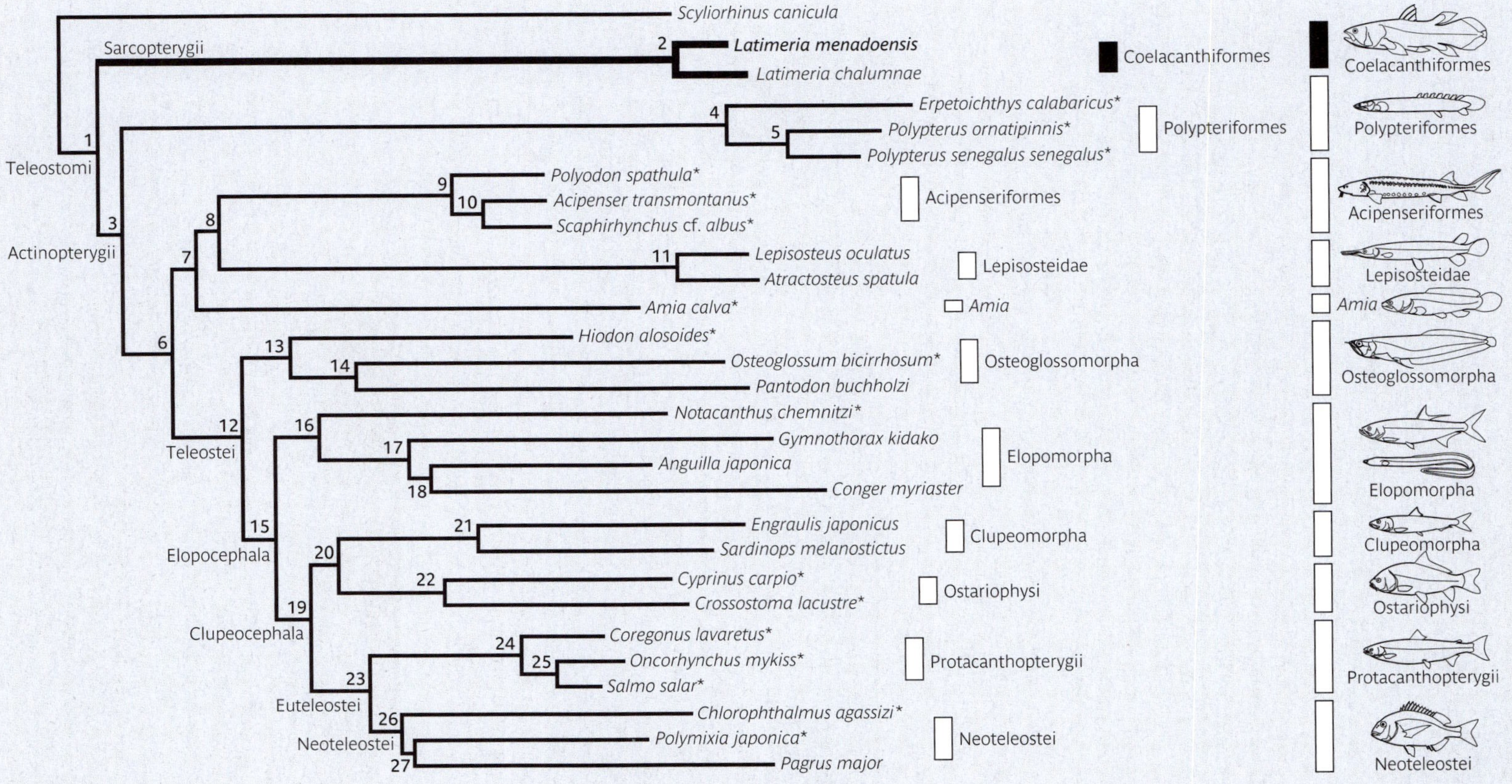

Figure 8.2 A molecular phylogeny of bony fish. The length of the branches represents the amount of genetic change estimated to have occurred along each of the lineages. From Inoue *et al*. (2005).

TECHBOX
8.1

Estimating branch lengths

KEYWORDS

rates across sites
transition
transversion
gamma distribution
base composition
distance
model selection
substitution rates

FURTHER INFORMATION

Although it's not about molecular branch lengths, the following review is a good discussion on choosing models and avoiding overparameterization (which, I am guessing, is the second longest word in this book, after methionine adenosyltransferase):

Ginzburg, L.R. and Jensen, C.X.J. (2004) Rules of thumb for judging ecological theories. *Trends in Ecology and Evolution*, Volume 19, pages 121–126.

RELATED TECHBOXES

TB 7.2: Maximum likelihood
TB 7.3: Bayesian inference

RELATED CASE STUDIES

CS 2.2: More moa (ancient DNA)
CS 8.2: Molecular detective (sources of infection)

Time, rate, and distance

The branch lengths of a molecular phylogeny usually represent the number of substitutions estimated to have occurred along each lineage, so each branch length is the result of both a particular rate of substitution and the amount of time the lineage has had to accumulate substitutions. This means that if you have a long branch in a phylogeny, you know that this sequence has accumulated many substitutions, but you don't know if that's because it has been diverging for a long time, or because it has a fast substitution rate. If you know the rate, you can convert a branch length to time. If you know the evolutionary time, you can convert a branch length to a rate.

Accuracy of branch-length estimation depends on the sample of substitutions. Branch-length estimates are not reliable at the shallow end: when there are too few substitutions, it is difficult to accurately estimate the true rate of change along a branch. Branch-length estimates are also not reliable at the deep end: when there have been too many substitutions, the historical signal is lost. The trick is to pick a sequence with an intermediate rate of substitution, relative to the timescale under consideration, and take the longest possible sequence you can.

The most straightforward measure of branch length on a molecular phylogeny is simply to count the number of differences between the sequences. This is known as the uncorrected distance, or Hamming distance. If there were no multiple hits, then this would be a perfect estimate of the number of substitutions that have occurred since two sequences diverged from each other. But, in reality, we must expect that some substitutions occur in the same sites as previous substitutions. Since we cannot directly observe overwritten changes, we must use a model to predict how many changes we have missed. Our model needs to include any factors that increase the probability of certain substitutions occurring. The values we give the parameters of the model are usually tailored to the particular dataset. We may choose a value based on some prior knowledge, for example if we happen to know the typical transition bias for our particular sequence. Or we may estimate these values directly from the data, either from a preliminary analysis of the data (e.g. estimating base frequencies from our alignment), or as part of the branch-length optimization (e.g. finding the values of parameters that optimize the likelihood of a given phylogeny; see **TechBox 7.2**).

Nucleotide substitution models

There are many different nucleotide substitution models, ranging from those that consider all possible base changes equally likely to those that have a separate rate parameter for each type of base change. Two common features of many nucleotide substitution models are base composition bias and transition/transversion ratio. Accounting for uneven base composition is an important part of the substitution probabilities. For example, if the sequences have a strong GC bias, we should expect changes to G or C to be more frequent than changes to A or T. Transitions are substitutions that exchange one pyrimidine (T and C) for another, or one purine (G and A) for another. Transitions are generally more frequent than transversions, in which a purine is exchanged for a pyrimidine or *vice versa*. Many

substitution models allow for the fact that transitions are more likely to occur than transversions. Some even allow for the different 'flavours' of transitions, by allowing one rate for pyrimidine transitions and one for purine transitions.

Amino acid substitution models

There are only four nucleotides, but there are 20 common amino acids, so protein sequences have a lot more possible substitutions than DNA sequences. But not all amino acid substitutions are equally likely. For example, if you look at the amino acid sequences in **Figure 8.22** (p. 308), you can see columns in which most species contain either a G or an A or an S or a T, which are all small, non-polar amino acids (see **TechBox 2.3**). It seems that in these positions, it is easier to exchange one small non-polar residue for another than it is to exchange it for a negatively charged amino acid such as D or E. If you are analysing protein-coding sequences you should take these differential amino acid substitution probabilities into account. There are a number of empirically determined matrices of amino acid substitution probabilities that have been determined from analyses of large numbers of protein alignments. These average substitution probabilities may not exactly match the evolutionary dynamics of your particular sequence, but some new methods allow substitution matrices to be tailored to particular datasets.

Rates across sites

Relatively few sequences have even rates of change across all sites. For many sequences, we observe that some sites more readily accept substitutions than others, and accounting for this bias is essential to accurate estimation of the number of substitutions that have been obscured by multiple hits. You might define rate categories based on your understanding of the evolutionary dynamics of the sequence under consideration, for example allowing first and second codon positions to have a different rate than third codon positions for a protein coding gene, or putting stem sites in a different rate category than loop sites for an rRNA gene. Or you could define a general model of rate variation, and use the estimation procedure to find the pattern of rates across sites that gives the highest likelihood. Many models of molecular evolution use a gamma distribution, for which a shape parameter (called alpha, α) changes the distribution of rates across sites. For example a low α value suggests that most sites are invariant with some sites having an intermediate or fast rate, but a high α value suggests that the majority of sites fall within an intermediate rate category.

How do you choose a model?

Different models can lead to very different estimates of branch length (see **Case Study 8.2**, p. 321). Clearly, choosing the best model is going to be a critical part of branch length estimation. You may instinctively wish to employ the model that you feel best represents the complexity of molecular evolution, with parameters for all possible biases in substitution rate. But, perhaps counter-intuitively, models with more free parameters can in some cases lead to greater errors due to 'overparameterization': the inclusion of so many free parameters that even random variance can be accounted for and nearly any possible answer can be entertained. Furthermore, the number of free parameters to be estimated greatly increases the computation time. So the aim is to choose the simplest adequate model, the one that does the maximum amount of explanatory work with the least amount of fuss.

You may pick a favourite model *a priori*, using your feel for the molecular evolutionary patterns of the sequences under consideration. Or you may use a formal statistical test to compare the 'fit' of different models to your data, selecting the one that best describes the patterns in the data with the fewest number of parameters. Many model tests ask whether adding a parameter provides a significant increase in the likelihood of the observed data. Whether you choose your model by instinct or statistics, you would do well to test the robustness of your assumptions by seeing if your conclusions hold when you analyse your data with a different model (**Case Study 8.2**).

the coelacanth has not had an obviously faster or slower rate of molecular change than any other fish lineage.

We also see this pattern of genomic change despite morphological stasis when we compare different coelacanths. In 1997, a coelacanth was spotted in an Indonesian fish market by two Americans on their honeymoon (one of whom happened to be a marine biologist). The original specimen was apparently sold and eaten, but after nearly a year of questioning fisherman and offering a reward, another coelacanth was brought to shore alive (**Case Study 8.1**). Over 10,000 kilometres from the African population, the announcement of the Indonesian coelacanth was nearly as surprising as Courtenay-Latimer's original discovery. The Indonesian coelacanth was virtually identical to the African ones, differing mainly in the colouring, not blue with flecks of white as described by Courtenay-Latimer, but brown with flecks of gold. However, when DNA sequences from the African and Indonesian coelacanths were compared, they were found to differ at around 4% of sites. This amount of genetic divergence would be expected from lineages that had been separated for millions of years (**Case Study 8.1**). The two populations of coelacanth have continued to accumulate molecular change despite the lack of morphological evolution.

CASE STUDY 8.1

Same but different: molecular divergence between African and Indonesian coelacanths

KEYWORDS

molecular dates
calibration
genetic variation
molecular taxonomy
mitochondria
substitution rates
confidence intervals

RELATED TECHBOXES

TB 2.4: DNA extraction
TB 6.2: What is a species?

RELATED CASE STUDIES

CS 6.1: Barcoding nematodes (DNA taxonomy)
CS 6.2: Keeping the pieces (DNA and conservation)

Holder, M.T., Erdmann, M.V., Wilcox, T.P., Caldwell, R.L. and Hillis, D.M. (1999) Two living species of coelacanths? *Proceedings of the National Academy of Sciences USA*, Volume 96, pages 12616–12620

"The fact that living coelacanths could escape detection in an area well studied by icthyologists for over 100 years is wonderful. It is a humbling and exciting reminder that humans have by no means conquered the oceans, and provides hope that 'Old Fourlegs' is more abundant and resilient that we initially dared hope." [1]

Background

The nature and distribution of the coelacanth population has been a subject of much investigation and debate. Although coelacanths have been caught along 4000 km of the African coast (particularly around the Comores archipelago), the African coelacanths appear to be have strikingly low genetic variation[2]. This genetic homogeneity has generated speculation about nature of the population. Is the population so small and inbred that genetic variation has been reduced? Are the African coelacanths a recently established population, or even 'waifs' that have drifted across the Indian ocean from the Indonesian population[2]? After all, Indonesian people reached Madagascar over a thousand years ago, so could the coelacanths have travelled on the same currents from the Pacific to Africa? Indonesian and African coelacanths are morphologically indistinguishable, apart from colour. But, given the lack of morphological change in the coelacanth lineage over 80 million years or more, we might expect them to be morphologically similar even if these two populations have separated for a long time.

Figure CS8.1 Arnaz Mehta Erdmann swimming with the first live specimen of the recently discovered Indonesian coelacanth.

Aim

Ideally, biologists would use a wide range of information to determine whether two populations are members of the same species or not, including distribution data, behavioural observations, or population biology. But coelacanths are rarely observed in their natural habitat, very little is know about their distribution, and this study had to be based on a single available Indonesian specimen. So in addition to measuring morphological parameters, DNA sequences from the Indonesian specimen were used to judge whether it was from the same population as the African coelacanths.

Methods

The first Indonesian coelacanth had been caught alive but badly injured. It survived long enough to be photographed and filmed[3]. As soon as it died, samples of gill tissue were preserved in ethanol. DNA was extracted and purified using phenol and chloroform (**TechBox 2.4**). Eighteen different primers were used to amplify DNA from a continuous stretch of the mitochondrial genome, providing nearly 5000 bases of DNA sequence that could be compared to the published mitochondrial genome of the African coelacanth (**TechBox 4.3**). They estimated the genetic distance between the Indonesian and African coelacanth mitochondrial sequences, using a Poisson distribution to infer the range of possible branch lengths that could explain the observed sequences differences. To estimate the amount of time that this level of divergence represents, the researchers needed a calibration rate, but they had no calibration dates that would allow them to estimate the rate directly for these sequences (**TechBox 8.1**). Instead, they used a published estimate of the average rate of molecular evolution in tetrapods for these genes, and inferred the date of the split between the two coelacanth mitochondrial genomes.

Results

There were a total of 185 nucleotide differences between the mitochondrial sequences of the Indonesian and African coelacanths (162 transitions and 23 transversions), plus 11 gaps (indels) in the non-protein-coding parts of the sequence (RNA-coding genes and the control region). Different regions of the sequence showed different rates of change. For example, the control region was 6.1% different between the sequences, whereas the structural RNA genes (rRNAs and tRNAs) were only 2.9% different. Likelihood ratio tests suggested that the divergence dates consistent with the observed sequences ranged between 6.3 and 4.7 million years, with an average of 5.5 million years. The researchers also tested the sensitivity of their date estimates to different substitution rates. If all of the sequences evolved as quickly as the control region, then the date estimate would be as young as 1.8 million years, but if the average rate is as slow as the ribosomal RNA genes then the date estimate would be as old as 11 million years.

Conclusions

Body measurements of the Indonesian specimen fell within the reported range for African coelacanths, with the only distinctive feature being the golden scale ornamentation. In contrast, the amount of genetic divergence between the Indonesian and African coelacanth sequences indicate it is unlikely to be part of the same interbreeding population, given the relative lack of genetic variation in the African coelacanths sampled thus far. This study demonstrates a sensible approach to molecular dating: use all the information you have to calibrate the substitution rate, consider confidence in the estimate by considering how different assumptions would change the estimate, and place the estimate in the context of other sources of information, such as biogeography, morphology, and palaeontology.

Limitations

This analysis was based on only one Indonesian coelacanth specimen; however, they were able to compare this to the observed level of divergence amongst individuals in the African population. More seriously, they have had to guess the calibration rate and if this is wrong, then the dates estimates could be inaccurate. Two other studies obtained very different dates by using different calibration rates. One assumed a rate of 1–2% per million years and obtained a date of divergence between African and Indonesian coelacanths of 1.2 to 1.4 million years ago[4]. The other estimated the coelacanth lineage rate from a calibration date of the split between coelacanths and other bony fish (which they placed at 450 million years ago) and obtained estimates of the divergence date between the coelacanths of 30 to 34 million years ago, with 95% confidence intervals of 24 to 44 million years[5].

Future work

Since its discovery, the African coelacanth population has been devastated by fishing to provide specimens for museums and collectors. It is imperative that the same fate does not befall the newly discovered Indonesian coelacanths. As long as genetic sampling is done sensitively, with the minimum amount of collateral damage, it could provide a valuable tool for monitoring the Indonesian population.

References

1. Erdman, M. quoted in Weinberg, S. (2000) *A Fish Caught in Time*. Harper Collins Publishers.
2. Schartl, M., Hornung, U., Hissmann, K., Schauer, J. and Fricke, H. (2005) Genetics: relatedness among east African coelacanths. *Nature*, Volume 435, page 901.
3. Erdmann, M.V., Caldwell, R.L. and Moosa, M.K. (1998) Indonesian 'king of the sea' discovered. *Nature*, Volume 395, page 335.
4. Pouyaud, L., Wirjoatmodjoc, I. Rachmatikac, A. Tjakrawidjajac, R. Hadiatyc, W. and Hadied, W. (1999) A new species of coelacanth. Genetic and morphologic proof. *CR Academic Science III*, Volume 322, pages 261–267.
5. Inoue, J.G., Miya, M., Venkatesh, B. and Nishida, M. (2005) The mitochondrial genome of Indonesian coelacanth *Latimeria menadoensis* (Sarcopterygii: Coelacanthiformes) and divergence time estimation between the two coelacanths. *Gene*, Volume 349, pages 227–235.

The story of the coelacanth illustrates an important point: the genome continues to record evolutionary history even when the fossil record stops and morphology is in stasis. The converse is also true: when morphological evolution accelerates, the genome continues, by and large, to steadily accumulate changes. In the time it has taken coelacanths to do precisely nothing (morphologically speaking) the Hawaiian honeycreepers have gone bananas (morphologically speaking). From an initial ancestral species that arrived in the newly-formed archipelago less than 10 million years ago, the honeycreeper lineage has produced a great variety of shapes and colours (**Figure 8.3**, p. 288). Some of these species are endemic to islands that are less than 1 million years old, suggesting a very rapid rate of speciation and divergence. But like the coelacanth, the honeycreeper genomes have continued to diverge at a steady rate. In fact, if you plot the amount of genetic divergence between honeycreeper species against the age of the island they come from, you get a surprisingly linear relationship (**Figure 8.4**, p. 289). These honeycreeper sequences seem to accumulate substitutions at a constant rate.

© Photo Resource Hawaii/Alamy

Chuck Babbit/istockphoto.com

Figure 8.3 The I'iwi (*Vestiaria coccinea*, left) and palila (*Loxioides bailleui*) are, like all Hawaiian honeycreepers, descended from a single ancestral species that probably colonized the islands less than 10 million years ago.

Pace of evolutionary change

Darwin proposed that evolutionary change is continuous (Chapter 5), but he never said it would occur at the same rate in all lineages, or uniformly across all periods of evolutionary history. We have seen that lineages can differ dramatically in their pace of morphological change and diversification. In 5 to 10 million years, the honeycreeper lineage in Hawaii has given rise to more than 50 different species with a wide variety of colours, shapes and ways of life. In 5 to 10 million years, the Indonesian and African coelacanths have stayed almost exactly the same.

George Gaylord Simpson (Figure 8.5) was a palaeontologist who was fascinated by such variation in the rate of evolution of different biological groups. Why did some lineages, like the honeycreepers, produce so many different forms, while others, such as the coelacanth, produced so few? Simpson coined the term 'tempo and mode of evolution' to encapsulate the way the pace (tempo) and type (mode) of evolutionary change can vary between lineages, over different periods in evolutionary history, or in different places. Simpson was interested in the way that palaeontology and genetics could be combined to shed light on the rates and mechanisms of evolutionary change.

To compare the tempo and mode across lineages or periods or places, you need to be able to estimate rates of evolutionary change. A rate is distance divided by time. So, for example, the rate at which you travel might be estimated as the number of kilometres you cover divided by the time in hours it took you to travel that far, to give a rate in kilometres per hour. To measure the rate of evolution, you need some measure of the amount of evolutionary change accumulated, and you need to know the period of evolutionary time over which the change occurred (**TechBox 8.1**, p. 282). Simpson's timescale came from the fossil record, allowing him to compare rates of change in taxa with a continuous fossil record. He used a variety of measures of evolutionary change. For example, he showed that the rate of change in the dimensions of horses' teeth had accelerated in some geological periods, and slowed down in others, and demonstrated that the rate of speciation had been several times faster in the horse lineage than it had been in ammonites.

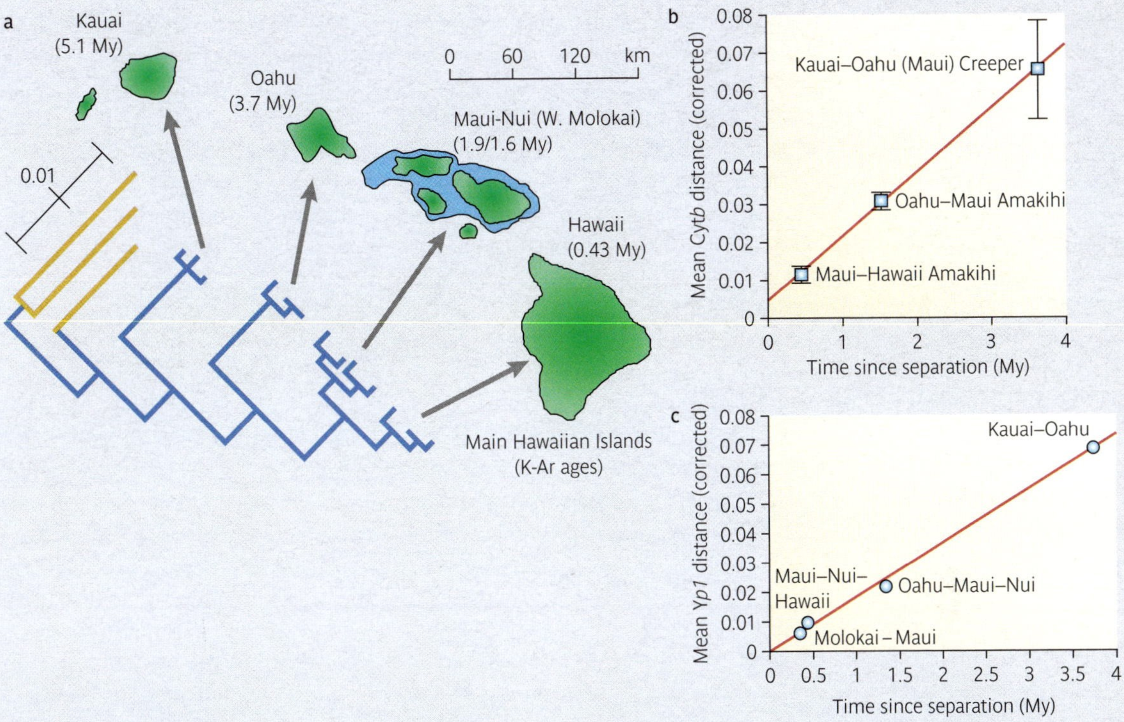

Figure 8.4 The volcanic origin of the Hawaiian islands has produced a chain of islands of increasing geological age (a). Species from the oldest islands tend to have the longest molecular branch lengths. Molecular date estimates for the age of honeycreeper, *Hemignathus* (b), and fruit fly, *Drosophila* (c), lineages from different islands produce a remarkably linear relationship between genetic divergence and time, when DNA distance is plotted against island age (see Fleischer *et al*. 1998).
From Bromham, L. and Penny, D. (2003) The modern molecular clock. *Nature Reviews Genetics*, Volume 4, pages 216–224.

Figure 8.5 George Gaylord Simpson.
Photo courtesy of Library, American Museum of Natural History. © Smithsonian Institute 2007.

A timescale for evolution

The fossil record is a rich and bountiful source of past history, providing the primary source of temporal information in evolutionary biology. But it is important to remember that fossilization is exceedingly rare. Most organisms that ever lived rotted away to nothing when they died. Only the occasional, lucky corpse encountered conditions that preserved its form after death. In fact, most biological lineages do not have a fossil record at all. For example, the Platyhelminthes (flatworms) are a diverse animal phylum. This lineage is over half a billion years old, and probably contains over 10,000 living species. Yet there are no identifiable flatworm fossils. Despite their antiquity, abundance, and diversity, flatworms are just the sort of small, soft-bodied creatures that tend not to leave fossils. And while there are many fossils of bacteria, they represent only a tiny fraction of the past and present diversity of bacteria (see **Case Study 4.1**).

A lineage will only be represented in the fossil record if some of its members happen to be captured in layers of sediment that remain undisturbed for long enough to turn into stone, and the stone survives the perturbations of geology long enough to be uncovered by a lucky palaeontologist, and the fossil contains features that allow it to be identified as a member of that lineage. So organisms that are unlikely to land in sediment (e.g. desert-dwelling shrubs), or unlikely to survive burial intact (e.g. soft, squishy worms), or unlikely to remain in undisturbed rock (e.g. birds on a volcanic island), or in a place under-explored by palaeontologists (e.g. a small mammal from the New Guinea highlands), or have few distinctive physical features (e.g. many bacteria) are a lot less likely to have an informative fossil record. What is the chance of an *E. coli* being preserved undamaged in sedimentary rock then being discovered by someone who can recognize it as a fossil and not just a spot on a rock?

In other words, the fossil record does not provide a complete or continuous record of evolutionary history. Instead, the fossil record is patchy. It's patchy in time (some periods are not recorded), in space (some areas are not recorded), and in its biological coverage (some lineages are not recorded). This does not in any way lessen the importance of fossils, which provide the primary source of information on evolutionary history. But it does make it difficult to interpret gaps in the fossil record. We have seen that gaps in the fossil record can be substantial. Some, such as the half-billion year absence of flatworms, may be unsurprising due to poor fossilizability. Others, such as the 80 million year gap in the coelacanth record, are harder to explain. If a lineage is missing from the fossil record, can we be sure that the species was absent from that place and time, or could it be the species was present but failed to leave fossils? When can we consider that absence of evidence is convincing evidence of absence?

There are two reasons why we are discussing the completeness of the fossil record in a book about molecular evolution. Firstly, there are many cases where it would be handy to have an alternative record of evolutionary history against which to compare the palaeontological record. The story in the genome provides a complement to the story in the rocks. Secondly, to use DNA sequences to investigate the tempo and mode of evolution, we need to be able to calibrate the rate of genomic change against a known evolutionary timescale. Since the fossil record provides the primary source of information on the appearance of different taxa throughout evolutionary history, most molecular phylogeneticists will, at some time, find themselves looking for appropriate fossil data with which to calibrate the rate of molecular evolution. Just as we need to develop a feel for the strengths and weaknesses of the molecular record, we should not use the palaeontological record uncritically.

All genomes carry a record of their evolutionary history. We have seen how the patterns of substitutions in DNA allow evolutionary relationships to be uncovered (Chapter 7). Now we will explore how we can use DNA data to estimate the timing of evolutionary events. Since molecular change accumulates continuously, we predict that two lineages that have a recent common ancestor will have fewer differences between their genomes than either has with a more distantly related lineage. More specifically, if we can estimate how many genetic changes have occurred since two lineages split, and if we know how fast such changes accumulate in these lineages, then we can use the amount of genetic change to predict when those two lineages diverged from each other (**Heroes 8**).

Inferring evolutionary time from molecular changes is a controversial topic. Some claim that 'molecular clocks' are usually right, and as many contend that they are usually wrong. In this chapter, we are going to focus on one of the most vigorously debated cases of molecular dating: the origin and diversification of the animal kingdom, an event often referred to as the Cambrian explosion. Estimating the diversification of the animal kingdom has been a difficult problem that has provided an important testing ground for improving molecular dating techniques, so it is a good case in point, to illustrate not only the advantages of molecular dating, but also the complications and problems. But the Cambrian explosion is not just an informative case study in molecular dating. It is also a critical test case for our understanding of tempo and mode of evolution. Many people have interpreted the Cambrian animal fossil record as the signature of a period of astonishingly high rates of change, and even as proof that unusual mechanisms of evolutionary change operate only at special times in life's history.

HEROES OF THE GENETIC REVOLUTION

8

Andrew Rambaut

Figure Hero 8 Andrew Rambaut and his son Hamish on Port Meadow in Oxford: a fine demonstration of genetic inheritance.

Photograph: Jo Kelly.

NAME

Andrew Rambaut

BORN

4th May 1971, London, United Kingdom

CURRENT POSITION

Royal Society University Research Fellow, Institute of Evolutionary Biology, University of Edinburgh, Scotland, United Kingdom

KEY PUBLICATIONS

Rambaut, A. (2000) Estimating the rate of molecular evolution: incorporating non-contemporaneous sequences into maximum likelihood phylogenies. *Bioinformatics*, Volume 16, pages 395–399.

Rambaut, A., Robertson, D.L., Pybus, O.G., Peeters, M. and Holmes, E.C. (2001) Human immunodeficiency virus – phylogeny and the origin of HIV-1. *Nature*, Volume 410, pages 1047–1048.

FURTHER INFORMATION

Andrew Rambaut's website is at: *http://tree.bio.ed.ac.uk/*

Andrew Rambaut's introduction to computational biology came about when he won a competition to create a 'biomorph' using code supplied with Richard Dawkin's 1986 book *The Blind Watchmaker*. This led him to study zoology at Edinburgh, then to a doctorate at the University of Oxford. He joined Paul Harvey's evolutionary biology group just as they were beginning to exploit the power of DNA sequence analysis to uncover patterns of evolution. Rambaut played a key role in developing software for detecting general evolutionary patterns from phylogenies, including programs for making statistically independent comparisons in evolutionary biology (CAIC) and for detecting changes in diversification rate of lineages over time (BiDe and EndEpi). He also wrote one of the most widely used programs for simulating the evolution of DNA sequences along phylogenies (SeqGen). While these methods are applicable to inferring evolutionary patterns from the phylogeny of any group of organism, Rambaut's research has generally focused on the application of these methods to understanding emerging diseases, such as HIV, dengue fever, SARS, hepatitis C, and West Nile viruses.

Rambaut has been at the forefront of many new methods in molecular dating. He developed one of the first variable-rate maximum likelihood dating methods, and applied it to estimating the timing of the radiation of animal phyla. He also devised TipDate which uses non-contemporaneous sample dates of sequences to estimate rates of molecular evolution along virus phylogenies. More recently, he codeveloped a new phylogenetic inference package, BEAST, that implements a range of new techniques in molecular dating within a Bayesian framework; for example, it allows various models of rate change along the phylogeny, and calibration dates can be treated as prior probability distributions rather than point estimates.

Despite an amazing track record of producing useful programs and interesting papers, Rambaut is sufficiently modest that he was somewhat perturbed by being described here as a 'hero', and asked me to rename this box 'useful people of the genetic revolution'. He also wants me to point that he is a friend of mine so, like all of the heroes in this book, my selection can hardly be described as unbiased. But I thought it important to include someone in the early stages of their career, and particularly someone whose contribution has been the production of freely available computer programs, which is an important aspect of progress in the field of molecular evolution.

Rambaut's programs begin as tools developed for his own research, but he generously makes his programs freely available for anyone who wishes to use them. This is a fine example of the way that academia should ideally operate, with the products of research and development being shared to speed the advancement of the field, and to allow any results to be tested independently by other scientists. Rambaut's programs are used in many of the studies reported in this book (e.g. **Case Studies 7.1** and **8.2**). He also wrote the alignment editor which was used to produce all the alignments in this book (Se-Al). His basic approach to programming seems to be to start with the acronym, think of a name that matches the acronym, then design the icon and the pull-down menus, and only then begin writing the code to make the thing work. Consequently, his programs (unlike so many academic software packages) usually have catchy names, are nice to look at, and easy to operate (though his enthusiasm rarely extends to writing a manual).

Rambaut is never happier than when he has a new problem to work on, or a new piece of computing equipment to play with. I have been fortunate enough to work with Andrew, and to take part in the rush of excitement that comes when he seizes on an idea for a novel method, or gets his hands on new data to explore, or his attention is brought to a questionable hypothesis to dismember. He is a constant innovator, so if you are analysing DNA sequences you would do well to keep an eye on his website for new programs that do neat stuff.

The Cambrian explosion

"Consequently, if my theory be true, it is indisputable that before the lowest [Cambrian] stratum was deposited, long periods elapsed, as long as, or probably far longer than, the whole interval from the [Cambrian] age to the present day; and that during these vast, yet quite unknown, periods of time, the world swarmed with living creatures."

Charles Darwin (1859) *Origin of Species*. John Murray

The fossil record of the animal kingdom starts with a bang. Rather than the gradual divergence of forms, many fundamentally different kinds of animals all appear in the fossil record more or less simultaneously. During the early- to mid- Cambrian period, from about 530 million years ago, the fossil record explodes with animal diversity. Here we see the first arthropods, with jointed legs, complex eyes, and a segmented body covered in a hard cuticle. There are molluscs, some protecting their soft bodies with shells, spikes, or scales. There are echinoderms with pentameral (five-pointed) symmetry. There are polychaete worms whose delicate hairy bodies are preserved in astoundingly fine detail. And there are a great variety of legged worms, some soft-bodied like today's velvet worms (**Figure 7.9**), others arrayed with formidable plates or spikes along their backs.

All of these diverse forms of animals are recorded in fine-grained shales that preserve a remarkable level of detail, even of the soft-parts of the animals (**Figure 8.6**). The fossils are of such high quality that in some cases it has been possible to examine the internal organs of ancient chordates, look through the crystalline eyes of a trilobite, or even infer the possible colours generated by refraction from the surface of polychaete body hairs. Despite being over half a billion years old, the fauna of

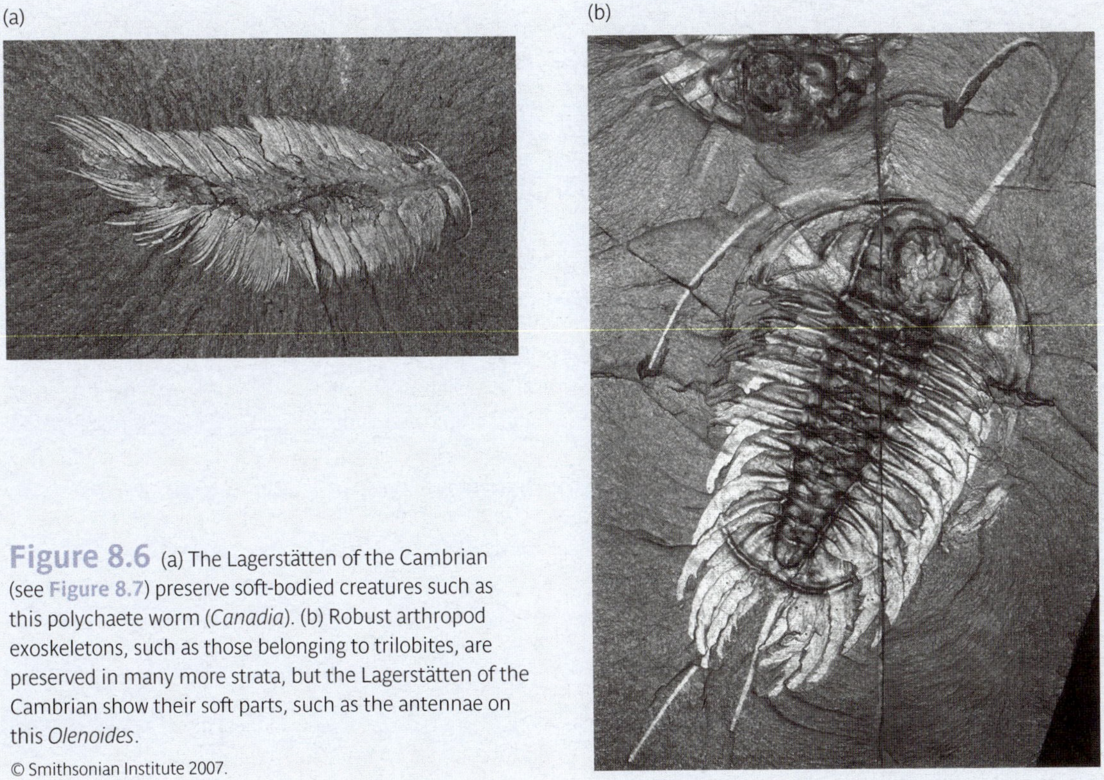

Figure 8.6 (a) The Lagerstätten of the Cambrian (see **Figure 8.7**) preserve soft-bodied creatures such as this polychaete worm (*Canadia*). (b) Robust arthropod exoskeletons, such as those belonging to trilobites, are preserved in many more strata, but the Lagerstätten of the Cambrian show their soft parts, such as the antennae on this *Olenoides*.

the Cambrian period is quite well known, due to the relatively high number of fossil beds with exceptional preservation (termed Lagerstätten; **Figure 8.7**, p. 294). This raises the possibility that the sudden explosion of animal forms in the early Cambrian is due to the unusually high capacity of the geological record of the period to record these forms.

Could it be that the animal kingdom had been evolving and diversifying for a long time, but none of the earlier forms have been recovered as fossils? Darwin was sure that the Cambrian explosion represented the start of the record of animal evolution, but not the start of the lineages themselves. Imagine turning on a video camera halfway through a party. Someone watching the tape the next day might guess that people arrived one by one in a relatively sober state, even though the recording suggests they miraculously appeared all at once in a state of combined hilarity. In Darwin's day, the fossil record was so clearly an imperfect and discontinuous record of evolutionary change that he could make a plausible and well-reasoned argument that the absence of earlier fossils did not indicate an absence of the animals themselves before the Cambrian.

But as more fossils were collected, gaps in the fossil record were gradually filled in. Animal-like fossils from before the Cambrian were discovered. These early animals are named after the Ediacaran hills in South Australia, where they were found by the geologist Reg Sprigg late in the day when the low angle of the sun caused even minor impressions in the rock to cast telltale shadows. He saw the faint shapes of jellyfish-like creatures pressed into the rust-coloured sandstone. Since then, fossils of Ediacaran animals have been found all over the world. They form a diverse collection of relatively simple multicellular organisms. They have no bones, no teeth, no legs, no eyes, no armour.

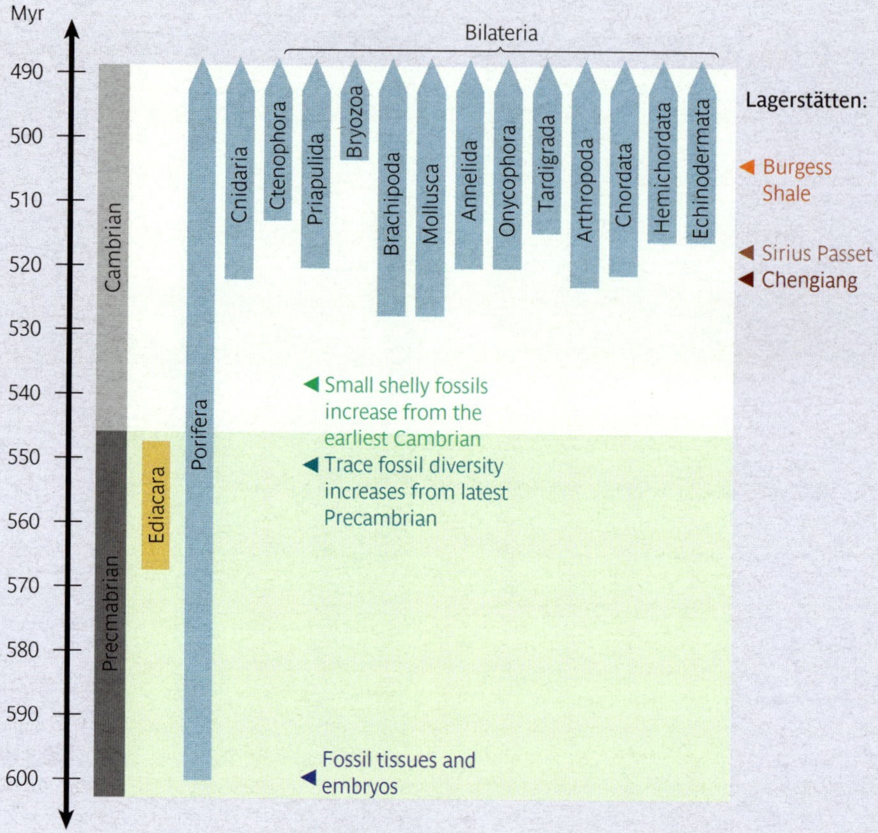

Figure 8.7 The known fossil ranges of around half of the bilaterian phyla begin in the early- to mid-Cambrian period. Note that the dates shown in this figure are approximate only, it is not intended to give precise ages for fossil beds or first appearances of phyla. This diagram is based on information from Valentine (2004; see Bibliography), and there is some debate about exactly when the earliest members of some phyla occur. For example, the fossil record of the Cnidaria (jellyfish, corals, and such like) is the subject of debate: some palaeontologists consider that various Ediacaran taxa are early cnidarians, others say the first unambiguous cnidarian fossil is Cambrian age.

The Ediacarans have been described as 'flat earthers', pudding-like creatures living a simple life, lounging around on the seafloor (Figure 8.8).

But, during the Cambrian period, the fossil record starts to change (Figure 8.7). Small animals with shells appear. Some of these shells seem to have bore-holes in them, suggesting the presence of specialized predators. Tracks show that animals were moving, and burrows show they began to colonize the sediment instead of just sitting on the ocean floor. By the mid-Cambrian, the diversity of forms display a range of ways of life: burrowing, grazing, hiding, swimming, eating, and being eaten. These new ways of life were characterized by more sophisticated morphological accessories such as appendages, sense organs, and armour.

These complex animals, with their different ways of life and bodily organization, can, by and large, be recognized as members of modern animal phyla. A phylum is the highest level of animal taxonomy, containing large collections of species that all share a fundamental

Figure 8.8 The Ediacaran fauna have variously been interpreted as ancestral animals, or modern metazoans, or a sister group to modern animals, a completely independent experiment in multicellularity, or even lichens or giant unicells. Debate continues as to whether the fauna contains early members of bilaterian lineages.

similarity of construction. For example, arthropods are characterized by having a hard, jointed exoskeleton, distinct body sections, and paired limbs (think of a spider or an ant; **Figure 8.9**, p. 296). Annelids are characterized by elongate segmented bodies with a mouth at one end and an anus at the other (think of an earthworm). Echinoderms have pentameral symmetry, no head or eyes, and a water vascular system that performs both as a circulatory system (instead of blood) and a hydraulic system (to operate their little tube feet).

These fundamentally different body organizations must have all ultimately originated from a single ancient animal ancestor. But the fossil record does not record a continuous series of animal forms linking the modern phyla, with lineages initially resembling each other then gradually becoming as distinct as the phyla we recognize today. The startling thing about the Cambrian explosion is that clearly identifiable members of almost all of the readily fossilizable animal phyla, such as arthropods, molluscs, and echinoderms, all appear in the fossil record during a period of little more than 10 million years (not surprisingly, at least a third of all animal phyla do not appear in the Cambrian because they have no fossil record at all).

So where did all these strikingly different body plans come from? There has been an ongoing debate about whether the Ediacarans represent ancestral forms of modern animals, whether they are a side branch from the early metazoan lineage, or an independent 'experiment' in multicellularity. Either way, the fossil record suggests a dramatic transition in form and function from the pudding-like Ediacarans to the sophisticated fauna of the Cambrian. This transition from simple, soft-bodied creatures to complex modern animals is not in itself surprising. Clearly the first multicellular animals must have been much simpler and less diverse than the animals of today. What is surprising is the speed of the transition. The Cambrian explosion is commonly cited to have taken only 10 million years to produce nearly all the major animal phyla. If this is true, then fundamentally different types of animals arose in the same period of time that it took for the Hawaiian honeycreepers to produce birds with different shaped beaks, or the coelacanths to remain exactly the same. The synchronous appearance of many animal phyla has led some biologists to doubt that this diversification could have been achieved by the gradual accumulation of many mutations, each of relatively small effect, by selection and drift. Instead, it has been proposed that

Figure 8.9 Animal phyla can have strikingly different body plans, for example Arthropoda (a), Annelida (b), and Echinodermata (c).

Reproduced courtesy of the National Estuarine Research Reserve Centre.

the rapidity of evolution in the Cambrian is due to special evolutionary mechanisms, such as the evolution of key developmental genes, or a consequence of particular conditions present in the Cambrian but never before or since.

As we saw in Chapter 7, when there are conflicting hypotheses for an evolutionary event, we need to make predictions, then test those predictions against observed data. To simplify this example, let's contrast just two alternative hypotheses for the origins of the animal phyla. One, which we will call the Cambrian explosion hypothesis, is that the pace of animal evolution in the early Cambrian was dramatically different to patterns of change seen after that time, with fundamentally different phyla originating in the time taken to produce slightly different species today. The other, which we will call the Precambrian slow-burn hypothesis, is that the diversification of the animal lineages only appears to be sudden because an earlier period of more gradual diversification was not recorded in the incomplete fossil record (**Figure 8.10**).

Clearly there are many different types of observations that could be used to test these hypotheses, using information from the geological and palaeontological records, from systematics and biogeography, from genetics and developmental biology. But, of course, this chapter is about estimating evolutionary time from molecular sequence data, so that is what we are going to focus on here. We want to know if the amount of genetic divergence between animal phyla is compatible with a divergence of lineages in the early Cambrian, or much earlier.

Figure 8.10

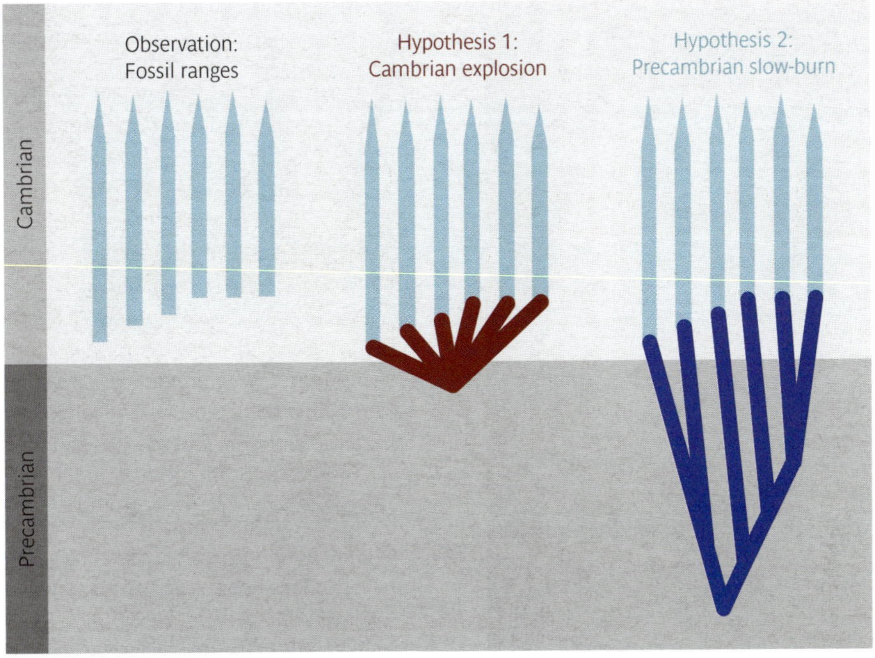

Molecular dating

> *Mutations happen, and such changes gather in the genetic code like bad memories stocking a guilty conscience: the effect is cumulative. These accreted mutations can provide a kind of clock, which can be reckoned in terms of millions of years if the right part of the genome is examined. There are 'fast' clocks and 'slow' clocks, and to try and look back into the Precambrian we need almost the slowest clocks of all, located in parts of the genome that are enormously conservative. We need to look for the genetic Collective Unconscious shared by all animals.*
>
> Richard Fortey (2000) *Trilobite! Eyewitness to Evolution*. Harper Collins

We saw in Chapter 7 that the longer two populations are separated the more substitutions we expect to see between their DNA sequences. So estimating the amount of sequence evolution not only gives a phylogeny that shows which populations are most closely related, it should also give an indication of just how closely related they are. If we knew how many substitutions accumulate per million years, then we could convert measures of substitutions to evolutionary time. The estimation of evolutionary age of lineages from molecular data is commonly referred to as molecular dating.

To contrast the Cambrian explosion and Precambrian slow-burn hypotheses (**Figure 8.10**), we want to know if the diversification of the animal phyla occurred during the Cambrian period or sometime before it. Obviously, we can't compare all the genomes that have descended from the animal radiation, but we don't need to, because the genome of any living species of animal can bear witness to their origins. If I compare homologous sequences from two living species from different animal phyla, then their last common ancestor was an animal that lived just before those phyla diverged. If the Cambrian explosion hypothesis is true, then all of the (bilaterian) phyla originated during one short evolutionary radiation, in the early- to mid-Cambrian. In this case, the date of divergence between any two sequences from different animal phyla should occur in

Figure 8.11

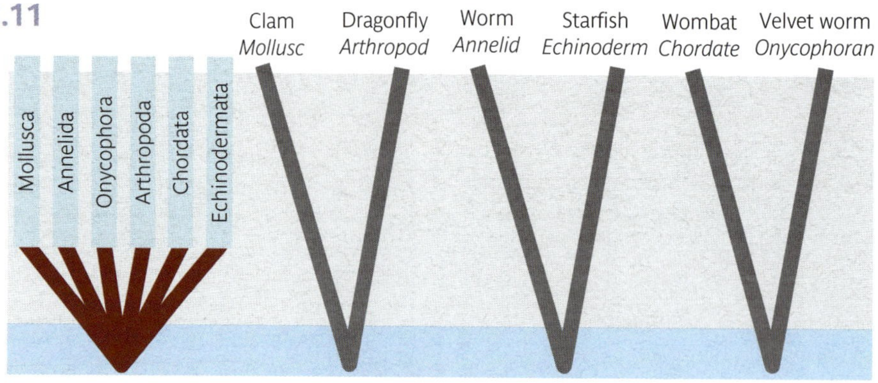

Figure 8.12 Two types of damselfly, a spreadwing and a bluet.

roughly the same time, presumably in (or just before) the early Cambrian (**Figure 8.11**).

But if the Precambrian slow-burn hypothesis is true, then the divergence date between any two animal phyla should be long before the early Cambrian. So estimating the date of divergence of any of the pairs shown in **Figure 8.11** (or any of millions of possible pairs of species from different phyla) provides a test of the Cambrian explosion hypothesis.

To estimate this divergence dates between phyla, we can select particular genes and compare them between representatives of different animal phyla. Clearly, this will need to be a fundamental and important sequence, so that it is present in the genomes of all animals, and evolves slowly enough that it is recognizably homologous between the species being compared. So, for example, we could align the methionine adenosyltransferase gene from some molluscs (perhaps a clam and a mussel) and some arthropods (say, two species of damselfly: a bluet and a spreadwing; **Figures 8.12** and **8.13**).

Chapter 6 explains the importance of alignment for evolutionary analysis of DNA sequences

Then we estimate the number of substitutions that have occurred between these sequences. Suppose that when we reconstruct the phylogeny, using whatever our favourite method is, we happen to get the following branch lengths (**TechBox 8.1**) (**Figure 8.14**).

Now we have an estimate of the number of substitutions have occurred in these sequences since they diverged from each other. So if we knew the rate at which these substitutions accumulated, we could work out how long it had been since their last common ancestor lived. But how are we going to work out the substitution rate?

We need some known time points that will allow us to calculate the number of substitutions that occur per million years. Happily, clams and mussels are just the sort of creatures that fossilize well. They live in or near sediments, they have a hard outer casing that is resistant

Figure 8.13

Mussel	Mytilus edulis	ATCACAGTATTCAGCTACGGAACATCGA
Clam	Nucula proxima	ATCACGGTTTTCAGCTATGGCACCTCTG
Damselflies	Lestes congener	ATCACCGTTTTTGACTATGGCACATCAA
	Enallagma aspersum	ATTACGGTTTTTGACTATGGGACATCAA

Figure 8.14

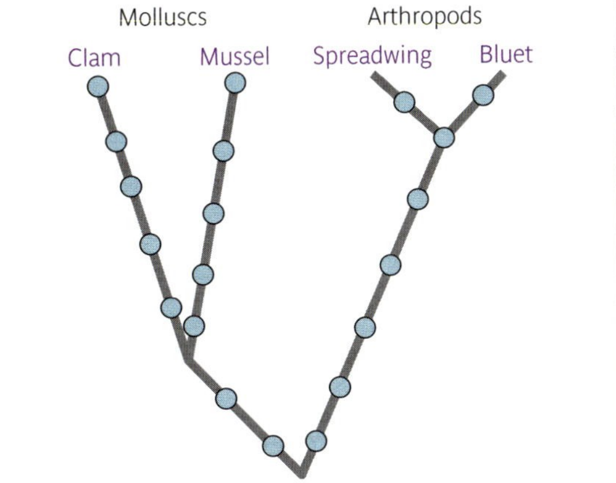

Figure 8.15

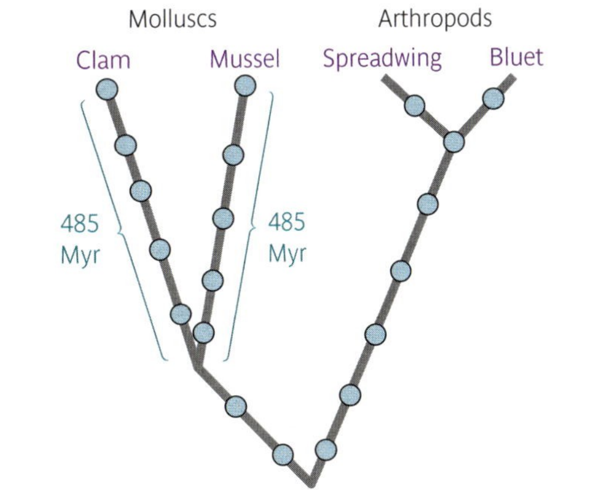

to stress and decay, and this outer casing contains a wealth of features that are used to distinguish different species. So, not surprisingly, these lineages have an informative fossil record, which suggests that the oldest known member of either lineage is at least 485 million years old. So the split between the two lineages must have occurred some time before this first fossil. So now we have a minimum age of the last common ancestor of the clam and mussel (**TechBox 8.2**, p. 300) (**Figure 8.15**).

In this case we can see that we have ten substitutions occurring between the clam and mussel sequences. Since each lineage has had at least 485 million years to evolve these differences, there has been approximately one substitution every 100 million years in this sequence, in these lineages.

So now we have a rate of substitution, calculated from a known date of divergence, that we can use to estimate our unknown divergence dates. We want to know the date of the split between the arthropods and molluscs, so we need to know how many substitutions have occurred between the two different phyla. Counting from an arthropod to a mollusc on this tree, there are 14 substitutions. If we expect approximately one substitution every 100 million years, then we would guess that it took a total of 1400 million years to accumulate these substitutions. In other words, if rates of substitution are roughly the same across the whole tree, then these data suggests that these phyla have been separated for around 700 million years (**Figure 8.16**).

Figure 8.16

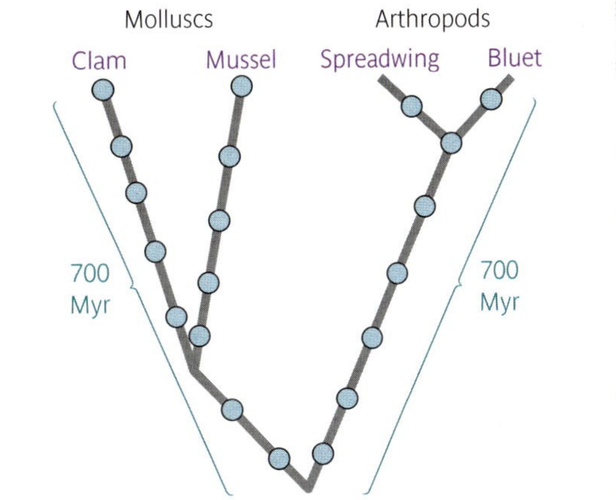

Well, that was easy, wasn't it?

TECHBOX 8.2

Calibration

KEYWORDS

fossil

ancient DNA

biomarker

island

divergence dates

maximum and minimum bounds

RELATED TECHBOXES

TB 8.1: Estimating branch length

TB 7.3: Bayesian phylogenetics

RELATED CASE STUDIES

CS 2.2: More moa (ancient DNA)

CS 8.2: Molecular detective (sample dates as calibrations)

> *How fast, as a matter of fact, do animals evolve in nature? That is the fundamental observational problem of tempo in evolution. It is the first question that the geneticist asks the paleontologist.*
>
> Simpson, G.G. (1944) *Tempo and Mode in Evolution*. Columbia University Press

Given that rates of molecular evolution can vary between lineages, substitution rates should, wherever possible, be estimated independently for different lineages. To estimate rates of substitution we need to be able to observe how many substitutions have accumulated in a known time period. Calibration is the use of any independent sources of information to infer the substitution rate for a particular dataset. The rate might be calibrated on a well-established rate for a related group of lineages (e.g. **Case Study 8.1**, p. 285), or it might be estimated from the data by calculating the number of substitutions that have occurred in a known time period (e.g. **Case Study 8.2**, p. 321). Both methods ultimately rely upon an independent source of temporal information, such as palaeontological specimens (e.g. the first recorded fossil of a particular lineage), biogeographic events (e.g. the splitting of two continents), ecological events (e.g. the coevolution of parasites with their hosts), or even the ages of the DNA sequences themselves (e.g. sample dates of viruses). On the whole, it is best to avoid using molecular dates derived from other studies to calibrate a molecular clock, as it introduces a worrying circularity to the date-estimation process and has the potential to magnify systematic errors.

If the rate of molecular evolution is uniform over all the lineages we wish to date, then we would need only one good calibration to estimate the rate of change. Many early molecular clock studies focused on finding a single, reliable fossil calibration. But if the rate of molecular evolution varies, then we are going to need as many calibrations as possible. And, just as we cannot expect molecular dates to be precise point estimates, we should not expect our calibrations to be without error either. So we will need to estimate confidence limits around the calibrations, and somehow use those confidence limits to inform our date estimation. These confidence limits may reflect the measurement error inherent in the calibration date. For example, geological clocks are sloppy clocks, like molecular clocks, so geochronological dates are usually reported with confidence intervals. But the error on calibration dates should reflect not just the confidence in the age estimate of the calibrating fossil, but also the confidence with which that fossil can be assigned to a node in the phylogeny.

To estimate the rate of molecular evolution, we need to know the date at which two or more lineages diverged from each other. In other words, calibration points are nodes in a phylogeny with a known age. But most calibrations do not mark the point in time when two populations split to become two genetically isolated lineages.

Imagine we want to estimate the rate of molecular evolution in two lineages, A and B. The earliest known fossil of either lineage is A4, which has all the defining features of the taxa in lineage A. We might calculate the rate of change in these lineages by assuming that the age of this fossil (which we will call T1) is a good approximation of the date of the split between lineages A and B.

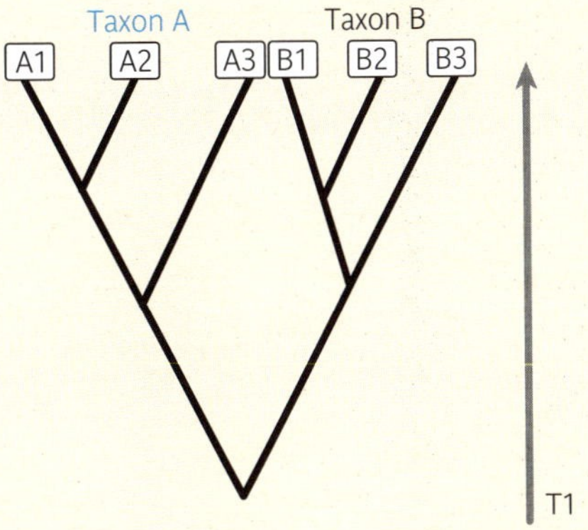

But does the first identifiable member of a taxon mark the split between lineages? The molecular divergence between lineages, marked by a branching point (node) in the phylogeny, occurs at, or soon after, the populations become reproductively isolated. But at this point, the populations may still be genetically nearly identical. Recognizable differences between lineages will evolve some time after that split. Furthermore, the defining characteristics of a taxon may not evolve all at once, but may accrue in a lineage over time. This means that when we use a fossil taxon to estimate the rate of molecular evolution, we need to think about how closely the age of the fossil matches the split between lineages.

In this case, we know that fossil A4 is in the crown group of lineage A, because it has all the defining features of taxon A. But we don't know what the interval of time is between the divergence of lineages A and B and the development of the crown characters of taxon A.

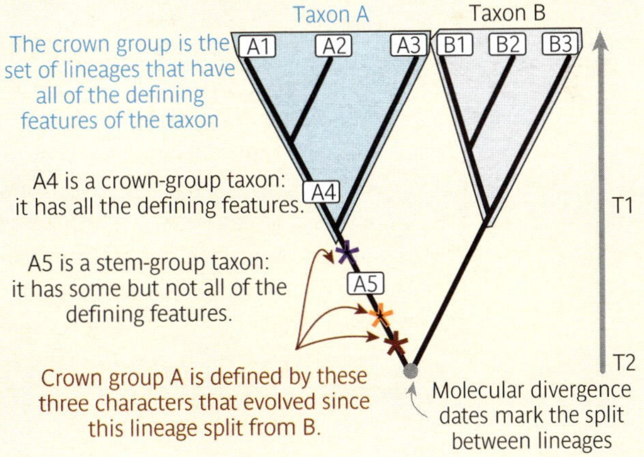

We might also have another fossil, A5, that has some but not all of the defining characteristics of taxon A, in which case we might assign A5 to the stem lineage of A. In some cases, particularly for taxa with a continuous fossil record, the oldest stem group fossil might be readily recognizable, reliably dated, and very close to the point of genetic divergence. But in other cases, it may be difficult to assign early members of the lineage to the correct stem lineage, or the earliest fossils may substantially post-date the split between the lineages. In such cases, the error on the age of the calibration itself may be overwhelmed by the uncertainty that arises from the assignment of the calibration to a point on a phylogeny.

Fossils are not the only kind of calibrations that can be used to estimate rates of molecular evolution. With all types of calibrations, it is important to remember that they represent different milestones in a lineage's evolutionary history, and most do not coincide exactly with the event we measure with molecular data, which is the genetic isolation of populations. To illustrate this point, we can imagine a mythical lineage of flightless beetles endemic to a particular island chain. Each island has one endemic beetle species. We want to be able to date the molecular phylogeny of beetles so that we can study the tempo and mode of the island radiation.

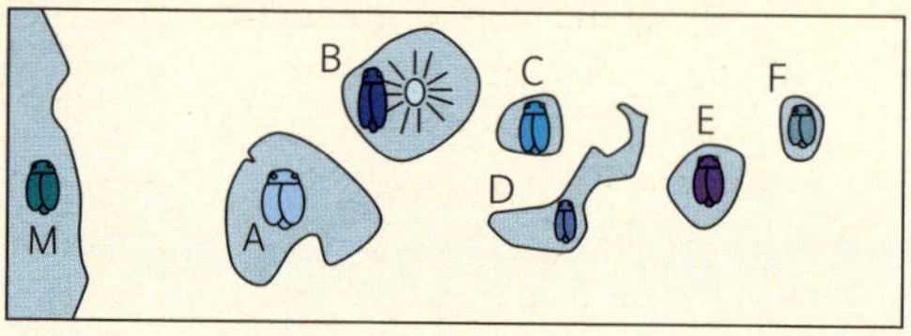

We are fortunate in this case to have a wealth of temporal information. We know that the island chain formed in a volcanic event approximately 10 million years ago. The first known fossil of the flightless lineage is preserved in the ashbed on island A from a volcano that erupted on island B 2 million years ago. There is ancient DNA from a subfossil preserved in a bog on island F, carbon dated to 50,000 years before present. There is also a biochemical calibration available from island C: sediments dated to 2.0 ± 0.8 million years ago contain a particular form of chitin that serves as a useful biomarker for the presence of this type of beetle. Combining all these observations, you might draw the calibrations on the phylogeny as follows.

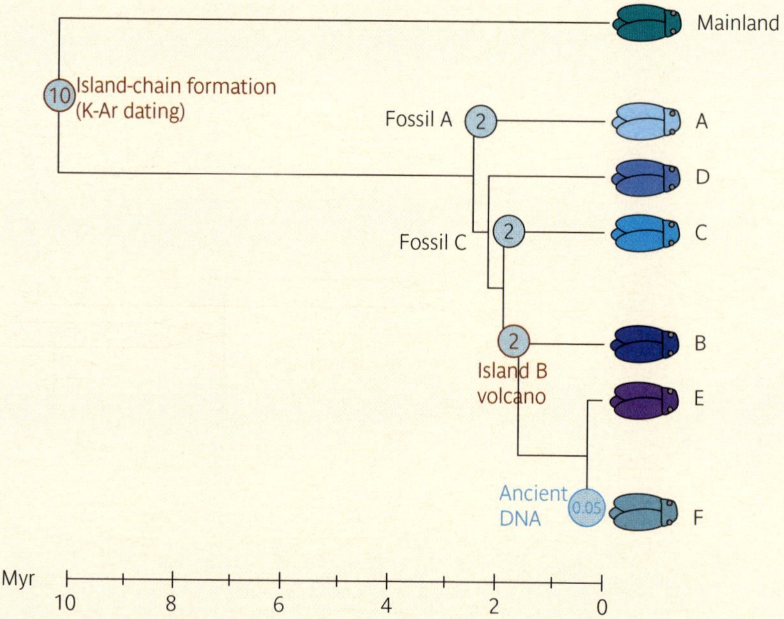

We might conclude, based on this information, that all of the island endemic beetles species date from around 2 million years ago, except for the outermost island which was only colonized recently. Thus we might propose that the radiation of these flightless beetles was driven primarily by rapid recolonization of islands after the recent volcano. But representing these calibrations as point estimates would be an oversimplification of the temporal information we have. Instead, we should consider how this information informs confidence ranges for divergence events on our phylogeny.

The island beetle lineage is unlikely to have split from the mainland population before the formation of the island, but it could have colonized any time after. In fact, the nearest mainland relative is capable of flight and is a regular visitor to the island, so the populations could interbreed even after the colonization of the island. At some point the island form became so distinct that it could not longer interbreed with the mainland immigrants, and from this point on, gene flow ceased and the island population began to accumulate in unique substitutions. So the island formation date provides information on the maximum age of the lineage, but it does not tell us when the genetic split between these lineages occurred. The fossil of the flightless form sets a minimum age on the flightless endemic clade: we know the flightless form was fully developed by the time this fossil was preserved, but it could have evolved any time before that point. So the split between the mainland and island beetles was probably after island formation at 10 million years ago, and definitely before the first distinct fossil of the island lineage at 2 million years ago.

The volcano on island B would probably have wiped out any beetles on island B at the time, though we can't be sure they didn't manage to persist in some sheltered cave somewhere. So we are fairly sure that species B is younger than 2 million years, but we don't know how much younger, and we can't rule out that it is older. The biomarker for the beetle lineage tells us that there were beetles on island C sometime between 2.8 and 1.2 million years ago, but it does not tell us which beetle lineage it was. As it happens, the biomarkers are the last trace of a population of species D that lived on C until it was wiped out by the volcano 2 million years ago. So we have placed this information on the wrong node on the phylogeny. It tells us nothing about the age of species C (which, as it happens, colonized island C 1 million years after the eruption). And finally, the ancient DNA date from the subfossil on island F represents the age of the sequence itself, not the origin of the lineage.

So we could draw a rather different picture of the diversification of these flightless beetles:

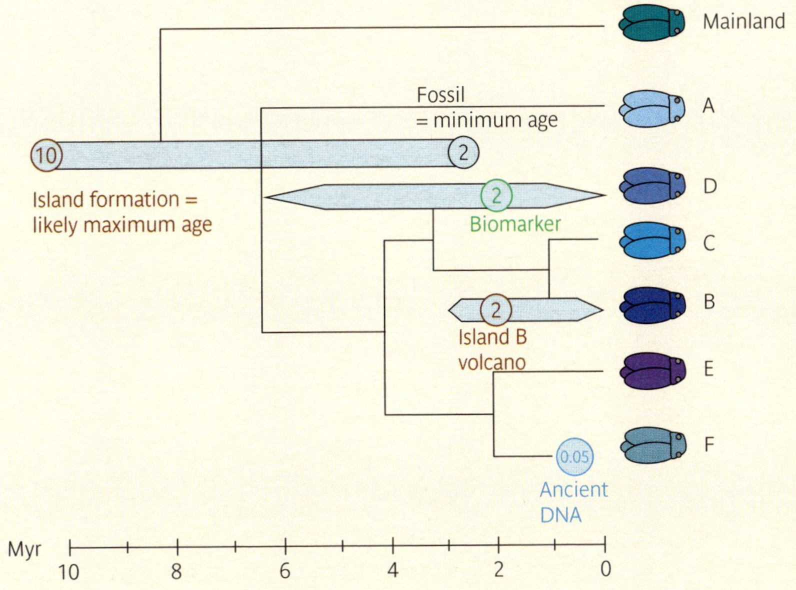

This phylogeny suggests a very different tempo and mode of island beetle evolution than the first one. Instead of a burst of diversification around 2 million years ago, there is a steady divergence into different lineages throughout the history of the island lineage. This alternative phylogeny does not support the hypothesis that the diversification of the flightless beetles was primarily driven by radiation into empty niches after the volcano on island B.

If you are wondering what on earth these mythical beetles can teach you, let me summarize. Some calibrations, such as island dates, are likely to represent maximum bounds on divergence dates. Some calibrations, such as fossils or biomarkers, tell you only that a lineages was present at a particular time, not when it originated, so provide probable minimum bounds on divergence dates. Some calibrations, like ancient DNA, represent the age of the sequence itself, not the date of divergence from a common ancestor. And all calibrations should be considered in light of the confidence with which we can assign nodes in the phylogeny to particular ranges of dates.

Calibrations rarely provide precise point estimates on the splits in a molecular phylogeny. Think about exactly what any given piece of information is telling you about the biological lineages they represent, and try to incorporate any uncertainty into molecular date estimation, either by repeating the analysis with different possible calibration dates, or by incorporating confidence limits on calibrations into the dating procedure. New molecular dating methods allow calibrations to be included with distributions of error around them, so that you can quantify the probability that the actual point of divergence takes a particular value. As with most things in biology, there are no magic bullets, no perfect solutions. Instead, in common with all the techniques we have discussed in this book, you should gather as much information as possible, consider that information in light of its margins of error, and weigh different sources of information up against each other.

In practice, estimating molecular dates is not nearly as straightforward as this example. That's because both of the things we need to know – the number of substitutions that have occurred and the rate at which they accumulate – are fairly tricky to estimate. First, let's consider the things that make estimating number of substitutions difficult, and then we will move onto the thorny problem of estimating substitution rates.

Estimating the number of substitutions

If we want to use molecular dating to test the hypothesis that the major animal lineages diversified in the Cambrian period, then the first thing we need to do is estimate how many substitutions have accumulated in each of the lineages since they diverged. The earliest molecular clock studies simply lined up the sequences and counted the differences between them. A count of observable differences, sometimes referred to as a Hamming distance, is a measure of similarity between sequences. But is it a fair measure of evolutionary time?

The problem with using Hamming distance is that it misses substitutions that have been overwritten by subsequent changes. The longer two sequences evolve independently, the more substitutions they will accumulate, and the more substitutions they accumulate, the greater the chance that substitutions will occur in the same sites as previous changes (multiple hits). If an A changes to a T and then to a G, then all we can see is the G, and we have no way of directly observing the bases that came before it. Multiple hits represent a loss of evolutionary signal, because they erase past evolutionary changes. If we are interested in using the number of substitutions between lineages as a measure of evolutionary time, then clearly we are going to

Figure 8.17 The Australian lungfish (*Neoceradotus forsteri*, also known as the Queensland lungfish) is an extreme case of an 'EDGE' species: Evolutionarily Distinct, Globally Endangered (see Isaac *et al.* 2007). Fossil evidence suggests this species has persisted since the Cretaceous, so it is probably the oldest known vertebrate species, and it closely resembles species that lived hundreds of millions of years ago. As one of the few surviving members of an ancient lineage of fish, the Queensland lungfish plays an important role in reconstructing the evolution of the vertebrates. However, this species only occurs naturally in two rivers: the Burnett River, which was dammed in 2005, and the Mary River, which the Queensland Government intends to dam in the near future. Many biologists fear that the destruction of its habitat could drive the Queensland lungfish extinct in the wild, representing a tragic loss of a unique species. It seems we have learned nothing from the loss of so many other irreplaceable species (such as the Thylacine; Chapter 6). You can find online petitions calling for moves to save the Queensland lungfish from extinction.

Figure 8.18

Latimeria_menadoensis	AACATCCGAAAGACACACCCACTAATTAAAATC
Latimeria_chalumnae	AACATCCGAAAGACACACCCGCTAATTAAAATT
Neoceratodus_forsteri	AATATCCGAAAAACACACCCGCTCCTAAAGATT

need a way of estimating how many substitutions we miss due to multiple hits.

> *The problem of multiple hits is introduced in Chapter 7*

Correcting for multiple hits

Since every new substitution that occurs has some chance of overwriting a previous substitution, the problem of multiple hits gets worse the more substitutions occur. Look at these mitochondrial DNA sequences from the Indonesian coelacanth (*Latimeria menadoensis*), the African coelacanth (*Latimeria chalumnae*), and the Australian lungfish (*Neoceratodus forsteri*: **Figures 8.17** and **8.18**).

The two species of coelacanth are probably separated by about 5 to 10 million years of evolution (**Case Study 8.1**, p. 285). In this small sequence, there are two nucleotides different between these species. The coelacanths last shared a common ancestor with the lungfish at least 400 million years ago. Yet the coelacanths differ from the lungfish at only eight nucleotides in this sequence. The lungfish–coelacanth split is 40 times older than the coelacanth–coelacanth split, but it has only four times as many substitutions. Why? Does this disprove that substitution rates are constant and clock-like, or suggest that the split between the lungfish and coelacanths is not as old as the fossil record would suggest?

A more likely explanation is that substitutions have continued to accumulate throughout the long history of the lungfish and coelacanth lineages, but that many of these substitutions overwrote previous changes. Since we can only observe the end points of the process, we cannot directly observe any past substitutions that have been erased. The suspicion that the surprisingly low number of differences between these sequences is due to saturation (past changes being lost) is supported by an inspection of the pattern of substitutions in these sequences. If you look at the alignment in **Figure 8.18**,

Figure 8.19

	123	123	123	123	123	123	123	123	123	123	123
Latimeria_menandoensis	AAC	ATC	CGA	AAG	ACA	CAC	CCA	CTA	ATT	AAA	ATC
Latimeria_chalumnae	AAC	ATC	CGA	AAG	ACA	CAC	CCG	CTA	ATT	AAA	ATT
Neoceratodus_forsteri	AAT	ATC	CGA	AAA	ACA	CAC	CCG	CTC	CTA	AAG	ATT

you will see that nearly all of the substitutions occur in every third position in the alignment (**Figure 8.19**).

Recall that protein-coding sequences are read in threes (codons; **TechBox 2.3**). There are more different codons than there are amino acids to be coded for, so most amino acids can be coded for by more than one three-base codon. If you look closely at the genetic code (**Figure 8.20**), you can see that changes in the third codon position often do not change the meaning of the codon. For example, the codon TCT codes for the amino acid serine. A mutation in the third codon position, from T to A or C or G, won't change the meaning of the codon, because TCT, TCA, TCG, and TCC all code for serine. But any change to the first or second position of this codon will change the amino acid – for example, changing the first T to A gives ACT which changes the meaning of the codon to threonine, changing the second C to A gives TAT which codes for tyrosine.

1st	2nd: T		C		A		G		3rd
T	TTT	F	TCT	S	TAT	Y	TGT	C	T
	TTC	F	TCC	S	TAC	Y	TGC	C	C
	TTA	L	TCA	S	TAA	Stop	TGA	W	A
	TTG	L	TCG	S	TAG	Stop	TGG	W	G
C	CTT	L	CCT	P	CAT	H	CGT	R	T
	CTC	L	CCC	P	CAC	H	CGC	R	C
	CTA	L	CCA	P	CAA	Q	CGA	R	A
	CTG	L	CCG	P	CAG	Q	CGG	R	G
A	ATT	I	ACT	T	AAT	N	AGT	S	T
	ATC	I	ACC	T	AAC	N	AGC	S	C
	ATA	M	ACA	T	AAA	K	AGA	Stop	A
	ATG	M	ACG	T	AAG	K	AGG	Stop	G
G	GTT	V	GCT	A	GAT	D	GGT	G	T
	GTC	V	GCC	A	GAC	D	GGC	G	C
	GTA	V	GCA	A	GAA	E	GGA	G	A
	GTG	V	GCG	A	GAG	E	GGG	G	G

Figure 8.20 The vertebrate mitochondrial genetic code. See **TechBox 2.3** for other versions.

In the majority of cases, you can change the third nucleotide of a codon without changing its meaning (a synonymous mutation), but changes to the first or second codon positions usually result in a change of amino acid (non-synonymous mutation). Since synonymous changes are essentially neutral, they won't be removed by negative selection. So, in protein-coding sequences, we expect a larger proportion of mutations occurring in third codon positions to become substitutions than those occurring in first and second codon positions. Since third codon positions have a higher substitution rate, they are more likely to accumulate multiple hits that obscure past changes. In this case, we can guess that the observed differences between the sequences in **Figure 8.18** do not reveal all the synonymous substitutions that have happened in the past.

Synonymous (silent) and non-synonymous (replacement) substitutions are introduced in Chapter 5

The practical result of saturation is that genetic distance does not always increase as a linear function of time. If you compare two sequences and find they differ at 4% of sites, you cannot assume that they diverged twice as long ago as two sequences that differ at 2% of sites. If we wish to use genetic distance to estimate evolutionary time, we are going to need to use the observable substitutions (Hamming distance) to estimate the number of substitutions that have really occurred between the two sequences. By definition, we can't observe these overwritten changes directly, so we are going to have to predict their occurrence from the substitutions we can see.

How can we account for changes that we can't directly observe? We can use a model of molecular evolution, which states the probabilities of different types of substitutions, to predict how many additional changes have occurred and been subsequently overwritten (**TechBox 8.1**, p. 282). Such a model allows you to say 'if I see this many changes, how many am I likely to have missed?' For example, if you observed only two substitutions across an alignment of 1000 nucleotide sites, you might conclude that, since there have been relatively few substitutions, the chances that one of them occurred on top of a previous substitution is pretty low. But if you find that 450 out of 1000 sites vary between sequences in your alignment, then you would expect that the chance of one of those substitutions overwriting a previous one is pretty high. Models of substitution probabilities provide a way of guessing how many changes you are likely to have missed due to multiple hits, and thus predicting how many actually occurred.

Variation in rates across sites

The alignment of coelacanth and lungfish sequences (**Figure 8.19**) illustrates the importance of considering variation in rates of substitution across different sites in an alignment. Knowing that some sites accumulate multiple hits at a faster rate than other sites has important implications for estimating amount of genetic change between two sequences. Imagine we observe 200 differences between some aligned sequences of 1000 bases, then we would consider that roughly a fifth of sites had undergone a substitution. But if we noticed that all of the 200 substitutions were in the third codon position, then we could see that the majority of the third codon positions had changed, and therefore we would expect that a large proportion of these overwrote previous changes. Similarly, if you observe only ten differences between two long sequences, you might assume a shallow divergence. But if you look closely at the sequences and find that all ten substitutions are in a particular region of the gene known to be unimportant for function, you might reassess your conclusion – perhaps there are only ten differences because only ten sites can change without destroying the function of the gene product. In this case, these ten sites might have been changing back and forth for aeons, erasing past history with every new substitution.

Chapter 5 explains why less important regions of a sequence have a greater rate of change

Allowing for difference in rate of change between sites in an alignment is critical for molecular dating. For example, one study estimated dates of divergence for major animal lineages from alignments of seven different proteins, taken from a range of modern animals. When they assumed that all sites evolved at the same rate, they got a date estimate of 573 millions of years ago. Since this date estimate corresponds with the earliest Ediacaran-type fossils, it was considered to support the contention that the early animal record is an accurate recorder of metazoan history. But when the authors used exactly the same data but a different assumption about patterns of evolution – that rates could vary between sites – they got a very different answer. By allowing rates to vary between sites, they estimated the bilaterian phyla diverged at 656 million years ago, long before the first body fossils of metazoans (**Figure 8.21**).

Figure 8.21

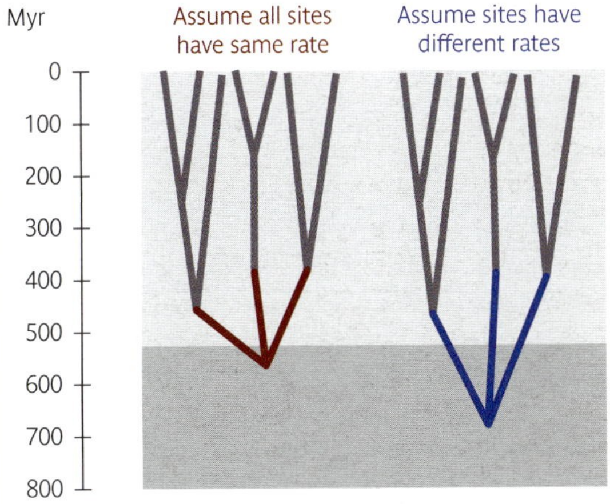

Why did allowing for different substitution rates across the sequence change the date estimate? We can see why when we look at the sequences used to produce these date estimates. Here is part of the amino acid sequence of one of the genes used in the study. In this small part of one alignment, which includes only half of the species used in the analysis, you can clearly see that some sites change more readily than others. Many of the amino acids have been preserved through over half a billion years of evolution, suggesting that they are strongly conserved by selection. Most of the changes occur in only half of the sites (**Figure 8.22**).

Sites that more readily accept substitutions are, *ipso facto*, more likely to undergo multiple hits than conserved sites. Therefore the average rate of change

Figure 8.22

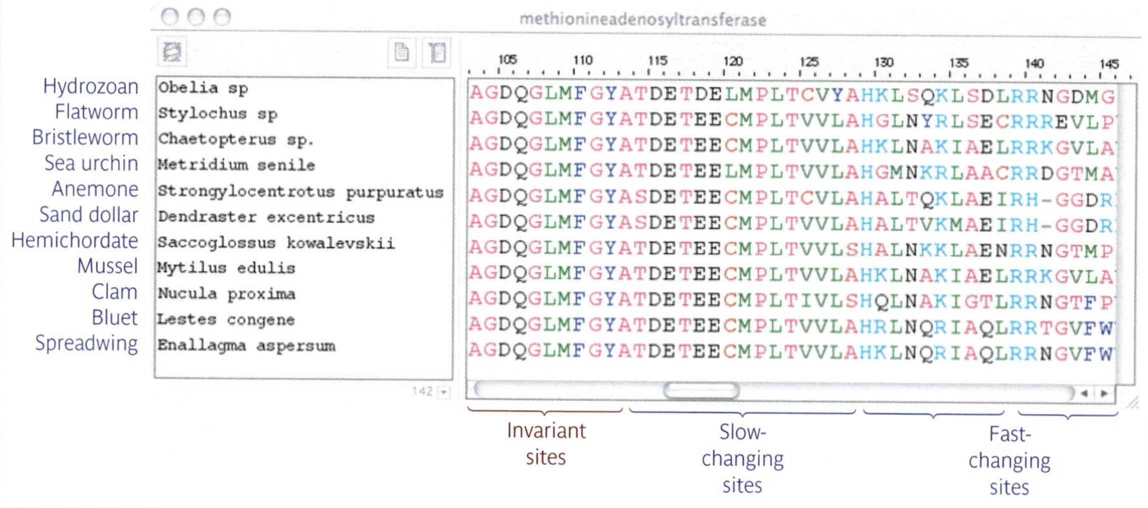

across all sites is not an accurate predictor of the percentage of sites that will have accumulated multiple hits. If we calculated an average rate across all sites, we would underestimate the number of substitutions that had occurred in the rapidly changing sites. And if we underestimate the number of substitutions that have occurred, then we will underestimate the time taken to accumulate these substitutions; therefore our estimated date of divergence will be too young. On the other hand, allowing for variation in rates between sites leads to the conclusion that far more substitutions were covered up by multiple hits. When these over-written changes are added to the estimate of genetic distance, the branch lengths are increased. Longer branch lengths move the common ancestor back in time.

Choosing the right sequence

Rates of change vary across the genome. When you choose a sequence to estimate genetic distance for molecular dating, you need to take care to choose a sequence that is evolving at a rate that is just right for the depth of divergence you are considering. Too slow, and there will not enough substitutions to tell a clear story. Too fast, and the historical story will be erased.

The causes of variation in rate of molecular evolution across the genome are discussed in Chapter 5

Consider this sequence from the control region of the mitochondria. The control region is known to have a rapid substitution rate so it is often chosen for distinguishing closely related populations or species. Here we have the control region sequence for three kiwis (*Apteryx*; **Figure 8.23**) and three moas (*Dinornis*; see **Case Study 2.2**), and we want to use this alignment get a picture of the patterns of divergence between all six species (**Figure 8.24**).

This particular alignment includes substitutions that distinguish the three kiwi sequences from each other. In this short sequence it seems that there are differences between the kiwi sequences at nearly 15% of sites. You can imagine that if you took a long enough sequence you would get a nice sample of informative substitutions for estimating the divergence time between these three kiwi species. But when we look at the same sequence in three moa species, there are no differences between these sequences. This particular short alignment would be no good for estimating the divergence between these moa species, because there are not enough substitutions. Conversely, this sequence would not be useful for estimating the relationships between kiwis and moas because there are too many substitutions between them. Half of all the nucleotides in this sequence differ between kiwis and moas, suggesting that the sequence has accumulated so many substitutions since the divergence between these two lineages

Figure 8.23 Kiwis (*Apteryx*), found only in New Zealand, are flightless birds related to emus and ostriches.

Reproduced courtesy of Phil Brown Photography.

that many of the sites will be saturated, and will have lost historical signal. This example illustrates that rather than there being 'good' and 'bad' sequences, we have to choose the sequence to suit the problem. This one sequence might have an ideal rate of change for investigating kiwi divergences, but too slow for investigating moa divergences, and too fast for investigating the split between kiwis and moas.

Sequence selection is not just a matter of choosing which gene you want to use. It may also involve choosing which sites you wish to compare within a gene. Here are sequences from four different animal phyla, for a gene frequently used to estimate the timing of the metazoan radiation (Figure 8.25, p. 310). These sequences are part of the 18s ribosomal RNA gene in Onycophora (velvet worms; Figure 7.9), Rotifera (wheel animalcules; Figure 6.3), Nematoda (roundworms; Case Study 6.1), and arthropods (here represented by a fly; Case Study 5.2).

Across a single, short sequence in the alignment, you can see evidence of variation in rate of substitution. In the first half of the alignment, there are substitutions in around 20% of sites. But in the second part of the alignment, over 90% of sites have substitutions. In this saturated region, we can no longer detect similarities that reveal the shared ancestry of these sequences. If we are interested in the date of divergence between these sequences, then the saturated sites are no help to us, because the number of differences between these sequences is no longer a good indicator of time since divergence. Since saturated sites hold no useful information for us, there is no point including them in our analysis. Indeed, including saturated sites could mislead our analysis. Therefore, it is prudent to exclude saturated regions of an alignment from any analysis aiming to estimate branch lengths that reflect evolutionary time.

Chapter 6 explains the importance of alignment to establishing homology

Another way to improve estimates of the divergence date between lineages is to take the longest possible sequences that you can. Whenever there is variation in

Figure 8.24

Kiwis	Apteryx_mantelli	CAATATGACTAGCTTCAGGCCCATTCATTCCCCGCGCACTACC
	Apteryx_rowii	CAACATGACTAGCTTCAGGATCATTCATTCCCCGCGCACTACC
	Apteryx_haastii	CAACATGACTAGCTTCAAGACCATTCATTCCCCGCGCACTACC
Moa	Dinornis_giganteus	GGTATGCGCTAGCTTCAGGAACCCTAAGTCCATATGTCATGCC
	Dinornis_novaezealandiae	GGTATGCGCTAGCTTCAGGAACCCTAAGTCCATATGTCATGCC
	Dinornis_struthiodes	GGTATGCGCTAGCTTCAGGAACCCTAAGTCCATATGTCATGCC

Figure 8.25

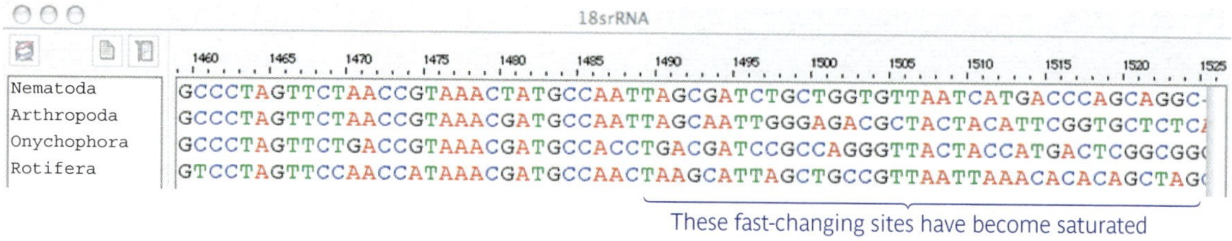

the quantity we want to estimate, we will have more confidence in our estimates if we have based them on a large sample size. If you wish to measure body size in a population of penguins, and you measure only two individuals, you cannot be sure that your measurement accurately reflects the population average. If you happened to sample two larger-than-average penguins, then your estimate will be higher than the true population average. In fact, the chance that the average size of your two penguin sample being the same as the overall average for the population is quite low. But now if you measure 60 penguins, your sample estimate is more likely to be close to the true population average. In a large sample, the occasional large individual is unlikely to derail your estimate, because it will probably be balanced by the occasional small one.

The same applies to using samples of substitutions to estimate the time frame of genetic divergence. In most cases, the sequence you choose to analyse is just a tiny sample of the genome, but you are hoping that the substitutions you observe are a representative sample of the differences between the genomes. If you choose a sequence that has few substitutions over the time period you are considering, then you could end up with an estimate much higher or lower than the real degree of divergence between these sequences. In **Figure 8.24**, the kiwi sequences differ from each by only one or two substitutions. This in itself is not enough to estimate the depth of divergence between these species. But if you took a sequence 50 times longer than this, then you might get 50 times as many substitutions, and the combined effect of all of these differences might be enough to give a clear picture of the patterns of divergence between them. Longer sequences are the main route to larger sample sizes, and larger sample sizes are the key to being able to develop and test hypotheses.

Finally, the accuracy with which you estimate genetic divergence depends critically on the model of molecular evolution (**TechBox 8.1**, p. 282). It is your model that allows you to correct for multiple hits and account for variation in rate of change between sites in a sequence. If your model does not adequately capture the patterns of substitution in your sequences, then you may fail to allow for many substitutions that have occurred and been overwritten, or you may falsely infer many more substitutions than have actually occurred. The problem of multiple hits will be exacerbated where relatively few sites in a sequence are able to change. If the free-to-vary sites change again and again, then they will lose historical signal, but this may not be obvious from a consideration of the number of substitutions across the whole alignment, because the sequences will differ only at a small number of variable sites.

Accuracy and precision

We have seen that too many substitutions can lead to inaccurate estimates of genetic distance due to the problem of multiple hits. Now let's consider the opposite problem: if there are too few changes between our sequences it will decrease the precision of our estimates of the rate of molecular evolution. Not surprisingly, estimates of substitution rate will be best when you have a lot of substitution events to consider.

Here is a hands-on example. Feel your pulse, and count the number of heart beats that occur in 2 seconds. Now I want you to calculate your heart rate, expressed as number of beats per minute. To estimate the number of heart-beats per minute, all you have to do is take the number of beats you felt in 2 seconds then multiply it by 30. The first time I tried this I measured two beats in 2 seconds. So my first estimate is that my heart-rate is 60 beats per minute. I took another 2-second measurement and this time I got three beats in 2 seconds, so my second estimate is 90 beats per minute. That's a 50% measurement error on a fairly regular heartbeat 'clock'. Now try using a bigger sample of heart beats to estimate your heart rate, by measuring the number of heart beats in 30 seconds. The first time I got 34 beats in 30 seconds, the next time 35. These result in estimates of 68 and 70 beats per minute respectively. That's less than 3% measurement error. So although I was measuring the same quantity (heart rate), and the absolute difference in the measurements was the same in both cases (samples differed by one heartbeat), the estimate based on a small sample of beats was less precise (had a greater margin of error) than the estimate based on a larger sample of beats.

The same principle applies to estimating the substitution rate. When there are few substitutions between two sequences, every new substitution makes a relatively large difference to the rate estimate. This would not be so bad if molecular change accumulated evenly. But the problem of estimating substitution rate from a small number of observable substitutions is made more difficult by the fact that, rather than ticking like a metronome, the molecular clock has an irregular tick rate. In other words, the molecular clock is a sloppy clock.

The molecular clock is a sloppy clock

A sloppy clock in one that does not tick regularly and evenly, like a metronome, but has an erratic tick rate, governed by chance events. Geochronological clocks, based on radioactive decay, are sloppy clocks. For example, the age of the Hawaiian islands (Figure 8.4, p. 289) was determined by potassium-argon (K-Ar) dating, which utilizes the radioactive decay of potassium isotopes (^{40}K) to argon isotopes (^{40}Ar) in a rock sample. Argon gas can escape from molten rock, but when the rock solidifies, any ^{40}Ar produced by decay of ^{40}K will be trapped in the rock sample. As more of the potassium in the rock decays to argon, the ratio of ^{40}K to ^{40}Ar will drop. At any given point in time, every atom of potassium has the same chance of decaying to argon, but it is impossible to say exactly when any given atom will release an electron and decay. While we cannot say precisely how many atoms will have decayed in a given time period and we certainly can't predict which atoms have decayed and which have not, we can estimate an average rate of decay and use it to convert measures of isotope frequencies to estimates of geological time.

Similarly, although we can describe an average rate of molecular change, and we can estimate the probability that a particular site will change in a given time period, we cannot say exactly which nucleotides in a DNA sequence will change and when. This is because the process of substitution is influenced by chance at many different stages: which mutations arise, when they arise, whether these mutations are lost from the population or go to fixation, whether the populations with particular substitutions persist or are wiped out. It is therefore not surprising that molecular change does not 'tick' regularly. Instead, the interval between substitutions varies.

To explore the implications of using a sloppy clock to estimate time, compare the distribution of changes along the two paths shown in Figure 8.26 (p. 312).

Both of these series of changes have the same overall rate of change (eight changes each in the same time period). So the average length of time between each change is the same in both cases. The difference is that in the blue line, the interval between changes is always the same, but the interval between changes varies on the red line. This is easier to see if we chop the lines up, cutting them at the point a change occurs. Each piece represents the interval between two changes. If we pile up the pieces we can compare the distribution of time intervals between changes (Figure 8.27, p. 312).

You can see that the blue line, with the regular tick rate, is divided into even pieces: the time interval between changes is always the same and therefore wholly predictable. But the sections of the red line have a range of different sizes, even though the average length of red pieces is the same as the blue pieces. So although we know the average time interval between changes on the red line, we could not predict with any certainty how long any particular interval was going to be.

Figure 8.26

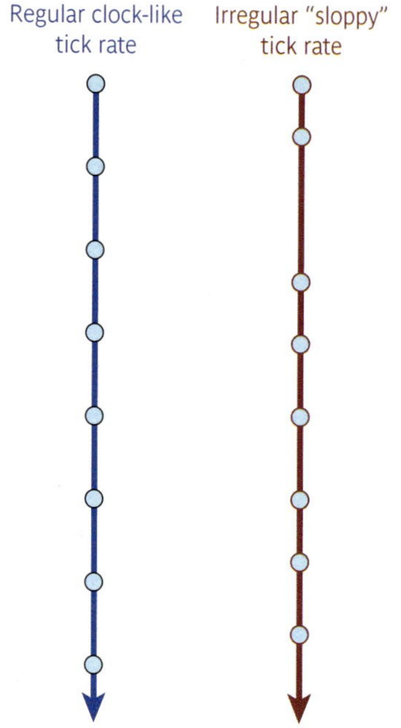

Figure 8.27

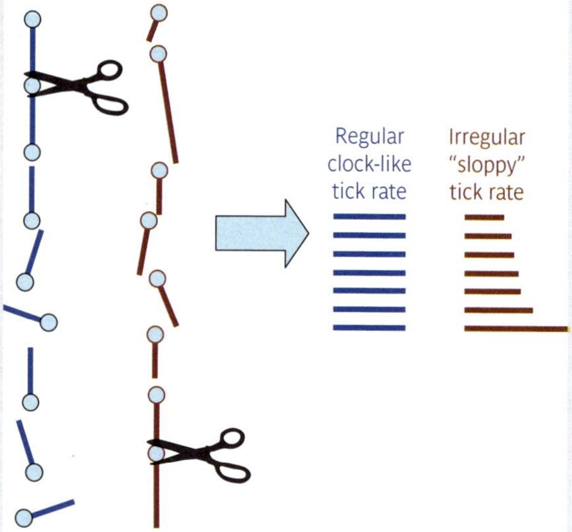

The point of this example is to illustrate that it is difficult to predict precise time from irregular changes. Each of the lineages below (**Figure 8.28**) has the same average rate of change, but, due to random variation in the interval between changes, they have accumulated different absolute numbers of changes in the same time period.

Figure 8.28

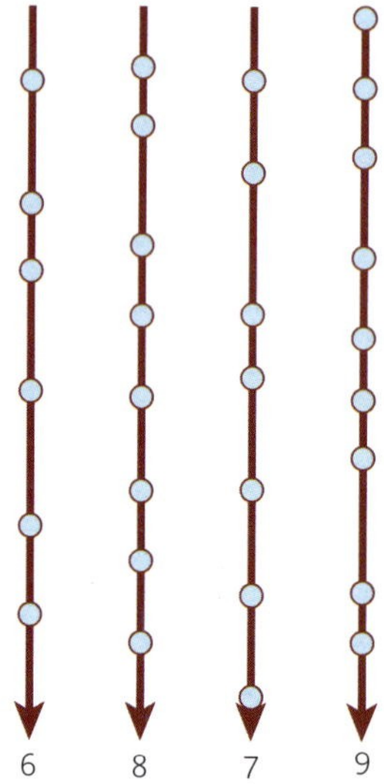

If we were to measure one branch with six changes, and one with nine changes, and we knew that they had the same substitution rate, we could not say for certain whether the difference in number of substitutions was due to time (one lineage older than the other) or whether the difference was due to chance (one just happened to have accumulated more substitutions in the same time period).

We need to take this sloppiness into account when we estimate genetic differences. Substitution models, which state the probability of a substitution happening at any site at any given time, have to allow for the fact that the chance elements in the substitution process cause variation in the time between substitutions. One way to do this is to presume that, although the exact substitution intervals are unknown, they are likely to follow a particular distribution of interval lengths. This distribution might be such that the majority of substitutions occur at an interval that is not very different from the average interval length. Intervals only slightly longer or shorter happen quite frequently, but intervals much longer or shorter occur occasionally.

Imagine cutting the lines in **Figure 8.28**, piling them up as we did before, then counting the numbers of intervals of each length. To make it easier to see, I have coloured the intervals according to their length.

Figure 8.29

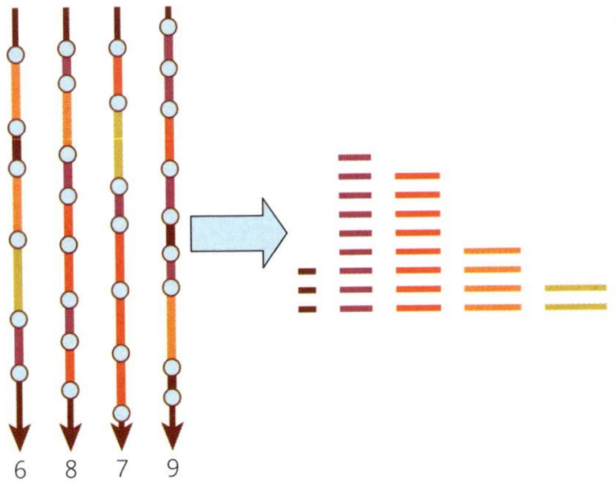

We could draw this as a distribution of interval lengths.

Figure 8.30

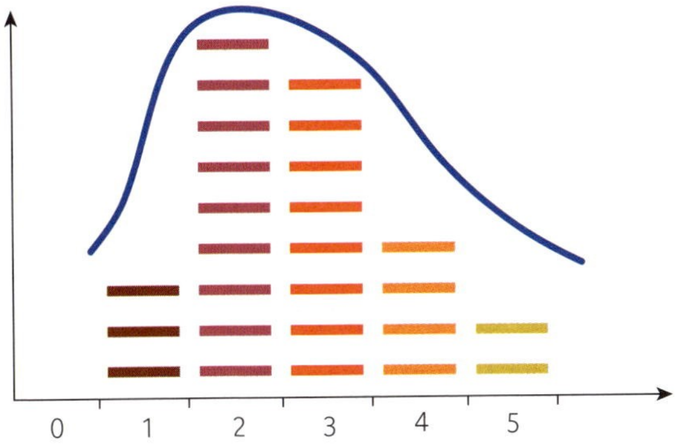

If we had a large enough sample of known interval lengths, then we might be able to describe a distribution that was typical of substitution patterns for our data. With this distribution, we can predict the scatter of substitution events in time along a lineage. We can also use this distribution to estimate the probability that a certain number of substitutions happened in a particular period of time (see **TechBox 8.1**, p. 282).

Dating with confidence

If the molecular clock is so sloppy, then how can anyone pretend they can estimate dates of divergence from sequence data? The sloppiness of the clock does not negate its usefulness. You might hope that this Christmas someone gives you a super accurate desk clock synchronized by satellite with an atomic time server. But in the meantime, you have to rely on the cheap watch you got last Christmas. Let's say that your cheap watch ticks on average once a second, but the ticks are not always exactly one second apart. In a 10-second interval, you might get ten ticks, or maybe eight ticks, or sometimes twelve. If you sat there for an hour and recorded the number of ticks per minute, then not many minutes would have exactly 60. But the average ticks per minute, taken over that whole hour, would be around 60. This clock would be imprecise, but it might be reasonably accurate. If the clock said it was 10:30, then, give or take 5 minutes, it might be more or less right. And, in the end, if you don't know what the time is, then even an imprecise clock tells you something useful.

If we know (or can guess) the distribution of interval times between substitutions, then we can predict the likely range of time periods that could have produced the number of substitutions we observe. If we observe six substitutions, and we know the distribution of intervals between substitutions follows the distribution given in **Figure 8.30**, then how long did it take those six substitutions to accumulate? Let's call the shortest interval length 1 million years' duration, and the longest 5 million years. I drew six intervals at random from the distribution, using a random-number generator (actually, I used the last column of numbers in the phonebook, but that's good enough for our purposes). Here are the first five branch lengths I produced by drawing at random from the distribution of substitution intervals. These branches have lengths of 20, 12, 17, 10, and 19 million years respectively.

Figure 8.31

Of course it would be possible to get a very short path (6 million years) or a very long path (30 million years), but these would be pretty unlikely to occur.

Figure 8.32

If we continue to draw random sets of six substitution intervals, we could build up a distribution of the possible lengths of time it would take to accumulate six substitutions. To speed this process up, I got a friend to write a little computer program that did the random sampling for me. Here is a distribution of 100,000 randomly sampled branch lengths.

Figure 8.33

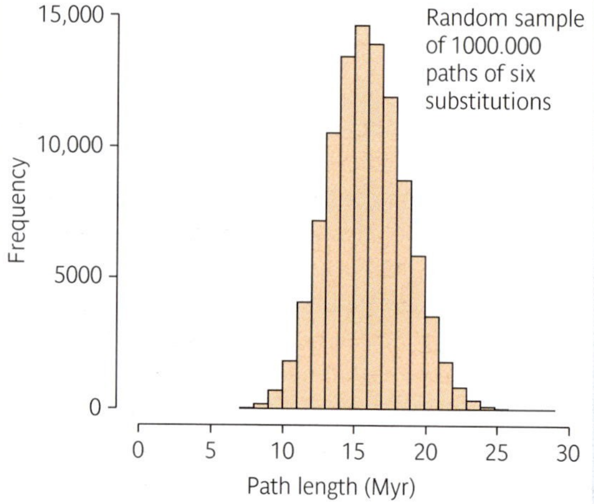

You can see that most of the branch lengths are clustered around the mean value of 16 million years. Although the random sampling occasionally produces a very short branch (the minimum here is 7 million years) or a very long branch (the maximum is 29 million years), 95% of the estimates lie between 11 and 22 million years. So if we observed a lineage with six substitutions, we could not say exactly how old it was, but we could say that, given what we know about this substitution rate for these sequences, the lineage was likely to be somewhere between 11 and 22 million years old. In this hypothetical case we based our distribution on observed substitution intervals. But in a normal situation we will not have that information. So most methods assume that substitution intervals vary according to some predefined distribution, the dimensions of which can be tailored to the data under consideration (see **TechBox 8.1**, p. 282).

A sloppy clock is better than no clock

The sloppiness of the clock prevents us from saying exactly how long it took to produce the observed number of substitutions. But we can use a distribution of possible path lengths to describe how likely the value is to fall between any given range of path lengths. In other words, we can define a confidence set: the range of values that we are quite confident the true estimate falls within. This is akin to the cheap watch example above, where we said the time was 10:30, give or take 5 minutes. We don't know for sure whether it's 10:25 or 10:35 or somewhere in between, but we are pretty certain it isn't lunchtime yet. Which brings us to an important point: even if you don't know the precise date of divergence, with confidence intervals you can reject certain values as being highly unlikely. In other words, even a sloppy clock can be used to test hypotheses about divergence times.

The sloppiness of molecular date estimates can be, to some extent, accounted for by methods that take the stochasticity of the substitution process into account. For example, one study that aimed to test the Cambrian explosion hypothesis quantified the uncertainty in date estimates for the origin of animal phyla by estimating confidence intervals using maximum likelihood (**TechBox 7.2**).

Like other studies, this one used DNA sequences from a range of different living animals to estimate the date of divergence between the phyla. A probabilistic model of molecular evolution allowed estimation of the range of possible date estimates that could be compatible with the observed differences between the sequences. This study also allowed for the variation in rate between lineages by estimating separate substitution rates for different phyla using a range of calibration dates. The results reflected the range of possible divergence times that could have produced the observed DNA sequence data, given an assumed pattern of sloppiness of the clock and allowing for variation in substitution rate between lineages. The confidence intervals were embarrassingly large (**Figure 8.34**).

Figure 8.34

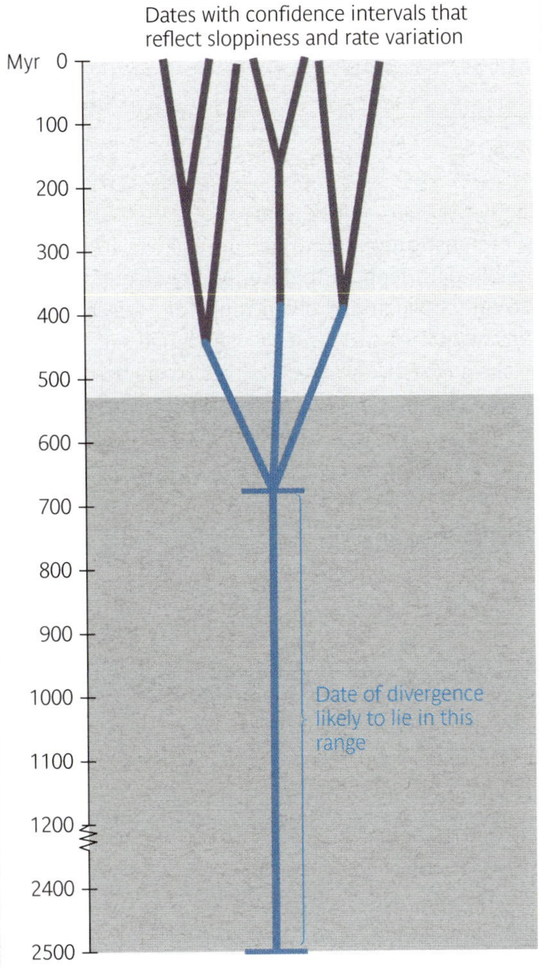

These estimates are horribly imprecise. The confidence intervals go so far back into the past that they cannot reject the possibility that the animal kingdom originated billions of years ago. But even imprecise estimates can be informative if they are accurate; that is, if the confidence intervals, however wide, contain the true value. In this case, even though the confidence intervals span over a billion years, they do not include the early Cambrian or late Precambrian. These estimates lack the precision that would allow us to say exactly when the animal kingdom diversified. But, if these estimates are accurate, then they can tell us when the animal kingdom didn't diversify: the results obtained using these data, methods and models of evolution suggest that these lineages must have diverged before the first recognizable animal fossils. Of course, if the results are inaccurate, due to problems with the data, methods, or assumptions, then we could be led astray.

For a similar approach on a much more recent evolutionary event see Case Study 8.1

So molecular dates are imprecise, but that doesn't necessarily make them unusable. Far more important is the question of whether molecular dates are inaccurate, in which case the confidence intervals, however wide or narrow, may fail to contain the true date of divergence. So far we have considered the imprecision arising from the sloppiness of the tick rate. But the pattern of substitutions is more than sloppy. Substitution rates can also vary consistently between lineages, and this lineage-specific variation in rate can generate inaccurate molecular date estimates.

Variation in the rate of molecular evolution

Given all that we have learned about molecular evolution thus far, we should not be the slightest bit surprised to find that the rate of molecular evolution varies between lineages. After all, factors that affect both the rate at which mutations are generated, and the proportion of these mutations that become substitutions, can vary between species.

In Chapter 3, we saw that DNA repair efficiency influences the mutation rate, and species can differ in their repair efficiency. Some of these differences are due to the presence or absence of whole repair systems: for example, rats and mice seem to lack some of the excision repair systems found in humans. Other differences may be due to the efficiency of repair systems. In particular, polymerase enzymes that copy DNA can vary greatly in their error rate. For example, the HIV genome is copied by reverse transcriptase which has a very high error rate, contributing to a mutation rate that is a million times higher than the mutation rate of its human host. The mutation rate can even vary within an individual: in mammalian cells, mitochondrial polymerases have a higher error rate than the major nuclear polymerases. Selection may play a role in shaping DNA repair rates in some species, allowing the evolution of higher or lower mutation rates. For example, in some cases treatment with antibiotics can favour 'mutator' bacteria that have less efficient repair, thus a higher rate of mutation, so are more likely to accidentally invent antibiotic resistance.

In Chapter 4, we saw that every time the genome is replicated there is a chance of acquiring copy errors. Anything that influences the number of DNA replications per unit time can influence the mutation rate. Some consequences of this fact are fairly obvious. Species with short generation times, such as mice, tend to have faster rates of molecular evolution than species that take longer to reproduce, such as humans, because they copy their genomes more often in any given time period. But some consequences of the copy number effect may be less obvious. Bird species that have more intense sexual selection have higher rates of molecular evolution. This is may be because males in strongly sexually selected species must produce more sperm, and the more sperm is produced, the more cell divisions the germline undergoes, so the more copy errors accrue per generation.

Male-biased mutation is explained in Chapter 4

In Chapter 5, we saw that population size influences the rate at which mutations go to fixation. So populations confined to islands tend to have a higher rate of fixation of mutations by drift than their wider-ranging mainland relatives (see **TechBox 8.3**). Bacteria normally have very large population sizes. But endosymbiotic bacteria live in small groups sequestered inside their host's body. These bacteria have much smaller effective population sizes, and higher substitution rates, than their free-living relatives. Effective population size can differ consistently between species, but it can also vary dramatically over time, as populations crash and expand. Fluctuations in population size are likely to cause changes in patterns and rates of substitution (**TechBox 5.3**).

In addition to these known effects on the rate of substitutions, there may be many other features of organisms that influence molecular evolution. For example, in **Case Study 7.2** it is reported that parasitic plants tend to have faster rates of molecular evolution than their non-parasitic relatives. It has also been shown that, for some phylogenies, the most species-rich lineages have longer molecular branch lengths. Since there are many possible influences on rates of molecular evolution, we expect variation in substitution rate to be quite common, even between closely related species. If rate variation is widespread, then this rather complicates our attempts to estimate evolutionary time from molecular data. Now we are going to look at three alternative approaches to estimating time, given we know that substitution rates can vary between lineages: selecting rate-constant sequences, estimating lineage-specific rates, or predicting rate changes along phylogenies.

Dating with variable rates

The earliest molecular clock studies were applied to sequences for which the rate of substitution seemed to have been constant in many lineages over a long time. Typically, these studies established rate constancy by plotting observed sequence differences between lineages of known divergence times (like in **Figure 8.4**, p. 289). If we know that the rate of molecular evolution has been the same in all lineages, then we only need a single calibration rate with which to date the rest of the tree. For example, **Case Study 8.2** (p. 321) describes how researchers found that the rate of substitution in part of the envelope gene of HIV was approximately the same for a diverse sample of sequences, so they decided to use this estimated rate to date the origin of sequences from a hospital outbreak of HIV infections. To estimate the date of the animal radiation, many researchers have sought to construct a database of sequences that evolve at a slow and steady rate, rejecting any genes that show significant rate variation between lineages. When the rate of molecular evolution in these apparently clock-like sequences is calculated using known

TECHBOX 8.3 Detecting rate variation

KEYWORDS

relative rates

ingroup

outgroup

likelihood ratio test

clock test

Akaike Information Criterion (AIC)

RELATED TECHBOXES

TB 8.1: Estimating branch length

TB 8.4: Molecular dating

RELATED CASE STUDIES

CS 7.2: Non-tree-like flowers (rate variation in phylogenies)

CS 8.2: Molecular detective (testing for rate variation)

Relative rates tests

Imagine that we want to investigate the tempo and mode of evolution of the raspberries (*Rubus* subgenus *Idaeobatus*), and we are particularly interested in the origin of the native Hawaiian raspberries, *Rubus hawaiensis* (common name 'ākala'; **Figure TB8.3a**). We have some good fossil pollen records from continental members of this subgenus, and we would like to use that information to calibrate the rate of molecular evolution in this group, and thus date the origin of the island endemic. We know that several factors could increase the rate of molecular evolution in island endemics, such as smaller population size, or relaxed selection, or adaptation to new niches. How can we check if the substitution rate is the same in all raspberry lineages?

Figure TB8.3a Conservation of the native Hawaiian raspberry (*Rubus hawaiensis*, left) is complicated by the introduction of non-native raspberries (such as *Rubus niveus*, right) which have invaded many of the islands ecosystems, particularly as the invasive species may hybridize with the locals. Attempts to use biological control agents against the invasive species may endanger the local raspberries. Should the native species be put under threat to enact a conservation programme aimed at restoring whole ecosystems?

Courtesy of Forest and Kim Starr (USGS).

The simplest test for rate variation is to estimate the amount of molecular change along two lineages since they last shared a common ancestor, then ask if one lineage has undergone more change than the other. Since we don't have a DNA sequence for the ancestor of the mainland and island lineages, we can't directly estimate the number of changes each has accumulated since they split. But we can compare the sequences of the living members of these lineages. For example, if we compare a mitochondrial DNA sequence between *Rubus hawaiensis* and its close relative, the salmonberry *Rubus spectabilis* (**Figure TB8.3b**), we might observe that there are six differences between them.

From this information alone, we cannot tell whether these lineages have the same or a different substitution rate, because, although we can see which positions in the alignment have undergone a substitution since these lineages last shared a common ancestor, we don't know on which lineages the substitutions happened. So the history of substitutions in these lineages could look like any of these (or other) reconstructions (**Figure TB8.3d**).

Figure TB8.3b The salmonberry (*Rubus spectabilis*), native to North America, is thought to be the closest relative of one of the Hawaiian endemic raspberries.

Figure TB8.3c

Rubus_hawaiensis	CAAAATCGAACCCACATCCCAGGTACCCTTACACCCTTTAA
Rubus_spectabilis	CAAAATCGAACCCACATTCAAGGTAGCCTTATATCATTTAA

Figure TB8.3d

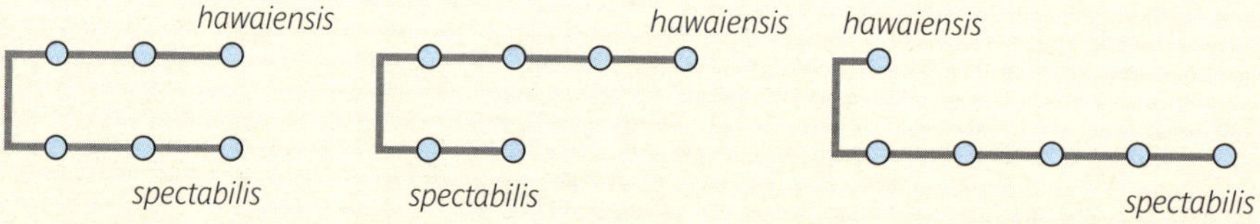

We can solve this problem by comparing the observed number of substitutions between each of these sequences and another more distantly related lineage, termed an outgroup. In this case we are going to include the Ceylon raspberry, *Rubus niveus* (**Figure TB8.3a**). The ingroup species (*hawaiensis* and *spectabilis*) share a more recent common ancestor than either does with the outgroup (*niveus*).

Figure TB8.3e

Rubus_hawaiensis	CAAAATCGAACCCACATCCCAGGTACCCTTACACCCTTTAA
Rubus_spectabilis	CAAAATCGAACCCACATTCAAGGTAGCCTTATATCATTTAA
Rubus_niveus	CAAAATCGAACCCGCATTCAAGTTAGCCTTATATCATTTTC

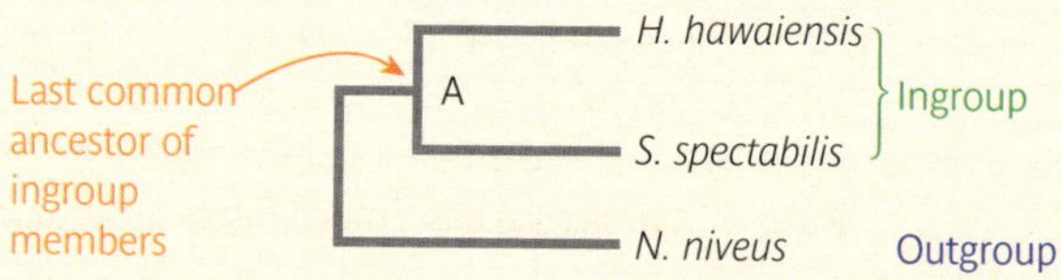

We want to use what we can measure directly (the differences between the three sequences) to estimate what we can't measure directly (the changes between each of the ingroup species and their last common ancestor). When we measure the distance from either ingroup to the outgroup, the contribution of substitutions on the lineage between the outgroup and the last common ancestor of the ingroup is the same for both species. Any difference in the distance from each ingroup species to the outgroup is due to substitutions that have accumulated in each of the ingroup lineages since their last common ancestor.

We want to be able to compare the length of the branch that connects A to H (we will call this distance dAH) to the length of the branch connecting A to S (dAS). To calculate dAH, first we measure the distance between both ingroup species (dHS), which in this case is six differences.

Figure TB8.3f

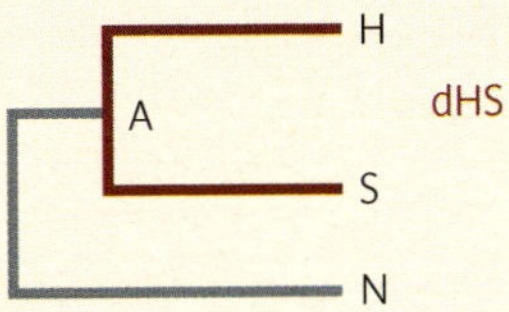

Then we measure the distance between H and the outgroup (dHN, ten differences):

Figure TB8.3g

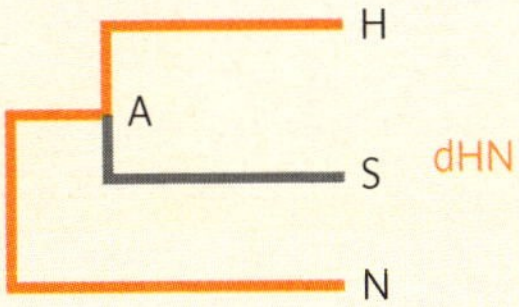

And then we measure the distance between N and the outgroup (dSN, four differences):

Figure TB8.3h

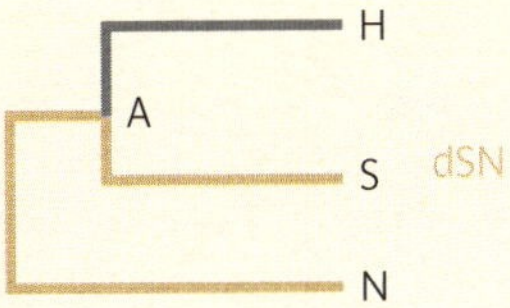

If you look at the diagrams in **Figures TB8.3f**, **g**, and **h**, you can see that dHS and dHN both cover the branch from A to H. By adding dHS to dHN then subtracting dSN, we get two times the branchlength from H to A.

Figure TB8.3i

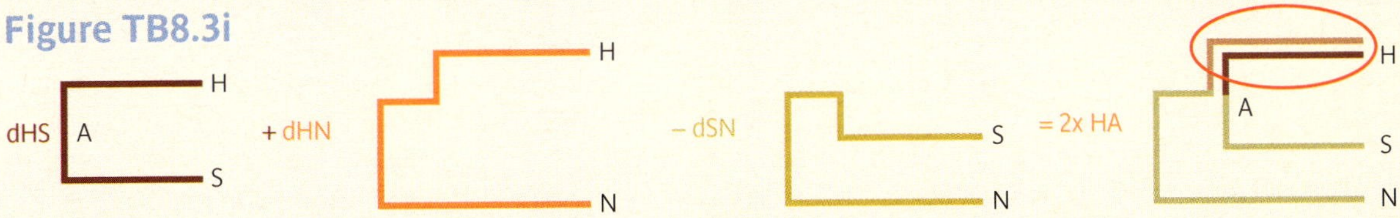

Using the differences between the sequences in **Figure TB8.3e**, we would estimate that there have been six substitutions along the HA lineage. When we repeat the process for the other ingroup lineages, adding dHS to dNS, subtracting dHS and dividing by two, we get a branch length of zero. We conclude that all six of the differences between *hawaiensis* and *spectabilis* occurred on the *hawaiensis* lineage. Thus we might conclude that the Hawaiian 'ākala' (*R. hawaiensis*) has a much faster substitution rate than the salmonberry (*R. spectabilis*).

Statistical tests for rate variation

There are two ways in which this approach is probably too simple. Firstly, just counting the observable differences between these sequences risks underestimating the number of substitutions that have actually occurred, as past substitutions may have been erased by multiple hits. If we fail to account for multiple hits, we may underestimate the number of substitutions that have occurred in either lineage, potentially masking the rate variation we are trying to detect. Secondly, even if the substitution rate is constant, there can be variation in the absolute number of substitutions along branches of the same age due to the stochasiticity of the substitution process. In other words, a difference in branch length does not necessarily imply a different underlying substitution rate. To judge whether any observed difference in branch length is more than we would expect from chance alone, we need to place our relative rates test within a statistical framework.

So we will need to estimate the branch lengths using a decent model of molecular evolution that allows for biases in the rates of different kinds of nucleotide substitutions, for variation in rates across sites, and for the sloppiness of the substitution process (see **TechBox 8.1**, p. 282). Then we need some way of testing whether these patterns of substitutions could be explained by the same underlying substitution rate, or whether we need to infer variation in rate between the lineages. We might contrast a model in which our three raspberry lineages all have the same rate to one in which the island lineage has a different rate to the mainland and outgroup lineages. If the model in which the island lineage has a faster lineage gives a higher likelihood than the model in which all lineages have the same rate, then we might conclude that the best description of our sequence data is that the island lineage has a higher substitution rate. However, adding an extra rate parameter to a model will usually increase the fit of that model to the data, because it will allow more variation in number of substitutions per branch to be explained. We want to make sure that the increase in fit of the model to the data is due to it capturing a significant biological pattern, not just random variation. There are various statistical tests that can be used to compare the goodness of fit of models that differ in the number of rate parameters, including the likelihood ratio test (**TechBox 7.4**) or the Akaike Information Criterion (AIC).

Power of tests for rate variation

As with any statistical test, the power of all of these 'clock tests' to detect rate variation depends on the data. If your alignment has relatively few substitutions – because you have a short alignment or recently diverged sequences or a slow average substitution rate – then it will be difficult to tell whether an excess of substitutions in one branch is indicative of a faster rate, or just the chance acquisition of a few more substitutions than the other branch. If a test has low power for a particular dataset then it may fail to detect rate variation. The important conclusion to draw from this is that if you fail to detect rate variation you cannot necessarily conclude that your sequences evolve in a 'clock-like' manner. Perhaps there is a lot of rate variation but you did not have the power to detect it. You might argue that any rate variation that cannot be detected can't possibly be enough of a problem to bias your analysis but alas this is not true: undetected rate variation can still cause consistent bias in molecular date estimates[1].

Tests of rate variation can be used not only in molecular dating studies, but also to investigate the tempo and mode of molecular evolution. For example, we might hypothesize that the Hawaiian raspberry has a faster rate of molecular evolution because it lives on an island. But, using only this one comparison, we cannot tell which aspect of the biology of *Rubus hawaiensis* is influencing its substitution rate, or even if the faster rate is due to an idiosyncratic event in the history of this lineage. But if we were to gather data from many different island endemic lineages, and compare each of them to their mainland relatives, and we found that, significantly more often than expected by chance, the island lineages had a faster rate of substitution, then we could start to believe we had seen a general pattern[2]. This pattern, in turn, may help us predict cases when rates will vary between lineages, and might ultimately lead to more biologically realistic models of rate change.

References

1. Bromham, L.D., Rambaut, A., Hendy, M.D. and Penny, D. (2000) The power of relative rates tests depends on the data. *Journal of Molecular Evolution*, Volume 50, pages 296–301.
2. Woolfit, M. and Bromham, L. (2005) Population size and molecular evolution on islands. *Proceedings of the Royal Society, Biological Sciences*, Volume 272, pages 2277–2282.

CASE STUDY 8.2

Molecular detective: using DNA to test sources of infection

KEYWORDS

molecular dating

relaxed clock

confidence intervals

hypothesis testing

RELATED TECHBOXES

TB 7.2: Maximum likelihood

TB 8.2: Calibration

RELATED CASE STUDIES

CS 7.1: Up the river (origins of HIV)

CS 8.1: Same but different (molecular dating)

de Oliveira, T., Pybus, O.G., Rambaut, A., Salemi, M., Cassol, S., Ciccozzi, M., Rezza, G., Gattinara, G.C., D'Arrigo, R., Amicosante, M., Perrin, L., Colizzi, V., Perno, C.F., Benghazi Study Group (2006) Molecular epidemiology: HIV-1 and HCV sequences from Libyan outbreak. *Nature*, Volume 444, pages 836–837

" *Assuming the presumption of innocence as a basis for a fair trial, it must be stated that, by any objective standard, there is no scientific evidence to convict anyone of deliberately infecting unfortunate Libyan children.* " [1]

Background

In March 1998, six new foreign staff members arrived at the Al-Fateh Hospital in Benghazi, Libya. Two months later, the hospital noted its first case of HIV-1 infection. Within 6 months of the arrival of the foreign staff, 111 children who had attended the hospital were found to be HIV-1 positive. An investigation by the World Health Organization (WHO) eventually revealed that 418 children had been infected by HIV-1 in the hospital outbreak, and many of these children were also infected with hepatitis C (HCV). A number of medical workers from the hospital were accused of involvement in the deliberate infection of the children. Six foreign medical workers were detained in prison in 1999, then sentenced to death in 2004. The death sentence was upheld in a retrial in 2007, though subsequently commuted to life imprisonment. Lawyers from an international charity who were representing the medical workers appealed to international experts to conduct an independent assessment of the evidence. This study was a response to that appeal.

Aim

Following the discovery of the Al-Fateh Hospital (AFH) outbreak, just over half of the infected children were sent to hospitals in Europe for assessment and treatment. The median age of these children at the time of diagnosis was only three and a half years old. Most of these children were asymptomatic, but some had begun to develop symptoms of AIDS. This international team of researchers had access to samples of HIV-1 from 44 of the children who were treated in Europe. If all of the children were infected by the foreign medical workers then you would expect that the divergence between their DNA sequences would be consistent with the genomes having a common ancestor in 1998 (unless they were deliberately infected with different strains of HIV, in which case the sequences from the children would occur in different parts of the phylogeny). These researchers used molecular phylogenetics and molecular dating to test this hypothesis.

Methods

HIV-1 RNA was extracted from plasma samples, amplified, and the *gag* gene sequenced. These sequences were blasted (**TechBox 3.4**) against GenBank (**TechBox 1.1**) and against the Los Alamos HIV sequence database (www.hiv.lanl.gov) to identify the most similar sequences. This provided a reference database of 56 sequences from Africa and Europe. The AFH and reference sequences were aligned automatically, then adjusted by hand (**TechBox 6.3**). The researchers tested which nucleotide substitution model best fitted the data (**TechBox 8.1**, p. 282), then used that model to estimate a phylogeny of the sequences using both maximum likelihood (**TechBox 7.2**) and Bayesian methods (**TechBox 7.3**). To estimate a substitution rate for this alignment, the researchers collated a reference set of the same genomic region from 48 HIV-1 sequences of known sample dates, spanning two decades. HIV-1 evolves so rapidly that sequences sampled in different years will be measurably different, so these sample dates could be used to estimate the substitution rates. They used both a strict clock and a Bayesian variable-rate method (**TechBox 8.4**, p. 327) to estimate rates of substitution, which did not vary greatly over the tree. This rate was used to estimate the age of the AFH cluster of sequences under a number of different epidemiological models (e.g. whether the population size of viruses stayed the same or grew exponentially over time) and different substitution models (e.g. allowing rates to vary across all sites, or specifying a codon model of variation in rates between sites).

Results

The AFH HIV-1 sequences all formed a distinct clade on the phylogeny, suggesting that the children had all been infected by a single strain of HIV-1. However, under nearly all of the models used, the date of origin of this clade of AFH sequences was estimated to be before the arrival of the foreign medical workers in March 1998. Furthermore, the confidence intervals on these estimates did not contain March 1998, so the probability that they could have originated from the foreign medical workers was considered to be effectively zero. Similar results were obtained for the hepatitis C (HCV) sequences (**Figure CS8.2a**).

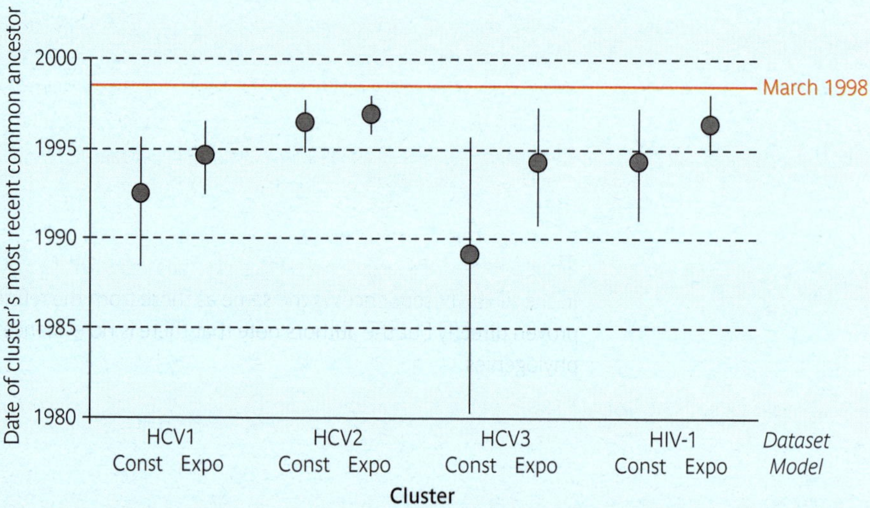

Figure CS8.2a Estimated dates of the most recent common ancestor for each cluster of hepatitis (HCV) and HIV-1 sequences taken from children infected at the Al-Fateh Hospital, derived under two different models of evolution: Const (which assumes constant population size) and Expo (which assumes exponential growth). The vertical lines represent confidence intervals.

 de Oliveira, T., Pybus, O.G., Rambaut, A., Salemi, M., Cassol, S., Ciccozzi, M., Rezza, G., Gattinara, G.C., D'Arrigo, R., Amicosante, M. (2006) Molecular Epidemiology: HIV-1 and HCV sequences from Libyan outbreak. *Nature*, Volume 444, pages 836–837.

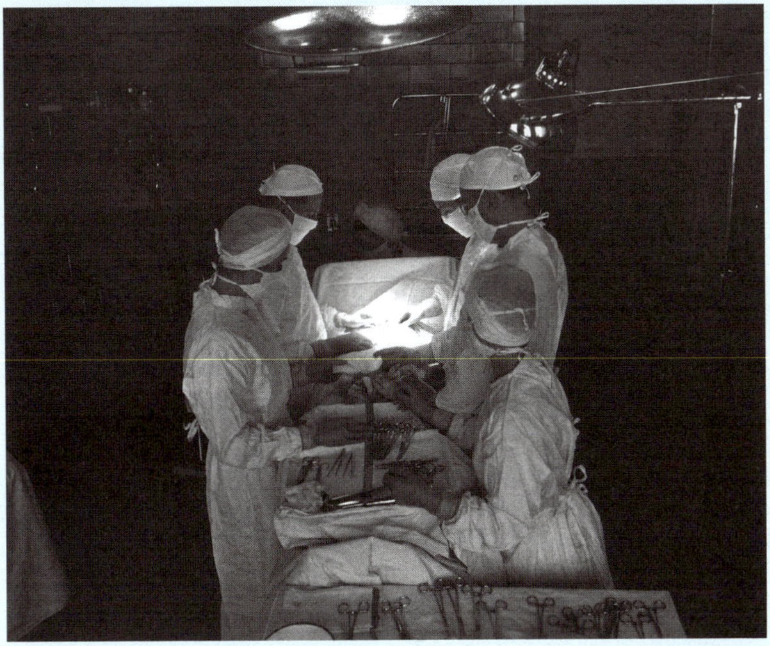

Figure CS8.2b Hygienic practices in hospitals, such as those shown in this photograph, reduce the chances of cross-infections between patients and healthworkers. Since a pioneering case demonstrated that a dentist had transmitted HIV to several of his patients (see Ou *et al*. 1992), DNA sequence analysis has been used in many different cases to investigate potential cases of iatrogenic transmission of HIV.

Reproduced courtesy of CDC/Barbara Jenkins, NIOSH; photographer: Roy Perry.

Conclusions

These results suggest that the Al-Fateh children were all infected with the same strain of HIV-1, but that this strain originated before the foreign medical workers arrived at the hospital. Those children for whom medical information was available had all had invasive medical procedures at the hospital, and several different international reports cite poor hygiene practices at the hospital as the probable cause of the infections. The high incidence of coinfection with HCV was also put forward as evidence of nosocomial (hospital-based) infection due to poor hygiene. The six accused medical workers were released in 2007 and allowed to return to their countries of origin, though their release was primarily the result of diplomacy rather than persuasive scientific evidence.

Limitations

The molecular date estimates rely on the assumption that the rate of molecular evolution in the AFH HIV sequences is the same as those from the reference database. This cannot be proven directly but the authors note that there is no evidence for such a rate change in their phylogenies.

Future work

DNA sequence analysis is being increasingly used as legal evidence in cases of alleged deliberate infection or negligence. Just as the application of DNA fingerprinting techniques in the courtroom has been confused by poor use of statistics and failure to explain the methods properly to legal professionals and jury members, so the use of molecular phylogenetic analysis in court cases will require careful attention to, and clear communication of, the confidence limits on molecular phylogenies and date estimates.

Reference

1. Hirsch, M.S. (2007) Justice in Libya? Let scientific evidence prevail. *Journal of Infectious Diseases*, Volume 195, pages 467–468.

dates of divergence, the molecular date estimates obtained for the radiation of the animal phyla are all much older than the Cambrian (ranging from 630 to 1200 million years; **Figure 8.35**).

Figure 8.35

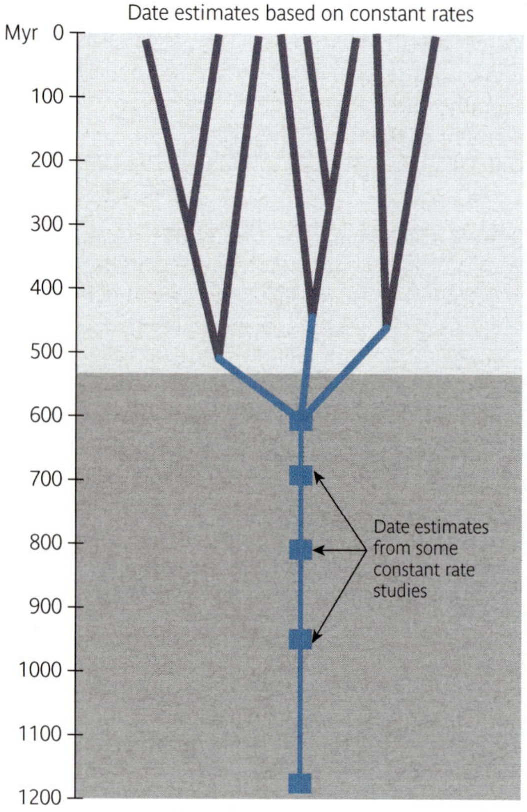

The problem with this approach is that the tests used to detect and reject sequences that vary in rate between lineages have relatively low power to detect rate variation (**TechBox 8.3**). This means that when they are applied to sequences with relatively few substitutions – such as short sequences with slow rates of divergence – then the tests may fail to reject all of the rate variable alignments. If rate variable sequences are included in a molecular dating study that relies on the assumption of rate constancy, then the calibration rate will not be an accurate reflection of the amount of time taken to produce the observed substitutions, so the date estimates could be inaccurate. Furthermore, these tests cannot detect concerted changes in substitution rates, so instances where all lineages increase or decrease in rate wont be rejected, even though such a pattern of rate variation can affect the accuracy of date estimates.

If we cannot rely on finding truly clock-like sequences, then we will need to find a way of allowing for rate variation in our date estimation procedures. One way of doing this is to use multiple calibration dates to estimate substitution rates separately for different parts of our phylogeny. If we want to estimate lineage-specific rates directly from our alignments, then we will need at least one calibration for every different rate. This may be tractable if rate changes are relatively rare, for example if each phylum tends to have a characteristic rate of molecular evolution. But we have seen that there are many factors that influence rate even between closely related species. In fact, significant rate variation exists at all levels of the animal kingdom, between species, between families, between phyla.

This brings us to two important conclusions. Firstly, constant rates of substitution may be a feature of some datasets, but clock-like molecular evolution should be regarded as the exception, rather than the rule. There are so many factors that influence the rate of substitutions that we should expect rate variation to occur in many, if not most, datasets. Secondly, if rates can vary consistently between species, then we should expect rates to evolve along lineages. For example, if we compare two species of snake, and we find that one has a shorter generation length than the other, then we do not expect that the difference in generation time evolved the instant that the two ancestral snake populations became reproductively isolated. Instead, we might assume that the two populations became gradually more distinct in generation time. We might also expect that substitution rates could have evolved in concert with generation time in these species: as the generation length became shorter in one species, its rate of substitution increased. If rates of molecular evolution evolve along phylogenies, just as other species traits do, then rates could be in a constant state of change over phylogenies.

The effect of generation time on substitution rates is explained in Chapter 4

If substitution rates evolve along phylogenies, then we would not be surprised to find that every branch in our phylogeny has a different rate of molecular evolution. If we wanted to estimate these rates directly from our sequence data, then we would need a calibration for every branch of the tree. This would be a marvellous way to proceed, but there are few groups that have such a well-known evolutionary history (and, of course, if we already know when all the lineages diverged then we don't really need molecular dates for that group). If we don't have calibrations for every branch in the tree, then we cannot estimate rates directly. The only way to allow for frequent rate change, if we cannot measure it directly, is to predict changes in rate.

We have already seen how substitution models can be used to predict the occurrence of substitutions that cannot be directly observed. The same principle can be applied to estimating changes in the substitution rate itself. Most recent molecular dating methods employ a model of evolution that expresses the probability of the substitution rate changing from one branch to the next. Then they use this model to evaluate alternative patterns of rate change along a phylogeny, finding the series of rate changes that maximizes some measure of the plausibility of the solutions. Once the rate changes along the tree have been predicted, then one or more calibrations can be used to convert the branch lengths into measures of evolutionary time.

This approach sounds simple but, in reality, a large number of different assumptions must be made in order to decide between different possible solutions. If we allow rates to change on any branch of the phylogeny, then when we estimate a certain number of substitutions between two lineages, we have to decide whether to reconstruct that branch length with a slow rate (therefore a deep divergence) or a fast rate (therefore a shallow divergence) or anywhere in between.

Figure 8.36

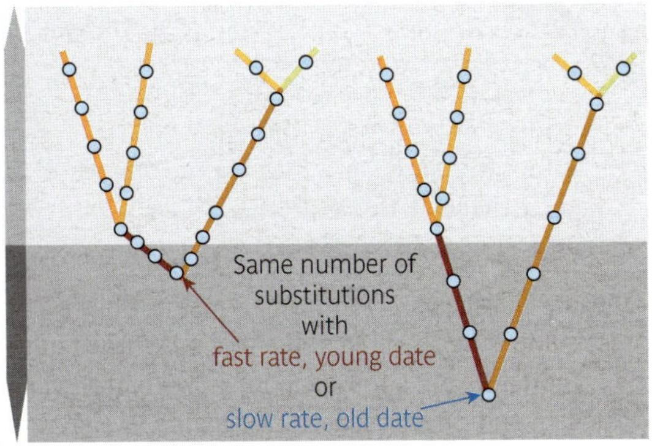

All molecular dating methods must make assumptions about the way substitution rates behave over phylogenies. In strict-clock methods, the assumption is that rates are the same over all lineages. In local-clock methods, the assumption is that defined lineages all share the same rate. Rate-variable molecular dating methods (sometimes referred to as relaxed-clock methods) also make assumptions about the way rates change over the phylogeny, in order to decide which pattern of rate changes best fits the observed data. Generally, these assumptions include some form of statement of the probability distribution of branching events along the phylogeny, and some kind of function that describes what kind of changes in rate are most likely (**TechBox 8.4**, p. 327). Molecular dating methods that allow changes in rate are very complex, although the casual user of molecular dating software may be largely unaware of all the statistical machinery that is being employed to produce a their date estimates.

The large number of different parameters included in these models makes it possible to come up with many alternative solutions by varying the assumptions made. So it is not surprising that when different research groups have applied rate-variable molecular dating methods to estimating the timing of the divergence of the animal phyla, they have come up with a range of estimates. Of course, these studies have also used different sequences, species, and calibrations, which may also add to the wide variation in the date estimates, which range from just before the Cambrian to a billion years ago (**Figure 8.37**, p. 326).

How can we tell which estimates are right and which are wrong? Since we don't know which model is right, the best approach is to explore the robustness of our results to different assumptions, to compare the estimates to other lines of evidence (such as the fossil record), and to consider the implications of the dates in light of what we know about evolution at the molecular and organismal levels (see **Case Study 8.2**, p. 321, and **TechBox 8.4**, p. 327). Most importantly, we must recognize the sources of error in molecular date estimates and use these to describe a range of possible date estimates, rather than aiming to produce a single magic number.

Did the Cambrian really explode?

“*The Cambrian, so beautifully documented . . . , was clearly a special time when animal life exploded into the wide range of forms that we associate with the modern faunas of today. So what does it mean to find molecular estimates of animal . . . divergences that are so much older? That is the challenge, both to paleontologists and phylogeneteticists.*"

Levinton, J. (2001) *Genetics, Paleontology, and Macroevolution*, 2nd edn. Cambridge University Press

At the present time, we cannot use molecular data to put precise date estimates on the divergence of the animal phyla. But this does not make molecular information useless. After all, we do not reject the fossil record when faced with evidence that it is less than perfect, such as the 80-million-year hiatus in the coelacanth record, or the total absence of flatworm fossils. Instead, palaeontologists work to derive as much information as possible from the palaeontological record, despite its obvious imperfections. We can take a similar approach to the imperfect record of history in the genome.

We know that there is information about the timing of animal diversification in the molecular record. We can see this in the short alignment shown in **Figure 8.13** (p. 299). Fossil evidence suggests that the two damselfly lineages have been separated for around 120 million years while the clam and mussel last shared a common ancestor at least 485 million years. Sure enough, there are far fewer differences between the more recently diverged damselflies that there are between the clam and the mussel. On the whole, we can reasonably expect that sequence alignments should be able to provide us with information about divergence times. But, as we have seen, molecular date estimates are usually imprecise and potentially inaccurate, so we must exercise caution when interpreting molecular evidence for the timescale for evolution, just as we exercise caution when interpreting tempo and mode of evolution from the fossil record.

This is not the place to review the debate about the nature, causes, and consequences of the metazoan radiation, discussions of which fill many greater volumes than this one. But we can summarize the evidence from molecular data as follows: when we compare DNA sequences from living members of different animal phyla, we find there are more substitutions between them than we would expect from only half a billion years of evolution, based on what we know about substitution rates in animal lineages. There are several possible explanations for this surprising observation. One is that phyletic lineages diverged long before the Cambrian, but we have no unambiguous evidence of their earliest history. To support this hypothesis, we would have to think of a good reason why members of bilaterian lineages have not been found, or not recognized, from the Precambrian geological record. An alternative explanation is that rates of molecular evolution were so much faster in earliest part of the animal radiation that molecular dates based on post-Cambrian rates greatly overestimate the origin of these lineages. To believe this hypothesis, we would need to come up with a good reason why all early animal lineages would have had systematically faster rates of molecular evolution than all of their descendants.

Figure 8.37

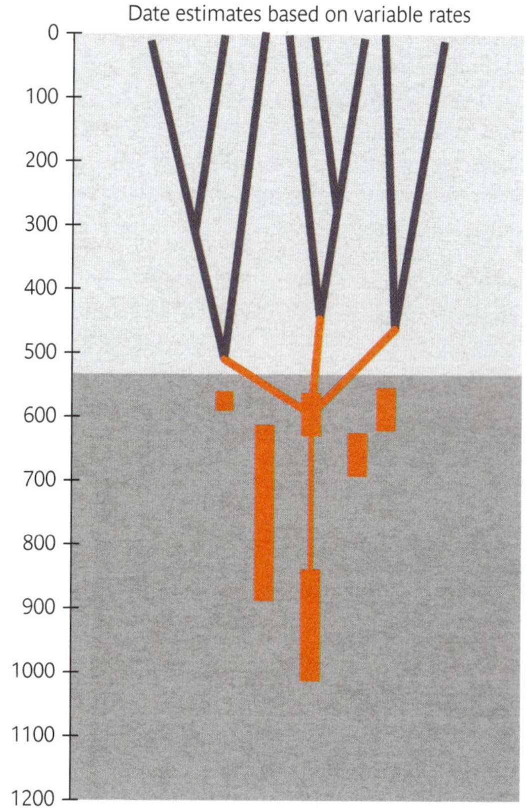

So often in biology, when there are two extreme hypotheses, the answer is eventually found to lie somewhere in between. Perhaps that is how different ideas

TECHBOX
8.4

Molecular dating

KEYWORDS

strict clock
local clocks
relaxed clocks
rate-variable dating
calibration
model
confidence interval

FURTHER INFORMATION

Welch, J.J. and Bromham, L. (2005) Estimating molecular dates when rates vary. *Trends in Ecology and Evolution*, Volume 20, pages 320–327.

RELATED TECHBOXES

TB 8.2: Calibration
TB 8.3: Detecting rate variation

RELATED CASE STUDIES

CS 8.1: Same but different (molecular dating)
CS 8.2: Molecular detective (estimating infection dates)

Molecular dating is the use of measures of genetic divergence to estimate the date of divergence of biological lineages. In order to estimate time from molecular sequence data, you need to be able to make assumptions about the rate of change in those sequences, and you need some way of calibrating the rate of molecular change against an absolute timescale (see **TechBox 8.2**, p. 300, for an explanation of calibration). There are a wide range of methods, all of which employ different assumptions about substitution rates. We will briefly review three broad approaches, but there are some methods that do not fit neatly into these categories.

Strict clocks are the simplest model of substitution rate evolution, resting on the assumption that all of the sequences in your alignment have the same underlying rate of substitution. In this case, we only need one calibration rate to date the whole phylogeny. Because it has high predictive power, this is the most powerful and useful molecular clock model. If rates are constant over the phylogeny, and we know the date of one node (branching point) in the phylogeny, then we can date all other nodes. One constant substitution rate of molecular evolution may provide a fair description of molecular evolution of some datasets (e.g. **Case Study 8.2**). But the rate of substitution will vary between lineages for many (perhaps most) datasets. If we apply an analysis that assumes constant rates to sequences that do not share the same average rate of change (whether we can detect that rate variation or not: **TechBox 8.3**, p. 317) then we will produce consistently misleading results.

Local clocks allow that lineages within a phylogeny can have different substitution rates, but the substitution rate is assumed to be constant within sets of lineages assigned to each rate category. Rates can then be directly estimated using multiple calibrations, at least one for each category of lineages, or can be predicted by an optimization procedure (similar to those outline in 'relaxed clocks' below). The rate categories might be defined by the availability of calibrations, or some other form of prior information (for example, free-living versus parasitic lineages), or lineages may be assigned to rate categories using an optimization approach. If you have sufficient calibrations to estimate rates across the phylogeny, and you are confident that rate variation across the phylogeny occurs mainly between lineages defined by different calibrations, then the local clock model provides a useful method for inferring node dates. But as with the strict clock model, if the assumption of rate constancy within rate categories is violated, then the date estimates will be inaccurate. Given that many different factors can influence the rate of substitution, it is reasonable to assume that the rate of molecular evolution can vary even between closely related lineages, which may invalidate the application of a local clock approach for many datasets.

Relaxed clocks are so called because the requirement for rates to be constant has been relaxed (not necessarily because you will feel relaxed while applying or interpreting them). The rate of substitution is allowed to vary between every node on the phylogeny. It would be impossible to directly estimate a substitution rate for every branch unless you had a calibration for every node (and if you did, you wouldn't need to estimate dates using molecular data). So these methods use a model of molecular evolution to predict changes in rate along the phylogeny. As with any statistical estimation procedure, the accuracy of prediction of rate changes will depend critically on how well the model fits the data. Current methods employ a variety of models of rate change, and new methods are frequently released.

Ideally we would develop a model of rate change based on a solid understanding of the dynamics of molecular evolution in the lineages under consideration. But, in most cases, we do not have a clear idea of what is driving rate variation, or how much rate variation to expect. Even if we did know what determined rate variation in a particular group, we might require a lot of extra biological information to predict patterns of rate change – like generation time or population size – which may not be available for all lineages in the phylogeny. So all current 'relaxed clock' methods use statistically tractable but biologically arbitrary models of molecular evolution. The best way to illustrate the importance of considering the influence of the model on the date estimates produced is to look at two studies that used the same data but different methods and produced very different date estimates.

Different methods give different results

Molecular date estimates for the metazoan radiation vary widely between studies, but since most studies use different sequences, species, calibrations, and methods, it is difficult to pin down exactly why the date estimates vary. However, two studies of the metazoan radiation used the same dataset but got different answers. One applied a 'local clock' method, using multiple calibrations to estimate rates for different animal lineages, and the other applied a 'relaxed clock' method, using a model of evolution to fit different rates to the molecular phylogeny. The local clock method produced estimates greater than 680 million years ago (see **Figure 8.34**, p. 315), suggesting that the Cambrian explosion of fossils does not mark the divergence of the major animal lineages[1]. The relaxed clock method produced date estimates between 498 and 616 million years, a result hailed by many biologists as supporting the Cambrian explosion of lineages and resolving the conflict between molecular and palaeontological dates for the animal radiation[2]. When faced with the same data, why did the local clock method reconstruct the branch lengths as a slow rate and deep date,

Figure TB8.4 A gratuitous photograph of a sea cucumber.

whereas the relaxed clock method reconstructed the branch lengths as a fast rate and recent date (Figure 8.37, p. 326)? The difference in date estimates can only be coming from the assumptions made in the analyses. For example, exploration of the results of the relaxed clock study show that the young date estimates rest critically on the birth–death model used to distribute the node heights of the phylogeny[3], which relied on the assumption of a constant rate of lineage speciation and extinction[2] throughout the phylogeny, and that the sequences used were a random sample of the tips of the phylogeny. This provided a statistically tractable model, but it is not derived from biological knowledge of the system; indeed, it is easy to demonstrate that speciation and extinction rates have not been constant, and that the sequences used are a biased sample of animal diversity.

In statistical estimation procedures there is a trade-off between adding parameters that capture important and consistent patterns in the data, and overfitting a complex model and producing inaccurate or imprecise results. Poor model fitting can lead to high confidence in the wrong results. Given that we rarely understand all of the forces acting on substitution rates for a given dataset, we have no way of checking whether our models are providing useful explanations of our data. As with all techniques outlined in this book, the best approach is to explore your data: test the influence of assumptions, compare to multiple lines of evidence, and consider all estimates in light of the confidence you have in your data and methods.

References

1. Bromham, L., Rambaut, A., Fortey, R., Cooper, A. and Penny, D. (1998) Testing the Cambrian explosion hypothesis by using a molecular dating technique. *Proceeding of the National Academy of Sciences USA*, Volume 95, pages 12386–12389.
2. Aris-Brosou, S. and Yang, Z. (2002) Effects of models of rate evolution on estimation of divergence dates with special reference to the metazoan 18S ribosomal RNA phylogeny. *Systematic Biology*, Volume 51, pages 703–714.
3. Welch, J.J., Fontanillas, E. and Bromham, L. (2005) Molecular dates for the 'Cambrian explosion': the influence of prior assumptions. *Systematic Biology*, Volume 54, pages 672–678.

about the tempo and mode of the diversification of the animals will be resolved. Some palaeontologists feel that a range of evidence points to an origin of animal lineages between 700 and 600 million years ago, based on evidence from fossilized embryos, possible animal traces, and suggestive biomarkers. Furthermore, this may coincide with the rise in multicellular forms of fungi and plants. While molecular estimates cover a bewilderingly large range of possible phylum divergence dates, the majority seem to fall around this period. Maybe, as we uncover more fossil evidence and learn more about molecular evolution, we might find that the fossil and molecular evidence is converging. However, even if the roots of the animal radiation lie deep in the Precambrian, there seems little doubt that there was a remarkable transformation across the Cambrian boundary. It is difficult to see how metazoan ancestors, possessed of shells, carapaces, teeth, or spikes, could have so comprehensively evaded being recorded in the many Ediacaran fossil faunas. Perhaps the molecular dates correctly identify the origins of the lineages, and the fossil record correctly illustrates the subsequent explosion of diversity and complexity.

The tempo and mode of the animal radiation is far from resolved. There remain many open questions. Molecular data may have some role in resolving these issues, but it is unlikely to provide all the answers. Hypothesis testing in evolutionary biology usually requires the combination of as many different lines of evidence, and the application of as many alternative modes of analysis, as possible. I am looking forward to seeing what emerges in the years ahead.

Conclusions

There is temporal information in DNA: the longer genomes evolve separately, the more different they will be. We cannot predict evolutionary time from molecular data with unswerving accuracy or deadly precision. However, in science we are accustomed to the idea of presenting estimates with confidence intervals that reflect our certainty that the true value lies within a range of estimates. For molecular dates, we should expect the confidence intervals to be quite wide, because there are so many things that can influence the number of substitutions that accumulate in a lineage over time. These confidence intervals can draw on two sources of inference. Firstly, we can use probabilistic models of molecular evolution to quantify the expected variation in substitution rate. Secondly, we can use our knowledge of the biological traits that influence the rate of molecular evolution to predict instances in which we expect rates to change. Sometimes our confidence intervals might be formally expressed as a distribution of values derived from statistical analysis (**Case Study 8.2**, p. 321), other times they may represent the upper and lower estimates obtained by varying the biological assumptions on which the date estimates are based (**Case Study 8.1**, p. 285).

Molecular dating will never replace the fossil record, which remains the primary source of information on the evolutionary past. But palaeontological dates are not available for all taxa. The majority of lineages have no fossil record, so we must find other ways to investigate the tempo and mode of evolution. Furthermore, the fossil record can give a misleading impression of the tempo and mode of evolution in some cases, so it would be handy to have an independent timescale to contrast it to and identify areas of discrepancy that require further investigation. We are liable to be led astray if we expect either the palaeontological or molecular record to always give us perfect point estimates of divergence dates.

Instead, the most robust approach to investigating the tempo and mode of evolution is to combine the strengths of both the palaeontological and molecular records, along with as many other sources of information as possible, and to constantly examine our findings in light of the assumptions we have made in our interpretation. Molecular dating has an important role to play in many areas of biology, so we need to develop realistic methods that provide honest confidence intervals on date estimates. Improving our understanding of the tempo and mode of molecular evolution will help us to unlock the wealth of information about biological history and evolutionary processes written in the genome.

Further information

The story of the rediscovery of the coelacanth is told in:

Weinberg, S. (2000) *A Fish Caught in Time*. Harper Collins.

A detailed account of all of the animal phyla, including their origins and diversity over time, is given in:

Valentine, J.W. (2004) *On the Origin of Phyla*. University of Chicago Press.

An accessible account of the importance of the Cambrian explosion in different views of how evolution proceeds can be found in:

Sterelny, K. (2007) *Dawkins vs Gould: survival of the fittest*. Icon Books.

You are a scientist

Or: what do I do now?

coda

"Therefore my success as a man of science, whatever this may have amounted to, has been determined, as far as I can judge, by complex and diversified mental qualities and conditions. Of these, the most important have been – the love of science – unbounded patience in long reflecting over any subject – industry in observing and collecting facts – and a fair share of invention as well as of common sense. With such moderate abilities as I possess, it is truly surprising that I should have influenced to a considerable extent the belief of scientific men on some important points."

C. Darwin, Autobiography. In Darwin, F. (1902) *The Life of Charles Darwin*. John Murray, London

What this chapter is about

You have learned how to collect data, either from GenBank or by extracting DNA from biological samples then amplifying and sequencing it. You have learned how to align the sequences, estimate the amount of evolutionary change between them, and use that information to construct phylogenetic trees and estimate evolutionary timescales. More importantly, you have learned to think about DNA sequence in an evolutionary context, comparing sequences and looking for patterns that reveal descent with modification by selection or drift. Now it is time to put all that you have learned into practice.

It is surprising, given the amount of time and paper devoted to exploring the issue, that there is no agreed definition of what science actually is. Many attempts have been made to define 'the scientific method', or to describe the motivations and behaviour of scientists in terms of philosophical principles or social rules. Perhaps we face the same problems defining science as we do defining species (Chapter 6). If different individuals or groups do science in different ways, our definition may need to be flexible, suited to the purpose and occasion for which we wish to apply it. I am not cavalier enough to wish to wade into this learned debate. All I wish to say is that being a scientist is about the way you do things. It is not dependent on your qualifications or employment. And that is particularly true of the ideas and techniques discussed in this book. The point of this little coda is to say that, whoever you are, dear reader, you should now feel happy to go out and start applying the basic principles you have learned in this book.

There are several important stages in most scientific investigations. First, you need to think of an issue that puzzles or interests you, or a problem that needs solving. Then, you need to devise some way of shedding light on that problem or idea. And then you need to communicate what you have learned to other people. You can do all of this. What's more, the field of molecular evolution makes all of these stages particularly accessible to you, because there is much that can be done without access to research budgets or specialist equipment.

One of the great principles of science is sharing. When a scientist reports research they have done, they must provide everything that an informed reader needs to verify those claims. In this way, science is fundamentally different to many other activities, such as commerce, where success may depend on keeping secrets, or the arts, where the outputs may be specific to a particular artist and not able to be reproduced by other practitioners. Obviously, as in any other field of human endeavour, there are selfish scientists who will not share their data, there are ego-driven scientists who wish to prevent others from making progress in the same field, and sadly there are, very rarely, dishonest scientists who cheat and make up their data. There are also cases where sharing data, methods, or materials may not be in the public interest (for example, genetic manipulation of dangerous pathogens). But on the whole, one of the marvellous things about molecular evolutionary biology is that almost all scientists working in the field make their data and methods freely available to anyone who wants to use them.

GenBank is a paragon of sharing (**TechBox 1.1**). Virtually every DNA sequence produced is submitted to this global database, and is then available to anyone in the world with access to an internet connection. This is a beautiful example of returning the fruits of scientific labours to the people: research dollars poured into DNA sequencing by governments and private enterprise produce a resource to which we all have equal access. New data are being generated at a phenomenal rate, greatly outstripping the rate at which they are being analysed. There are many fascinating questions in evolutionary biology that could be answered simply by playing with sequences available on GenBank.

Not only are the data freely available, but so are many of the computer programs used to analyse those data. Some software for molecular genetic analysis is produced by companies that aim to make a profit. But all of programs I currently use in my research are produced by academics in universities or research organizations, and almost all of these are made freely available to anyone who wants to download them. In addition, there are an increasing number of publicly funded server-based applications that not only provide the software but also run the analysis for you. This is handy, because it means that you do not have to own a computer with sufficient capacity for analysis, as long as you can access an internet connection. Again, the provision of software and servers for molecular genetic analysis is an admirable case of research funded by governments returning a resource to anyone in the community who wishes to use it.

Scientific research is worth little if the results are not communicated to anyone else. This is why most researchers publish the results of their work in one form or another, usually in a recognized scientific journal. If you are interested in a particular area, the first thing you should do is read about what others have been doing in this field. Review articles, which summarize the state of play in a particular field, are often the most useful place to start. Unfortunately, scientific papers can be difficult to read. If you find your head swimming with jargon, or you feel like crying with frustration at your inability to understand the methods in a particular paper, take heart in the fact that many scientists feel the same. We all struggle at some time to understand a paper, or to stop our eyes from closing while perusing some less than riveting details. But even if you just read the titles and abstracts of papers, you will begin to get a feel for the current issues and the approaches people are taking to address them (an abstract is a short summary of the aims and results of the research).

You can use public search engines such as PubMed or GoogleScholar to search the scientific literature for papers on a particular topic, or by specific authors. While many journals still require a (prohibitively expensive) subscription to access the whole paper, you can usually read at least the abstract, and increasing numbers of scientific journals are making their contents freely available over the internet. You will be able to access a wider range of journals if you are a member of a library with journal subscriptions. Try visiting the webpages of scientists working in the field, as they may have summaries of their research or copies of their papers to download. You can even write to scientists and ask for a reprint (copy) of their paper if you cannot get the paper by any other means.

The public availability of scientific literature, molecular sequence data, analytical software, and servers for molecular genetic analysis means that, unlike many fields of scientific endeavour, research in molecular evolution is technically open to anyone who can gain access to a computer with an internet connection. Evolutionary biology is, it must be admitted, predominantly in the hands of the professionals: academics employed at universities, scientists working in research organizations. Science is most easily practised within professional communities, partly because of the availability of resources such as laboratories and libraries, but more importantly because of the availability of like minds to discuss ideas with. But evolutionary biology was founded on the investigations of passionate and dedicated amateurs. And the post-genomic age makes participation in evolutionary research accessible to anyone passionate enough to apply themselves to it.

This book will not give you everything you need to undertake research in molecular evolution. But I hope it gives you a starting point. Now the thing to do is get out there and get your hands dirty: think of a question and work out how you could answer it, contrast hypotheses and think of how you would test them, delve into the data and see what you find. Do not fuss about whether your question is too big or too small, as long as it is a question you find interesting, and you think you can see a way of shedding light on it. The only way to start is begin with something, anything, and see where it leads you. Good luck. I hope you have fun.

Appendix I

The genetic code

1st	2nd: T				C				A				G				3rd
T	TTT	F	Phe	Phenylanaline	TCT	S	Ser	Serine	TAT	Y	Tyr	Tyrosine	TGT	C	Cys	Cysteine	T
	TTC	F	Phe		TCC	S	Ser		TAC	Y	Tyr		TGC	C	Cys		C
	TTA	L	Leu	Leucine	TCA	S	Ser		TAA	–		*Stop*	TGA	–		*Stop*	A
	TTG	L	Leu		TCG	S	Ser		TAG	–		*Stop*	TGG	W	Trp	Tryptophan	G
C	CTT	L	Leu		CCT	P	Pro	Proline	CAT	H	His	Histidine	CGT	R	Arg	Arginine	T
	CTC	L	Leu		CCC	P	Pro		CAC	H	His		CGC	R	Arg		C
	CTA	L	Leu		CCA	P	Pro		CAA	Q	Gln	Glutamine	CGA	R	Arg		A
	CTG	L	Leu		CCG	P	Pro		CAG	Q	Gln		CGG	R	Arg		G
A	ATT	I	Ile	Isoleucine	ACT	T	Thr	Threonine	AAT	N	Asn	Asparagine	AGT	S	Ser	Serine	T
	ATC	I	Ile		ACC	T	Thr		AAC	N	Asn		AGC	S	Ser		C
	ATA	I	Ile		ACA	T	Thr		AAA	K	Lys	Lysine	AGA	R	Arg	Arginine	A
	ATG	M	Met	Methionine	ACG	T	Thr		AAG	K	Lys		AGG	R	Arg		G
G	GTT	V	Val	Valine	GCT	A	Ala	Alanine	GAT	D	Asp	Aspartic acid	GGT	G	Gly	Glycine	T
	GTC	V	Val		GCC	A	Ala		GAC	D	Asp		GGC	G	Gly		C
	GTA	V	Val		GCA	A	Ala		GAA	E	Glu	Glutamic acid	GGA	G	Gly		A
	GTG	V	Val		GCG	A	Ala		GAG	E	Glu		GGG	G	Gly		G

This is the 'universal genetic code', the most commonly used in the widest range of organisms (see **TechBox 2.3**). The colours of the one-letter codes represent the chemical properties of the amino acids: small non-polar (G, A, S, T; orange), hydrophobic (C, V, I, L, P, F, Y, M, W; green), polar (N, Q, H; magenta), negatively charged (D, E; red), positively charged (K, R; blue). This is the scheme used by Lesk (*Introduction to Bioinformatics*, 2005, Oxford University Press) but there are many other possible amino acid colouring schemes (see for example, *www.bioinformatics.nl/~berndb/aacolour.html*). The amino acid colours are the ones used in Se-Al (Rambaut, A. (1996) *Se-Al: Sequence Alignment Editor*, available at *http://evolve.zoo.ox.ac.uk/*). There are many slight variants on the genetic code. Listed below are some of the alternative codon translations used in various genomes. Note that this is not an exhaustive list, and that the alternative codes may not be used by all organisms in that taxon. Some organisms do not use all codons in their code, and it is reasonably common for some organisms to use alternative start codons. This information is taken from the NCBI website.

Alternative nuclear codes

Blepharisma (a common ciliate protist), nuclear code: TAG = Q
Ciliate (protists such as *Paramecium*): TAA = Q; TAG = Q
Dasycladacea (green algae): TAA = Q; TAG = Q
Hexamita (a flagellated protozoan that may cause 'hole in the head' disease in fish): TAA = Q; TAG = Q
Euplotids (ciliate protists): TGA = C
Mycoplasma (bacteria): TGA = W

Alternative mitochondrial codes

Arthropods, nematodes, molluscs: AGA = S; AGG = S; ATA = M; TGA = W
Ascidian: AGA = G; AGG = G; ATA = M; TGA = W
Chlorophycea (green algae including *Volvox*): TAG = L
Echinoderm: AAA = N; AGA = S; AGG = S; TGA = W
Fungi: TGA = W
Protozoa: TGA = W
Scenedesmus obliquus (a unicellular green alga): TCA = stop; TAG = L
Trematode: TGA = W; ATA = M; AGA = S; AGG = S
Vertebrate: AGA= stop, AGG = stop, ATA = M, TGA = W
Yeast: ATA = M; CTT = T; CTC = T; CTA = T; CTG = T; TGA = W

Appendix II

Sequences used in this book

This is a list of the accession numbers of sequences used as illustrations in this book. If you want to retrieve these sequences, go to GenBank (most easily accessed through the Entrez search engine), and type in these accession numbers. See TechBox 1.1 for an explanation of GenBank and Accession numbers. Note that the part of the alignment shown in the figures will be a very small percentage of the entire sequence.

Figure CS1.1: Blob sequences

Description from GenBank entry	Accession
Physeter catodon NADH dehydrogenase subunit 2 (*nad2*) gene, partial cds; mitochondrial (note: obtained from Elsa Cabrera of the Chilean Centro de Conservacion Cetacea, Los Muermos, Chile)	AY582746
Physeter macrocephalus mitochondrial genome	AJ277029

Figures 2.2, 2.7, 4.15, 4.16, 4.17, 5.23, 5.24: RNA polymerase beta

Description from GenBank entry	Accession
Homo sapiens polymerase (RNA) II (DNA directed) polypeptide B	NM_000938
Rattus norvegicus similar to DNA-directed RNA polymerase II 140 kDa	XM_21402
Drosophila melanogaster CG3180-PA (RpII140) mRNA, complete cds	NM_057358
Neurospora crassa strain OR74A	XM_324476
Oryza sativa (japonica cultivar-group) chromosome 10, section 62 of 77 of the complete sequence – partial sequence	AE017108
E. coli RNA polymerase beta subunit (*rpoB* and *rpoC*) genes, 3′	ECORPOBCY

Figure 4.19: Bacillus

Description from GenBank entry	Accession
Bacillus anthracis strain GJ-2 RNA polymerase beta subunit (*rpoB*)	AY169514
Bacillus anthracis strain GJ-1 RNA polymerase beta subunit (*rpoB*)	AY169513
Bacillus anthracis strain Army RNA polymerase beta subunit (*rpoB*)	AY169512
Bacillus anthracis strain ATCC 14185 RNA polymerase beta subunit (*rpoB*)	AY169511
Bacillus anthracis strain Sterne RNA polymerase beta subunit (*rpoB*)	AY169510
Bacillus thuringiensis strain IMSNU 10051 RNA polymerase beta	AY169538
Bacillus mycoides strain KCCM 40260 RNA polymerase beta subunit (*rpoB*)	AY169540
Bacillus cereus strain ATCC 9634 RNA polymerase beta subunit (*rpoB*)	AY169515

Figures 5.25, 5.26: alpha globin

Description from GenBank entry	Accession
Homo sapiens haemoglobin, alpha 1 (HBA1), mRNA	NM_000558
Homo sapiens haemoglobin alpha 2 (HBA2) gene, complete cds	DQ499017
Human alpha-2 pseudogene of alpha-globin gene cluster	X03583
Homo sapiens haemoglobin, zeta (HBZ), mRNA	NM_005332
Homo sapiens HBZP pseudogene, exons 1, 3, 2 and complete cds	HUMHBA3

Figure 5.25, 5.28: HIV Nef

Description from GenBank entry	Accession
HIV-1 isolate Q23–17 from Kenya, complete genome	AF004885
Human immunodeficiency virus type 1 (HXB2), complete genome; HIV1/HTLV-III/LAV reference genome	K03455
Human immunodeficiency virus type 1 (HXB2), complete genome; HIV1/HTLV-III/LAV reference genome	AF067155
HIV-1 isolate 02CM.0016BBY from Cameroon gag protein (*gag*) and pol protein (*pol*) genes, partial cds; vif protein (*vif*), vpr protein (*vpr*), tat protein (*tat*), rev protein (*rev*), vpu protein (*vpu*), and envelope glycoprotein (*env*) genes, complete cds; and nef protein (*nef*) gene, partial cds	AY371158

Human immunodeficiency virus type 1 proviral mRNA for partial GAG protein, partial POL protein, partial ENV protein and partial NEF protein isolate 95CM-MP255	AJ249236
HIV-1 isolate CRF01_AE/B from Thailand gag protein (*gag*) gene, complete cds; pol protein (*pol*) gene, partial cds; and vif protein (*vif*), vpr protein (*vpr*), tat protein (*tat*), rev protein (*rev*), vpu protein (*vpu*), envelope glycoprotein (*env*), and nef protein (*nef*) genes, complete cds	AF516184
Human immunodeficiency virus type 1 gag polyprotein (*gag*), pol polyprotein (*pol*), vpr protein (*vpr*), vif protein (*vif*), rev protein (*rev*), vpu protein (*vpu*), env polyprotein (*env*), and nef protein (*nef*) genes, complete cds	L39106
HIV-1 isolate X397 from Spain, complete genome	AF423756

Figures 6.4 to 6.12: marsupial and placental 12S rRNA

Description from GenBank entry	Accession
Thylacinus cynocephalus 12S ribosomal RNA gene, complete sequence; mitochondrial gene for mitochondrial product	TCU87405
Dasyurus geoffroii 12S ribosomal RNA gene, mitochondrial gene for mitochondrial rRNA, complete sequence	AF009891
Sminthopsis psammophila 12S ribosomal RNA gene, mitochondrial gene for mitochondrial RNA, complete sequence	AF088974
Notoryctes typhlops 12S ribosomal RNA gene, complete sequence; mitochondrial gene for mitochondrial product	NTU21179
Canis familiaris mitochondrial 12S rRNA gene	MICF12S
F.domesticus mitochondrial 12S rRNA gene	MIFD12S
Talpa europaea complete mitochondrial genome	Y19192
Sorex cinereus voucher FMNH 159789 12S ribosomal RNA gene, partial sequence; mitochondrial	AY691825

Figure 8.13: clam, mussel, and damselfly

Description from GenBank entry	Accession
Mytilus edulis methionine adenosyltransferase mRNA, partial cds	AY580273
Nucula proxima methionine adenosyltransferase mRNA, partial cds	AY580233
Lestes congener methionine adenosyltransferase mRNA, partial cds	AY580226
Enallagma aspersum methionine adenosyltransferase mRNA, partial	AY580212

Figures 8.18, 8.19: coelacanths and lungfish

Description from GenBank entry	Accession
Latimeria menadoensis mitochondrion, complete genome	NC_006921
Latimeria chalumnae mitochondrial DNA, complete genome, isolate: Kigombe-9	AB257297
Neoceratodus forsteri mitochondrion, complete genome	AF302933

Figure 8.22: methionine adenosyltransferase

Description from GenBank entry	Accession
Obelia sp. KJP-2004 methionine adenosyltransferase mRNA, partial cds	AY580240
Stylochus sp. KJP-2004 methionine adenosyltransferase mRNA, partial cds	AY580254
Chaetopterus sp. KJP-2000 methionine adenosyltransferase mRNA, partial cds	AY580185
Metridium senile methionine adenosyltransferase mRNA, partial cds	AY580247
Strongylocentrotus purpuratus methionine adenosyltransferase mRNA, partial cds	AY580282
Dendraster excentricus methionine adenosyltransferase mRNA, partial cds	AY580198
Saccoglossus kowalevskii methionine adenosyltransferase mRNA, partial cds	AY580278
Mytilus edulis methionine adenosyltransferase mRNA, partial cds	AY580273
Nucula proxima methionine adenosyltransferase mRNA, partial cds	AY580233
Lestes congener methionine adenosyltransferase mRNA, partial cds	AY580226
Enallagma aspersum methionine adenosyltransferase mRNA, partial cds	AY580212

Figure 8.24: kiwi and moa mitochondrial control region

Description from GenBank entry	Accession
Apteryx mantelli mitochondrion, partial genome	AY016010
Apteryx australis rowii haplotype Okarito 5 ATPase subunit 8 and ATPase subunit 6 genes, complete cds; mitochondrial genes for mitochondrial products	AY150600
Apteryx haastii mitochondrion, complete genome	NC_002782
Dinornis giganteus mitochondrion, complete genome	NC_002672
Dinornis novaezealandiae specimen-voucher CM_Av30497 mitochondrial control region, partial sequence	AY299875
Dinornis struthoides specimen-voucher CM_Av8872 mitochondrial control region, partial sequence	AY299874

Appendix III

The geological timescale

Eon	Era	Period	Epoch	Age (millions of years ago)
Phanerozoic	Cenozoic	Quaternary	Holocene	0.01
			Pleistocene	1.8
		Tertiary – Neogene	Pliocene	5.3
			Miocene	23
		Tertiary – Palaeogene	Oligocene	33
			Eocene	55
			Palaeocene	65
	Mesozoic	Cretaceous		145
		Jurassic		199
		Triassic		251
	Palaeozoic	Permian		219
		Carboniferous – Pennsylvanian		318
		Carboniferous – Mississippian		359
		Devonian		416
		Silurian		443
		Ordovician		488
		Cambrian		542
Precambrian	Proterozoic			2500
	Archean			4000
	Hadean			4600

This chart gives the approximate age of the lower boundary of some of the major stratigraphic intervals (see www.stratigraphy.org). Note that the ages of these boundaries are periodically revised, so it is always best to consult the most recent published timescales.

Glossary

Since this book is intended to be accessible to entry-level students with a minimal background in evolutionary biology or genetics, this glossary provides definitions of some words and phrases that may be familiar to some readers but mysterious to others. It is not an exhaustive dictionary of biological terms used in the book, but contains explanations of a number of key concepts that can be confusing or ambiguous. I have bundled together related terms to make meanings clearer, so many terms in the glossary are cross-referenced to other entries. This cross-referencing doesn't imply equivalence. For example you may look up 'zygote' and be referred to 'gamete': these words mean very different things but are explained in the same paragraph. I have not included terms that occur only in one place in the text, and are defined at first mention (for example, 'affine gap penalty' is used only in **TechBox 6.3**, where its meaning is explained), so if you don't find a word in the glossary, try looking in the index for the page on which it is first mentioned. Finally, a word of warning: as with any language, the meaning of scientific words can vary between users. The purpose of this glossary is to explain terms as they are used in this book. You may find some words included here that are used in a different sense by other authors.

3′ (three-prime) and **5′** (five-prime) refer to the two different ends of a polynucleotide strand. The 5′ end terminates with a phosphate molecule which is attached to the 5′ carbon of the sugar molecule. The 3′ end terminates with a sugar, the third carbon atom of which has a free hydroxyl group which forms a phosphodiester bond with the phosphate of a newly added nucleotide. This is why DNA synthesis always proceeds in the 5′ to 3′ direction – new bases can be added to the 3′ end (carbon with free hydroxyl group), not to the 5′ end (phosphate) (**TechBox 2.2**).

454 sequencing: see pyrosequencing

Acquired characteristics are modifications to phenotype that occur during an individual's lifetime. Some early evolutionary thinkers believed that acquired characteristics of the parent could be inherited by its offspring. Weismann was one of the first to propose that acquired characteristics could not be inherited by an individual's offspring, so did not contribute to descent with modification (evolution). See germline.

Active site: see enzymes

Adaptation can refer to the process whereby a trait has evolved, or to the end product of this process. An adaptation is a feature of an organism that suits it to a particular way of life, often described as increasing the 'fit of an organism to its environment'. The recognition of adaptations does not depend upon the mechanism of their generation, for example a clear treatise on adaptation was given by William Paley in 1802 (i.e. long before Darwin's theory of evolution was published) where he described how living organisms are notable for having features that are 'formed . . . for the purpose which we find it actually to answer'. However, natural selection is the only known mechanism for adaptation.

Adenine (A) is one of the four DNA bases that make up the 'alphabet' of the genetic code. See base.

Afrotheria: see mammal

AIDS: see Human Immunodeficiency Virus (HIV)

Alignment is the process of arranging homologous sequences so that comparisons can be made between the character states (nucleotides or amino acids) that have descended from a shared ancestral character. See **TechBox 6.3**.

Allele is derived from 'allelomorphs' (a term coined by William Bateson, 1902) meaning alternative versions of a trait that are transmitted independently and segregate randomly, so the term allele predates molecular genetics, and can be applied to alternative forms of any heritable trait. In molecular genetic terms, alleles are variations on the possible genetic sequence at a particular locus, that are present in the same population. Since diploid individuals carry two copies of every locus, they may be heterozygous (carry two different versions – alleles – of the same locus, one on each chromosome) or homozygous (two copies of the same allele). Alleles at closely linked loci will tend to be inherited together: sets of linked alleles form a haplotype.

Alpha globin: see haemoglobin

Amino acids are the building blocks of proteins. There are 20 common amino acids, which vary from each other in the functional groups attached to the central carbon atom. The sequence of bases in a protein-coding gene specifies an exact sequence of amino acids when it is translated.

Ammonites are an extinct group of cephalopods (the group of molluscs that includes octopuses and squid), typically with spiral, coiled shells. Ammonites are important for biostratigraphy of some marine geological strata.

Anaemia is a disease that results from insufficient oxygen being transported in the blood. This may have a genetic cause (for example, a haemoglobinopathy) or an environmental cause (for example, iron deficiency).

Analogy: see homology

Ancient DNA (aDNA) is, in the broad sense, any DNA sample derived from non-fresh material. Examples mentioned in this text include DNA extracted from museum specimens (pinned flies, preserved thylacines), frozen samples (woolly mammoths), and subfossils (moa bones). DNA decays over time, and currently the oldest reliable DNA sequence is around 400,000 years old. Because the amount and quality of DNA in a sample reduces with age, ancient DNA extraction and DNA amplification relies on sterile techniques to prevent the sample DNA being overwhelmed by contaminating DNA from other sources (Case Studies 2.2 and 5.2).

Annotation is the addition of genetic information about a DNA sequence to its entry in a database, for example the gene name, chromosomal location, or regulatory and coding regions of the sequence. Large-scale sequencing projects tend to use automated annotation to identify sequence features by similarity to a reference database of sequences of known function (TechBox 1.1).

Anopheles: see malaria

Anthrax is an infectious disease of livestock and humans, caused by the bacterium *Bacillus anthracis*. It is transmitted by spores which can enter the body through the lungs (causing pulmonary anthrax), the digestive system (causing gastrointestinal anthrax) or the skin (causing cutaneous anthrax) (Figure 4.20).

Antimutator: see mutator

Arthropoda is a phylum of animals, including spiders, insects and crustaceans, characterized by segmented bodies, paired jointed limbs and an exoskeleton (a hard covering over the body) made of chitin.

Artiodactyl: see mammals

Association studies aim to identify genetic markers that are significantly associated with a particular trait, such that individuals with that marker have a higher chance of having that trait than individuals without the marker. Association studies may target a particular informative pedigree, or may be applied to samples from a population (for example, a study based on a biobank) (TechBox 3.2).

Autapomorphy is a trait unique to one lineage from a group of related lineages. Because it is not shared with other related lineages, an autapomorphy cannot be used to determine phylogenetic relationships between lineages.

Bacillus: see anthrax

BACs (bacterial artificial chromosomes) are circular DNA molecules into which can be inserted a DNA sequence from another organism (say, several hundred kilobases of the human genome), which can then be replicated in bacterial cells.

Bacteriophage (also known as phage) are viruses that infect bacteria. Because phage have simple genomes and are amenable to laboratory experiments, phage genetics has played an important role in the development of molecular genetics (e.g. see TechBox 2.1)

Barnacles are arthopods in the class Cirripeda: underneath the calcareous plates that form their 'shell', most barnacles have typical aspects of the arthropod body plan such as segmented legs and antennae. The larvae swim then settle on a surface, but the adult form is immobile, cemented to the surface. There is a wonderful illustration of barnacle life cycles on *www.mesa.edu.au/friends/seashores/barnacles.html*.

Bases are small biomolecules that make the four 'letters' of the DNA 'alphabet': adenine (A), cytosine (C), guanine (G), and thymine (T). RNA molecules have a slightly different alphabet, using uracil (U) instead of T. A and G are purines (double-ring bases) and T and C are pyrimidines (single-ring bases). A nucleotide consists of a base attached to a sugar molecule which is attached to a phosphate. DNA is a polynucleotide because it is made of series of nucleotides joined together by phosphodiester bonds which link the sugar of one nucleotide to the phosphate of the next nucleotide in the chain. The bases of one polynucleotide can form hydrogen bonds with the bases of another, linking two nucleotide strands. Watson–Crick pairing rules state that A pairs with T, and G pairs with C (in RNA molecules U pairs with A as T does in DNA). In DNA, it is these bonds between bases that bind the two strands of the double helix together. In RNA, base pairing can bind one part of the polynucleotide to another to generate secondary structure (TechBox 2.2).

Bayesian inference is a form of statistical inference, which begins with a set of prior beliefs (prior probabilities), then uses the observed data to modify those beliefs in the face of the evidence (posterior probabilities). See TechBox 7.3.

Beta globin: see haemoglobin

Bilaterian can refer broadly to any animal having bilateral symmetry, such that if you cut it down the plane of symmetry you would get two similar halves (imagine slicing a cat in half from between the ears to the tail, and you would get two similar cat-halves). In this book, the term bilatarian is used to refer to a clade of animals that includes all members of the superphyla Lophotrochozoa (including molluscs and annelids), Ecdysozoa (including arthropods and nematodes), and Deuterostomes (including echinoderms and chordates). The bilaterian clade excludes the older animal lineages of Cnidaria (jellyfish and their kin) and Porifera (sponges).

Biodiversity is a contraction of the phrase 'biological diversity', and is intended to capture the variation in numbers and kinds of living organisms. Often, biodiversity is measured by summing the total number of species found in a given area; however, some biodiversity measures are more sophisticated, for example including the distribution of organisms among different types, or capturing the total amount of evolutionary divergence as represented by phylogenetic branch length.

Biogeography is the study of the geographical distribution of organisms.

Bioinformatics is, broadly speaking, the application of computational and statistical techniques to the analysis of biological data, but the common usage of the word applies more narrowly to statistical analysis aimed at detecting patterns in large collections of molecular sequence data.

biostratigraphy: see geochronology

BLAST is a bioinformatic method for searching a database of sequences for the closest match to a query sequence. See **TechBox 3.4**.

Body plan refers to a basic level of physical organization in animals. The body plan is commonly interpreted to be laid down early in development, and to represent fundamental differences between deep lineages, such as phyla.

Calibration, in molecular phylogenetics, is either the process of converting some measure of genetic distance or branch length to time using a known date of divergence, or it is the date of divergence used in such a calculation. See **TechBox 8.2**.

Carnivorans are mammals from the order Carnivora, which includes cats, dogs, bears, weasels, and seals. Not to be confused with the term 'carnivore' which refers to any meat-eating animal (or plant).

Chain-termination sequencing method: see Sanger method

Chance: see random

Character is a broad term meaning any discrete heritable trait possessed by an organism, that may vary independently of other such characters. Each character has a number of different character states (possible variants of that character). Systematics is based on the identification of informative character states that define sets of related lineages: see homology.

Chargaff's rules describe the relative amounts of the four bases in DNA molecules. Chargaff's first rule states that the amount of A equals the amount of T and the amount of G equals the amount of C. This observation was important for the discovery of Watson–Crick pairing. Chargaff's lesser known second rule is that the relative amounts of the four bases differ between species, a phenomenon now often referred to as base composition bias. See bases.

Chiasma (plural: chiasmata): see crossing-over

Chloroplasts are a type of plastid, which are organelles found in photosynthesizing cells of plants and algae. Chloroplasts capture the energy from light and use it to drive a series of reactions that result in chemical energy storage. Chloroplasts, like mitochondria, were originally derived from free-living bacteria, and retain a small circular genome. Chloroplast DNA sequences are frequently used in plant systematics.

Chromosomes are cellular structures containing the genome, consisting of nucleic acids and supporting proteins (such as histones). Some small genomes are circular, such as those of bacteria, mitochondria, and chloroplasts. Larger genomes tend to be divided into a number of linear packages. Diploid cells carry two copies of the genome, on pairs of homologous chromosomes that each contain the same genes (but may have different alleles). Before cell division, each chromosome replicates to produce two sister chromatids, held together at a region called the centromere, which is important for proper segregation of chromatids into daughter cells at cell division.

Clade represents a group of related lineages united by descent from a common ancestral lineage. Phylogenies generally describe sets of nested clades, each defined by sharing a common ancestral node. Clades, defined by the set of all lineages descended from a specific node in a phylogeny, may be used as the basis of taxonomic grouping under the cladistic framework for systematics.

Cloning has several broad meanings in molecular genetics. Cloning may refer to any method for producing copies of a particular sequence, usually by inserting the sequence into a cultured cell. Cloning cells refers to the production of a colony of cells derived from a single cell. Cloning also refers to the production offspring that are genetically identical (or near identical) to a single parent. Naturally occurring clones may be produced by asexual reproduction or when a single embryo splits to form identical twins; artificial clones may be produced by transferring the nucleus of an adult cell into an anucleate embryo. See **TechBox 6.4**.

Coding region generally refers to a DNA sequence that codes for an amino acid sequence – that is, the exons of a protein-coding gene. However, the term is sometimes used more broadly to refer to all of the transcribed DNA in the genome, whether it codes for proteins or RNA molecules, as a contrast to non-transcribed 'junk DNA'.

Codon is a three-base sequence that specifies a particular amino acid (or a stop codon, which signals the end of a peptide). When a messenger RNA transcript is translated into a protein sequence at the ribosome, each codon is matched to the complementary recognition sequence on the correct transfer RNA molecule, which joins the corresponding amino acid to the growing peptide chain. See translation (**TechBox 2.3**).

Confidence intervals refer to a range of likely estimates for a parameter, reflecting the precision of the parameter estimate.

Copy errors are changes in the DNA sequence that arise from imperfect replication of a nucleotide strand. Every time the genome is copied there is a small but finite chance that the sequence in the copy will not be exactly the same as in the parent strand. The more times DNA is copied the more errors will accumulate (Chapter 4).

Crick, Francis (1916–2004): one of the co-discovers of the structure of DNA, who also made significant contributions to understanding how the genome specifies phenotype, for example predicting and decoding aspects of the genetic code, and formalizing the central dogma. See **Heroes 4**.

Crossing-over is the exchange of genetic material between homologous chromosomes, resulting in recombination (disruption of linkage) between genetic markers. Unequal crossing over, which can occur when repeat sequences become misaligned, results in the net loss of sequences from one chromatid and the net gain of sequences in the other. The crossover points are referred to as chiasmata (singular: chiasma) (Figure 3.6).

Cryptic species are those that do not show any detectable differences from one or more other species, despite evidence of consistent and sustained reproductive isolation from other such populations. Cryptic species are often detected by observing a significant level of genetic divergence from otherwise similar populations of a particular organism.

Cryptozoology is the study of elusive, and probably illusory, organisms, which are considered by a minority to exist but for which there is no conclusive proof.

Cytosine (C) is one of the four DNA bases that make up the 'alphabet' of the genetic code: see bases.

Darwin, Charles (1809–1882), though he was not the first to describe evolution, essentially created the science of evolutionary biology through his rigorous investigation of the evidence for the transformation of species over time, and by providing a plausible mechanism for evolutionary change (natural selection).

Degenerate refers to redundancy of information. The genetic code is said to be degenerate because many different codons can specify the same amino acid (a non-degenerate code would have exactly one codon for each amino acid). Degenerate primers are a set of primers with slightly different sequences, used when the exact target sequence is unknown (a non-degenerate primer would bind to one specific sequence only).

Deoxyribose: see sugar

Dicot is a plant from one of two major groups of flowering plants. The dicots include most flowering trees, daisies, strawberries and so forth. The other major group is the monocots which include the grasses and palms.

Dideoxy sequencing method: see Sanger method

Dinornis: a genus of moa.

Diploid cells have two copies of each chromosome. For example, most human cells are diploid, carrying two copies of each of the 22 autosomal chromosomes (each copy containing the same genes but potentially carrying different alleles) and two versions of the sex chromosomes (two Xs or an X and a Y). Human germ cells (sperm and eggs) are haploid, carrying one copy of each of the autosomes and one sex chromosome.

Divergence date is the point in time when a single interbreeding population became separated into two separate lineages. The divergence date marks the age of the last common ancestor of a clade, but usually predates the evolution of distinct characteristics that define members of the descendant lineages (**TechBox 8.2**).

Diversification is the generation of many distinct lineages from an original ancestral lineage. More formally, diversification rate of a clade is defined as the net gain in lineages over a time period as a result of the addition of lineages by speciation and the loss of lineages by extinction.

DNA amplification is the production of many copies of a DNA sequence. The polymerase chain reaction (PCR) is the most common means of amplifying a DNA sequence. See **TechBox 4.2**.

DNA barcoding refers, in the broad sense, to identifying members of a species using DNA analysis. More specifically, DNA barcoding refers to the hope that the base sequence at a single locus (for example, a mitochondrial gene sequence) may be able to act as a universal means of species identification.

DNA fingerprinting refers to any method for genotyping individuals, revealing a unique combination of alleles carried by an individual that can be used to distinguish them from all other individuals in the population. DNA fingerprints were originally based on restriction length fragment polymorphism (RFLPs), but now more commonly based on a set of microsatellite loci, or single nucleotide polymorphisms (SNPs) at defined polymorphic loci (Chapter 3).

DNA hybridization refers to the technique of combining DNA from multiple sources, heating to separate the double-stranded helices into single strands, then cooling so that the single-strands reanneal into double strands. When a helix forms between DNA strands from different sources, the strength of binding between the strands is proportional to number of complementary bases that match between the sequences. The amount of heat required to separate the hybridized strands will be proportional to the number of matched bases between the strands, so the melting temperature reflects the similarity between DNA sequences from different sources. In this way, DNA hybridization can be used as a measure of the amount of difference between the two genomes (**Case Study 4.1**).

dNTP (deoxyribonucleotide triphosphate) is the form of nucleotide used in DNA synthesis, so dNTPS must be added to DNA amplification reactions such as the polymerase chain reaction (PCR).

Dominant alleles create the same phenotypic effect whether present in the homozygous or heterozygous state, so it only takes one copy of the allele to generate the trait.

Drift is a change in allele frequencies in a population from one generation to the next due to incomplete sampling of the alleles in the parent generation. Drift refers specifically to random fluctuations in allele frequencies across generations (not changes in frequency due to selection, migration, or mutation). The effect of drift on allele frequencies is most pronounced for neutral alleles, or in populations with small effective population size.

Drosophila is a genus of fruit flies commonly used in genetic experiments, as they can be reared in large numbers in the laboratory and have short generation times. *Drosophila melanogaster* was developed as a model genetic organism by Thomas Hunt Morgan in the early part of the twentieth century. Early genetic studies were aided by the very large chromosomes in the salivary glands of fruit flies, making inheritance of chromosomal regions observable under the microscope.

Echidna: see monotreme

Echinoderms are members of the animal phylum Echinodermata, including starfish, sea cucumbers, sea urchins, crinoids, and brittlestars.

Echolocation is the detection of objects using reflected sound. Animals such as whales, bats, and some birds (e.g. cave swiftlets) emit sounds then use the echoes of these sounds from solid objects to build a picture of their surroundings.

Edge refers to the line connecting two nodes in a phylogeny (including the lines connecting an internal node to a tip). This term is derived from mathematical graph theory, but is used by biologists as a synonym for ‘branch’.

Effective population size (N_e) represents the number of parents that contribute alleles to the next generation. This number is generally much smaller than the total number of breeding individuals in the population, because generally not all individuals in a population successfully reproduce, and even those that do may not pass all of their alleles on to viable offspring. See **TechBox 5.3**.

Empirical refers to observation or measurement: an empirical estimate is derived from the data. Theoretical refers to predictions made from consideration of hypotheses; theoretical estimates are derived from the predictions of a model.

Endangered species are species that are considered to be threatened with extinction. Endangered is also a defined category under the IUCN red list, which is a scheme for classifying species according to the perceived likelihood of extinction, drawing on information from the global population size of the species, extent of the species distribution, and the rate of decline in distribution and population. There are three recognized levels of extinction threat under the IUCN classification, vulnerable, endangered, and critically endangered.

Endemic species are found only in the area under consideration and nowhere else. Therefore endemism is considered with reference to a specific area.

Endonuclease is an enzyme that cuts a DNA or RNA molecule in the middle of the strand by breaking the phosphodiester bonds holding adjacent nucleotides together. An exonuclease removes nucleotides from the end of a nucleotide strand.

Enzymes are proteins that perform a specific catalytic role, changing the form of some substrate by actively making or breaking chemical bonds. The active site of an enzyme is the regions that binds to a substrate and performs some catalytic function. Proteins with enzymatic functions usually have a name ending in 'ase', for example DNA polymerase catalyses the formation of a polymer of nucleotides to make a DNA polynucleotide.

Epitope is the part of a molecule recognized by the immune system, which can form antibodies specific to that epitope.

Escherichia coli (*E. coli*) is a species of bacterium found in the human digestive system (and also present in the environment). Like *Drosophila*, *E. coli* is a model organism in genetics research.

ESTs (expressed sequence tags) are produced by sequencing the messenger RNA content from a cell, so represent the sequences being transcribed by that cell. Genes may been identified by comparing sequences from ESTs to genomic data or by hybridizing ESTs to chromosomes to determine the location of the gene.

Eucalyptus (gum trees) is a speciose genus of trees, found in Oceania, forming a predominant part of many Australian ecosystems and an increasingly common timber species around the world. The smell of eucalyptus oil often makes Australians feel nostalgic.

Eugenics in the broad sense refers to the active intervention in human reproduction to achieve a change in the frequency of heritable traits within a particular population. However, some people apply the word eugenics only to the selection of desirable traits whereas others include programmes aimed at reducing the incidence of heritable disease.

Eukaryote cells have their genome contained within a distinct organelle called the nucleus. Most eukaryotes also have mitochondria. Animals, plants, fungi, and protists are eukaryotes. The genome of prokaryotes, the archaebacteria and bacteria, is not enclosed in a nucleus.

Eusocial species live in social groups that show a high degree of co-operation amongst individuals, with reproductive division of labour (a small number of individuals have offspring which the rest of the colony help to raise) and often specialization to different tasks (food gathering, defence, etc). Examples of eusocial animals include naked mole rats, honey bees, and termites.

Evolutionary tree: see phylogeny

Exhaustive search considers all possible states: in phylogenetics, an exhaustive search evaluates every possible phylogeny for a set of taxa. An heuristic search uses some strategy to consider a sub-set of all possible states, with the aim of finding the optimum state without having to evaluate every possibility; in phylogenetics, a heuristic search makes a partial exploration of tree space, using a search strategy directed to finding trees that provide the best explanation of the data (e.g. with a higher likelihood or posterior probability).

Exons are the parts of protein-coding genes that specify an amino acid sequence. Introns are regions of protein-coding genes, which may contain regulatory elements but are excised from the messenger RNA transcript before translation.

Exonuclease: see endonuclease

Expression: see gene expression

Extinction occurs when the last member of a species, or other distinct taxon, dies. Extinction represents the loss of a unique lineage (or a unique set of alleles).

Fisher, Ronald Aylmer (1890–1962) was a key figure in the development of population genetics, as well as establishing many key statistical techniques such as maximum likelihood (and whenever you use 0.05 as a cut-off for statistical significance, think of Fisher). Fisher demonstrated that even alleles with a very small advantage over other alleles in the population could go to fixation by selection in a large, randomly mating population. Fisher was a committed eugenicist, believing that selective breeding was an important tool in improving human health and prosperity (Figure 5.29).

Fitness can be defined in many ways (see **TechBox 5.1**), but usually reflects the relative reproductive success of different types of individuals in a population. In this book, we consider that the fitness of an allele (or genotype) is a reflection of its relative selective advantage or disadvantage relative to the other alleles in the population, and thus influences the chance of that allele going to fixation.

Fixation of an allele occurs when the frequency of that allele in a given population reaches 1, so that all members of the population carry the same allele at a particular locus and there is no polymorphism at that locus. See also substitution.

Flanking sequence refers to the DNA sequence either side of the locus of interest.

Frameshift mutations result from the insertion or deletion of bases from a protein-coding sequence in multiples other than three, which disrupts the way that the codons are read from the subsequent sequence, changing the amino acid specified by the rest of the sequence.

Gamete is a haploid reproductive cell, such as sperm or eggs. In diploid organisms, gametes are produced by meiosis. Gametes from two parents fuse to form a diploid zygote, which develops into an embryo.

Gap: see indel

Gene is a surprisingly slippery term, sometimes used in the abstract sense to indicate an independent heritable trait, sometimes used to describe a DNA sequence with particular features that allow it to be transcribed, sometimes to indicate a variant of a heritable trait (allele), and sometimes as a short-hand for any locus in the genome. Mendel was the first to describe the action of genes – discrete inherited units of heritable information – but the term 'gene' was not coined until 1909 (by Wilhelm Johanssen, who also introduced the words 'phenotype' and 'genotype' in order to distinguish the heritable component of variation from that caused by the environment). Genes began to take on a physical reality when, in 1910, the great fruit fly geneticist Thomas Hunt Morgan showed that particular genes were located on specific chromosomes; however, the nature of the gene was unknown. In the 1940s, genes were defined as the units of hereditary information, each of which produces a particular protein. With the discovery of the genetic code, a gene could be recognized as a particular DNA sequence that carries the information needed to produce a gene product, such as a peptide or RNA molecule. Perhaps we need different terms to describe different aspects of genes, as proposed by Seymour Benzer who, in the 1950s, invented the terms 'cistron' (unit of function), 'recon' (unit of recombination), and 'muton' (unit of mutability). Or perhaps it doesn't matter so long as the meaning of 'gene' is clear from context. Thomas Hunt Morgan declared, in his Nobel Prize acceptance speech in 1933: 'At the level at which the genetic experiments lie it does not make the slightest difference whether the gene is a hypothetical unit, or whether the gene is a material particle'.

Gene expression is the process whereby the information in the genome is converted into biological structures and processes. Typically this involves the recognition of regulatory elements by the transcription machinery, which makes an RNA transcript of the gene.

Gene family is a set of related genes, generated by duplication in an ancestral genome. Related copies of the same gene within the same genome are referred to as paralogues, to distinguish them from orthologues (related genes in different lineages). In other words, paralogues are produced by gene duplication, orthologues are produced by speciation (lineage divergence). For example, alpha globin 1 and 2 in the human genome are paralogues, but human alpha 1 and chimp alpha 1 are orthologues.

Gene flow describes the movement of alleles from one population to another, typically by migration of individuals or through hybridization where populations meet.

Gene pool: see population

Gene regulation is the control of gene expression, so that the gene product is produced in appropriate amounts when and where it is needed, in response to external or internal signals.

Generation time effect refers to the prediction that lineages that have a shorter generation turnover time (that is, the time it takes for an embryo to become and adult and produce another embryo) should have a higher mutation rate, because their genomes are copied more often per unit time and therefore are expected to accumulate more DNA copy errors. The generation time effect has been observed for DNA sequences in vertebrates.

Genetic code is the set of 64 possible three-base codons that correspond to 20 amino acids. See **TechBox 2.3**. The universal genetic code is not actually universal, but is the code used in the majority of genomes. However, there are about a dozen known variations on the code, usually differing from the universal code by one or several codons. The code is said to be degenerate because multiple codons specify the same amino acid, and these synonymous codons are effectively interchangeable without altering the protein made from the sequence.

Genetic drift: see drift

Genetically isolated: see population

Genetic marker is an allele that allows a certain piece of DNA to be identified. Markers can be used to track the inheritance of linked alleles.

Genome refers to all of the DNA inherited as a coherent set. The human nuclear genome is arranged on 23 different chromosomes. A diploid cell contains two copies of the nuclear genome (46 chromosomes total);

haploid gametes contain only one copy of the nuclear genome (23 chromosomes). Most human cells also contain multiple copies of the mitochondrial genome, which is transmitted to the next generation in the cytoplasm of the egg cell.

Genome-wide scan is a survey of a large set of genetic markers, distributed throughout the genome, to detect any markers that are significantly associated with a particular trait (found in more people with the trait than expected by random sampling alone).

Genomic system is short-hand for the core hereditary system of nucleic acids plus proteins common to all life on Earth. I must admit that I made up this term while I was writing this book, so it is unlikely to appear anywhere else.

Genotype can refer to the total genetic information carried by an organism (as distinct from its phenotype); the specific alleles carried by an individual at a particular locus; or to the process of determining the alleles carried by an individual (see DNA fingerprinting).

Genus is a group of related species, defined within a taxonomic hierarchy. In bionomial nomenclature, genus is the first part of a species' formal name.

Geochronology is the discipline in earth sciences concerned with determining the absolute or relative ages of geological strata (layers of rock, minerals, or sediments with defined age ranges). The most common dating techniques make use of the decay of radioactive elements such as uranium. Many rock types cannot be dated directly (such as sedimentary rocks) so their age is inferred by measuring the age of underlying and overlaying strata, or by the presence of fossils of characteristic species (a process known as biostratigraphy).

Germline cells can pass genetic information to future generations. Soma (body) cells do not copy their genomes to the next generation. In most animals, the germline is set aside early in development as the gamete-producing cells; all other body (soma) cells die when the individual dies. The situation is less clear cut in many plants, where the gametes can be formed from body tissues, and parts of the adult body can reproduce vegetatively (grow a new individual from a piece of the parent's body).

Googol is 10^{100}, or ten duotrigintillion, or a one followed by 100 zeros.

Gradualism refers to a model based on the cumulative effect of many small changes. Darwinian evolutionary theory expects that most evolutionary change is gradual, achieved by a series of substitutions of alleles each of relatively small (or no) phenotypic effect. Gradualism does not refer to speed of change (which may vary over time or between lineages) and does not discount periods of rapid change.

Guanine (G) is one of the four DNA bases that make up the 'alphabet' of the genetic code: see bases.

Guthrie test (heel-prick test) is the practice of taking a blood sample from newborn babies by pricking the baby's heel and blotting the blood onto a card. The blood samples are then analysed for evidence of a range of metabolic disorders, including phenylketonuria. In most cases, the Guthrie card is stored with the baby's and mother's names, and the place and date of birth. Stored Guthrie cards represent a biobank, and have been used in medical research, in court cases, and to identify victims of a terrorist attack.

Haeckel, Ernst (1834–1919) was a doctor, scientist, and artist who used comparative development and anatomy to illuminate evolution.

Haemoglobin is a blood protein, made of two alpha globin chains and two beta globin chains, that transports oxygen around the body. Haemoglobinopathies are disorders arising from mutations in the various globin genes, for example thalassaemia which results from underproduction of one of the globin chains (Figure 5.8).

Haldane, John Burdon Sanderson (known as JBS; 1892–1964) was a biochemist, physiologist, and evolutionary biologist, one of the key contributors to the formation of the neo-Darwinian synthesis. Haldane was also a great popularizer of science, publishing many readable and entertaining books and essays, and, as a committed socialist, he wrote a regular column for *The Daily Worker*. His larger-than-life personality and outrageous behaviour generated countless stories and legends, retold with relish (and probably improved) by his student John Maynard Smith. (Figure 5.5).

Hamilton, William Donald (Bill) (1936–2000) was a key figure in the development of the 'gene-centred' view of evolution, developing ideas of kin selection to explain the evolution of eusociality, and evolutionary arms races to explain the evolution of sex as a means of generating genetic variability (particularly as an adaptation to parasitism).

Hamming distance is a term from information theory that describes the number of positions at which two strings of symbols differ from each other. For two aligned nucleotide or amino acid sequences, the Hamming distance is the observed number of differences between them.

Haploid: see diploid

Haplotype is a set of linked alleles that are usually inherited together as a unit.

Helicase is an enzyme that separates the two strands of a DNA helix so it can be replicated.

Heterozygote is the term for a diploid organism that has two different alleles at the locus of interest, whereas a homozygote carries two copies of the same allele. The heterozygosity of a population is the proportion of heterozygous at a particular locus, so is a measure of the genetic variability of a population. Severe reduction in effective population size is likely to lead to inbreeding (mating between relatives), which is expected to reduce the heterozygosity of a population because an increasing proportion of the population carry alleles copied from the same recent ancestor. Sustained reduction in population size will thus ultimately reduce the number of polymorphic loci.

Heuristic search: see exhaustive search

Higher taxa: see taxon

Homologue: see gene duplication

Homology refers to similarity by descent. If you wish to uncover evolutionary relationships, it is important to distinguish homologies (which are evolutionary signal) from analogies (which represent noise, because they do not reflect descent). Homologous traits are similar because they were copied from the same ancestral trait. Analogous traits are superficially similar traits that have been arrived at independently in different lineages (i.e. same ends from different starts). Some analogies are the result of convergent evolution, whereby similar selection pressures promote the evolution of the same (or similar) features in different species: for example, many different insect lineages have evolved the same amino acid changes that confer resistance to insecticides (see **Case Study 5.2**). Other analogies may result from chance, for example a particular site in a DNA sequence may happen to acquire an adenine at the same position in two different lineages. Sequence alignment is the process of arranging DNA or protein sequences so that homologous sites, originally copied from the same position in an ancestral gene, can be compared in order to uncover evolutionary patterns.

Homozygote: see heterozygote

Horizontal gene transfer (HGT) is the movement of DNA between individuals other than the transmission of the genome from parent to offspring. Examples include the bacteria taking up genes from the genomes of distantly related lineages, the transfer of genes from the mitochondrion to the nucleus, the formation of recombinant virus genomes, or the movement of DNA from parasite to host (see **Case Study 7.2**). Horizontal gene transfer produces non-tree-like signal in sequence data.

Human Immunodeficiency Virus (HIV) is a retrovirus that infects cells of the immune system, reducing the effectiveness of the immune response and causing Acquired Immune Deficiency Syndrome (AIDS) which increases vulnerability to infections and particular cancers.

Huntington's disease is a neuromuscular disorder caused by inheritance of a single dominant allele. The *Huntington disease* gene contains a trinucleotide repeat region which causes disease if it contains more than a threshold number of repeats.

Huxley, Thomas Henry (1825–1895) was a doctor and scientist, who used comparative anatomy to investigate and illustrate evolution. Huxley is famous for his vociferous promotion of Darwin and evolutionary theory, but he was also a tireless champion of scientific research and science education.

Hybridization is used in two senses in this book. Firstly, it refer to the production of an hybrid individual who has inherited alleles from two distinct populations or species. Secondly, in molecular genetic terms, the word hybridization refers to joining together nucleotide strands from different sources by complementary base pairing.

Hypothesis is a proposed explanation for observed data, which can then be tested by experiments or observations. In some cases, the relative level of support for different hypotheses is contrasted. In other cases, the emphasis may be on falsifying a particular null hypothesis by showing that the observed data could not have been produced if the null hypothesis was true. The null hypothesis is a statistical tool used to generate patterns of data expected if the process of interest does not operate: if the null hypothesis cannot be rejected, then the pattern in the data could have been produced without the process of interest operating.

Inbreeding results from mating between relatives, either by non-random mate selection, or random mating within a consistently small population, which is expected to result in lost of genetic variability and increasing homozygosity, which may increase the incidence of heritable recessive diseases.

Indel stands for insertion or deletion, where one or more nucleotides are added or taken away from a sequence. Indels are represented by gaps in an alignment.

Information is a very tricky word to define, but here we will simply consider two senses in which the genome can be said to contain information. Firstly, parts of the genome contain information in a semantic sense: genes provide a set of instructions for making RNA and protein molecules that have specific function. This information is transferable from the genome to the cytoplasm, and between one generation and the next, through complementary base pairing. Secondly, the genome contains information in the sense that related sequences are likely to be more similar, so the genome contains a record of its own evolutionary history.

Informative pedigree: see pedigree

Interbreeding population: see population

Intron is a DNA sequence within a protein-coding gene that does not code for an amino acid sequence. Introns are transcribed but then removed from the processed messenger RNA transcript. Most intron sequences have a rapid substitution rate, compared to the exons, and are sometimes used to estimate the assumed neutral rate of substitution.

Invariant sites are columns in an alignment where all sequences have the same nucleotide. Note that invariant is with respect to a particular collection of sequences and does not imply the site could not change in some other sequence.

In vitro means a process carried out in the laboratory, not within a living organism (Latin for 'in glass', presumably referring to test tubes). *In vivo* ('in a living thing') is a process occurring in a living cell, or within an organism.

IUCN red list: see endangered species

Junk DNA is a term coined to described DNA sequences in the genome that have no apparent function in building or maintaining the organism. This term is not used much these days as it has become apparent that some of the non-genic DNA in the genome codes for other things, such as regulatory elements or small RNA molecules, and much of it is derived from viruses or transposable elements, so whether it is considered functionless or not might depend on who it is functioning in aid of. A more useful term might be 'intergenic DNA' which refers to all of the sequences between genes. See also non-coding DNA.

Kingdom is one of the highest taxonomic categories. Various numbers of kingdoms have been recognized, though most current schemes include six kingdoms: animals, plants, fungi, protists (single-celled eukaryotes), bacteria, and archaebacteria.

Label, in molecular genetics, generally refers to a chemical added to a molecule to allow its detection, for example a radioactive label (in traditional Sanger sequencing) or a fluorescent label (in pyrosequencing).

Lagerstätten are fine sedimentary deposits that record an unusually high degree of fossil detail and diversity.

Leading strand of DNA runs 3′ to 5′ so during DNA synthesis its complement is constructed from 5′ to 3′ in one continuous strand. But the lagging strand runs 5′ to 3′, and DNA polymerase cannot build a new strand in the 3′ to 5′ direction, so the complementary strand must be built from short strands (Okazaki fragments) made 'backwards' along the lagging strand from 5′ to 3′.

Likelihood ratio test compares the likelihood of observing a particular dataset under two (or more) alternative hypotheses.

Lineage is the line of descent, drawn as a branch on a phylogeny, linking a series of populations all descended from a common ancestor.

Linkage relates to the joint inheritance of alleles that are physically connected together on a chromosome. Linked alleles may form identifiable haplotypes.

Locus (plural: loci) is a broad term meaning any location in the genome. Locus is often used to refer to an independent heritable unit of genetic diversity, such as a polymorphic site in the sequence, a microsatellite region, or a gene.

Lysis is the rupture of the cell membrane to spill the cell's contents.

Macroevolution: see microevolution

Malaria is a disease caused by infection by the single-celled protist *Plasmodium*, which infects over 500 million people every year and kills at least a million people annually (particularly young children). Malaria is characterized by fever, headache, and vomiting, and is transmitted by the bite of an *Anopheles* mosquito. There are at least five different species of *Plasmodium* that can cause malaria, and around 30 different species of *Anopheles* that transmit the infection.

Mammals are a class of animals named for their unique milk glands (mammae), which are modified sweat glands that exude a specialized substance to feed young offspring. The Class Mammalia includes three main groups, Prototheria (monotremes), Metatheria (marsupials), and Eutheria (placentals). Another obvious mammalian feature is hair, which has an important role in thermoregulation in many species (even apparently hairless species may grow hair; for example, baby dolphins are born with a moustache). Unlike other vertebrates, in mammals the jawbone is made of a single bone, and two bones from the jaw have become the distinctive hammer and anvil bones in the mammalian inner ear.

Marker: see genetic marker

Marsupials are one of three major lineages of mammals (the others are monotremes and placentals), including possums, kangaroos, and koalas, characterized by development of young in a pouch on the outside of the body.

Matthewson, John (1973–), a brilliant philosopher of biology, originally trained as a medical doctor and now completing a doctorate in the philosophy of ecology, focusing on the way that models are used in biology.

Maximum likelihood is a statistical technique for comparing the plausibility of different hypotheses. In phylogenetics, maximum likelihood is used to compare the probability that a given phylogeny would have produced the observed sequence alignment, given a particular model of molecular evolution.

Maynard Smith, John (known as JMS; 1920–2004): I asked Kim Sterelny (a philosopher of biology) to write a definition of JMS and this is what he wrote: 'One of the most brilliant, and certainly the most sane, of the great UK twentieth century biologists, who largely built modern evolutionary biology. His most distinctive contribution was to incorporate game theory within evolutionary biology' (**Figure 5.4**).

Meiosis: see mitosis

Melt refers to heating a DNA sample to break the hydrogen bonds that form between the complementary bases on each strand, resulting in a single-strands of DNA rather than double helices. The melting temperature (temperature needed to convert the majority of the double-stranded DNA in the sample to single strands) will depend on a number of factors, including the GC content of the DNA: since GC pairs have three hydrogen bonds but AT pairs only two, it takes more energy to melt GC-rich DNA.

Mendel, Gregor (1822–1884) revealed the particulate nature of genetic inheritance by conducting breeding experiments on pea plants, so was the first quantitative geneticist. The significance of this work became apparent only after Mendel's death, when it became the basis of the neo-Darwinian synthesis. Incidentally, Mendel suffered from severe exam anxiety, which you may find comforting next time you are facing an exam.

Messenger RNA (mRNA) is an RNA polynucleotide made as a complementary copy of a gene in the nucleus, which is then processed and transported to the cytoplasm where it is translated into amino acid sequence on the ribosome.

Metagenomics is the analysis and interpretation of DNA from environmental samples, typically applied to characterizing bacterial communities, using a range of techniques such as DNA hybridization (see **Case Study 4.1**) or shotgun sequencing.

Methylation is the addition of a methyl group to a protein or nucleotide chain. DNA is methylated at cytosines; in mammalian genomes, the majority of CpG sites (where a C is next to a G) are methylated. Methylation plays a role in gene expression (e.g. gene silencing), and the methylation state of genes may be inherited.

Microarray is a solid surface (usually a chip) onto which are fixed a series of polynucleotides. When a sample of labelled DNA or RNA is added to the microarray, it will hybridize to any complementary sequences on the chip, and the label will reveal the position of the matching sequence on the array, thus allowing the identity of the sample sequence to be determined.

Microevolution is descent with modification by changes in the frequency of heritable variants in a population over generations. Mechanisms of microevolution include selection and drift. Macroevolution refers to evolutionary patterns that can be observed by comparing different lineages. Most biologists consider that macroevolutionary phenomena arise from microevolutionary processes; that is, that microevolution and macroevolution are different views of the same underlying process. For example, speciation is a microevolutionary process involving division and genetic change within populations, but patterns of species richness are observed comparing the relative numbers of species produced by different lineages. Taking a macroevolutionary perspective involves asking whether there are any interesting evolutionary patterns observable only when considering patterns in species and other higher taxa.

Microsatellites are loci in the genome where the same short nucleotide sequence (typically less than ten bases long) is repeated multiple times. The number of repeats has a high mutability, so microsatellite loci are often useful as genetic markers to detect within-population variability, and for genotyping individuals.

Microsporidia are unicellular intracellular parasitic eukaryotes, once thought to be one of the oldest lineages of eukaryotes on account of the simplicity of their genome and phenotype, but now considered to be highly simplified fungi.

Mismatch is a base pair other than those specified by Watson–Crick pairing rules. When the incorrect bases are opposite each other in a DNA helix, they cannot bond properly. Mismatch repair detects these incorrect base pairs, and excises one or more bases on one of the strands, and replace them with the matching nucleotide sequence.

Mitochondria (singular: mitochondrion) are energy-generating organelles in eukaryotic cells, originally derived from a symbiotic prokaryotic cell, and retaining a small circular genome with a small number of functional genes. Mitochondrial DNA is usually, though not exclusively, maternally inherited.

Mitosis is the normal process of cell division, whereby each daughter cell gets an identical (or nearly identical) copy of the genome of the parent cell. Meiosis is a special form of cell division associated with the production of gametes in sexually reproducing organisms. In meiosis, a diploid cell undergoes two rounds of cell division, one in which each daughter cell receives one copy of each chromosome, then each divides again to produce four haploid gametes. Unlike mitosis, the products of meiosis are genetically non-identical.

Moa were flightless birds native to New Zealand, from the ratite family which includes kiwis, emus, and ostriches. Moa went extinct approximately one thousand years ago, not long after the arrival of humans in New Zealand, probably due to hunting pressure.

Mobile genetic elements are any DNA sequences capable of moving from one part of the genome to another, also known as transposable elements.

Model is an abstract description of the behaviour of a system, which can be used to interpret observed patterns or predict future occurrences.

Molecular clock has two meanings. Firstly, it may refer to constant rates of molecular evolution in all lineages under consideration. Secondly, it may it may refer to the method of molecular dating, where the amount of divergence between sequences is used to estimate the age of the last common ancestor of those sequences. These two concepts are not equivalent: the molecular clock assumption underlies various analytical techniques (not just estimating dates), and molecular dating does not necessarily involve assuming constant rates of molecular evolution. See **TechBox 8.4**.

Monocot: see dicot

Montremes are a lineage of egg-laying mammals including the platypus and echidnas found in Australia and Papua New Guinea. Monotremes can hunt for invertebrates using electrodetection, as their soft beaks pick up the electromagnetic radiation given off by moving animals. Platypuses are one of the few mammal species that are poisonous, as male platypuses are able to inject venom using a spur on each ankle.

Morgan, Thomas Hunt (1866–1945) was a pioneering geneticist who established the use of the fruit fly, *Drosophila*, in genetical research, and used them to demonstrate that genes, localized to chromosomal regions, were the agent of heritability.

Morphology typically refers to the observable characteristics of an organism, though sometimes the term is used synonymously with phenotype.

Multiple hits occur when two or more nucleotide substitution events have occurred at the same site in a DNA sequence, so that only the last substitution is directly observable. Multiple hits obscure the true number of substitutions that have occurred in a sequence. See saturation.

Mutagen is any agent capable of causing mutation.

Mutation is a permanent, heritable change to the information in the genome.

Mutator and **antimutator** refer to variation in mutation rate between members of a population. A mutator has a higher-than-average mutation rate, antimutators have a lower-than-average mutation rate.

Natural selection: see selection

N_e: see effective population size

Nearly neutral theory is a modification of the neutral theory to encompass a continuum of mutation effects. In particular, the nearly neutral theory predicts that the fixation probabilities of 'nearly neutral' mutations, which are slightly advantageous or slightly deleterious, can contribute significantly to patterns of molecular evolution (**TechBox 5.2**).

Negative selection: see selection

Neo-Darwinian synthesis describes the body of ideas developed primarily in the first half of the twentieth century linking the evolutionary theories of Darwin and his contemporaries to Mendelian genetics, demonstrating the power of natural selection to drive change in allele frequencies, thus vindicating the hypothesis of gradualism. Key players in the development of the neo-Darwinian synthesis include Fisher, Haldane, and Wright who forged a mathematical framework for population genetics, combining the effects of selection and drift on allele frequencies. The neo-Darwinian synthesis was extended by researchers such as Theodosius Dobzhansky, who provided lab-based evidence of mutation and substitution in *Drosophila*, and G. G. Simpson, who demonstrated the compatibility of Darwinian mechanisms with observed patterns in the fossil record.

Neurospora crassa is a haploid fungus that, like *Drosophila* and *E. coli*, is a classic model organism for research in genetics. It can be grown on simple medium in the laboratory, and produces arrays of sexual spores contained within a body that allows the products of segregation to be easily detected.

Neutral alleles are variations of a trait that are functionally equivalent, such that none is favoured by selection over the others. Since the frequency of neutral alleles is not influence by selection, they fluctuate due to random events (see drift). It is important to note that neutral alleles are not necessarily phenotypically silent: observable phenotypic variants may be selectively equivalent. Neutral sites are positions in the genome where all nucleotides states are selectively equivalent, so that the substitution rate at those sites should reflect the mutation rate. Neutral theory suggests that the majority of observed substitutions were fixed by drift, not by selection (see TechBox 5.2).

Nodes are terminal points of branches (edges) in a phylogeny, such that each branch is defined by two nodes. Internal nodes are the branching points where a lineages splits to give rise to two or more descendant lineages. The tips of a phylogeny, which in molecular phylogenies usually represent the sequences from the alignment, are also nodes. The root node is the primary branching event at the base of a rooted phylogeny.

Noise: see signal

Non-coding is a confusing term, because sometimes it is used to refer to any DNA sequence that does not code for a protein (including RNA genes), sometimes it is used to represent all non-transcribed DNA sequences (including regulatory elements), and sometimes it means apparently functionless DNA (previously known as 'junk DNA').

Nuclear DNA: see nucleus

Nucleotides are the basic units of nucleic acids. Each nucleotide consists of one base (A, C, T, or G) bound to a sugar molecule which is bound to a phosphate molecule.

Nucleus is a membrane-bound organelle in eukaryote cells that contains the nuclear DNA. The nuclear DNA is the primary genome of a eukaryotic cell, typically packaged into multiple chromosomes.

Null hypothesis (or null model): see hypothesis

Okazaki fragments: see leading strand

Oligonucleotide is a short nucleotide strand; usually refers to a artificially created polynucleotide.

Onycophora is a phylum of adorable little caterpillar-like animals, otherwise known as velvet worms or peripatus.

Open reading frame (ORF) is a DNA sequence that does not contain a stop codon so has the potential to be translated into a continuous amino acid sequence.

Organelles are structural subunits of cells that have specific functions, for example mitochondria and chloroplasts.

Orthologues are homologous copies of a sequence produced by the divergence of lineages: see gene family.

Oryza sativa is the scientific name for the most commonly cultivated species of rice.

Overparameterization (overfitting) is the inclusion of too many parameters in a model, which may increase the fit of the model to the data, but decreases the explanatory or predictive power of the model by making the model explain one specific dataset (potentially describing noise rather than signal) rather than provide a general explanation of a pattern or process.

Panspermia is the hypothesis that life on Earth was ultimately derived from elsewhere in the universe (an idea not supported by many biologists).

Paralogues are copies of genes produced by gene duplication: see gene family.

Parsimony, as applied to phylogenetic reconstruction, is the principle that the most reasonable phylogeny is the one that requires the inference of the smallest number of evolutionary changes.

Pauling, Linus (1901–1994) pioneered the study of the structure and evolution of proteins, and won two Nobel prizes, one for chemistry and one for peace.

p-distance is the Hamming distance divided by the number of sites compared.

Pedigree is a family history, usually drawn as a branching diagram with male family members represented as squares and females as circles. In genetics, pedigrees are used to uncover patterns of inheritance, for example to identify carriers of an allele associated with a particular trait. An informative pedigree is one in which the known phenotypes and family relationships provide sufficient information to allow the genetic basis of a trait to be identified.

Peptides are chains of amino acids, held together by peptide bonds which are covalent bonds between the amino group of one amino acid and the carboxyl group of the next amino acid. Proteins are formed of single or multiple peptides, which usually adopt a particular three-dimensional structure determined by the amino acid sequence.

Phage: see bacteriophage

Phenetics, as applied to phylogeny reconstruction, is the principle that the relationships between lineages can be determined by analysis of measures of similarity.

Phenotype generally refers to the observable properties of an organism, particularly its morphology and behaviour. However, phenotype is sometimes used more broadly to encapsulate all of the results of gene expression, including development and metabolism, in order to provide a contrast to genotype, which refers only to the genetic information contained in the genome. Phenotype often refers to a 'partial phenotype', the expression of a trait or traits of interest, rather than to the entire morphology.

Phenylketonuria (PKU) is a heritable metabolic disorder caused by a recessive allele, homozygotes for which cannot metabolize phenylalanine, which then builds up in the nervous system and brain causing onset of mental retardation during childhood.

Phosphates are one of the three units in a polynucleotide strand: see nucleotide.

Phosphodiester bonds join nucleotides to make a polynucleotide chain, through a covalent bond between 5′ phosphate group of one nucleotide to the 3′ hydroxyl group of the sugar of the next nucleotide, catalysed by a polymerase enzyme.

Phylogeny is a representation of the evolutionary history of biological lineages as a nested series of branching events, each marking the divergence of two or more lineages from a common ancestral lineage. Phylogenetics is the discipline of inferring phylogenies – there are many different phylogenetic techniques including those based in maximum likelihood and Bayesian inference.

Phylum (plural: phyla) is a taxonomic category: kingdoms are divided into phyla.

Placental mammals, more formally known as Eutheria, are one of three major lineages of mammals (the others being monotremes and marsupials). The placentals have been split into three major clades, largely based on molecular phylogenies: the Afrotheria (including the elephants, hyraxes, dugongs, aardvarks and others); Laurasiatheria (including artiodactyls, cetaceans, carnivorans, horses, rhinos and others); and Euarchontoglires (including rodents, bunnies, primates, bats, and others).

Plankton refers to the microscopic organisms, including animals, plants, and algae, that float in surface waters.

Platypus: see monotreme

Polymerase is an enzyme that catalyses the formation of phosphodiester bonds that bind nucleotides together in a DNA or RNA molecule.

Polymerase chain reaction (PCR) is a laboratory method for amplifying DNA, in which DNA is heated to separate the double helices, cooled so primers bind to specific sequences in the single-stranded DNA, then heated with polymerase which makes complementary strands starting from the primer sequences. The cycle is repeated many times, resulting in exponential increase in the number of copies of the amplified sequence (TechBox 4.2).

Polymorphism occurs when there is variation within a population for a given trait. At the molecular genetic level, polymorphism is the presence of two or more alleles for a given locus in a population.

Popper, Karl (1902–1994) was an influential philosopher of science who described approaches to hypothesis testing in science, most famously that scientific hypotheses can be refuted when the facts are shown to be against them, but cannot be conclusively proven. In the Popperian view of science, a well-supported hypothesis is one that has survived repeated attempts at refutation.

Population, in evolutionary genetics, represents an interbreeding set of the individuals capable of combining genetic material to produce offspring. Populations are kept distinct from all other such populations by reproductive isolation mechanisms that prevent the formation of hybrid offspring with parents from different populations. These isolating mechanisms may prevent potential parents coming together (for example, distance between populations or behavioural differences in mating rituals), or may prevent the development or reproduction of offspring resulting from hybrid crosses (such as genetic incompatibility). Note that a population may have 'fuzzy borders' if there are low levels of genetic exchange between connected populations, through occasional hybridization or rare migration events. Some populations maintain genetic distinctness even in the face of ongoing genetic exchange with other populations.

Positive selection: see selection

Posterior probability, **prior probability** see Bayesian inference

Prokaryotes (bacteria and archaebacteria) are single-celled organisms without a nucleus or mitochondria: see eukaryote.

Promoter is a sequence located near a gene to which RNA polymerase binds to begin transcription of the gene.

Proofreading is an exonuclease function possessed by some polymerase enzymes that removes incorrect nucleotides from the end of the newly synthesized DNA strand.

Proteins are a type of organic molecule made of linear chains of amino acids linked together by peptide bonds. The sequence of amino acids, which determines a protein's characteristics, is specified by the nucleotide sequence of a gene. The linear chain then takes on a three-dimensional structure due to interactions between non-adjacent amino acids in the chain: this protein folding either happens spontaneously, or is assisted by chaperone proteins. Different proteins may combine together to make a functional unit (e.g. haemoglobin). Proteins perform a wide variety of essential functions in organisms, including structural (e.g. keratin which forms nails, claws, and hair), enzymatic (e.g. protease digests proteins in food), metabolic (e.g. haemoglobin takes oxygen around the body), and defence (e.g. immunoglobins which protect the body from infections). Proteins are also essential for DNA replication (e.g. polymerase makes a new nucleotide strand to match an existing template strand).

Protein-coding sequences are parts of the genome that can be transcribed and translated into an amino acid sequence.

Pseudogene is a non-functional version of a gene, produced by gene duplication or by inactivating mutations within a previously function gene. Because they are not subject to selection, pseudogenes are assumed to accumulate substitutions at the neutral rate.

Purines are the double-ring bases, adenine and guanine.

Pyrimidines are the single-ring bases, thymine, cytosine, and uracil.

Pyrosequencing is an alternative to traditional Sanger sequencing. Rather than producing a series of fragments, each ending at a particular nucleotide in the sequence, pyrosequencing uses light reactions during DNA synthesis to report the identity of each nucleotide added to the growing strand. 454 sequencing is a particular parallel pyrosequencing technology that allows very large amounts of sequencing to be done in a short time period. 454 technology has been used for whole genome sequencing, and producing partial genome sequences from ancient DNA samples such as Neanderthal and woolly mammoth (**TechBox 1.2**).

Quagga is a subspecies of zebra that went extinct when the last individual died in captivity in an Amsterdam zoo in 1883.

Random processes are undirected with respect to a particular outcome. Therefore it is important to define 'random with respect to what?'. A process that is random with respect to particular outcomes (e.g. mutations are random with respect to fitness) may be biased towards certain other outcomes (e.g. mutations are more likely to occur in certain places in the genome). In this book, 'random' is used synonymously with stochastic and chance.

Recessive alleles affect phenotype only when in the homozygous state, not when heterozygous. Recessive diseases are only evidence when the carrier has a 'double-dose', carrying a copy of the disease allele on both homologous chromosomes.

Recombination is the exchange of genetic material between chromosomes or between genomes: see crossing-over. Recombinant DNA has DNA from two different sources, thus breaking up normally linked genes or alleles.

Reduction/division: see meiosis

Redundancy: see degenerate

Reinforcement is selection for reproductive isolation mechanisms that reduce the incidence of hybridization between members of different populations.

Relaxed clock may refer in the broad sense to any phylogenetic method that does not rely upon the assumption of a molecular clock (constant rate of change in all lineages), but usually refers to the set of phylogenetic methods that allow substitution rates to vary across the phylogeny according to a predefined model of rate change.

Repeat sequences contain multiple tandem (side-by-side) copies of the same nucleotide sequence, which are prone to change in repeat number. Microsatellites loci contain short nucleotide repeats: see microsatellite.

Replacement substitutions change the amino acid sequence of the resulting protein.

Restriction enzyme is an endonuclease that recognizes a specific nucleotide sequence and cuts the DNA strand wherever that sequence occurs. Restriction fragment length polymorphism (RFLP) is variation between genomes in a population in the number or location of a particular restriction sequences, so that when DNA is digested with a specific restriction enzyme, individuals will be characterized by different fragment numbers and lengths which can be separated on a gel: see DNA fingerprinting.

Retroviruses replicate by using reverse transcriptase to make a DNA copy of their RNA genome, and the DNA copy is then inserted into the host's genome. The inserted retroviral genome, referred to as a provirus, contains regulatory elements that co-opt the host's transcription machinery, which makes RNA copies of the viral genome. These RNA viral genomes can act as transcripts for the production of viral proteins, or as genomes to be packaged into new virus particles. Endogenous retroviruses are viral genomes (or the remains thereof) embedded in the host genome and inherited by offspring when the genome is replicated.

Reverse transcriptase is a polymerase enzyme that makes a complementary DNA strand from an RNA template. Retroviruses use reverse transcriptase to make the provirus which is inserted into the host genome: see retrovirus.

Ribose: see sugar

Ribosomal RNA (rRNA) is a component of the ribosome, the cellular organelles that are responsible for translating the information in a messenger RNA transcript to the amino acid sequence of a peptide. Transcription of rRNA genes produces an RNA polynucleotide which then adopts a secondary structure by complementary base pairing between specific parts of the sequence. Some rRNA genes are commonly used in phylogenetics, including the mitochondrial-encoded 12S rRNA and the nuclear-encoded 18S rRNA (also referred to as small-subunit RNA, or SSU) and 28S rRNA (also called the large-subunit RNA or LSU). The 'S' refers to the size of the molecule produced by the gene, measured in Svedburgs, which reflects the rate of sedimentation under centrifugation – the larger the S value the bigger the particle. rRNA genes have different patterns of molecular evolution than protein-coding sequences, for example indels are more common, and rates of change may vary between stem and loop regions of the sequence (**Figure 6.14**).

Ribosome: see ribosomal RNA

RNA polymerase is an enzyme that makes a complementary RNA strand to match a DNA template strand.

Sanger method is the standard approach to DNA sequencing, where DNA is replicated in the presence of modified nucleotides that halt synthesis when incorporated, producing fragments of different lengths which can be separated on a gradient. See **TechBox 1.2**.

Saturation is the loss of evolutionary information in a sequence through multiple hits which erase the signal of descent.

Schizophrenia is a psychiatric diagnosis covering a wide range of symptoms and signs, characterized by disordered thought and perception, usually including delusions and hallucinations, and often social withdrawal.

Scientific literature describes research articles published in 'scholarly' (professional science) journals, rather than in the popular press. Articles in the scientific literature should fully describe the research so that, technically speaking, any suitably experienced and equipped person could repeat the research. Articles published in the scientific literature should have been through peer review, where the article is first sent to a number of independent scientists who evaluate the article (usually anonymously) in order to detect any flaws in the research.

Segregation is the division of the diploid genome into haploid gametes in meiosis. Segregation refers more specifically to the separation of alternative alleles for one or more loci into gametes so that they can be inherited separately. Some geneticists refer to polymorphic alleles as 'segregating in the population' (or segregating in a particular pedigree).

Selection (natural selection) is change in the frequency of an allele as a result of its influence on its own chances of being passed to the next generation. Alleles under positive selection have an increased chance of being included in the next generation relative to other alleles in the population. In a large population, positive selection will usually result in the fixation of an allele so that it replaces all other variants in the population (though this will not always be the outcome of positive selection: see Chapter 5). Alleles under negative selection have a decreased chance of being included in the next generation: any alleles that cause reduction in their carriers' chances of reproduction will tend to reduce in frequency until lost from the population. Because negative selection results in the removal of alleles from the population, it commonly results in the conservation of sequences.

Selective sweep occurs when selection on a specific locus affects the frequency of alleles at neighbouring loci. Alleles in loci that are linked to a locus under positive selection may be swept to fixation along with the selected allele, regardless of their own effects on fitness. Selective sweeps result in the reduction in nucleotide diversity at loci linked to a trait under strong positive selection, detected through the over-representation of particular haplotypes in a population.

Signal is patterns in the data created by the process of interest. Noise is random variation resulting from stochastic processes not from the process of interest. The signal-to-noise ratio of a dataset reflects how clearly those data reflect the process of interest, therefore whether the data can be used to discriminate between alternative hypotheses.

Silent changes alter the nucleotide sequence of the genome but do not have a noticeable effect on phenotype. For example, synonymous changes to the nucleotide sequence of a protein-coding gene do not change the amino acid sequence of the resulting protein, so they are phenotypically silent. Note that 'silent' is not exactly the same as 'neutral'. Silent changes may have a fitness cost or benefit (for example, synonymous codon bias suggests that there may be an efficiency gain in using certain codons), and non-silent changes may be effectively neutral (for example, a non-synonymous change may have a noticeable effect on phenotype that has no implications for fitness).

Simpson, George Gaylord (1902–1984) was a palaeontologist who contributed to the codification of the neo-Darwinian synthesis by showing how palaeontological patterns could be interpreted in light of Darwinian gradualism, natural selection, and population genetics.

Simulated data are data produced artificially, rather than observed from natural systems. In phylogenetics, data are simulated according to a particular model of molecular evolution (e.g. a substitution model used to 'evolve' a set of sequences from a single ancestral sequence). The simulated data may be used to test the effectiveness of methods of phylogeny reconstruction, or to generate an expected pattern of substitutions to use as a null hypothesis against which another hypothesis can be tested. See **TechBox 7.4**.

Single nucleotide polymorphism (SNP) is a locus in the genome where the nucleotide sequence differs between members of a population. See **TechBox 3.3**.

Soma: see germline

Species is one of the lowest levels in the taxonomic hierarchy, given a scientific name consisting of *Genus species* (e.g. *Homo sapiens*). The precise definition of species is a matter of debate, and there are many alternative species concepts that set out criteria for delineating species: see **TechBox 6.2** for a fuller discussion. An undescribed species is a set of organisms that probably form a distinct species but have not been given a formal taxonomic description with a scientific species name: it is likely that the majority of living species have not yet been described.

Spencer, Herbert (1820–1903), the originator of the phrase 'survival of the fittest' to describe natural selection (a phrase adopted by Darwin in later editions of the *Origin of Species*), who used evolutionary principles to inform models of social change.

Substitution is the loss of all alternative alleles in a population but one, so that all members of the population carry the same allele at that locus. Substitution should not be confused with mutation, which changes the genetic information in a single individual. A mutation may become a substitution if it increases in frequency over generations until it reaches fixation.

Sugar, in the broad sense, is any carbohydrate molecule. Ribose is a pentose (five-carbon) sugar, with four carbons arranged in a ring with oxygen, and four hydroxyl (OH) groups. Deoxyribose is the version of this sugar found in DNA, in which one of the hydroxyl groups is replaced with a hydrogen.

Synonynmous codon: see codon

Systematics is the study of the relationships of living things, nowadays most commonly conducted in a phylogenetic framework.

Taxonomy is the practice (and the output) of classifying biological diversity. Taxonomy divides organisms into defined units, known as taxa (singular: taxon). A taxon could refer to any division of organisms into groups, whether species, subspecies, phyla, etc. The phrase 'higher taxon' generally refers to taxonomic levels above species (or above the level referred to). See **TechBox 6.1**.

Theoretical: see empirical

Thylacine (Tasmanian tiger) refers to a lineage of marsupial predators, originally distributed throughout Australia and Papua New Guinea. Thylacines disappeared from the Australian mainland before European settlement (probably more than a thousand years ago, possible due to competition with introduced dingos), but survived on the island of Tasmania until the first half of the twentieth century.

Thymine (T) is one of the four DNA bases that make up the 'alphabet' of the genetic code: see bases. A thymine dimer (also called thymidine dimer) forms when two adjacent thymines bond together, rather than pairing with the A on the opposite strand.

Topology is the branching order of a phylogenetic tree. Trees that contain the same taxa but differ in the way those taxa are connected together are said to differ in topology. Topology is sometimes used to distinguish the branching order from information about branch-lengths: two phylogenies that depict the same relationships between taxa but show different amounts of change along lineages have the same topology but differ in branch length.

Trait is generally used synonymously with character to indicate a discretely heritable aspect of phenotype.

Transcription is the production of a complementary RNA strand from a DNA sequence in the genome. This RNA transcript can move from the nucleus to the cytoplasm, where it may be directly involved in cellular processes (for example, ribosomal RNAs and transfer RNAs) or translated into a protein sequence at the ribosome (messenger RNA).

Transfer RNA (tRNA) is a small RNA chain, often depicted as having a clover-leaf secondary structure, that carries amino acids to the ribosome for protein synthesis. Each type of tRNA carries a particular amino acid and has an anticodon that can pair to the complementary codon in messenger RNA, ensuring that the messenger RNA is accurately translated into a specific amino acid sequence.

Transition is a mutation (or substitution) involving the exchange of one pyrimidine for another, or one purine for another. A transversion changes a purine to a pyrimidine or vice versa. In most sequences, transitions are more common than transversions, so the transition–transversion ratio is an important component of many substitution models.

Translation is the conversion of information from the nucleotide sequence of a messenger RNA molecule into the amino acid sequence of a peptide, which occurs on the ribosome.

Transposable elements: see mobile genetic elements

Tree-like data can be represented by a branching diagram. Non-tree-like data does not follow a simple hierarchical branching pattern.

Uncorrected distance: see Hamming distance

Unequal crossing over: see crossing over

Uracil: see base

Velvet worm: see onycophoran

Vulnerable species: see endangered species

Wallace, Alfred Russell (1823–1913) was a British naturalist who independently discovered the principle of natural selection, and who made a major contribution to establishing the field of biogeography.

Weismann, Friedrich Leopold August (1834–1914) placed Darwin's evolutionary theory within the framework of heritability and development: see **Heroes 2**.

Wright, Sewall (1889–1988) was one of the three key figures in the development of population genetic theory in the first half of the twentieth century (the others being Haldane and Fisher), promoting the importance of drift in determining allele frequencies and embedding the concept of the adaptive landscape in evolutionary biology (Figure 5.4).

Zygote: see gamete

Bibliography

This is not an exhaustive list of the sources of information used to write each chapter, nor is it a suggested reading list. Instead, I have listed only those studies or sources that are referred to specifically in the text, or directly quoted, so that interested readers can follow up the sources of the observations referred to in each of the chapters. References given elsewhere have not been repeated. This is why some chapters have few references – because they discuss information widely available in a number of printed and online sources – and some have more – because the reader would need to know the particular publication details to find this particular information.

Chapter 1

Gatesy, J. and O'Leary, M.A. (2001) Deciphering whale origins with molecules and fossils. *Trends in Ecology and Evolution*, Volume 16, pages 562–570.

Hillis, D.M., Mable, B.K., Larson, A., Davis, S.K. and Zimmer, E.A. (1996) Nucleic acids IV: Sequencing and cloning. In *Molecular Systematics*, 2nd edn. Edited by D.M. Hillis, C. Moritz and B.K. Mable. Sinauer Associates, Sunderland, Mass.

Lyrholm, T., Leimar, O. and Gyllensten, U. (1996) Low diversity and biased substitution patterns in the mitochondrial DNA control region of sperm whales: implications for estimates of time since common ancestry. *Molecular Biology and Evolution*, Volume 13, pages 1318–1326.

Lyrholm, T., Leimar, O., Johanneson, B. and Gyllensten, U. (1999) Sex-biased dispersal in sperm whales: contrasting mitochondrial and nuclear genetic structure of global populations. *Proceedings of the Royal Society of London Series B Biological Sciences*, Volume 266, pages 347–354.

Milinkovitch, M.C. (1995) Molecular phylogeny of cetaceans prompts revision of morphological transformations. *Trends in Ecology and Evolution*, Volume 10, pages 328–334.

Pierce, S.K., Massey, S.E., Curtis, N.E., Smith, G.N., Olavarri, C. and Maugel, C. (2004) Microscopic, biochemical, and molecular characteristics of the Chilean Blob and a comparison with the remains of other sea monsters: nothing but whales. *Biological Bulletin*, Volume 206, pages 125–133.

Richard, K.R., Dillon, M.C., Whitehead, H. and Wright, J.M. (1996) Patterns of kinship in groups of free-living sperm whales (*Physeter macrocephalus*) revealed by multiple molecular genetic analyses. *Proceedings of the National Academy of Sciences USA*, Volume 93, pages 8792–8795.

Whitehead, H. (2002) Estimates of the current global population size and historical trajectory for sperm whales. *Marine Ecology Progress Series*, Volume 242, pages 295–304.

Chapter 2

Duncan, R.P., Blackburn, T.M. and Worthy, T.H. (2002) Prehistoric bird extinctions and human hunting. *Proceedings of the Royal Society of London Series B Biological Sciences*, Volume 269, pages 517–521.

Maynard Smith, J. (1989) Weismann and modern biology. *Oxford Surveys in Evolutionary Biology*, Volume 6, pages 1–12.

www.dnai.org/timeline

www.mendelweb.org

Chapter 3

Asif, M.J. and Cannon, C.H. (2005) DNA extraction from processed wood: a case study for the identification of an endangered timber species (*Gonystylus bancanus*). *Plant Molecular Biology Reporter*, Volume 23, pages 185–192.

Carpen, J.D., Archer, S.N., Skene, D.J., Smits, M. and Schantz, M. (2005) A single-nucleotide polymorphism in the 5′-untranslated region of the hPER2 gene is associated with diurnal preference. *Journal of Sleep Research*, Volume 14, pages 293–297.

Dawkins, R. (1982) *The Extended Phenotype*. Oxford University Press, Oxford.

Deguilloux, M.-F., Pemonge, N.-H. and Petit, R.J. (2004) DNA-based control of oak wood geographic origin in the context of the cooperage industry. *Annals of Forest Science*, Volume 61, pages 97–104.

Eyre-Walker, A. (2006) The genomic rate of adaptive evolution. *Trends in Ecology and Evolution*, Volume 21, pages 569–575.

Goossens, B., Graziani, L., Waits, L.P., Farand, E., Magnolon, S., Coulon, J., Bel, M.-C., Taberlet, P. and Allaine, D. (1998) Extra-pair paternity in the monogamous Alpine marmot revealed by nuclear DNA microsatellite analysis. *Behavioural Ecology and Sociobiology*, Volume 43, pages 281–288.

Guelbeogo, W.M., Grushko, O., Boccolini, D., Ouedraogo, P.A., Besansky, N.J., Sagnon, N.F. and Costantini, C. (2005) Chromosomal evidence of incipient speciation in the Afrotropical malaria mosquito *Anopheles funestus*. *Medical and Veterinary Entomology*, Volume 19, pages 458–469.

Hanger, J.J., Bromham, L., McKee, J.J., O'Brien, T.M. and Robinson, W.F. (2000) The nucleotide sequence of koala (*Phascolarctos cinereus*) retrovirus (KoRV): a novel type-C retrovirus related to gibbon ape leukemia virus (GALV). *Journal of Virology*, Volume 74, pages 4264–4272.

Mi, S., Lee, X., Li, X.P., Veldman, G.M., Finnerty, H., Racie, L., *et al.* (2000) Syncytin is a captive retroviral envelope protein involved in human placental morphogenesis. *Nature*, Volume 403, pages 785–789.

Rosenberg, S.M. (1997) Mutation for survival. *Current Opinion in Genetics and Development*, Volume 7, pages 829–834.

Tafti, M., Maret, S.P. and Dauvilliers, Y. (2005) Genes for normal sleep and sleep disorders. *Annals of Medicine*, Volume 37, pages 580–589.

Toh, K.L., Jones, C.R., He, Y., Eide, E.J., Hinz, W.A., Virshup, D.M., Ptacek, L.J. and Fu, Y.-H. (2001) An hPer2 phosphorylation site mutation in familial advanced sleep phase syndrome. *Science*, Volume 291, pages 1040–1043.

Wexler, N.S., Lorimer, J., Porter, J., Gomez, F., Moskowitz, C., Shackell, E., *et al.* (2004) Venezuelan kindreds reveal that genetic and environmental factors modulate Huntington's disease age of onset. *Proceedings of the National Academy of Sciences USA*, Volume 101, pages 3498–3503.

Zane, L., Bargelloni, L. and Patarnello, T. (2002) Strategies for microsatellite isolation: a review. *Molecular Ecology*, Volume 11, pages 1–16.

Chapter 4

Read, T.D., Salzberg, S.L., Pop, M., Shumway, M., Umayam, L., Jiang, L., *et al.* (2002) Comparative genome sequencing for discovery of novel polymorphisms in *Bacillus anthracis*. *Science*, Volume 296, pages 2028–2033.

Chapter 5

Beehler, B. (1983) Lek behavior of the Lesser Bird of Paradise. *Auk*, Volume 100, pages 992–995.

Bromham, L. and Leys, R. (2005) Sociality and rate of molecular evolution. *Molecular Biology and Evolution*, Volume 22, pages 1393–1402.

Clark, R. (1968) *J.B.S: the Life and Work of JBS Haldane*. Hodder and Stoughton, London.

Crow, J.F. (1988) Wright, Sewall – Obituary. *Genetics*, Volume 119, pages 1–4.

Darwin, F. (1902) *The Life of Charles Darwin*. John Murray, London.

Dawkins, R. (1989) *The Selfish Gene*, 2nd edn. Oxford University Press, Oxford.

Evans, T.A., Wallis, E.J. and Elgar, M.A. (2002) Making a meal of mother. *Nature*, Volume 376, page 299.

Kerr, B., Riley, M.A., Feldman, M. and Bohannan, B.J.M. (2002) Local dispersal promotes biodiversity in a real-life game of rock–paper–scissors. *Nature*, Volume 418, pages 171–174.

Kohn, M. (2004) *A Reason for Everything: natural selection and the English imagination*. Faber and Faber, London.

Maynard Smith, J. (1998) *Evolutionary Genetics*. Oxford University Press, Oxford.

Medawar, P. (1986) *Memoir of a Thinking Radish: an autobiography*. Oxford University Press, Oxford.

Muller, H.J. (1950) Our load of mutations. *American Journal of Human Genetics*, Volume 2, pages 111–176 (reprinted in Ridley, M. ed. (1997) *Evolution*. Oxford Readers Series, Oxford University Press, Oxford).

Nussbaum, R.L., McInnes, R.R. and Willard, H.F. (2004) *Thompson and Thompson Genetics in Medicine*. Saunders, Philadelphia, Penn.

Wood, E.T., Stover, D.A., Slatkin, M., Nachman, M.W. and Hammer, M.F. (2005) The β-globin recombinational hotspot reduces the effects of strong selection around HbC, a recently arisen mutation providing resistance to malaria. *American Journal of Human Genetics*, Volume 77, pages 637–642.

Zanotto, P.M., Kallas, E.G., de Souzaa, R.F. and Holmes, E.C. (1999) Genealogical evidence for positive selection in the nef gene of HIV-1. *Genetics*, Volume 153, pages 1077–1089.

Chapter 6

Bromham, L. and Leys, R. (2005) Sociality and rate of molecular evolution. *Molecular Biology and Evolution*, Volume 22, pages 1393–1402.

Cameron, S.A. and Mardulyn, P. (2001) Multiple molecular data sets suggest independent origins of highly eusocial behaviour in bees (Hymenoptera: Apinae). *Systematic Biology*. Volume 50, pages 194–214.

Duffy, J.E., Morrison, C.L. and Ríos, R. (2000) Multiple origins of eusociality among sponge-dwelling shrimps (*Synalpheus*). *Evolution*, Volume 54, pages 503–516.

Gomez, A., Serra, M., Carvalho, G.R. and Lunt, D.H. (2002) Speciation in ancient cryptic species complexes: evidence from the molecular phylogeny of *Brachionus plicatilis* (Rotifera). *Evolution*, Volume 56, pages 1431–1444.

Keeling, P.J. and Fast, N.M. (2002) Microsporidia: biology and evolution of highly reduced intracellular parasites. *Annual Reviews in Microbiology*, Volume 56, pages 93–116.

Moritz, C. (1993) The origin and evolution of parthenogenesis in the *Heteronotia binoei* complex: synthesis. *Genetica*, Volume 90, pages 269–228.

Storz, J.F., Sabatino, S.J., Hoffmann, F.G., Gering, E.J., Moriyama, H., *et al.* (2007) The molecular basis of high-altitude adaptation in deer mice. *PLoS Genetics*, Volume 3, page e45.

Thompson, G.J. and Oldroyd, B.P. (2004) Evaluating alternative hypotheses for the origin of eusociality in corbiculate bees. *Molecular Phylogenetics and Evolution*, Volume 33, pages 452–456.

Villanueva-G., R., Roubik, D.A. and Colli-Ucán, W. (2005) Extinction of *Melipona beecheii* and traditional beekeeping in the Yucatán peninsula. *Bee World*, Volume 86, pages 35–41.

Chapter 7

Andersson, M., Page, D.C. and de la Chapelle, A. (1986) Chromosome Y-specific DNA is transferred to the short arm of X chromosome in human XX males. *Science*, Volume 233, pages 786–788.

Beddall, B.G. (1957) Historical notes on avian classification. *Systematic Zoology*, Volume 6, pages 129–136.

Bromham, L., Eyre-Walker, A., Smith, N.H. and Maynard Smith, J. (2003) Mitochondrial Steve: paternal inheritance of mitochondria in humans. *Trends in Ecology and Evolution*, Volume 18, pages 2–4.

Dawkins, R. (2004) *The Ancestor's Tale: A Pilgrimage to the Dawn of Life*. Weidenfeld & Nicholson, London.

Page, R.D.M. and Holmes, E.C. (1998) *Molecular Evolution: A Phylogenetic Approach*. Blackwell, Oxford.

Roques, P., Robertson, D.L., Souquiere, S., Apetrei, C., Nerrienet, E., Barre-Sinoussi, F., Muller-Trutwin, M. and Simon, F. (2004) Phylogenetic characteristics of three new HIV-1 N strains and implications for the origin of group N. *AIDS*, Volume 18, pages 1371–1381.

Sereno, P.C. (1999) The evolution of dinosaurs. *Science*, Volume 284, pages 2137–2147.

Swofford, D.L., Olsen, G.J., Waddell, P.J. and Hillis, D.M. (1996) Phylogenetic inference. In *Molecular systematics*, eds. Hillis, D.M., Moritz, C. and Mable, B.K, pages 407–514. Sinauer Associates, Sunderland, Mass.

Chapter 8

Bromham, L., Rambaut, A. and Harvey, P.H. (1996) Determinants of rate variation in mammalian DNA sequence evolution. *Journal of Molecular Evolution*, Volume 43, pages 610–621.

Darwin, C. (1859) *On the Origin of Species by Means of Natural Selection, or the Preservation of Favoured Races in the Struggle for Life*. John Murray, London.

Ellegren, H. (2007) Characteristics, causes and evolutionary consequences of male-biased mutation. *Proceedings of the Royal Society of London Series Biological Sciences*, Volume 274, pages 1–10.

Fleischer, R.C., McIntosh, C.E. and Tarr, C.L. (1998) Evolution on a volcanic conveyor belt: using phylogeographic reconstructions and K-Ar based ages of the Hawaiian islands to estimate molecular evolutionary rates. *Molecular Ecology*, Volume 7, pages 533–545.

Inoue, J.G., Miya, M., Venkatesh, B. and Nishida, M. (2005) The mitochondrial genome of Indonesian coelacanth *Latimeria menadoensis* (Sarcopterygii: Coelacanthiformes) and divergence time estimation between the two coelacanths. *Gene*, Volume 349, pages 227–235.

Isaac, N.J.B., Turvey, S.T., Collen, B., Waterman, C. and Baillie, J.E.M. (2007) Mammals on the EDGE: conservation priorities based on threat and phylogeny. *PLoS ONE*, Volume 2, page e296.

Knoll, A.H. and Hewitt, D. (2008) Phylogenetic, functional and geological perspectives on complex multicellularity. In *Major Transitions in Evolution Revisited*, eds B. Calcott and K. Sterelny, Vienna Series in Theoretical Biology. MIT Press, Cambridge, Mass.

Levinton, J. (2001) *Genetics, Paleontology, and Macroevolution*, 2nd edn. Cambridge University Press, Cambridge.

Ou, C.Y., Ciesielski, C.A., Myers, G., Bandea, C.I., Luo, C.C., Korber, B.T.M., *et al.* (1992) Molecular epidemiology of HIV transmission in a dental practice. *Science*, Volume 256, pages 1165–1171.

Peterson, K.J., Lyons, J.B., Nowak, K.S., Takacs, C.M., Wargo, M.J. and McPeek, M.A. (2004) Estimating metazoan divergence times with a molecular clock. *Proceedings of the National Academy of Sciences USA*, Volume 101, pages 6536–6541.

Thomas, J.A., Welch, J.J., Woolfit, M. and Bromham, L. (2006) There is no universal molecular clock for invertebrates, but rate variation does not scale with body size. *Proceedings of the national Academy of Sciences USA*, Volume 103, pages 7366–7371.

Valentine, J.W. (2004) *On the Origin of Phyla*. University of Chicago Press, Chicago, Ill.

Weinberg, S. (2000) *A Fish Caught in Time*. Harper Collins Publishers, London.

Welch, J.J. and Bromham, L. (2005) Estimating molecular dates when rates vary. *Trends in Ecology and Evolution*, Volume 20, pages 320–327.

Woolfit, M. and Bromham, L. (2003) Increased rates of sequence evolution in endosymbiotic bacteria and fungi with small effective population sizes. *Molecular Biology and Evolution*, Vollume 20, pages 1545–1555.

Sources of date estimates for Figures 8.34 to 8.37:

Aris-Brosou, S. and Yang, Z. (2002) Effects of models of rate evolution on estimation of divergence dates with special reference to the metazoan 18S ribosomal RNA phylogeny. *Systematic Biology*, Volume 51, pages 703–714.

Aris-Brosou, S. and Yang, Z. (2003) Bayesian models of episodic evolution support a late precambrian explosive diversification of the metazoa. *Molecular Biology and Evolution*, Volume 20, pages 1947–1954.

Ayala, F.J., Rzhetsky, A. and Ayala, F.J. (1998) Origin of the metazoan phyla: molecular clocks confirm palaeontological estimates. *Proceedings of the National Academy of Sciences USA*, Volume 95, pages 606–611.

Blair, J.E. and Hedges, S.B. (2005) Molecular phylogeny and divergence times of deuterostome animals. *Molecular Biology and Evolution*, Volume 22, pages 2275–2284.

Bromham, L., Rambaut, A., Fortey, R., Cooper, A. and Penny, D. (1998) Testing the Cambrian explosion hypothesis by using a molecular dating technique. *Proceedings of the National Academy of Sciences USA*, Volume 95, pages 12386–12389.

Douzery, E.J.P., Bapteste, E., Delsuc, F. and Philippe, H. (2004) The timing of eukaryotic evolution: Does a relaxed molecular clock reconcile proteins and fossils? *Proceedings of the National Academy of Sciences USA*, Volume 101, pages 15386–15391.

Gu, X. (1998) Early metazoan divergence was about 830 million years ago. *Journal of Molecular Evolution*, Volume 47, pages 369–371.

Lynch, M. (1999) The age and relationships of the major animal phyla. *Evolution*, volume 53, pages 319–325.

Peterson, K.J. and Butterfield, N.J. (2005) Origin of the Eumetazoa: Testing ecological predictions of molecular clocks against the Proterozoic fossil record. *Proceedings of the National Academy of Sciences USA*, Volume 102, pages 9547–9552.

Peterson, K.J., Lyons, J.B., Nowak, K.S., Takacs, C.M., Wargo, M.J. and McPeek, M.A. (2004) Estimating metazoan divergence times with a molecular clock. *Proceedings of the National Academy of Sciences USA*, Volume 101, pages 6536–6541.

Wang, D.Y.-C., Kumar, S. and Hedges, S.B. (1999) Divergence time estimates for the early history of animal phyla and the origin of plants, animals and fungi. *Proceedings of the Royal Society of London Series Biological Sciences*, Volume 266, pages 163–171.

Index

N